Internet of Things

Technology, Communications and Computing

Series Editors

Giancarlo Fortino, Rende (CS), Italy

Antonio Liotta, Edinburgh Napier University, School of Computing, Edinburgh, UK

The series Internet of Things - Technologies, Communications and Computing publishes new developments and advances in the various areas of the different facets of the Internet of Things.

The intent is to cover technology (smart devices, wireless sensors, systems), communications (networks and protocols) and computing (theory, middleware and applications) of the Internet of Things, as embedded in the fields of engineering, computer science, life sciences, as well as the methodologies behind them. The series contains monographs, lecture notes and edited volumes in the Internet of Things research and development area, spanning the areas of wireless sensor networks, autonomic networking, network protocol, agent-based computing, artificial intelligence, self organizing systems, multi-sensor data fusion, smart objects, and hybrid intelligent systems.

** Indexing: *Internet of Things* is covered by Scopus and Ei-Compendex **

More information about this series at http://www.springer.com/series/11636

Naercio Magaia · George Mastorakis ·
Constandinos Mavromoustakis ·
Evangelos Pallis · Evangelos K. Markakis
Editors

Intelligent Technologies for Internet of Vehicles

 Springer

Editors
Naercio Magaia
LASIGE, Departamento de Informática,
Faculdade de Ciências
Universidade de Lisboa
Lisbon, Portugal

George Mastorakis
Department of Management Science
and Technology
Hellenic Mediterranean University
Agios Nikolaos, Lakonia, Greece

Constandinos Mavromoustakis
Department Computer Science
University of Nicosia
Nicosia, Cyprus

Evangelos Pallis
Department of Electrical
and Computer Engineering
Hellenic Mediterranean University
Heraklion, Greece

Evangelos K. Markakis
Department of Electrical
and Computer Engineering
Hellenic Mediterranean University
Heraklion, Greece

ISSN 2199-1073   ISSN 2199-1081   (electronic)
Internet of Things
ISBN 978-3-030-76495-1   ISBN 978-3-030-76493-7   (eBook)
https://doi.org/10.1007/978-3-030-76493-7

This Springer imprint is published by the registered company Springer Nature Switzerland AG
The registered company address is: Gewerbestrasse 11, 6330 Cham, Switzerland

Introduction

The Internet of things (IoT) has appeared as a novel and ground-breaking technology aiming to provide Internet connectivity to every device that has computation, communication, and storage capability. Numerous smart applications, e.g., smart cities, are a result of this phenomenon of connecting smart devices with the Internet. In order to provide secure and safe transportation, IoT has evolved into a novel concept, also known as the Internet of vehicles (IoV), where smart vehicles equipped with computing, communication, and storage capabilities communicate with each other and with surrounding infrastructure via vehicle-to-everything (V2X) connectivity. The latter enabled the exchange of critical information among smart vehicles, including sudden lane changes, steep curves, black ice warnings, and traffic accident avoidance, therefore assisting drivers in extreme and difficult situations.

IoV has an enormous potential to provide secure and safer transportation as it enables vehicles to make the correct decision promptly. In addition, IoV will also have intelligence and learning capabilities to anticipate the vehicular users' intentions. However, IoV-based objects produce a vast volume of information frequently called big data, which traditional algorithms and schemes cannot easily process. Emerging vehicular environments will enable efficient handling and processing of big data that is generated by the associated objects for effective communication and cooperation among them.

In addition, artificial intelligence (AI) will play a crucial role in such environments, enabling smarter services and applications, business processes, and social interaction among the IoV objects. In this respect, great potential exists for advanced technological solutions combining AI and IoV. AI mechanisms applied in IoV environments can be exploited, for instance, in smart cities, to analyze drivers' actions via data gathered or exchanged by enabled V2X vehicles' electronic control unit (ECU) and microcontrollers. AI will play a significant role in future IoV applications and infrastructures by providing insights from collected data. This capability will enable identifying patterns and will allow operational predictions with higher accuracy in short time periods. AI applications for smart vehicles will also allow the development of new services, reduce possible risks, and increase efficiency by predicting failures that are usually non-detectable by humans.

The future of implementing AI-powered IoV infrastructures depends on the effective solutions to several technical challenges that such paradigms introduce. Some initiatives try to incorporate AI schemes in IoV environments, but new frameworks have to be defined. As a matter of fact, the convergence of both has the potential benefits of building extensive ecosystems for increasing the number of automated services and their value.

This book gathers recent research works in emerging AI methods for the convergence of communication, caching, control, and computing resources in emerging IoV infrastructures. In this context, the book's major subjects cover the analysis and the development of AI-powered mechanisms in future IoV applications and architectures. It addresses the major new technological developments in the field and reflects current research trends and industry needs. It comprises a good balance between theoretical and practical issues, covering case studies, experience and evaluation reports, and best practices in utilizing AI applications in IoV networks. It also provides technical/scientific information about various aspects of AI technologies, ranging from basic concepts to research-grade material, including future directions.

Research Solutions

The Chapter **"The Fundamentals and Potential of the Internet of Vehicles (IoV) in Today's Society"** provides a modern perspective of IoV, categorizing and synthesizing its techniques. The authors point out its similarities to IoT where the things are connected vehicles, focusing on their connectivity and ability to exchange information. Through IoV and mobile communication technologies, drivers and passengers are connected through solutions integrated into data, device, and operations management via the implementation of secure and unified network access considering mass data collection and analysis. A plethora of services can explore such data if the information and automotive technologies are fully integrated.

The development of autonomous vehicles is seen as a solution to many of today's societal problems such as traffic congestion, road accidents, theft prevention, and air pollution. The Chapter **"Intelligent Approaches for Fault-tolerance in Radio Communication of Autonomous Vehicles"** proposes novel strategies and methods to ensure fault-tolerance in autonomous vehicles' communication systems. This was motivated by the fact that radio technology offers a great contribution to the developments of such vehicles but continues suffering from several shortcomings such as security breaches, radio interference, heterogeneity of protocols, vulnerability to climate changes, among others. Nonetheless the impact of these shortcomings being minimal in everyday life, it can be very disastrous for autonomous vehicles where the loss of transmission, or misinterpretation of received data, can lead to a global malfunction or even create a domino effect that will spread throughout the entire system.

With the deployment of the fifth-generation (5G) communication technology, characterized by a wider bandwidth and an increased computational capability, it is now possible to develop more complex services that require low latency, such as public safety AI-based applications. The Chapter **"AI-based Traffic Queue Detection for IoV Safety Services in 5G Networks"** investigates the deployment of an application that recognizes the formation of traffic queues on the highway through the analysis of video streams, by detecting the traveling vehicles and tracking their movement to understand when a traffic jam is occurring. This was enabled by using 5G multi-access edge computing (MEC) that reduces the latency thanks to computational resources being located closer to the user. The authors used convolutional neural network (CNN) to detect vehicles. The third version of You Only Look Once (YOLO) was selected due to its trade-off between accuracy and real-time computation. In addition, the Simple Online and Real-time Tracking (SORT) algorithm was used to identify the direction of the traffic flows and then to calculate when the vehicles are slowing down or stopping by either measuring the number of stationary vehicles or vehicles' traveling time within a region of interest.

The Chapter **"Internet of Vehicles—System of Systems Distributed Intelligence for Mobility Applications"** presents the IoV concept, technologies, and applications used to realize intelligent functions, optimize vehicle performance, control, and decision-making for future electric, connected, autonomous, and shared (ECAS) vehicles mobility scenarios. The concept addresses the convergence of the edge intelligence embedded in the vehicles based on AI technologies with the cooperative, collaborative intelligence distributed into the IoT devices and edge computing infrastructure federated with the hierarchical cognitive processes and analytics in physical, network, infrastructure, and data spaces. The chapter advances the latest architectural concepts for ECAS vehicles. It proposes an IoV 3D multi-layered architecture that combines AI, edge computing and connectivity as part of the functional layers while integrating the system properties and trustworthiness properties into the overall architecture to provide efficient new mobility applications and services. The proposed system of systems concept for IoV applications allows for distributed intelligent functions to be embedded into the edge and cloud infrastructure for ECAS vehicles to provide a computing, processing, and intelligent connectivity continuum for IoV applications and services.

There are different scenarios with different network requirements to be considered in an IoV. For instance, safety messages require a low latency network, meanwhile infotainment services demand a high bandwidth one. Due to such requirements, the network should allocate appropriate resources to accomplish the desired service. One of the promising technologies that is leveraged to fulfill such a goal is network slicing. The underlying infrastructure is divided into multiple slices, each equipped with required resources to meet specific needs. To slice the network, the infrastructure should be controllable to allow a central unit to guide the slicing process. Therefore, software-defined networking (SDN) is utilized to decouple the control plane from the data plane, allowing a separate unit to take control. The

Chapter "**Cross Network Slicing in Vehicular Networks**" discusses vehicular SDN slicing and how to boost it with intelligent capabilities leveraging recent advances in machine and deep learning.

SDN has the capability of designing a flexible programmable IoV network that can foster innovation and reduce complexity. SDN-enabled IoV devices can be controlled seamlessly from an external server (i.e., a controller), which can be located in the cloud and have computational resources to run resource-intensive algorithms, hence making intelligent decisions. The Chapter "**Toward Artificial Intelligence Assisted Software-Defined Networking for Internet of Vehicles**" describes the benefits of integrating SDN in IoV and reports recent advances. It presents an AI-based architecture and an automatic configuration method with which SDN can be deployed automatically in IoV without any manual configuration. The experimental evaluation is performed on a publicly available European testbed using an emulator for wireless SDN networks. It is conducted for automatic configuration of SDN in IoV network's topologies and for data collection in SDN-enabled IoV. The experimental results show the effectiveness of the proposed automatic configuration method.

The rapid development of intelligent transportation systems (ITS) and computational systems boosted the development of smart vehicle applications. IoV's V2X enables vehicles to communicate with public networks and interact with the surrounding environment. It also allows vehicles to exchange information, besides collecting information from other vehicles and roads. However, existing smart IoV systems' applications face many challenges related to different problematic issues such as big data interconnection with IoV, cloud network, data processing, and efficient communication between many different vehicle types to optimum data processing decision on or off board. AI technology with machine learning (ML) mechanisms offers smart solutions that can improve IoV network efficiency. For example, data processing decisions at various layers, i.e., onboard units (OBUs), fog, or cloud level, are among the problems that need ML algorithms. ML mechanisms can resolve other critical issues: time, energy, rapid IoV topology changes, quality of experience (QoE) optimization, and channel modeling. The Chapter "**Machine Learning Technologies in Internet of Vehicles**" provides theoretical fundamentals for ML models, algorithms in IoV applications, and future directions.

IoV has become an important revolution of ITS. It became an emerging research area as its need has increased tremendously. With a great number of applications available, besides the intention to improve the quality of life and services, the application of AI techniques would dramatically enhance the overall performance of the IoV network. The Chapter "**Deep Learning Approaches for IoV Applications and Services**" discusses deep learning (DL) networks as a type of ML with the influence of neural networks (NN) used in IoV, where great amounts of unlabeled data are processed, classified, and clustered. DL network approaches, i.e., convolutional neural networks (CNN), recurrent neural networks (RNN), deep reinforcement learning (DRL), are briefly introduced here, in addition to reviewing their ability to obtain better performing IoV applications.

Reducing vehicle energy demand has been the subject of academic research for several years. Distributing energy equally and efficiently among vehicle wheels addresses challenges that may have an impact on energy consumption. Therefore, the Chapter "**Intelligently Reduce Transportation's Energy Consumption**" proposes a novel system integrating smart panels. The effects of the road's change of gradient and anomalies motivated the authors to work on more torque at the wheel joints that demand more energy procurement. However, the supply affects all the wheels equally, which the engine interprets as more consumption. The authors approach the emerging issue by developing a Vehicle to smart-panel (V2SP) system in fluid dynamics. They implemented an algorithm that could run in real-time on the grid based on fundamental mathematics equations of fluid flow from Navier–Stokes that will be integrated on panels and interact accordingly with a vehicle's operational system. The objective is the distribution of the appropriate energy among wheels based on their needs.

Since an IoV is a heterogeneous environment involving information exchange between system components such as road infrastructure devices, vehicular embedded sensors, and other vehicular elements, devices communications and data exchange needs to be secure, efficient, and transparent to achieve the platform's goals. Blockchain technology is proved to provide decentralization, immutability, security, and transparency properties due to the features of its distributed ledger. The Chapter "**Blockchain-based Internet of Vehicle**" introduces the integration of the blockchain technology into IoV networks to support the essential data exchange and storage requirements such as decentralizing, security, and transparency. The authors describe blockchain terminologies and how they can be adopted in IoV environments. Then, they explain how it supports IoV data sharing, IoV trust, and verification, besides exploring the most popular blockchain-based IoV applications developed.

Vehicle guidance systems are considered key to improve the capacity and safety of transport systems. Information and communication technologies enable both vehicle speeds and distances to be optimized without being limited to human reaction times. As the major two competing topological approaches for communication networks exhibit specific advantages and disadvantages in different applications, there is still no standardized solution. Established encryption methods have proven to be insecure or lack real-time capabilities when used in distributed automation systems. The only proven secure concept for encryption—perfect security—has not been employed so far due to practical shortcomings. Meeting existing standards, the Chapter "**Vehicle Guidance System based on Secure Mobile Communication**" presents a communication architecture for vehicle guidance systems allowing for perfectly secure encryption and observing real-time requirements for wireless communication. Its core components are a central instance authenticating all participants, generating and distributing the required keys, and a transmission infrastructure based on relay stations. Different sensitivity analyses show that one-time pad cryptography can keep up with or even outperform the advanced encryption standard in the presented use case. The keys required for a sufficiently long operating time can be stored on common storage media.

With the rapid development of ITS, autonomous vehicles (AVs) are becoming one of the most anticipating means of transport. However, as their complexity increases, it is intuitive to consider the existence of more possible attacks and higher potential risks. For example, tampering with the in-car sensors or hacking into any vehicle's electronic control units (ECUs) could severely affect the driving performance or even cause life-threatening situations to users. Moreover, since AVs are part of IoV, the security of the intra- and inter-vehicular communication links should also be carefully studied. The Chapter **"Attack Models and Countermeasures for Autonomous Vehicles"** provides a comprehensive taxonomy for attack surfaces and countermeasures for the defense to identify and mitigate the security risks involved in AV holistically. Specifically, four different attack surfaces are defined, namely ECUs, sensors, intra-, and inter-vehicular communication links. For each of the attack surfaces, various common attack vectors are discussed in detail. Subsequently, the authors also survey the latest major existing work for defending the attacks on each surface.

The Chapter **"VASNET Routing Protocol in Crisis Scenario based on Carrier Vehicle"** tackles one of the most important disadvantages that takes place in the vehicular ad hoc network (VANET) during a crisis scenario. It summarizes five VANET routing protocol categories, where the common criteria among them are the requirement of infrastructure, i.e., road-side unit (RSU) or base station. The role of the RSU is to provide an internet connection. Packets are uploaded to the RSU, which, in turn, upload them to the Internet, thus making them available for future download. In a crisis scenario, one or more RSU might be disconnected from the Internet, which results in a disconnected network. The latter will lead to the failure of packet upload and timeout, leading to packet loss. The Chapter **"VASNET Routing Protocol in Crisis Scenario based on Carrier Vehicle"** proposes a protocol to ensure a successful packet delivery with disconnected RSUs. The network is considered a set of sensors that periodically upload data to the RSU, which uploads them to the Internet. The proposed protocol considers that vehicles are assumed to be capable of short-range wireless communication and can collect data from a nearby RSU. The main target of the proposed protocol is to ensure a successful data transfer during the time a certain vehicle, known as the carrier vehicle, is within the communication range of an RSU.

Delay-tolerant networks (DTNs) are networks where there are no permanent end-to-end connections; that is, they have a variable topology, with frequent partitions in the connections. Given the dynamic characteristics of these networks, routing protocols can take advantage of dynamic information, such as the node's location, to route messages. Vehicular delay-tolerant network (VDTN) routing is referred to as the hybridization of DTNs with VANETs, which mobilizes both knowledge-based and geography-based forwarding techniques. Numerous shortages are stated in existing VDTN routing protocols in both modes, such as inaccurate location information and uncontrolled congestion due to bundle flooding. The Chapter **"Hybrid Swarm-based Geographic VDTN Routing"** introduces a hybrid routing strategy combining a swarm-inspired algorithm, namely the firefly algorithm, to enhance the decision-making of finding better next

Store-Carry-and-Forward (SCF) relay vehicles by the use of geographical forwarding for better localization of closer vehicles to the destination. Therefore, the flooding of bundles is controlled by the movement of fireflies in early routing stages. Then, a reliable geographic routing is followed to better track closer SCF vehicles toward the bundle's destination.

Geolocation-based routing protocols choose the node that moves closer to the location of the message destination as the message carrier. However, such protocols suffer from obsolete location information due to node mobility and network partitions. The Chapter "**SnLocate: A Location-based Routing Protocol for Delay-Tolerant Networks**" proposes an epidemic-based decentralized localization system (i.e., DTN-Locate) and a hybrid location-based routing (i.e., SnLocate) approach. The former is used to disseminate node's localization information while the latter to create and route multiple copies of a message, using geographic mechanisms to disseminate them. In addition, a novel distributed contention mechanism is also proposed. The performance evaluation shows that the SnLocate protocol has a higher delivery rate and lower latency than other geographic and non-geographic routing protocols considered.

The Chapter "**New Ambient Assisted Living Technology: A Narrative Review**" presents a narrative literature review study of the acceptability of new assistive and information technology, including IoT technologies and AI, by older adults (i.e., 65 or older). The study followed a careful search strategy in specific databases and was based on inclusion/exclusion criteria and keywords. The search strategy resulted in 28 articles, which reflected the research aim, and was reviewed based on an interpretive approach and critically appraised by the critical assessment skills program's guidelines"The sentence 'The search strategy resulted in...' is misiing closing quote. Please check.` –>. This study explored both assistive and information technology and looked for overarching reasons why older adults may accept new technology. The results showed that older adults accept technology when they have a good sense of control over the devices and their lives: the technology being useful, having characteristics that are not threatening, and other compromising factors such as financial cost, restricting health conditions, and inappropriate physical environment are not present. Based on these findings, the authors proposed the N-ACT principles whereby technology developers should consider users' needs, adjustable technology and personalized service, users' control over technology and their lives, and trust in technology.

The Chapter "**Piloting Intelligent Methodologies for Assisted Living Technology Through a Mixed Research Approach: The vINCI Project in Cyprus**" presents the procedure and results of piloting vINCI, i.e., a new ambient assisted technology. vINCI aims to enhance older adults' active life and quality of living by measuring end-users' physical, psychological, and social state and providing them with information and feedback about any necessary corrective measures they need to take. To achieve this, vINCI has been based on microservice architecture and integrated IoT monitoring technologies and AI. In order to ensure that end-users would accept and use such a set of technologies, this study employed a mixed research methodology to understand any acceptability factors. The results indicated that clarity of instructions, the comfort of technology, ease to use, use-fulness, and a sense of safety, control, familiarity, and normalization were very important features of vINCI, which could cause participants to accept and use such technology. The study highlights the importance of a mixed research method for gauging acceptability to ensure that the end-users experience with new technology during pilots is fully captured and understood.

Conclusion

The book is intended for researchers, practitioners in the field, engineers, and scientists involved in designing and developing protocols and AI applications for IoV-related devices. It can also be used as the recommended textbook for under-graduate or graduate courses. The intended audience includes college students, researchers, scientists, engineers, and technicians in the field of IoV technologies and issues related to the associated AI-enabled applications. As the book covers a wide range of applications and scenarios where IoV technologies can be applied, the material covered is readable and a solid base for introduction, penetration into the comprehensive reference material on advanced communication and networking concepts in the related research field. It can also be a reference for selection by the audience with different but close to the field backgrounds. Finally, it adopts an interdisciplinary approach, and its form reflects both theoretical and practical approaches to be targeted by multiple audiences.

Contents

Emerging Trends of AI and IoV

The Fundamentals and Potential of the Internet of Vehicles (IoV) in Today's Society

Reinaldo Padilha França, Ana Carolina Borges Monteiro, Rangel Arthur, and Yuzo Iano

Abstract The IoT concept comes from products that go from the refrigerator to the washing machine, even smart sensors and smart devices that can be connected and regulated by mobile devices. In this sense, as the automotive industry continues to accelerate towards connected cars, more Internet of Vehicles (IoV) solutions, which technology would be similar to the IoT that is already known, but focused on connectivity in vehicles. Thus, the IoV is the possibility for vehicles to connect and exchange information, it is the possibility that vehicles communicate with each other automatically, consisting of a subarea of IoT applied to automobiles. Through IoT technology and mobile technology, people have the feeling of being connected all the time, and through IoV, this possibility is transformed into reality in solution integrated into data, device, and operations management through the implementation of secure and unified network access considering mass data collection and analysis. Representing the innumerable possibility of services can be explored with data and with Information Technology fully integrated with Automation Technology. Besides, IoV is an ecosystem that has an extensive field of intelligent transport systems, with a variety of traffic control solutions, transport management, emergency telecommunications cluster, secure, efficient, and ubiquitous wireless connections. Therefore, this chapter has the mission and objective of providing an updated overview of IoV, it is worth mentioning the novelty of this manuscript is in dealing with the approach to the theme focusing on the role of this technology in modern perspectives, categorizing and synthesizing the potential of the technique.

R. P. França (✉) · A. C. B. Monteiro (✉) · R. Arthur · Y. Iano
School of Electrical and Computer Engineering (FEEC), University of Campinas – UNICAMP, Av. Albert Einstein – 400, Barão Geraldo, Campinas, SP, Brazil
e-mail: padilha@decom.fee.unicamp.br

A. C. B. Monteiro
e-mail: monteiro@decom.fee.unicamp.br

R. Arthur
e-mail: rangel@ft.unicamp.br

Y. Iano
e-mail: yuzo@decom.fee.unicamp.br

© Springer Nature Switzerland AG 2021
N. Magaia et al. (eds.), *Intelligent Technologies for Internet of Vehicles*, Internet of Things,
https://doi.org/10.1007/978-3-030-76493-7_1

Keywords IoT · IoV · 5G · 4G · Vehicles · Network · Connectivity

1 Introduction

With the spread of 5G, which promises speeds up to 50 times faster than those obtained with 4G, one of the sectors that will benefit the most from the high speed and reliability rates of the fifth generation of the mobile data network will be the industry automotive, with vehicles connected to the internet, a trend that has been called IoV (the Internet of Vehicles) [1–3].

The need for mobile connectivity and automotive safety, through applications such as connected automated steering, access to services anywhere, and integration with smart cities and transport. Numerous services can be explored with data in hand and with Information Technology fully integrated with Automation Technology [4, 5].

The Internet for Vehicles (IoV) is the possibility for vehicles to connect and exchange information. It is a subarea of the Internet of Things (IoT) applied to automobiles. Through it is possible for vehicles to communicate with each other automatically. Allowing a driver to be driving on a given highway and receiving information about the traffic of other vehicles coming in the opposite direction. This can cause the vehicle to suggest a new route to the driver [5–7].

In the future, we will have autonomous vehicles circulating on the streets, like the prototypes created by Google. But for this point to be reached, we need to go through the Vehicle Internet first, ensuring a reliable connection with low latency. IoV has entered a period of rapid growth, with the integration and development of network, electronic, and information communication technologies, together with the manufacture of automobiles, providing new services for consumers [8–10].

High-speed 4G mobile networks are excellent platforms for intelligent Vehicle Internet, but IoV will be able to expand in a broader environment and act more securely, reliably, efficiently, and flexibly with the advent of 5G [11, 12].

With respect to vehicle connectivity, the major chipmakers have focused on areas such as communication systems, digital entertainment, advanced driver assistance, and radar solutions. It is a true mix of technologies, bringing together IoV, automated steering, artificial intelligence, sensing, multimedia, and advanced chip processing, to create a comprehensive and integrated automotive processor solution. To assist with driving, modules are now available that include vision-based driver assistance systems (V-ADAS). This technology uses versatile visual processing units (VPU) to handle large amounts of dynamic images with low power consumption [11–13].

Through the exchange of information between vehicles, it will be possible to anticipate various issues related to traffic, such as slowdowns in the route, interruptions, accidents, rain, fog, among other aspects. In addition, when a vehicle is stopped on the road, it will be able to communicate with nearby vehicles informing the problem to seek help, such as a lack of fuel or mechanical failure. In the second case, there may already be a direct connection with the concessionaire to arrange for repairs on-site, if it is not possible to perform it via the Internet [13–15].

With data collected in real-time by the vehicle's sensors, driving safety is enhanced. Connected vehicles will allow a large amount of data about vehicles to be collected (and sent over the internet), which can monitor resources such as brake systems, tire pressure, and even in cases of vehicle theft, with vehicles tracking, with data being shared with insurers or even with the authorities [15–18].

And the systems also learn from the driver and traffic situations, adopting machine learning technologies to improve accuracy and speed of detection, object recognition and to track resources that allow the possibility to "see" lanes, vehicles, pedestrians, and others moving items, with multiple lens calibration and panoramic monitoring [19, 20].

IoV Intelligent is expected to establish its roots in 4G and evolve with 5G. Key technologies supported by 5G and we have already achieved advances in channel coding, Massive MIMO, virtualization and network slicing, and precise positioning. All of these achievements will design a new user experience with intelligent IoV [1–3, 11, 12, 21].

As the automotive industry continues to accelerate toward connected vehicles, more Vehicle Internet (IoV) solutions are emerging through a network, 5G, and Internet of Things (IoT) advance. Traditional vehicle companies must take on additional roles in addition to vehicle manufacturers and evolve into automotive service providers. With IoV and increased connectivity, the development of innovative applications in the industry now features management capabilities for data, interconnections, operations, and security, opening up to application platform interfaces (APIs) for third-party solutions [1–3, 11–13, 21, 22].

The Internet for Vehicles brings a panorama of accident reduction, which should happen in the near future, once the fleet is renewed, to allow a wide communication between vehicles. The technology in automobiles is already helping to reduce accidents and, with the adoption of the Internet for Vehicles, it should decrease even more [1–7].

Therefore, this chapter has the mission and objective of providing an updated overview of the Internet of Vehicles (IoV), addressing its evolution and fundamental concepts, showing its relationship as well as approaching its success, with a concise bibliographic background, categorizing and synthesizing the potential of the technique.

2 Methodology

This research study was developed based on the analysis of scientific papers and scientific journal sources referring to the Internet of Vehicles (IoV) area, aiming to gather pertinent information regarding thematic concerning evolution and fundamental concepts of technology. Thus, it is also possible to boost more academic research through the background provided through this study.

This survey performs the bibliographic review of the main research of scientific papers and manuscripts related to the topic of Internet of Vehicles, published in renowned bases in the last 5 years, such as Scopus, Web of Science, EI-Compendex, Google Scholar, Springerlink, SciELO, IEEE Xplore and more.

The differential of this manuscript in dealing with its approach with respect to the topic presented through current examples of the applicability of the IoV concept, focusing on the role of this technology in modern perspectives. In the same sense that it is also motivated to boost more academic research through the background supplied through this study.

3   Internet of Vehicles (IoV) Concept

The user's experience in purchasing products has been encouraging organizations and industry to change the concept and design of offers and products, evaluating that the modern user has a tendency to pluggable products into apps with the ability to monitor, collect and obtain insights from your consumption and performance information during use. In this aspect and connected, the Internet of Things (IoT) emerged as a technology capable of exploiting these characteristics [1–7, 23].

In the same vein, the telecommunications and automotive sectors are developing and promoting communication solutions, supporting standardization, and accelerating the commercial viability of mobile connectivity and automotive security needs. Through intelligent applications, through technologies such as IoT, which provide connected automated steering, access to services anywhere, integration services with smart cities, in the same vein as solutions that encourage intelligent transport [1–7, 24].

The IoT concept (Fig. 1) comes from products that go from the refrigerator to the washing machine, even smart sensors and smart devices that can be connected and regulated by mobile devices. In the automotive industry, the need is increasingly present, since it needs to establish a dialogue and mutual relationship between the vehicle and IoT technology, in order to understand, more and more, the experience of users in the use of services embedded in vehicles [1–7, 24].

In this sense, as the automotive industry continues to accelerate towards connected vehicles, more Vehicle Internet (IoV) solutions, which technology would be similar to the IoT that is already known, but focused on connectivity in vehicles (Fig. 2). As well as designs, technologies, and applications emerging through network advance, the 5G trend among others [1–3, 11, 12, 24].

5G is the fifth generation of mobile networks, developed to be the successor to the 4G network, which promises higher connection speeds and data download, offering higher data connection speed for smartphones and mobile devices. However, in practice, 5G is being designed for the reality of IoT and derivatives, since this technology needs speed, coverage, and latency. What in turn measures the time it takes between a command executed until it is sent over the network and executed, with less latency is essential, which respects the management of autonomous vehicles or to operate any IoT by IoV [1–3, 11, 12, 24].

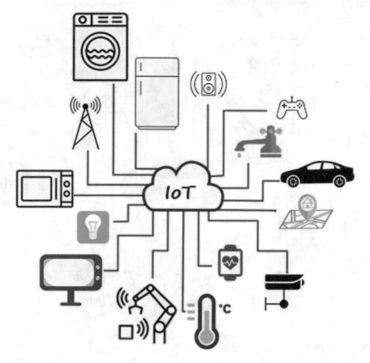

Fig. 1 IoT illustration

Even with a historical perspective, the maximum and average speeds have increased considerably since 2G, considered as the first digital network, a successor to the 1G (analog) network. Still considering the 3G, which allowed the transfer of data at high speed, without using any wire. Although it is important to note that 4G is still being improved, which can reach higher speeds from that. 4G generally considers an average latency of 50 ms, while the 5G trend promises latency of around 5 ms [25, 26].

Over time, more and more electronic devices are becoming smarter, which in the face of this evolution requires cloud services and network resources for communication to exist. In this respect, mobility which is the technology to connect at any time and in any place, which in the face of this need, IoT technology provides the best global connectivity solution for devices aiming to meet the needs of machines and vehicles that are different connectivity needed to connect people [6, 8, 9].

The IoT connectivity solution can for example preload IMSI (International mobile subscriber identity) on devices, listing a number that uniquely identifies all users of a cellular network, consisting of a permanent and unique identifier that identifies a mobile subscriber to the network. This helps to centralize Connectivity Management with the data rate, instantly experiencing the performance of connected vehicles taking on additional functions. In addition to vehicle manufacturers (automotive industry) having the ability to evolve into automotive service providers, through IoV

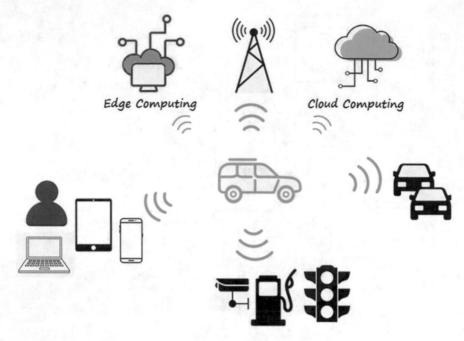

Fig. 2 IoV illustration

and increased connectivity, coupled with the development of innovative applications featuring interconnection, operations, and security features, data management, yet pondering the openness to application platform interfaces (API) for third-party solutions, for example [1–7, 27–29].

Thus, the Internet of Vehicles (IoV), is the possibility for vehicles to connect and exchange information, it is the possibility that vehicles communicate with each other automatically, consisting of a subarea of IoT applied to automobiles. Evaluating that the data coming from these intelligent sensors and vehicular systems can still be extracted by the vehicle in an automated way, from standardized communication protocols for the vehicular area [1–7, 27–29].

3.1 Types of Vehicle Networks

In-Vehicle Networking (In-Vehicle Networking) allows applications through controllers that allow the management of information traffic with interfaces through a reduced physical medium, possibly a serial bus, but it has the capacity to enable the multiplexing of this information. The use of an intra-vehicular network offers benefits such as requiring a much smaller number of wires and connectors, reducing material and installation costs, even considering the possibility of sharing sensors available on the network in different measures. The uniformity of intra-vehicle network designs offers several standards, such as CAN (Controller Area Network), SAE (Society of Automotive Engineers standard), VAN (Vehicle Area Network) among others [30–33].

Inter-vehicle networks are understood as the junction of the vehicle's internal networks, which connect the vehicles and all other entities (nodes) connected to them. Representing that the vehicles are equipped with a set of systems for sending and receiving signals, allowing each vehicle to know, for example, its position, speed, and even the direction of other vehicles of a certain group, still allowing for some joint decision making or individually [34, 35].

Just as IoV is actually an intelligent system for managing the traffic network using advanced sensors, performing interaction formed by motor vehicles and fixed infrastructure located on the side of streets or roads, wireless communication, and data processing. These networks are different from other wireless networks, providing that information is collected and analyzed through intelligent sensors installed in vehicles or terminals, mainly due to the nature of its nodes, which have high mobility and trajectories that follow the limits of public access roads, and thus transmitted by the mobile network to a management platform. Making technology capable of managing a transit system for example, so that it is more efficient, and safe, both from the perspective of vehicles and pedestrians [36, 37].

3.2 IoV Applications

Still evaluating that in IoV, it is possible to use On-Board Diagnostics (OBD), consisting of a self-diagnosis system available in most vehicles, capable of collecting a series of data in real-time, such as the status of the electronic injection light, or even problem diagnosis codes (DTC, i.e., Diagnostic Trouble Codes), or engine rotation, speed, coolant temperature, among other aspects [38, 39].

With IoV, GPS navigation (formed by three segments: space, control, and user) allows, through artificial satellites, a better obtaining of information about the geographical location anywhere on the Earth's surface and at any time of the day [40, 41].

The IoV also allows the recording of vehicle data in real-time, similar to OBD, which represents a solution that helps computers to identify objects, record data, and thus monitor the entire vehicle information chain, with more speed, information (by recording and re-recording data) and therefore accuracy [5–7, 38, 39]

The telemetry box, which is a fleet management system that collects the data generated by each vehicle, with IoV allows a fleet manager, for example, to evaluate and detect business optimization opportunities. In this aspect, the collection of these data, storage, and availability in a management system, is essential so that through analysis, these data can be used as a foundation for strategic decisions [42, 43].

Still reflecting on the control of traffic lights, which requires a combination of technologies such as IoT sensors, high-capacity communication networks, such as

4G or even the 5G trend, and the V2R, V2X, V2V, V2S system, which coordinate all this context as well as among other applications [1–3, 11, 12, 44].

IoV is a set of technologies and applications that uses wireless, mobile communication, which is an open and converged network system based on the coordination of vehicles, humans, and devices. In order to provide connectivity between vehicles of all types, between vehicles and road infrastructure, and between vehicles, infrastructure, and wireless devices, with the aim of improving safety, mobility, and the environment. In which the recognition, sensing, collection, transmission, and processing of information about people, vehicles of the most diverse types, communication network, and even the road traffic infrastructure, using information and communication technologies (ICT), IoT devices, and processing technologies [1–3, 11, 12, 45].

IoV allows a driver of a vehicle driving on a highway to receive information about the traffic of other vehicles coming in the opposite direction, which allows the driver and even the vehicle's onboard system to suggest a new route to the driver. In this context, IoV can work in different ways, since in more modern vehicles, it is possible to access the internet on the panel itself, via 3G/4G, still pondering the use of cellular connection service, still evaluating the hotspot technology of wi-fi for network distribution internally, according to the existing technology in the vehicle, and the application scenario. This allows the connection of smartphones, tablets, notebooks, or even any other mobile device with internet access through the vehicle [1–3, 11, 12, 46].

Through IoT technology and mobile technology, people have the feeling of being connected all the time, and through IoV, this possibility is transformed into reality in private or even collective vehicles such as buses, trains, subways. What enables the trend of that automakers can offer more and more vehicle options with this factory technology. Realizing more and more the expectation that this context will become even more a common reality with the advancement of autonomous vehicles and even the 5G internet [1–3, 11, 12, 46].

Assessing that the access to the internet in a vehicle goes far beyond the meaning of the possibility of accessing maps, or listening to music by streaming, or even videos by streaming to passengers. The objective is on autonomous vehicles, stating that the connection will enable the operation without the need for a driver.

3.3   Information Technology Integrated with Automation Technology for IoV

The increasingly present trend of 5G technology should be the main lever for data in vehicles to be collected in real-time. Enabling automakers to obtain quality data, which allows the performance of their products and consequently the satisfaction of their customers; still pondering the concessionaires that carried out reviews, recall, and preventive services in a scheduled and precise manner. In other words, it will allow the formation of a complete automotive ecosystem, through IoV, including members such as vehicle manufacturers, telecommunications infrastructure providers, telecommunications operators, IoT chip and sensor suppliers, and even automotive parts suppliers [1–3, 11–20, 47].

The IoV solution integrates data, device, and operations management through the implementation of secure and unified network access, considering flexible device adaptation and mass data collection and analysis, while still significantly contributing to the generation of potential new sources of data recipe [1–3, 11–20, 48].

However, vehicles must be able to collect and provide this data, a necessary factor to be implemented. Data and information obtained through sensors related to oil pressure and tires, speed control on autopilot, parking, reversing, and driving cameras. Still pondering about a multimedia center obtaining route and congestion data, which, through the use of Artificial Intelligence (AI), will interact with drivers proposing diversion routes for these situations [50, 51].

What represents the innumerable possibility of services can be explored with data and with Information Technology fully integrated with Automation Technology. Reflecting on IoV will contribute to the safety of drivers through the exchange of information between vehicles, making it possible to anticipate various issues related to traffic, such as slowdowns in the route, interruptions, accidents, rain, fog, among other situations, among others. Still considering the situation in which a vehicle is stopped on a certain highway, it can communicate with nearby vehicles informing the problem, and requesting help, such as lack of fuel, among others. In cases of mechanical failure, the vehicle will have conditions and properties to make a direct connection with the concessionaire to arrange for repairs on-site, if it is not possible to perform it via the internet [11–20, 48–51].

In addition, IoV is an ecosystem that has an extensive field of intelligent transport systems, with a variety of traffic control solutions, transport management, emergency telecommunications cluster, secure, efficient and ubiquitous wireless connections, as well as intra-vehicle devices such as 4G multi-mode modules, on-board diagnostics (OBD) and wi-fi for vehicles, among other information services [3, 18, 34, 38, 39].

Through IoV solutions it will be possible to build a series of secure platforms, open and shared between terminals and the cloud and environments for intelligent transport, meeting the needs of drivers and passengers aspects related to security, control, and information. What brings a panorama of accident reduction allowing a wide communication between vehicles, through the technology applied in automobiles collaborating for the reduction of accidents [51, 52].

Although 4G mobile networks are great platforms for IoV, the solution will be able to expand in a broader environment and act more securely, reliably, efficiently, and flexibly with the advent of 5G. Considering that IoV will be able to identify and solve gaps and potential needs in services such as Telematics, traffic efficiency, and automotive safety [1–3, 11–20, 47, 48].

Still evaluating that the automotive industry will be one of the sectors that will benefit the most from high-speed rates, advances in channel coding, Massive MIMO, network virtualization, and slicing, and precise positioning, and 5G reliability in a mobile data network. And in that direction through IoV, it will be possible to design a new user experience with intelligent transport in key technologies supported by 5G [1–3, 11–20, 47–49].

Through vehicle connectivity, the chipmaker industry will focus on areas such as communication systems, digital entertainment, advanced driver assistance, and radar solutions, or even vision-based driver assistance systems (V-ADAS) to assist in the direction, using versatile visual processing units (VPU) dealing with large amounts of dynamic images, with low energy consumption, among other applications. Providing a mix of technologies bringing together IoV, automated steering, AI, IoT sensing, multimedia, and advanced chip processing, creating a comprehensive and integrated automotive processor solution [1–3, 11–20, 47–49, 53].

By obtaining this large volume of data on vehicles collected in real-time, by the vehicle's IoT sensors and sent over the internet, also guaranteeing the total anonymity of the driver's data, since it is not appropriate to know the routes of a given user. Through this data, it is possible to monitor resources such as brake systems, tire pressure, and even in cases of vehicle theft, allowing tracking, with data being shared with insurers or even with the related authorities [11–20, 47–49, 54].

In the same perspective of the potential through the data, it is possible, that the systems can learn from the driver and the traffic situations, adopting the machine learning technology (sub-area of the AI) improving the precision and the speed of detection, the object recognition and even enabling resources that "see and understand" lanes, vehicles, pedestrians and other moving items. What generates more security in maneuvers, for both drivers, cyclists, and even pedestrians, forming an advanced communication system that allows more security to be made available in cities, with multiple lens calibration and panoramic monitoring [11–20, 47–49, 51–53].

Anyway, through IoV, innumerable services can be provided, since through technology the automotive segment is transformed to meet the challenge of lowering their operating costs at the same time that they need to launch increasingly innovative products. Representing that the automobile industry advances towards advanced power steering and autonomous driving, allowing more safety on the streets, avoiding the cause of accidents, through the interaction between vehicles, between the infrastructure and the possibility of identifying pedestrians and cyclists on the streets, in addition to the possibility to detect accidents in order to avoid them, avoiding further accidents. What is possible with the vehicle system gradually replacing human initiatives, paving the way for these technological advances through the IoV infrastructure [1–3, 11–20, 47–49].

4   IoV Communication Networks

They are integration between three types of communication networks Inter-Vehicle, Intra-Vehicle, and Vehicle-Internet, which define as a large-scale distributed system, formed by motor vehicles and fixed infrastructure, or with wireless communication between vehicles, highways, people, and the internet. These networks have high mobility and trajectories that follow the limits of public access roads, interconnected through communication protocols, and standardized data [30–35, 55].

4.1 V2V (Vehicles with Vehicles)

In this technology (Fig. 3), each vehicle is able to perceive the presence of other vehicles using a communication protocol based on radio waves. The technology uses radio frequency to establish connectivity between vehicles, in order to prevent accidents, making it possible to send and receive relevant information in real-time, such as position, speed, accident alerts, bad weather, or catastrophes, blitzes, and information about traffic in general [33, 44, 56].

With V2V it is possible, for a vehicle to send information to the other regarding potential problems that are outside the range of the vehicle's sensors and, therefore, outside the scope of operation of the system. This communication can work either through a radio frequency, such as Dedicated Short-Distance Communication (DSRC), or over a mobile network. And through data collection, countless possibilities for applying the technology, are alert to vehicles in the blind spot of drivers, frontal collision warning, indication to abort some overtaking, among many others [33, 44, 56, 57].

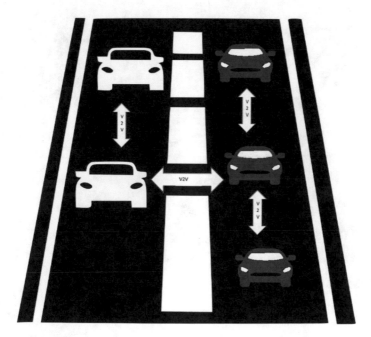

Fig. 3 V2V illustration

V2V networks have a great potential to prevent accidents, reducing the number of deaths and injuries in traffic, however, it is worth mentioning that the collection of data by these devices, must respect privacy, protecting information about drivers [33, 44, 56, 57].

Still evaluating that systems using V2V technology can be both passive (only receive information) and active (bidirectional communication). Through the development of this technology, vehicles will be able to establish a solid connection that will allow advanced functions, such as file sharing, among other features [33, 44, 56, 57].

4.2   V2R (Vehicles with Infrastructure)

Through this technology (Fig. 4) the vehicle can exchange information in real-time with a traffic center, allowing the verification of the vehicle's location and informing when a certain traffic light will close or open. However, in a "vehicle-infrastructure" communication it is a one-way street, i.e., the vehicle obtains information from a central, and nothing else. Even so, if there is an exchange of information, the system for determining the positions of vehicles can make the traffic more flexible, for example, helping the passage of an ambulance [44, 51, 58].

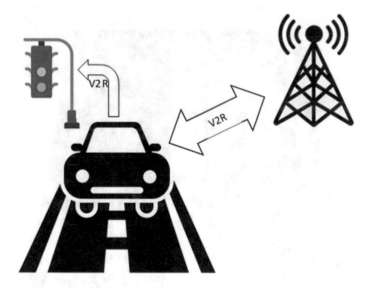

Fig. 4 V2R illustration

4.3 V2X (Vehicles with Internet (Cloud))

In this technology, it is a set of protocols based on Wireless Local Area Network (WLAN), it is a concept for connected, autonomous, and assisted vehicles, which allows the transmission of data from a vehicle, to any system that has the ability to affect such vehicle or vice versa. The technology makes the vehicle transport people in a safe and faster way, resulting from the ability to communicate with other automobiles or even with infrastructures be buildings, smart traffic lights, traffic cameras, among others in a context of intelligent transport, ensuring expressive gain in performance, with regard to the direction [7, 33, 44, 47, 54, 55].

The V2X technology (Fig. 5) serves as a reference for technological evolution in vehicles, evaluating that it gave rise to two technological standards, the DSRC (Dedicated Short Range Communications) and the C-V2X (Cellular Vehicle to X), which use the same initial concept, that C-V2X is an evolution of the DSRC. V2X encompasses specific forms of communication, such as V2V (Vehicle-to-Vehicle), V2I (Vehicle-to-Infrastructure), V2P (Vehicle-to-Pedestrian), and even V2N (Vehicle-to-Network), so it is robust [7, 33, 44, 47, 54, 55, 59, 60].

This is a technology that defines communication from the vehicle to everything around it, i.e., from the vehicle to any element in traffic, capable of perceiving all agents involved in traffic to ensure more efficiency. From the application point of view, V2I is for transport infrastructure, V2P is for pedestrians, V2N is for a network, and V2V is for vehicles [7, 33, 44, 47, 54, 55, 59–61].

In a scenario of intelligent transport, with most vehicles connected, since the technology will generate a series of data about the highway that vehicles are driving, it will be possible to know how traffic conditions are very easily, helping big cities to

Fig. 5 V2X illustration

reduce traffic jams. Still reflecting that the main focus of V2X is to analyze and work together with other vehicles, but pedestrians and other active elements of traffic such as traffic lights, cyclists, still considering other variables may also be involved more participative with the evolution of the concept [7, 33, 44, 47, 54, 55, 59–61].

4.4 VANETs (Vehicular Ad Hoc Networks)

IoV is a broad concept, of which VANET is a subset, which is the technology focused on the integration of computing and communication technologies with the objective of enabling transport, efficiency, and improving city services. Technology is a technological evolution in the means of communication present in automobiles, based on crossing the communication barrier existing between different vehicles [62, 63].

The technology is characterized as a network formed dynamically and randomly covering a restricted geographical area, meaning that communication, performed only between driver and vehicle, can be carried out between vehicles. However, one of the biggest challenges in its adoption is scalability. Considering that the ad-hoc characteristic of VANETs undermines the provision of global cooperative and sustainable services, since, in order to solve the traffic problems of large cities, it is necessary to cover very large areas and perform complex correlations between vehicles, pedestrians, and the traffic environment [62, 63].

Still pondering about the protocols created for wireless networks are not appropriate for vehicular networks, and to maintain a connected network on a large scale, since there are challenges with respect to scalability, i.e., a high number of nodes, and high mobility of them. Or even pondering the lack of guarantee of connectivity to the Internet and incompatibility with mobile devices and with the cloud computing storage and processing model, are other important limitations in VANETs. Faced with these limitations, networks in which the nodes are composed of the vehicles themselves, such as "vehicle-to-vehicle" (V2V) and road infrastructure, such as "vehicle-to-road" (V2I), are the best answer [33, 44, 47, 54, 55, 59–63].

5 IoV Architecture

The IoV network architecture is generally heterogeneous, consisting of an infrastructure distributed along the route, communicating with vehicles via a wireless connection, in order to verify the validity of the received message. Which can be categorized into five types of vehicle communication: vehicle-to-vehicle (V2V), vehicle-to-infrastructure (V2I), vehicle-to-roadside unit (V2R), vehicle-to-sensors (V2S), and vehicle-to-pedestrians (V2P) [7, 33, 44, 47, 54, 55, 59–63].

In this sense, each vehicle is equipped with a tamper-proof device (TPD) consisting of a device with the function of verifying the driver and jointly maintaining the keys safely, and even plays a role in decoding the data during execution. Since through IoV, encrypted data is transmitted, which by nature must be difficult to compromise, which is obtained with the TPD to obtain computer security against and copy of the target program and even illegal analysis. While communicating with other vehicles and RSUs (Roadside Unit) over wireless connections, periodically transmitting safety-related messages to nearby vehicles via the DSRC (Dedicated Short Range Communications) protocol, a wireless protocol typically used in V2R or V2V, or even other types of short-range dedicated communications [1, 64, 65].

The IoV represents a prominent instantiation of the IoT in an ITS (Intelligent Transport System), designed to support a wide range of safe and unsecured applications for intelligent traffic management. Allowing for efficient and sustainable traffic, performing intelligent vehicle control, environmental monitoring, and even obtaining driving experiences through the exchange of V2I, V2R messages, thereby enabling safer, more comfortable, and pleasant travel and traffic for both drivers and passengers [7, 33, 44, 47, 51, 54, 55, 59–61, 66].

However, it is worth mentioning that there is no well-defined layered architecture for IoV, evaluating the models found in the literature, which in general derive from a three-level architecture. Consisting of the first level for data acquisition, the second relating to communication, and the third corresponding to applications [47, 51, 54, 55, 59–61, 67].

Still reflecting that positioning and location recognition technology is the main technology of IOV, which through the proper implantation of devices such as RFID readers, which is a short-range communication technology through RFID tags that can be read automatically by sensors, still pondering traffic lights, video cameras, and other wireless range technologies. Or even corresponding location recognition algorithms, using AI technology or Machine Learning, Deep Learning and derivatives, which can be used in vehicles in operation [11–13, 15, 50, 51, 68].

And in this context, one of the central objectives of IoV is the construction of an intelligent system to improve the quality of life or driving quality of drivers, and even aiming to increase the quality of the digital experience or the digital service provided to the user.

Still considering the IoV information generated by the vehicles, they can be categorized into two parts: information on board monitoring the vehicle status, with respect to speed, engine parameters, brake status, among other characteristics obtained directly from the IoT sensors in the vehicle. At the same time that considering the route information related to the events that happen on the road, which can be either of sensors onboard (respective the distance between vehicles, objectives of the hidden point of direction, video of the pilot camera, among others) or even the from others via IoV connections related to traffic light status, brake notification of the neighboring vehicle, traffic map, among other features) [33, 44, 47, 54–57, 59–63].

5.1   Advantages and Benefits of IoV

In everyday life, accessing the internet in the vehicle facilitates people's lives and contributes to advantages related to an in-vehicle connection with respect to intelligent traffic management, intelligent dynamic information service, and intelligent integration of vehicle control networks through IoV [1–3, 11, 12, 24, 69, 70].

In this sense, it facilitates the location and identification of the best route via the internet in the vehicle by GPS, since due to the existing satellite tracking devices working connected to the internet by digital devices, it is possible to identify the best route to reach a destination without wear and tear traffic jams, or even blocked traffic due to accidents [1–3, 11, 12, 24, 40, 41, 58].

Through IoV, various services such as collision warnings, informational entertainment, or even sound entertainment when listening to music playlists through streaming platforms, detecting traffic congestion, or even finding establishments close to your location, such as restaurants, parking lots, and even gas stations, i.e., route planning are provided, making human traffic trips more convenient and enjoyable. Still related to the savings in the use of the mobile data package of the users' cell phone plan, who, in general, use these resources for travel traffic maps. From the passenger's point of view, IoV provides greater distraction and entertainment for the family during long journeys (Fig. 6), with respect to passengers being able to entertain themselves during the trip without spending their own data package [1–3, 11, 12, 18–24, 34, 40, 41, 58].

Also relating the benefits of IoV concerning cognitive intelligence in reliable decision making, through the efficient use of resources and great potential to implement the intelligence in the transport system. A factor that is essential for the success of autonomous steering systems, which is an improvement on the traditional IoV architecture. Or even reflecting on the potential of IoT in the identification of mechanical problems, with respect to the most modern vehicles, since in these vehicles, the technological system of the automobile itself has the capacity to identify failures in

Fig. 6 Entertainment IoV vehicle illustration

Fig. 7 Maintenance IoT-IoV vehicle illustration

operation and signal to the driver, which through addition, preventive maintenance can be triggered (Fig. 7) [8, 14, 71].

Still relating to cognitive intelligence, it is fundamental for cognition in the intra-vehicle network, which is related to drivers, passengers, smart devices, among others, inter-vehicle network with respect to the adjacent intelligent vehicles, and even its presence is in the external network to the vehicle, i.e., in the road environment, cellular network, edge nodes (vehicles), remote cloud, among others [8, 14, 71].

IoV enables a more accurate perceptual capacity, providing information and programming strategies for the entire transport system. What is noticed through the presence of cognitive computing in autonomous steering systems, through the cognitive cycle of perception, training, learning, and feedback (AI algorithms). In view of the learning capacity of these vehicles effectively improved, through a more reliable and complete decision-making process for autonomous vehicles [8, 14, 15, 50, 51, 68, 71].

Also reflecting on the greater stimulus on application drivers, making this transport market even more competitive, with a better offer related to every detail and benefits, and other facilities that the internet in the vehicle can bring, made available to the passenger during the trip making the difference. How to offer the possibility of watching films and series during the journey, in good stability and quick connection. With the traffic apps, it is possible to identify the best route to reach the destination, avoiding accidents and traffic jams, allowing passengers to arrive on time and without delays. This improves the quality of the trip, transforming the service into a highly differentiated offer [8, 14, 15, 50, 51, 68, 71].

Still reflecting on the context of Smart Cities, which through IoV it is possible to provide smart transportation through real-time traffic information, increasing the safety of drivers and pedestrians, consequently increasing efficiency and reducing traffic problems through advanced navigation [3, 8, 14, 15, 18, 21, 50, 51, 68, 71].

5.2   Challenges for Implantation of IoV Structure

Internet infrastructure with software updates, privacy, and cybersecurity and coverage in remote areas is required. In other words, the successful implementation of IoV depends on factors, ranging from the interconnection between different sectors such as energy, environment, transport, automotive manufacturing, and others, to technologically sustained developments in the lower and upper layers considering a stack of protocols of communication [51, 71, 72].

In IoV vehicle data, in general, is organized into flows, even considering that each vehicle carries a small data flow, a large flow is merged into a cloud due to the high frequency and large scale of the fleet, for example. What requires the processing of vehicle data flow in a timely manner, and consequently errors in data transmission, occurring when an emergency message is released to provide traffic recognition, and vehicles need to increase their channel transmission capacity ensuring additional coverage and mitigating possible accidents, which represents another important challenge for the consolidation of IoV. This can result in channel congestion and unnecessary power consumption due to improper transmission configuration [51, 71, 73].

Still reflecting on the challenges related to the scalability of an industrial IoV system where this system needs to be highly scalable, dealing with the large scale of data; the need for an effective trade-off between real-time requirements and the accuracy of data-centric services, when processing data with low-value density and low quality on a large scale; and even the need for high reliability to support critical security services such as emergency assistance, for example [51, 71, 74].

The IoV sustainability factor is a constant challenge, which can be achieved not only by the use of pollution-free vehicle systems, but concurrently by maintaining traffic safety or even preventing vehicle accidents or collisions [51, 71, 75, 76].

With respect to digital security and privacy, IoV systems, in general, must be efficient against cyber-attacks of authentication and identification, availability, routing, confidentiality, or even data authenticity, among many others [16, 51, 71, 73, 75, 76].

Still reflecting on security factors, the adequate collection of information, it is necessary requirements with operational functions and management functions that perform authentication identifying the vehicle node, the collecting node, and even the big data center (use of this technology for IoT due to the volume of data); allow confidentiality of information sent to the appropriate entity; carry out the integrity of messages against modification or destruction, and even allow authorization to guarantee that only authorized nodes access digital resources [16, 51, 71, 73, 76–79].

6   Discussion

The development of smart cities is a strong driver of IoV, since technology is efficient to connect, vision systems, autonomous driving, anti-collision, and entertainment.

Because through its buildings, houses, devices, and people are able to communicate through connection via wireless networks.

Still evaluating that transportation represents an economic factor in the world economy, which has been driving IoV with respect to rapid growth, with the integration and development of network, electronic, and information communication technologies, together with automotive manufacturers, providing new services for consumers.

In a Smart City, an intelligent transport system, with the IoV model, requires traffic of information from several devices, forming a complex communication system. Pondering the management of Smart Cities is not a simple collection of devices, yet evaluating the advancement of intelligent building technology, the integrated design focused on centralized management of the information system and the network of devices for the control of lighting and air quality, for example. This requires highly integrated systems, extensive communication technologies, as well as the ability to provide comprehensive network connectivity for IoV.

Or even the use of Big Data, Cloud Computing, and IoT in smart cities has a greater dependency than simply the integration of systems, but also the intelligent optimization of efficient management while saving energy. In this regard, wireless technologies are essential in V2V and V2X communications, as they allow ample connectivity and especially the intrinsic mobility of vehicles and other devices.

While implementing an IoV platform, including the collection, storage, analysis, and services related to vehicle data, they must load driving data, including instant locations and vehicle engine states, allowing control over vehicle maintenance, while still reflecting that this data can be used in real-time and offline vehicle management applications.

In the context of Smart Cities, quality of life is guaranteed by solid municipal infrastructure, water, electricity, and gas supply, and even considering roads and traffic as fundamental aspects in everyday life. In this sense, in order to improve efficiency and operational safety, investments are needed to update all these facilities with more intelligent resources, in general, using IoT technology.

With respect to intelligent traffic, the traffic light functions as an intersection control center. While in contact with vehicles approaching the intersecting roads, this can determine which closing time is most appropriate for each of them. Due to the sensors in the vehicle and the multimedia channels that provide information about direction, and route, considering a fully connected traffic connection. Still pondering that this system may even inform the ideal approach speed for each vehicle, which can appear on the vehicle panel, making a useful technology or even serving as a parameter for autonomous vehicles.

Still estimating the use of the various wireless technologies existing today, mainly those applicable to IoT, such as NBIoT, LoRA, Sigfox, or even LTE-M, which has properties related to the low power consumption on the device, long-range, and low transfer rate.

Through IoV it is still possible to employ a vision-based driver assistance system (V-ADAS), using machine learning technology (AI derivations) that improve the accuracy and speed of detection, object recognition, and even supporting the detection of lanes, vehicles, pedestrians, among others. Allowing motion analysis, multiple lens calibration, and panoramic vehicle monitoring, i.e., a true traffic assistant for drivers. It is also worth mentioning in this context, the use of versatile vision processing units (VPU) dealing with large volumes of dynamic visual data, providing a high level of functionality, with low energy consumption.

Along with the data collected in real-time by the vehicle's sensors, driving safety can be improved through a traffic center that receives information from vehicles approaching a point with traffic lights. In this sense, parameters such as speed and intention of direction (direction of the crossing of the road or corner) are informed, allowing these systems to make calculations to optimize the flow. Consequently, when receiving guidance on the approach speed and the lane to be occupied, the traffic light may open (green light) when a certain vehicle arrives or even providing a shorter stopping time. Research shows that systems like this can double the traffic capacity at intersections or, in some cases in cities, even eliminate the need for traffic lights. Still reflecting on emergency scenarios, traffic lights can be activated, for example, freeing the highways for ambulances or fire engines. However, intelligent traffic lights are a premise for optimizing traffic in large urban centers, but they are extremely relevant for autonomous vehicle traffic.

From an environmental perspective, remote data collection with respect to intelligent street lighting, air pollution measurement, and intelligent traffic control, can be managed more efficiently by saving energy, creating a more environmentally friendly society.

Another important factor is the behavioral aspect, i.e. an educational effect, in view of the traffic systems guiding the speed of approach of the vehicles, if the drivers do not respect the instructions. Still considering the integration of social aspects in IoV related to the vehicles' ability to establish social, static (between vehicles of the same automobile company) or dynamic (developed through V2V), with each other in an autonomous way, considering the restrictions from the owner.

With the emergence of social IoV, powerful applications such as transportation and logistics management, multimedia services, friend locators, content sharing, among others, can be widely applied. Since it solves many problems regarding limited connectivity between vehicles, limited connectivity between vehicles and road infrastructure, heterogeneity and privacy, and even effective content distribution, among other aspects.

Vehicle networks have characteristics that differ from most IoT applications, such as the need for very low latency for critical applications, reliability, and connectivity even in high-speed situations. Through these characteristics, the perception of network traffic status and road circumstances in real-time, enables decisions derived from analytical technologies such as machine learning (ML) and deep learning (DL). Both derivations of AI, helping the mechanism cognitive of resources to conduct a more effective control over vehicles increasing the efficiency of information sharing within vehicular networks.

However, for contexts like this to become increasingly real and allow the consolidation of IoV, it is necessary to overcome challenges related to high-speed mobility, delay sensitivity, i.e., the delay in communication in milliseconds, because in the face of the existence of congestion, accidents can occur; still correlating connectivity, information privacy, and resources.

IoV has the properties necessary to boost automotive maintenance, by sharing information about oil changes directly from the vehicle systems, this data is only shared with vehicle dealers or workshops, analyzing the data helping in possible repairs and maintenance of vehicles. Or even in another scenario where the vehicle's IoV system sends data that is running out of oil allowing the owner to place the order online, linking with manufacturers who will have information about the vehicle's conditions, being able to prepare products and maintain customers' vehicles.

Finally, with IoV technology, still combining AI, communication, and IoT sensing, with multimedia technology and data processing technology, it is possible to create a complete and highly integrated automotive solution allowing vehicle manufacturers to develop products that significantly improve the experience of intelligent consumer travel.

7 Conclusions

IoT is the interconnection of networks of physical, connected, and intelligent devices, or even vehicles, buildings, and other items incorporated in electronics, software, sensors, actuators, among others. What through network connectivity allows these objects to collect and exchange data by intelligently changing various existing research areas to new topics, including smart health, smart home, smart industry, and even smart transportation. Based on intelligent transport it is evolving as a new research and development theme from IoV.

The evolution of intra-vehicular networks with standards and protocols has the capacity to allow greater capacity and performance in data transmission together with the addition of the diversity of sensing systems and autonomous systems, representing an advance in-vehicle technologies. Still pondering the IoV technology, it allows the sharing of information between vehicles, roads, and the transportation environment, helping to guide and supervise vehicles, together with the provision of multimedia services and mobile internet.

In this respect, IoV is a complex integrated network system connecting vehicles, people in and around vehicles, intelligent on-board systems, and even various cyber-physical systems in urban environments. On the other hand, the slow evolution in telecommunications networks and road transport infrastructure. Also relating the evolution and development of autonomous vehicles, mobile technologies such as 5G, and even the improvement of Cloud Computing technologies, such as Edge and Fog.

In this sense, the scope of application of IoV is broad, since technology is of essential importance in the development of smart cities, allowing vehicles, connected smart devices, and even buildings and individuals to remain connected to each other. What from the point of view of connectivity, this goes beyond smartphones, implying in all sectors of society, which with respect to the road sector, noted for its economic and social importance, consists of the foundation of a technological revolution, having the IoV as a fundamental part for the development of cities that are increasingly intelligent and connected.

IoV is a technology that goes beyond telematics, reaching vehicle networks and intelligent transport by integrating these vehicles, IoT sensors, and mobile devices in a global network, enabling various digital services to be offered related to vehicle and transport systems, reaching people on board and around vehicles. Still reaching manufacturers of vehicles and automotive equipment creating new business models, offering digital services such as telemetry, Wi-Fi, infotainment, among others, in addition to monetizing data collected from accessibility and visualization of its users.

Despite the implementation of IoV, it still faces many challenges related to flexible and efficient connections, a guarantee of the quality of service, simultaneous requests for support, the security of data and information, and adequate information transmission. Through IoV technology it is possible to make vehicles more modern and dynamic, allowing passengers and drivers to have the possibility and benefits brought by the connection and all the practicality of using the technology.

8 Trends

Cooperation between connectivity technologies will enable IoV to bring together even more capabilities allowing for car design, engineering, and manufacturing and mobile software technology to further fuel the electric vehicle market [80].

Still pondering dynamic road signs for intelligent transport systems, which display the status of highways in real-time, toll rates, lane closures, and travel times, automatically transmitted from IoT sensors installed in vehicles [81].

IoV applied around logistics and transport can provide an active IoT ecosystem when mobile connectivity and vehicle networks serve vehicles and containers equipped with sensors. Creating a bidirectional channel for cloud technology, sharing, and receiving data, allowing location tracking to assisted and autonomous activation, noting the characteristic that IoV has to be able to transform the entire transport chain [82].

Assessing the context of commercial fleets, remote communication systems capable of analyzing vehicle performance and driver behavior, managing fuel consumption, through Telematics and tracking, allowing for optimized route planning and vehicle maintenance based on almost real-time data. IoV has capabilities through a set of computer services that allow vehicles to have their data recorded in detail, using it to guide other connected vehicles to brake and accelerate more efficiently in the face of road conditions [83].

IoV will provide an even more global scale with a focus on close-up, making it economically viable to use IoT sensors in vehicles, aiming at autonomous vehicles to become part of people's daily lives, with integrated cameras and several other features and properties in an automated and remote way [3].

IoV has properties to overcome mobility challenges, representing a powerful tool, related to highly congested urban centers, among which it is worth highlighting the reduction of travel time, the improvement in the safety and quality of roads, traffic instruments, and the transport service public transportation. Still considering the prioritization of public transport over private, as well as the expansion of accessibility of public space with the encouragement of walking by pedestrians and bicycles, on safer roads. Considering the time spent in traffic that reduces productivity and prevents the increase in GDP in developing countries, such as Brazil, which has a shortage in the supply of quality public transport and high rates of traffic accidents [84].

Still reflecting on the properties of IoV from the user's point of view in helping to provide information to the population about traffic in different locations in a city, offering types of services enhanced through data from social media and platforms that allow the monitoring traffic conditions by the users themselves, or even on situations that may impact traffic in the city [85].

However, it is worth mentioning the prospect of a secure environment against cyber-attacks and data loss related to the existence of numerous smart devices and IoT sensors connected to the network, leading to an abundance of possible attack vectors through this immense volume of data, and for that reason, security is needed at several levels, including containment of the IoT networks themselves, which is critical to reducing the risks of cybercrime, which can be achieved through joint use with Blockchain technology [86].

References

1. Priyan, M.K., Usha Devi, G.: A survey on internet of vehicles: applications, technologies, challenges and opportunities. Int. J. Adv. Intell Paradigms **12**(1–2), 98–119 (2019)
2. Sharma, S., Kaushik, B.: A survey on internet of vehicles: applications, security issues & solutions. Veh. Commun. **20**, 100182 (2019)
3. Hamid, U.Z.A., Zamzuri, H., Limbu, D.K.: Internet of vehicle (IoV) applications in expediting the implementation of smart highway of autonomous vehicle: A survey. In: Performability in Internet of Things, pp. 137–157. Springer, Cham (2019)
4. Sinha, N.: Emerging technology trends in vehicle-to-everything connectivity. In: 2019 Wireless Telecommunications Symposium (WTS). IEEE (2019)
5. Dabboussi, A.: Dependability approaches for mobile environment: Application on connected autonomous vehicles. Université Bourgogne Franche-Comté, Diss (2019)
6. El-hajj, M., et al.: A survey of internet of things (IoT) authentication schemes. Sensors **19**(5), 1141 (2019)
7. Storck, C.R., Duarte-Figueiredo, F.: A 5G V2X ecosystem providing Internet of vehicles. Sensors **19**(3), 550 (2019)
8. Lu, H., et al.: The cognitive internet of vehicles for autonomous driving. IEEE Netw. **33**(3), 65–73 (2019)
9. Nanda, A., et al.: Internet of autonomous vehicles communications security: overview, issues, and directions. IEEE Wirel. Commun. **26**(4), 60–65 (2019)
10. Sanchez-Iborra, R., et al.: Empowering the internet of vehicles with Multi-RAT 5G network slicing. Sensors **19**(14), 3107 (2019)

11. Zhou, H., et al.: Evolutionary V2X technologies toward the Internet of vehicles: challenges and opportunities. Proceedings of the IEEE 108.2 (2020): 308–323.
12. Wang, Ye., Liu, Z., Deng, W.: Anchor generation optimization and region of interest assignment for vehicle detection. Sensors **19**(5), 1089 (2019)
13. Kim, Y., et al.: Low-power RTL code generation for advanced CNN algorithms toward object detection in autonomous vehicles. Electronics **9**(3), 478 (2020)
14. Chen, M., et al.: Cognitive internet of vehicles. Comput. Commun. **120**, 58–70 (2018)
15. Wang, J., et al.: Internet of vehicles: Sensing-aided transportation information collection and diffusion. IEEE Trans. Vehicular Technol. **67**(5), 3813–3825 (2018)
16. Kong, Q., et al.: A privacy-preserving sensory data sharing scheme in Internet of Vehicles. Future Gener. Comput. Syst. **92**, 644–655 (2019)
17. Qiu, T., et al.: Community-aware data propagation with small-world feature for internet of vehicles. IEEE Commun. Mag. **56**(1), 86–91 (2018)
18. Teng, H., et al.: A novel code data dissemination scheme for Internet of Things through mobile vehicle of smart cities. Future Gener. Comput. Syst. **94**, 351–367 (2019)
19. Kumar, S., et al.: Delimitated anti jammer scheme for Internet of vehicle: Machine learning-based security approach. IEEE Access **7**, 113311–113323 (2019)
20. Wang, W., et al.: Vehicle trajectory clustering based on dynamic representation learning of internet of vehicles. IEEE Trans. Intell. Transp. Syst. (2020)
21. Phan-Huy, D.-T., et al.: Adaptive massive MIMO for fast-moving connected vehicles: it will work with predictor antennas! In: WSA 2018; 22nd International ITG Workshop on Smart Antennas, VDE (2018)
22. Sadiku, M.N.O., Tembely, M., Musa, S.M.: Internet of vehicles: an introduction. Int. J. Adv. Res. Comput. Sci. Softw. Eng. **8**(1), 11 (2018)
23. Zhang, J., Letaief, K.B.: Mobile edge intelligence and computing for the internet of vehicles. Proc. IEEE **108**(2), 246–261 (2019)
24. Adly, A.S.: Integrating vehicular technologies within the IoT environment: a case of Egypt. In: Connected Vehicles in the Internet of Things, pp. 85–100. Springer, Cham (2020)
25. Patel, S., Shah, V., Kansara, M.: Comparative study of 2G, 3G and 4G. Int. J. Sci. Res. Comput. Sci. Eng. Inf. Technol. **3**, 55–63 (2018)
26. Mishra, A.R.: Fundamentals of Network Planning and Optimisation 2G/3G/4G: Evolution to 5G. John Wiley & Sons, Hoboken (2018)
27. Starsinic, M.F.: Connecting IMSI-less devices to the EPC. U.S. Patent No. 9,930,613 (2018)
28. Mijić, D., Varga, E.: Unified IoT platform architecture platforms as major IoT building blocks. In: 2018 International Conference on Computing and Network Communications (CoCoNet). IEEE (2018)
29. Wu, Yu., Fang, X., Wang, X.: Mobility management through scalable C/U-plane decoupling in IoV networks. IEEE Commun. Mag. **57**(2), 122–129 (2019)
30. Lee, T.-Y., Lin, I.-A., Liao, R.-H.: Design of a FlexRay/Ethernet gateway and security mechanism for in-vehicle networks. Sensors **20**(3), 641 (2020)
31. Lokman, S.-F., Othman, A.T., Abu-Bakar, M.H.: Intrusion detection system for automotive Controller Area Network (CAN) bus system: a review. EURASIP J. Wirel. Commun. Netw. **2019**(1), 184 (2019)
32. Kuang, X., et al.: Intelligent connected vehicles: the industrial practices and impacts on automotive value-chains in China. Asia Pacific Bus. Rev. **24**(1), 1–21 (2018)
33. Mishra, A., et al.: Novel VPDU based framework in V2X communication for Internet of Vehicles (IoV) and performance evaluation of VAN in urban scenario. In: 2020 5th International Conference on Communication and Electronics Systems (ICCES). IEEE (2020)
34. Ang, L.-M., et al.: "Deployment of IoV for smart cities: applications, architecture, and challenges. IEEE Access **7**, 6473–6492 (2018)
35. Sahbi, R., Ghanemi, S., Djouani, R.:A network model for internet of vehicles based on SDN and cloud computing. In: 2018 6th International Conference on Wireless Networks and Mobile Communications (WINCOM). IEEE (2018)

36. Alouache, L., et al.: Survey on IoV routing protocols: security and network architecture. Int. J. Commun. Syst. **32**(2), e3849 (2019)
37. Yang, H., Xie, X., Kadoch, M.: Intelligent resource management based on reinforcement learning for ultra-reliable and low-latency IoV communication networks. IEEE Trans. Veh. Technol. **68**(5), 4157–4169 (2019)
38. Awathare, A.A., Malathi, P., Gaikwad, N.G.: Development of an offline simulation tool to test the on-board diagnostics software For BS-VI. i-Manager's J. Softw. Eng. **12**(3), 21 (2018)
39. Ramaguru, R., Sindhu, M., Sethumadhavan, M.: Blockchain for the internet of vehicles. In: International Conference on Advances in Computing and Data Sciences. Springer, Singapore (2019)
40. Sharma, T.P., Sharma, A.K.: Heterogeneous-internet of vehicles (Het-IoV) in twenty-first century: a Comprehensive Study. In: Handbook of Computer Networks and Cyber Security, pp. 555–584. Springer, Cham (2020)
41. Wang, X., et al.: Parallel internet of vehicles: ACP-based system architecture and behavioral modeling. IEEE Internet Things J. **7**(5), 3735–3746 (2020)
42. Hasenburg, J., et al.: Managing latency and excess data dissemination in fog-based publish/subscribe systems. In: 2020 IEEE International Conference on Fog Computing (ICFC). IEEE (2020)
43. Guillen, M., et al.: The use of telematics devices to improve automobile insurance rates. Risk Anal. **39**(3), 662–672 (2019)
44. Storck, C.R., Duarte-Figueiredo, F.: A survey of 5G technology evolution, standards, and infrastructure associated with vehicle-to-everything communications by internet of vehicles. IEEE Access **8**, 117593–117614 (2020)
45. Shen, X., Fantacci, R., Chen, S.: Internet of Vehicles. Proc. IEEE **108**(2), 242–245 (2020)
46. Huang, X., et al.: Secure roadside unit hotspot against eavesdropping based traffic analysis in edge computing-based internet of vehicles. IEEE Access **6**, 62371–62383 (2018)
47. França, R.P., et al.: Improvement of the transmission of information for ICT techniques through CBEDE methodology. In: Utilizing Educational Data Mining Techniques for Improved Learning: Emerging Research and Opportunities, pp. 13–34. IGI Global (2020)
48. Padilha, R., et al.: Betterment proposal to multipath fading channels potential to MIMO systems. In: Brazilian Technology Symposium. Springer, Cham (2018)
49. Raza, N., et al.: Social vehicle-to-everything (V2X) communication model for intelligent transportation systems based on 5G scenario. In: Proceedings of the 2nd International Conference on Future Networks and Distributed Systems (2018)
50. Chen, P.-Y., Cheng, S.-M., Sung, M.-H.: Analysis of data dissemination and control in social internet of vehicles. IEEE Internet Things J. **5**(4), 2467–2477 (2018)
51. Dai, Y., et al.: Artificial intelligence empowered edge computing and caching for internet of vehicles. IEEE Wirel. Commun. **26**(3), 12–18 (2019)
52. Ning, Z., et al.: Joint computing and caching in 5G-envisioned Internet of vehicles: a deep reinforcement learning-based traffic control system. IEEE Trans. Intell. Transp. Syst. (2020)
53. Darwish, T.S.J., Bakar, K.A.: Fog based intelligent transportation big data analytics in the internet of vehicles environment: motivations, architecture, challenges, and critical issues. IEEE Access **6**, 15679–15701 (2018)
54. Chen, L.-W., Ho, Y.-F.: Centimeter-grade metropolitan positioning for lane-level intelligent transportation systems based on the internet of vehicles. IEEE Trans. Industr. Inf. **15**(3), 1474–1485 (2018)
55. Wu, D., et al.: Similarity aware safety multimedia data transmission mechanism for Internet of vehicles. Future Gener. Comput. Syst. **99**, 609–623 (2019)
56. Sun, G., et al.: Security and privacy preservation in fog-based crowdsensing on the internet of vehicles. J. Netw. Comput. Appl. **134**, 89–99 (2019)
57. Ni, Y., et al.: Toward reliable and scalable internet of vehicles: performance analysis and resource management. Proceedings of the IEEE 108.2 (2019): 324–340.
58. Wang, X., et al.: A city-wide real-time traffic management system: enabling crowdsensing in social Internet of vehicles. IEEE Commun. Mag. **56**(9), 19–25 (2018)

59. Limbasiya, T., Das, D.: IoVCom: reliable comprehensive communication system for internet of vehicles. IEEE Trans. Depend. Secure Comput. (2019)
60. Ivanov, I., et al.: Cybersecurity standards and issues in V2X communications for Internet of Vehicles, pp. 46–56 (2018)
61. Senouci, O., Aliouat, Z., Harous, S.: MCA-V2I: a multi-hop clustering approach over vehicle-to-internet communication for improving VANETs performances. Future Gener. Comput. Syst. **96**, 309–323 (2019)
62. Ning, Z., et al.: Mobile edge computing-enabled Internet of vehicles: toward energy-efficient scheduling. IEEE Netw. **33**(5), 198–205 (2019)
63. Tanwar, S., et al.: Tactile Internet for autonomous vehicles: Latency and reliability analysis. IEEE Wirel. Communi. **26**(4), 66–72 (2019)
64. Atzori, L., et al.: Towards the implementation of the Social Internet of Vehicles. Comput. Netw. **147**, 132–145 (2018)
65. Cui, J., et al.: Privacy-preserving authentication using a double pseudonym for internet of vehicles. Sensors **18**(5), 1453 (2018)
66. Senouci, O., Harous, S., Aliouat, Z.: A new heuristic clustering algorithm based on RSU for internet of vehicles. Arab. J. Sci. Eng. **44**(11), 9735–9753 (2019)
67. Ligo, A.K., Peha, J.M.: Cost-effectiveness of sharing roadside infrastructure for Internet of Vehicles. IEEE Trans. Intell. Transp. Syst. **19**(7), 2362–2372 (2018)
68. Jain, B., et al.: A cross-layer protocol for traffic management in Social Internet of Vehicles. Future Gener. Comput. Syst. **82**, 707–714 (2018)
69. Fan, K., et al.: Permutation matrix encryption-based ultralightweight secure RFID scheme in internet of vehicles. Sensors **19**(1), 152 (2019)
70. Yu, X., et al.: Dynamic testing for RFID based on photoelectric sensing in Internet of Vehicles. In: Smart Innovations in Communication and Computational Sciences, pp. 319–327. Springer, Singapore (2019)
71. Zhang, D., et al.: Novel reliable routing method for engineering of internet of vehicles based on graph theory. Eng. Comput. (2018)
72. Qian, Y., et al.: Secure enforcement in cognitive internet of vehicles. IEEE Internet Things J. **5**(2), 1242–1250 (2018)
73. Butt, T.A., et al.: Privacy management in social internet of vehicles: review, challenges and blockchain-based solutions. IEEE Access **7**, 79694–79713 (2019)
74. Vivek, S., et al.: "Cyberphysical risks of hacked internet-connected vehicles. Phys. Rev. E **100**(1), 012316 (2019)
75. Joy, J., Rabsatt, V., Gerla, M.: Internet of Vehicles: Enabling safe, secure, and private vehicular crowdsourcing. Internet Technol. Lett. **1**(1), e16 (2018)
76. Smida, K., et al.: Software-defined Internet of Vehicles: a survey from QoS and scalability perspectives. In: 2019 15th International Wireless Communications & Mobile Computing Conference (IWCMC). IEEE (2019)
77. Tan, J., Gong, Li., Qin, X.: Global optimality under Internet of Vehicles: strategy to improve traffic safety and reduce energy dissipation. Sustainability **11**(17), 4541 (2019)
78. Gružauskas, V., Baskutis, S., Navickas, V.: Minimizing the trade-off between sustainability and cost-effective performance by using autonomous vehicles. J. Cleaner Prod. **184**, 709–717 (2018)
79. Habib, M.A., et al.: Security and privacy based access control model for internet of connected vehicles. Future Generat. Comput. Syst. **97**, 687–696 (2019)
80. Mondal, S., Nandi, D., Bera, R.: V2X communication test bed for smart electrical vehicle with 5G IOV technology. In: 2020 URSI Regional Conference on Radio Science (URSI-RCRS). IEEE (2020)
81. Ejaz, W., et al.: IoV-based deployment and scheduling of charging infrastructure in intelligent transportation systems. IEEE Sensors J. (2020)
82. Liu, D., et al.: Cold chain logistics information monitoring platform based on Internet of Vehicles. In: 2019 International Conference on Intelligent Transportation, Big Data & Smart City (ICITBS). IEEE (2019)

83. Hu, W.-C., et al.: Optimal route planning system for logistics vehicles based on artificial intelligence. J. Internet Technol. **21**(3), 757–764 (2020)
84. Cheng, J., et al.: Accessibility analysis and modeling for IoV in an urban scene. IEEE Trans. Vehicular Technol. **69**(4), 4246–4256 (2020)
85. Khan, Z., Koubaa, A., Farman, H.: Smart route: Internet-of-Vehicles (IoV)-based congestion detection and avoidance (IoV-Based CDA) using rerouting planning. Appl. Sci. **10**(13), 4541 (2020)
86. Kim, S.K.: Enhanced IoV Security Network by using Blockchain Governance Game (2019). arXiv preprint arXiv:1904.11340

AI-Enabled IoV Applications and Systems

Intelligent Approaches for Fault-Tolerance in Radio Communication of Autonomous Vehicles

Yazid Benazzouz, Oum-El-Kheir Aktouf, Rachid Boudour, and Jerry Gao

Abstract The development of autonomous vehicles is seen as a solution to many of today's societal problems such as traffic congestion, road accidents, theft prevention and air pollution. Radio technology offers a great contribution to these developments but continues to suffer from several shortcomings such as security breaches, radio interference, heterogeneity of protocols, vulnerability to climate changes, *etc.* The impact of these shortcomings is minimal in everyday life but can be very disastrous for autonomous vehicles where the loss of transmission, or misinterpretation of received data, can lead to a global malfunction or even create a domino effect that will spread throughout the entire system. Hence, we need intelligent strategies and methods to ensure fault-tolerance in the communication systems of autonomous vehicles.

1 Introduction

Automation has seen a growing interest in many different industrial sectors with a general goal to improve productivity. Regarding automated driving systems, piloting (control?) is nearly all carried out automatically, such as in drones, commercial flights, and space rockets. Vehicle automation is being implemented gradually, and there are varying degrees of automation, as presented in [1]. The standards organization Society of Automotive Engineers (SAE) International has defined 6 levels of automation.

- In Level 0 (No Automation), all aspects of the driving task are in the hands of the driver.

Y. Benazzouz (✉) · R. Boudour
Embedded system laboratory, Badji Mokhtar University, Annaba, Algeria

O.-E.-K. Aktouf
Institut polytechnique de Grenoble, Grenoble, France
e-mail: oum-el-kheir.aktouf@lcis.grenoble-inp.fr

J. Gao
Computer Engineering Department, San José State University, San Jose, USA

© Springer Nature Switzerland AG 2021   33
N. Magaia et al. (eds.), *Intelligent Technologies for Internet of Vehicles*, Internet of Things,
https://doi.org/10.1007/978-3-030-76493-7_2

- In level 1 (Driver Assistance), the automobile includes some built-in capabilities to operate the vehicle, such as assisting the driver with tasks like steering or acceleration/deceleration.
- At level 2 (Partial Automation), two or more automated functions work together to relieve the driver from some control operations. The driver must remain fully engaged with the driving task.
- In Level 3 (Conditional Automation), the driver still has to keep his eyes on the road, ready to take over at a moment's notice, but the vehicle can handle certain parts of the trip on its own – mainly highway driving.
- Level 4 (High Automation) suggests that the vehicle can essentially do all the driving, but the driver can intervene and take control if needed.
- However, Level 5 (Full Automation) considers humans as just passengers; a completely automated vehicle can perform all driving functions under all conditions.

Giant steps have already been taken for almost all levels and we believe that the full driving automation will soon become a daily reality. While the idea of using algorithms to totally control cars may sound challenging, users need time to fully embrace this new technology. The main consideration is that self-driving cars will be totally accepted if their levels of security and safety are the same as human driving ones [2]. The global traffic fatality risk associated with human errors is already high. Autonomous cars must improve safety on the road to reach an acceptable, or even better safety rate. The achievement of that objective would enhance our quality of life, the most glaring example of which is vehicle traffic management, especially in big cities, because it increases emissions, wastes time and productivity, and increases stress.

The current development of artificial intelligence is already beginning to actively contribute to the self-driving vehicles business. Some ideas [3] include package delivery vehicles, smart AI-based systems that provide data on available parking spots, locations and times to park, iSee (an AI and deep learning based system which allows cars to learn from data and negotiate any and all types of traffic conditions), and AI based Corner Cameras to identify people or objects situated in blind corners of the roads. Perhaps, one of the most fundamental questions, as reported by Buttice in [2], is: how can we determine the ability of AI compared to a human at not crashing when things go sour, such as when the weather is bad or when the user must drive on a steep slope or dirt road, or when a pedestrian unexpectedly steps into the road? Right now, we cannot – at least, not in a reliable way. Furthermore, the situation may get worse if hacking attempts (even failed ones) can tamper with the delicate controls of autonomous vehicles.

However, one major issue in autonomous vehicles design is that they are not yet capable of perfectly measuring risk and making decisions when they are confronted with an unusual situation for which they were not prepared. The need for autonomous vehicles to share information and to perceive their environment using a wide range of sensors, leads to exponential growth of the number of situations and therefore a risk of occurrence of unexpected and fuzzy situations. Loss of communication, even for a short time, is a prime cause of confusing situations, particularly if the message is necessary to autonomous vehicles to make a decision, for example sharing mutual localization to avoid collision.

Indeed, radio communication is a central element for autonomous vehicles in determining their position and sharing data with the infrastructure and other vehicles in the area. However, it could also be a source of faults and insecurity. Indeed, due to its complexity, dealing with loss of communication raises a number of challenges because the systems will not be prepared to react with the right actions when these situations occur. To achieve a quality of reactivity comparable to that of humans, or even better, three possible artificial intelligence (AI)-based approaches can be explored.

- The first one is learning, but it is difficult to implement because learning takes time. Putting a vehicle in a real environment requires a minimum of guarantees. In this case, learning will be complementary and not essential. Moreover, the evaluation of the quality of learning is often problematic after the vehicle is put into circulation. Furthermore, the question arises that different vehicles will have acquired different experiences.
- The second approach that seems to be more practical, is the collection and integration of expert knowledge. In this way, it is possible to integrate this knowledge into the vehicles via expert, probabilistic and planning systems.
- The last approach is to map the environment and generate possible scenarios of emergency situations while giving the vehicle the possibility to acquire new knowledge from past experience.

In this chapter, loss of inter-vehicle communication will be used as a case study to show how to deal with unexpected situations using the second AI-based approach (expert-knowledge-based approaches). This case study will serve as a tutorial example for the presentation of AI-based techniques. The next section highlights the problem to put it into the context. Section 3 will then present radio technologies and their usage in autonomous vehicles, as well as a fault review and existing fault-tolerance approaches. Then Sect. 4 explores artificial intelligence solutions to provide fault-tolerance mechanisms for wireless reliability during inter-vehicle messaging. In addition, the chapter will present the case study that will show how AI could be used to resolve disruptions in inter-vehicle communications (Sect. 5). Finally, before concluding this chapter, some thoughts about AI application to radio communication fault management are provided in Sect. 6, while future challenges and research opportunities are summarized in Sect. 7.

2 Problem Overview

Fault-tolerance has been extensively studied in the literature in an effort to address failures within a system. If the system is in interaction with other entities, internal faults should be known by all entities in the interaction to prevent the propagation of the fault to other systems, or to cause the occurrence of operating challenges

Fig. 1 Illustration example

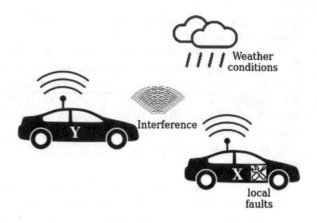

that could lead to damage. This is even truer in a vehicle-to-vehicle (V2V) context. Vehicles have internal faults that should be known by other adjacent vehicles to avoid hazardous situations. Short range wireless technology represents the preferred means of communication but using wireless technology in real-time safety-critical applications is a challenging task. Wireless technologies cannot provide guarantees that transmitted packets will actually be delivered on time [1]. Furthermore, the loss of wireless messages can have a very negative impact so putting the passengers lives in danger.

Let us consider the example of Fig. 1. If a vehicle (X) has steering issues and lacks the communication capabilities to inform the nearest vehicles, this may lead to an accident. Now, if we suppose that the vehicles are using a broadcast protocol to inform others about local faults, loss of steering in this case, the vehicle (Y) which is almost directly behind, can use this information to take evasive maneuvers to avoid accidents.

Inter-vehicles communication has the advantage of providing enhanced safety for autonomous vehicles. However, the implementation of this solution presents some challenges, amongst others, the optimization of the communications to ensure an efficient continuous exchange of information between vehicles; and the robustness with regards to other factors affecting this exchange like bad weather and wireless interference.

This study addresses the wireless propagation of local faults. There are three situations that will be used as a tutorial for this chapter:

- (a) A local fault occurs and a message reporting the fault is shared with the surrounding vehicles.
- (b) A local fault occurs, and the message is lost.
- (c) No fault message is shared with the surrounding vehicles.

The dynamic nature of the system composed by an autonomous vehicle and its surrounding vehicles makes it very hard to handle such faults. In the next section, we review the main features of radio communication technologies in autonomous vehicles and some existing fault-tolerance solutions. We will see throughout the chapter how AI approaches can enhance radio communication fault-tolerance in autonomous vehicles.

3 Radio Technology Usage in Autonomous Vehicles

This section is a short survey of radio communication technology that is used in the area of autonomous vehicles. Its goal is to highlight both the issues and the difficulties when using this technology.

3.1 Radio Technologies in AV

The real-time nature of data issued from sensors and inter-vehicle communications in autonomous vehicles creates the need for high speed communication protocols and technologies. In addition, due to the criticality of AV, these inter-vehicle communications must be reliable and secure [4]. VANET (Vehicular Ad-hoc NETwork) is the representative protocol for inter-vehicle communications. It encompasses different communication types, known as V2X communications:

- V2V: Vehicle-to-Vehicle.
- V2I: Vehicle-to-Infrastructure.
- V2P: Vehicle-to-Pedestrian.
- V2N: Vehicle-to-Network.

It is worth mentioning that some studies are currently investigating the introduction of the promising ICN (Information-Centric Networking) technologies in inter-vehicle communications. ICN is expected to provide better security, scalability, and content distribution. However, this brings many challenges and issues [5] that are out of the scope of this chapter. Other technologies for inter-vehicle communications exist, such as low power technologies, IEEE 802.11 family technologies, and base station driven technologies, just to mention a few of them. So, in this chapter, we focus on the VANET protocol and more precisely, on the V2V communication.

3.1.1 VANET

VANET is the declination of MANET (Mobile Ad-hoc NETwork) on traffic road vehicle communications with their environment (other vehicles, pedestrians, *etc.*). VANET has been introduced into vehicles because of the advancement in the design of vehicles that has made them endowed with sensors, computing, and processing capabilities, as well as communication functions. In VANET, vehicles are considered as being wireless nodes [4]. This section presents VANET architecture and its main features. Further details can be found in [6].

Basics

VANET is an open network where nodes are free to join and leave the network. A network is formed by moving vehicles and other devices that can connect to each other over a wireless medium. A VANET network can be considered as being completely connected as each node in the network receives information sent by all the other nodes. The main objectives through the communications are to increase the efficiency of the network and to ensure road safety conditions, based on exchanged information. Indeed, the two main applications of VANET are driver assistance and information dissemination. The main features of VANET are [6]:

- Dynamic topology, that changes frequently because of moving nodes.
- Intermittent connectivity, where two communicating devices can be disconnected at any time because of the dynamic topology of the underlying network.
- Mobility patterns: mobility patterns are deduced from traffic rules and help in designing specific routing protocols.

VANET architecture

The architecture of VANET encompasses different plans: node or vehicle level, network level, road level, and even cloud level for vehicular cloud systems.

- Road and network level architecture:
At this level, the main components found are the ones necessary for allowing safe traffic management, such as road infrastructure (lights *etc.*) and vehicle and traffic management centers. Road Side Units (RSUs) are components connected to the backbone network and fixed on the road in different positions so facilitating communication between vehicles.
- Node level architecture:
Vehicles compliant with the VANET protocol are equipped with radio interfaces enabling the formation of short-range wireless ad hoc networks among On-Board Units [OBUs) [7]. In addition, they have different sensors allowing the gathering of data from the environment (LIDAR, GPS, Ultrasonic radar).

3.1.2 Fault Models in Inter-vehicle Communication

If we focus on the V2V communication, the main faults and/or errors that could affect the inter-vehicle communication are: communication drops [8], failures in message delivery, contention, and broadcast storm [9] and also link disruption [10]. We need to distinguish between dense traffic and sparse density traffic to establish the most likely fault model. In dense traffic, contention and broadcast storms are the most observed faulty behaviors. In sparse traffic, failures in message delivery and communication drops occur more frequently. Link disruptions are mainly due to the highly dynamic environment. As these fault models are related to the communication behavior between vehicles, it is worth noting that AI techniques can play an interesting role for communication fault detection and/or diagnosis. Indeed, most AI techniques can analyze patterns of common behavior and then raise alarms if relevant [11]. The goal is mainly to proactively detect and diagnose the cause of abnormal network behavior and to propose, and if possible, take corrective actions [11, 12].

3.2 The General Fault-Tolerance Process

Fault-tolerance is usually performed through a process that includes: fault detection; fault diagnosis; and fault recovery. Fault detection is performed by checking the system's state and/or behavior related data. Any deviation from the expected state or behavior can be due to a fault occurrence, and is determined by testing, monitoring, or any other approach that enables the deviation from the expected system state or behavior to be determined. These are usually statistical approaches, like the Principal Component analysis method [13].

Fault diagnosis aims at locating the root cause of the faulty behavior. It generally consists in determining the faulty item within the system, whether it is a variable, a message, or an operation. Fault diagnosis is performed with classification methods (simple pattern classification, Bayes classification, neural networks, *etc.*) or inference methods (approximate reasoning, hybrid neuro-fuzzy systems) [13].

Recovery actions are performed by fault masking or system reconfiguration [14]. Fault masking is possible when redundancy is used. So, fault-free spare components are used and allow system correct functioning, despite the occurrence of faults in some components. Fault recovery can be performed through backwards or forwards recovery. Backwards recovery allows the system to return to a previous correct state and continue executing. It needs regular checkpoints which can be costly in terms of memory usage. Forwards recovery is used mainly in real-time systems. It consists in bringing the system into a new fault-free state, from which the system continues executing.

From this very brief overview, we can say that AI techniques are mainly used in fault detection and fault diagnosis steps of the general fault-tolerance process. Some works, such as [11, 12], have also investigated the fault recovery step using AI techniques. With the booming of research work in AI, we are expecting that AI techniques will also be increasingly used in fault recovery and system reconfiguration after fault diagnosis [15]. Before considering more deeply AI-based fault-tolerance approaches in V2V (Sect. 4), Sect. 3 below provides an overview of some representative traditional fault-tolerance solutions that have been developed for V2V.

3.3 Existing Fault-Tolerant Radio Technologies in AV

Many approaches have been proposed for making V2V communications safe and fault-tolerant. A complete survey of these approaches would need a chapter by itself and is out of the scope of this study. However, it is worth mentioning some of the fault-tolerant technologies. So, in this section an overview of some fault-tolerant technologies for V2V communication in autonomous vehicles is provided, without claiming completeness. Two classifications of fault-tolerant approaches for autonomous vehicles have been encountered. The first one is based on the existence or not of a central entity, and the second one considers passive/active approaches.

3.3.1 Central vs Distributed Approaches

In [16], the authors present a peer-to-peer approach for ensuring V2V communications. Their objective is to overcome the limitation of centralized structures which present the single-point-of failure issue. In the proposed model, vehicles are considered as relays for information exchanges. The authors consider a distributed consensus algorithm between nodes (vehicles) to implement fault-tolerance and improve communication quality between vehicles. Nodes are not considered to be malicious.

Other distributed approaches rely on in-network data aggregation and majority vote. In this scheme, neighboring vehicles agree on a given information (*e.g.*, the average speed) and sign the corresponding aggregate to be disseminated using less bandwidth [17]. Authors in [10] deal with the link disruption issue and develop MOCA (Mechanism for connectivity management in Cognitive vehicular networks). This mechanism periodically evaluates values related to some observable parameters that indicate the communication channels quality. Thus, the MOCA solution copes with the highly dynamic nature of VANET and adapts the communication link between two neighboring nodes.

In [18], the authors develop an autonomous decentralized system architecture that supports online extensibility, online maintainability, and fault-tolerance that are critical to setup a system with continuous operation. The implementation details of this architecture are presented, focusing on the mobility feature of the underlying network and the improvement of end to end communications.

Extensibility of fault-tolerance solutions in VANET is also addressed in [9]. In this work, the authors adapt the multi-hop broadcast-based communication scheme used in VANET for disseminating safety messages, to the actual network state (dense or sparse network) thus allowing reliable and timely delivery of safety messages. The proposed protocol selects specific nodes for the broadcast allowing to reduce the network load in terms of messages.

Fault-tolerant path planning between connected vehicles is considered in [19]. In this work, the authors present a cooperative approach where vehicles share their path plans to allow new path calculation. In case of communication faults that prevent correct plan transmission the authors propose a Bayesian-based approach for allowing a vehicle to cope with lacking information from other nodes and determine to the best, the most suited paths. In [20], authors adapt classical shortest path calculation in order to allow nodes (vehicles) to find the most reliable path to destination, based on the underlying street map. In this work, a reliability rating or weight of each street edge is provided using traffic monitoring data. Obtained weights allow to calculate the most reliable paths.

3.3.2 Passive vs Active Approach

Fault-tolerant approaches for autonomous vehicles can also be categorized as passive or active [21]:

- In the passive approach, the design of the communication system involves robust components in such a way that the communication is made resilient against faults by design.
- In the active approach, redundancy techniques and fault detection and identification methods are used to handle faults that may occur, so allowing the communication system to be fault-tolerant.

Passive approaches are costly as designing robust components that can reliably adapt to real-time environment is a hard topic. Furthermore, relying on passive approaches necessitates frequent component updates, which is not convenient in case of hardware components for example. This is why, most of the proposed approaches and those reviewed in Sect. 3.3.1. range in the active approaches class. This is clearly justified by the nature of VANET which needs real-time and proactive methods for dealing with communication issues between vehicles. Nevertheless, passive approaches can still be found in some works, like in [22] where a solution based on the combination of active and passive replications is presented. This solution is developed in the general scheme of multi-agent systems and applied to VANET as a case study. A general discussion on active and passive replication approaches for fault-tolerance in IoT systems is presented in [23].

3.3.3 Miscellaneous Approaches

Other existing works on fault-tolerant approaches for V2V communications and the VANET protocol in general have focused mainly on the modelling part related to faults and fault management. These works are mainly theoretical and provide a good basis for modelling and evaluating fault-tolerance approaches. For instance, topology-based and graph-based approaches are presented in [24] and [25] to model the communication networks, and especially, faulty, and unreliable communication links. In [26], a probabilistic approach is used for dealing with the problem of fault-tolerant distributed sampling towards uniform probabilistic distribution in dynamic multi-hop wireless networks. In [8], the authors tackle the issue of modelling several fault types in V2V communications. In [27], the authors discuss the specific phase during which manned and unmanned vehicles coexist, and how to deal with traffic safety.

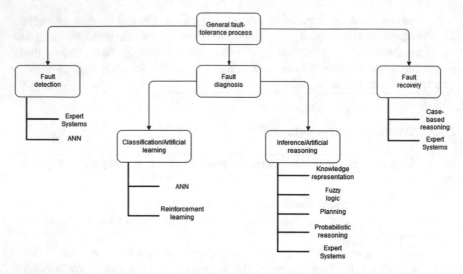

Fig. 2 AI-based techniques for fault-tolerance

Some other works like in [28] develop specific and cutting-edge solutions. The authors propose to use Commercial Unmanned Aerial Vehicles (UAVs) or what are commonly referred to as drones in cooperation with VANET to improve the reliability of the data delivery by coping with the high mobility of nodes in VANET and bridging the communication gap whenever it is possible. Although interesting, this solution can be costly by adding extra equipments and generating additional messages.

In the next section, we introduce the different ways AI can be used as a tool for enhancing fault-tolerance in V2V communications, by reviewing basic AI techniques and showing through illustrations, how these techniques can contribute to better fault-tolerant V2V communications.

4   Artificial Intelligence as a Tool for Fault-Tolerance

AI-based fault-tolerance approaches have the advantage of benefiting from traditional AI methods to provide intelligent process automation for fault-tolerance using, for example, logical reasoning and models extracted from data.

This section highlights a selection of artificial intelligence techniques described in the literature and how to apply them to improve reliability of autonomous vehicles. Figure 2 represents different AI approaches by type of faults. They are split into two different sections: artificial reasoning and artificial learning. Emphasis will be put on enhancing fault-tolerance based on safety information sharing, and inter-vehicle wireless communication.

4.1   Artificial Reasoning

Artificial Reasoning is a central topic that explores models of human thinking. For clarity, artificial reasoning is defined as a group of techniques and methods that aim to automate the human reasoning process based on logic reasoning, terms and concepts that make sense in a given field of application. Among these techniques we can find knowledge representation, expert systems, fuzzy systems, probabilistic reasoning, and planning. The ability of computer models to reason like humans is something that can be debated, like Addis did in his book "Natural and Artificial Reasoning" [29].

4.1.1   Expert Systems

An expert system is defined as a computerized, interactive, and reliable tool that uses both facts and heuristics to solve complex decision-making problems. It is based on knowledge acquired at the highest level of human intelligence and expertise. Expert systems are mostly involved in system diagnostics and management. For example, they have been used to assist a mechanic service in diagnosing vehicle faults [30], such as faults of automotive engines. An expert system [31] provides fault diagnosis of engines by providing systematic and step-by-step analysis of the failure symptoms and offering maintenance or service advice. In telecommunication networks [32], an expert system has been used for network failure management *i.e.,* a group of functions for the detection, diagnosis, recovery, and correction of shortcomings of the elements that form the network.

Example. We consider the example of Sect. 2 and, more precisely, the problem tagged (a). In this example, an internal fault such as loss of brakes, engine-fire, loss-of-steering or flat-tire occurs at vehicle **X**. A message is sent to all surrounding vehicles containing a description of the fault. Vehicle **Y** receives the message and triggers the fault manager which is an expert system to determine what is the best action the vehicle should accomplish. The principle is to summarize the drivers knowledge and imitate their reaction when the same kind of fault occurs. This can be determined by survey design, or simulation, to develop an expert system. The following rule issues a request to slow-down if the vehicle which has a loss of brakes is at the front-end with regards to the current vehicle position. It is assumed that each autonomous vehicle is able to perceive its surrounding vehicles. The rules are written in CLIPS (C Language Integrated Production System).[1] This is an expert system shell from NASA. The full example is detailed in Sect. 5.

```
(defrule brakes-emergency ''rule 1''
    (and (emergency (type loss-of-brakes))
         (vehicle (location front-end)))
```

---

[1] http://www.clipsrules.net/.

```
=>
(and (assert (response (action slow-down)))
     (printout t ''Slow-down'' crlf )))
```

4.1.2 Knowledge Representation

Hinchey [33] puts an emphasis on knowledge representation (KR) concepts and their role in the self-adaptation behavior. To achieve fault-tolerance, representation techniques [34] provide the means of describing the system operation to ensure the continuity of execution when the primary process fails. A hot standby sparing process can immediately take over the system operation based on the information summarized and stored in the system's structure. This could be, for example, a checkpoint strategy in case of software redundancy or a complete knowledge redundancy for hardware redundancy.

Example. The description in 4.1 gives only a partial view of the expert system. A real implementation includes a **Knowledge Base** which is a collection of facts. Each time a fault message is received by an autonomous vehicle, its content is transformed into facts and inserted into the knowledge base. Then, the inference engine will apply rules to the facts and resolve any rule conflictions whenever multiple rules are applicable to a particular case. The knowledge base is kept consistent by considering the life-time of the facts. For instance, cf. Figure 1, vehicle **X** is broadcasting a message about loss of brakes. Surrounding vehicles receiving the message will process and eliminate it when the right actions are provided by the inference engine.

```
(assert (emergency
     (type loss-of-brakes)))
```

Using CLIPS syntax, the list of facts are presented as follows:

```
(facts)
f-0     (initial-fact)
f-1     (emergency (type loss-of-brakes))
f-2     (vehicle (location front-end))
f-3     (emergency (type flat-tire))
```

4.1.3 Fuzzy Logic

This is an area of active research in artificial intelligence domain. Shahnaz *et al.* [35] provide an overview and a recent perspective on decision-making in a fuzzy environment.

The concept of fuzzy logic has been used to overcome the shortcomings of classical logic; i.e., the inability to manipulate data representing subjective or unclear concepts. It was designed to allow the computer to determine distinctions between data, which are neither true nor false, somewhat similar to the process of human reasoning, *e.g.*, slow speed.

A general implementation of fuzzy systems contains 4 parts:

- Rules base (Rules) contains all the rules and "if-then" conditions proposed by the experts to control the decision-making system.
- Fuzzification allows the conversion of inputs into fuzzy sets. For example, inputs such as weather conditions: temperature, pressure, wind speed, level of radio interference, *etc.* are measured by sensors and transmitted to the control system for processing.
- An inference engine helps in determining the degree of correspondence between fuzzy inputs and rules. On the basis of the matching percentage, it determines which rules should be implemented according to the given input field. Then, the applied rules are combined to develop control actions.
- Defuzzification is performed to convert fuzzy sets to exact values. Fuzzy logic helps to cope with uncertainty in engineering and is mainly robust because no precise input is required.

Fuzzy logic has been applied to various fields, from control theory to artificial intelligence. In [36], a two layer fuzzy logic system is used to monitor vehicle operation to detect and compensate for any faults or abnormal system behavior. The lower layer is to detect the presence of a faulty system, and the higher layer is used to pinpoint which system is faulty. The fuzzy rules used in the lower layer are generated from experience and acquired data and are used in the form of IF-THEN statements.

Example. The expert system proposed in 4.1 and 4.2 is not enough to model all types of data faults. For example, suppose for any reason vehicle **X** has a low fuel tank. If the vehicle stops suddenly this could result in an accident. A practical way to share this fault with surrounding vehicles is to use fuzzy data in the expert system. This can be expressed using FuzzyClips,[2] a fuzzy logic extension of the CLIPS language. The fuel tank level is represented by the fuzzy membership function of Fig. 3. There are two actions the artificial intelligence driver can take when a fault message is received by autonomous vehicle **Y**: slow-down or overtake of Fig. 4. Both are defined as fuzzy depending on the fuel level of vehicle **X**. If it is low it might be safe to slow-down rather than to overtake.

```
(deftemplate fuel
  0 100 gallons
  (
   (low (Z 15 30))
   (average (pi 25 50))
   (full (S 70 85))
   (empty extremely low)
```

---

[2]https://github.com/rorchard/FuzzyCLIPS.

```
    )
  )

  (deftemplate Act
    0 100 move
    (
      (slow_down (Z 40 60))
      (overtake (S  40 60))
    )
  )

  (deftemplate message
      (slot vehicle_id)
      (slot fuel)
      (slot location)
  )

  ;fuzzify the inputs

  (defrule fuzzify
    (message (vehicle_id ?a) (fuel ?b) (location ?c))
   =>
    (assert (fuel (pi 0 ?b )))
  )
  ;; defuzzify the outputs
  (defrule defuzzify1
  (declare (salience -1))
  ?f <- (Act ?)
  =>
  (bind ?t (moment-defuzzify ?f))
  (printout t ``action--> '' ?t crlf))

  (defrule slow_down
      (fuel low)
    =>
    (and (assert (Act slow_down))
         (printout t ``Slow-down'' crlf))
  )

  (defrule overtake
      ( or
         (fuel average)
         (fuel full)
      )
    =>
```

```
(assert (Act overtake))
)
```

The reception of a fuel fault message is converted into the following fact.

```
;(assert (fuelValue 10))

(deffacts fuzzy-fact
   (message (vehicle_id mz) (fuel 10) (location ahead))
)
```

The value of the fuel level, 10, will be evaluated as being low and will trigger the action **slow-down**.

```
f-0       (initial-fact) CF 1.00
f-1       (message (vehicle_id mz) (fuel 10) (location ahead))
          CF 1.00
f-2       (fuel ???) CF 1.00
( (10.0 0.0) (10.0 1.0) (10.0 0.0)   )

f 3       (Act slow_down) CF 1.00
( (40.0 1.0) (42.5 0.9688) (45.0 0.875) (47.5 0.7188)
(50.0 0.5) (52.5 0.2812) (55.0 0.125) (57.5 0.03125)
(60.0 0.0)   )
```

Fig. 3 Matching of fuzzy fuel facts

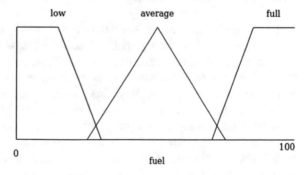

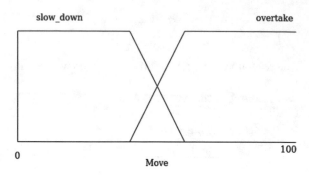

Fig. 4 Matching of fuzzy actions facts

4.1.4 Planning

Planning means developing an action plan to achieve objectives. Unlike other problems, such as classification, planners provide guarantees on the quality of the solution. Two examples of planning methods are search-based problem-solving and hybrid propositional logic. As our interest lies in the unusual or unexpected situation issue, the approach proposed in [34] uses planning and learning to cope with unexpected real-time situations. Because it is not possible to predetermine and code all possible situations, the idea was to store rules instead of code, and dynamically generate plans based on real-time situations. Then, by identifying an unexpected situation that occurs once, the unexpected situation can be included as one of the expected situations to be planned ahead in the future.

4.1.5 Probabilistic Reasoning

Artificial intelligence enables the modelling of intelligent behavior through logical reasoning. Probability theory provides a consistent basis for how belief should change in light of partial or uncertain information. Typically, network representations are used to group facts into structures in the form of causal chains and inheritance hierarchies. In the first place, Bayesian methods provide a formalism for reasoning about partial beliefs under conditions of uncertainty [37, 38].

After distinguishing the possible values of the random variables (*i.e.,* their spaces), we judge the probabilities of the random variables having their values. We ascertain probabilities concerning relationships among random variables that are accessible to us. We can then reason with these variables using Bayes' theorem to obtain probabilities of events of interest. The next example illustrates this idea.

Example. Many things may cause harmful interference to radio autonomous vehicles inter-communications, such as direct satellite services, wireless speakers, and an abundance of other wireless devices. For example, to evaluate the possibility of losing periodic messages during communication, an uncertainty mechanism could be used as in the following. The example below assumes that vehicles are equipped with a radio interference sensor and periodic messages about internal faults are broadcast. If, after a certain period of time, the message is not received by vehicle **Y** of Fig. 1, an uncertainty value about wireless message loss is calculated. Using a radio interference sensor detector, the value is converted to a fact and added to the knowledge base and then processed by the uncertainty rule.

```
(defrule certainty-rule
    (declare (CF 0.7))
    (radio_interference high)
  =>
    (assert (message  loss))
)


(deffacts fuzzy-uncertainty-fact
    (radio_interference high) CF 0.9
)
```

If the interference is measured as high then the certainty of message loss is about 0.63.

```
f-0      (initial-fact) CF 1.00
f-1      (radio_interference high) CF 0.90
f-2      (message loss) CF 0.63
```

4.2   Artificial Learning

Artificial intelligence learning brings together learning techniques and methods that often draw on the processes of nature. A particular example is the theory of artificial neural networks, which seeks to mimic the functioning of the human brain. Learning enhances system reliability by the ability to construct fault models, using fault data collection offline, that can be validated using simulation and then online real time operations.

4.2.1   Artificial Neural Networks

Artificial neural networks (ANN) simulate the functioning of the biological neuron by an activation function to build complex networks, linking neurons in different configurations. These networks are then trained on large volumes of data in order to capture the knowledge carried by this data. A large volume of data is needed to allow the neural network to correct the weights of the connections between neurons so that they best reflect the data being used. Once the expected accuracy and quality of learning is achieved, the network is then used to produce an estimator of the new data.

Deep learning models are much more resource-intensive and often, several configurations have to be re-tried by modifying, for example, the number of layers and the type of activation functions, *etc.* to produce a good model. They are used when large volumes of data are available and their structure is difficult to manipulate with the usual learning algorithms, such as linear regression, decision trees, support vector machines (SVMs), naive Bayes, *etc.*

Artificial neural networks are used as fault detectors by estimating changes in process model dynamics [39]. According to [40], neural networks are used for fault detection and identification (FDI). They have the advantage of preventing usage of sensor redundancy, although the required computational power could be higher if multiple estimators are used. The authors demonstrated through this work that it is possible to replace a bank of neural networks with a single one, supporting 70 fault scenarios, to reduce computational cost and to use simplified programming. In contrast to the Kalman Estimators, neural networks present an increased False Alarm Rate (FAR) mainly due to a very small residual that remains after fault estimation. Nevertheless, ANNs are widely used since they can work without having precise and formal knowledge of the system.

Example. An example of Neural networks usage for fault-tolerance could be taken from [41]. The idea was to monitor and approximate any abnormal behavior using neural networks. In the same way, it is possible to generate decision rules which summarize the human driving experience to decide what action to do when certain faults occur in a real-time situation for an autonomous vehicle. All data for deep learning models could be generated by simulation by including different parameters such as fuel level, weather conditions, radio interference, and so on. The following rule means that during simulation, and in 70% cases, the driver overtakes when the fuel level of the vehicle in front is low.

```
(defrule down
   (declare (CF 0.7))
   (fuel low)
  =>
  (assert (Act overtake))
)
```

4.2.2 Reinforcement Learning

Reinforcement learning is the use of observed rewards to learn an optimal (or near-optimal) policy to adopt in a given environment [42]. This can be difficult to put into practice for autonomous vehicles because any error can be very costly. However, the model can be produced by simulation and then deployed in real situations.

Particularly, reinforcement learning can be used to create robust fault models by simulating different kinds of faults. Positive reinforcement occurs when there is a positive impact on the actions taken by the autonomous driver. However, too much reinforcement can lead to over-optimization and will have a bad impact on some specific situations when other driving moves are a better choice. Negative reinforcement occurs when a propagated fault leads to a bad situation, an accident for example. This type of reinforcement provides sufficient means to avoid some actions that are more appropriate for a given situation, but because of an unpredictable event, an accident happened. Let us take the following example.

Example. Instead of generating decision rules from simulated driving data (cf. example 4.5), it is possible to train a reinforcement model by giving appropriate "rewards" in case a bad move is engaged after fault propagation. This way, the program adapts itself to improve its piloting. Assume the driver can perform one of three possible actions: **slow-down**, **gain speed**, **overtake**; and a unique fault "low fuel level" of vehicle **X** of Fig. 1. In this case, both actions: **overtake** and **slow-down**, will have positive rewards but only one of them, which might not be the better choice, is considered.

5   Case Study

The case study examines the interaction between two autonomous vehicles. The goal is to investigate a reliable artificial intelligence solution to wireless message loss during V2V interaction, case (b) of the example presented in Sect. 2. For this case study, and to make things easy to understand, it is assumed that vehicles labeled **X** and **Y** are equipped with sensors measuring radio wireless interference, and that they operate under the following weather conditions: wind **W**, rain **R**, snow **S**, wind and rain **WR**. Moreover, the use case supposes 4 types of vehicle faults: loss of brakes, engine-fire, loss-of-steering and flat-tire. Figure 5 summarizes the reliable solution which is instantiated from the fault-tolerance process of Sect. 3.2. In addition to a fault propagation model, the solution employs two artificial intelligence methods: Bayesian inference and expert system. The critical part of this process is the instant between fault detection and fault handling by diagnosis and recovery. Some faults might already cause damage before other vehicles handle a propagated fault. The sooner the fault is detected and processed, the higher is the chance to avoid an accident.

V2V fault's propagation is based on **simplex transmission**. It is a one-way only communication standard that broadcasts information about internal vehicle faults to other autonomous vehicles. This information may only travel in one direction, the sender does not know what happens at the receiver. All the decisions to handle the propagated faults are taken at the receiver side. The only assumption made in the method is that a vehicle is able to identify surrounding vehicles using, for example, video recognition technology which is more robust and secure than radio commu-

Fig. 5 Artificial intelligence
based fault-tolerance
solution

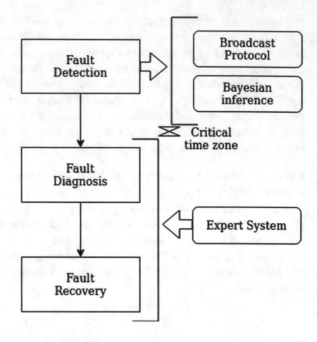

nication. To this end, a list of surrounding vehicles is kept continuously updated.
A vehicle using this protocol starts by periodically sending a keep alive message if
there is no internal fault *i.e.,* I am OK!. If a fault occurs a message with the fault
and vehicle ID and position is sent to surrounding vehicles. The analysis of possible
situations gives the following:

- Any vehicle receiving a keep alive message continues broadcasting its own keep
  alive message to surrounding vehicles.
- When a fault message is received, the expert system diagnoses the fault and triggers
  appropriate actions.
- If no keep alive message is received and the vehicle is always in the surrounding list,
  then the loss of wireless message is evaluated using Bayesian inference and based
  on the environmental conditions. If there is no loss fault, the vehicle continues
  broadcasting its own keep alive message. But in case the loss of wireless message
  is confirmed by Bayesian inference, then the expert system will trigger appropriate
  actions to avoid a potential accident.

5.1 Propagated-Fault Detection

The broadcasting protocol is a state machine model representing the interaction
between vehicles **X** and **Y**, depicted by Figs. 6 and 7. It is normal that the protocols
are equivalent on both sides. The symbols **!** and **?** followed by the type of the message

Fig. 6 Vehicle X

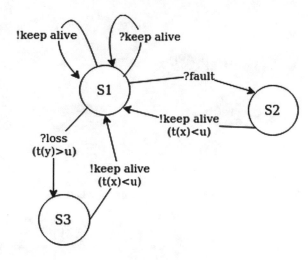

Fig. 7 Vehicle Y

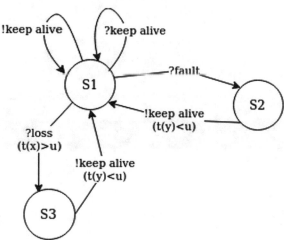

indicate if the message is being transmitted or received, respectively. The protocol uses a timeout value **u**. Each sender should ensure sending a live message before the timeout, else other vehicles, **Y** in this case, might resolve this fact as a message loss. The determination of the timeout value is closely related to the need of a balance between fault detection and fault handling, addressed previously in this section.

The probability of wireless message loss is computed using a Bayesian network (BN) represented by Fig. 8. BN is composed of three variables (weather conditions, radio interference and loss accuracy nodes) and their conditional dependencies (directed edges) which, together, form a directed acyclic graph (DAG). A conditional probability table (CPT) is associated with each node. It contains the conditional probability distribution of the node given its parents in the DAG. The conditional probability table (CPT) of accuracy node is initialized at the beginning. It indicates

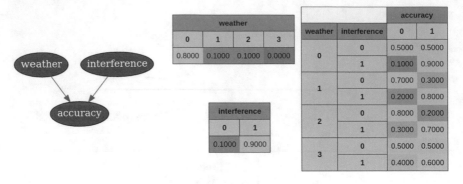

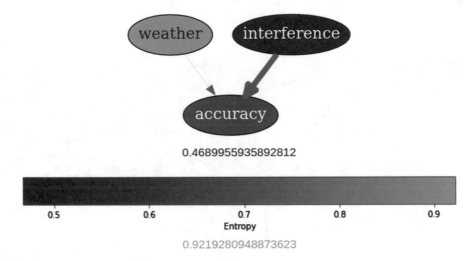

Fig. 8 Bayesian network for wireless message loss

0.4689955935892812

0.9219280948873623

Fig. 9 Bayesian network entropy

the contribution of weather and interference nodes to wireless message loss. It can be determined statistically, or by a human expert, before deploying the system in autonomous vehicles. In contrast, the weather and interference nodes are determined at run-time and injected to the Bayesian network to determine the probability that a wireless message is lost. A running example is given by CPT tables of Fig. 8. Weather conditions are named respectively: WR (0), W(1), R(2) and S(3). Interference states are low (0) and high (1), and the final accuracy has 2 states: not lost(0) and lost (1).

This example is developed using pyAgrum.[3] It is a Python library dedicated to Bayesian Networks and other Probabilistic Graphical Models.

```
<accuracy:0> :: 0.172 /<accuracy:1> :: 0.828
```

---

[3] https://pyagrum.readthedocs.io/.

The probability obtained by doing inference in a Bayesian network can be used to reach a decision. The wireless message is lost with a probability of 0.828. Figure 9 shows the impact of the weather condition and radio interference on the wireless message loss. The entropy calculated by pyAgrum shows the big contribution of radio interference information.

Even though, the Bayesian network itself does not recommend a decision. This brings us back to the expert systems in the next section.

5.2 Fault Diagnosis and Recovery

The preceding section shows how to compute the conditional loss probability of wireless messages by sensing the autonomous vehicle environment. Expert system is still a feasible option for decision making in the fault diagnosis and recovery context. The following CLIPS code summarizes the decisions of the smart driver for this case study. The knowledge of the expert reflects the decision that a human driver can take if a similar problem happens.

```
(deftemplate emergency
     (slot type))

(deftemplate response
     (slot action))

(deftemplate vehicle
     (slot location))

(defrule brakes-emergency ''rule 1''
   (and (emergency (type loss-of-brakes))
        (vehicle (location ahead)))
  =>
  (and (assert (response (action slow-down)))
       (printout t ''emergency call to free the road'' crlf )))

(defrule steering-emergency ''rule 2''
   (and (emergency (type loss-of-steering))
        (vehicle (location ahead)))
  =>
  (and (assert (response (action hold on)))
       (printout t ''Slow-down'' crlf )))

(defrule engine-emergency ''rule 3''
   (or  (and (emergency (type insufficient-engine-power))
```

```
        (vehicle (location ahead)))
        (and (emergency (type engine-caught-fire))
        (vehicle (location ahead))))
   =>
   (and (assert (response (action take left)))
        (printout t ''Slow-down'' crlf )))

(defrule tire-emergency ''rule 4''
   (and (emergency (type flat-tire))
        (vehicle (location ahead)))
   =>
   (and (assert (response (action slow-down)))
        (printout t ''Slow-down'' crlf )))
```

The insertion of the facts below requests a smart driver to slow-down.

```
(assert (vehicle
    (location ahead)))
(assert (emergency
    (type flat-tire)))
```

6 Thoughts About AI Application to Wireless Propagated Fault Management

Artificial intelligence based on machine learning models uses huge volumes of data and it is incapable of reasoning. Its models are black boxes. Thus, it is difficult to interpret or to verify their own validity and safety. Conversely, decision support systems require the definition of policies, and need human expert knowledge to reason. It seems that the mixture between learning, reasoning, and uncertainty constitutes the foundation of fault analysis for major systems. Autonomous behavior, such as autonomous driving requirements cannot be ensured by only one method. A behavior is a complex system where errors and risks cannot be excluded.

7 Future Challenges and AI Opportunities Toward Reliable Autonomous Vehicles

Monitoring faults in autonomous vehicles is a complex and difficult task due to changes in operating conditions, external disturbances, vehicle to vehicle variability, *etc.* According to [34], two aspects reveal the deficiency in fault-tolerance consideration in most autonomous applications: (1) how to react to unexpected situations, and (2) how to provide continuous uninterrupted control of the process in the event of computer failure without putting human health at risk, *e.g.* when the expert system portion fails due to hardware failure. AI can play an important role in modelling, interpreting, and analyzing complex autonomous systems behavior, as well as their environments. AI can also offer foundations for fault recovery solutions that ensure a safe and secure functioning of autonomous systems despite the occurrence of faults.

8   Conclusion

It is admitted that autonomous vehicles will captivate our future. Today, there are many attempts and experiments to provide a safe solution for using them. The problem with that is to believe autonomous safety is largely supported. If we refer to the safety systems in airplanes, history has proven that whatever safety management system is put in place, faults still occur and sometimes lead to dramatic accidents. Safety and fault-tolerance for autonomous vehicles should be a continuous process, even after commissioning. Actual advancement in artificial intelligence could help to resolve many of them. Indeed, the autonomous process is one of the core processes of artificial intelligence, widely reported by the literature. This chapter has shown that it is possible to combine one, or many, proven methods to tackle very hard problems, such as loss of wireless critical messages. Some precursor methods that are judged useless, such as expert systems, can even make a major contribution to safety enhancement.

References

1. Brown, D.S.: Driverless cars: Levels of autonomy (Oct 2019). https://www.techopedia.com/driverless-cars-levels-of-autonomy/2/33449. Accessed 01 June 2020
2. Buttice, C.: Hacking autonomous vehicles: Is this why we don't have self-driving cars yet (Feb 2019). https://www.techopedia.com/hacking-autonomous-vehicles-is-this-why-we-dont-have-self-driving-cars-yet/2/33650. Accessed 01 June 2020
3. Pal, K.: The 5 most amazing AI advances in autonomous driving (Apr 2018). https://www.techopedia.com/the-5-most-amazing-ai-advances-in-autonomous-driving/2/33178. Accessed 01 June 2020
4. Wang, J., Liu, J., Kato, N.: Networking and communications in autonomous driving: a survey. IEEE Commun. Surv. Tutor. **21**(2), 1243–1274 (2019)
5. Din, I.U., Kim, B.S., Hassan, S., Guizani, M., Atiquzzaman, M., Rodrigues, J.J.P.C.: Information-centric network-based vehicular communications: overview and research opportunities. Sensors **18**(11), 3957 (2018)
6. Tomar, R., Prateek, M., Sastry, G.H.: Vehicular adhoc network (VANET) - an introduction. Int. J. Control Theor. Appl. **9**(18), 8883–8888 (2016). https://hal.archives-ouvertes.fr/hal-01496806
7. Mchergui, A., Moulahi, T., Alaya, B., Nasri, S.: A survey and comparative study of QoS aware broadcasting techniques in VANET. Telecommun. Syst. **66**, 253–281 (2017)

8. Lu, X.Y., Shladover, S.E., Kailas, A., Altan, O.D.: Messages for cooperative adaptive cruise control using V2V communication in real traffic. In: Hu, F. (ed.) Vehicle-to-Vehicle and Vehicle-to-Infrastructure Communications - A Technical Approach, Chapter 8. CRC Press (2018)

9. Oliveira, R., Montez, C., Boukerche, A., Wangham, M.S.: Reliable data dissemination protocol for VANET traffic safety applications. Ad Hoc Netw. Elsevier Ltd. **63**, 30–44 (2017)

10. Silva, C., Nogueira, M., Kim, D., Cerqueira, E., Santos, A.: Cognitive radio based connectivity management for resilient end-to-end communications in VANETs. Comput. Commun. **79**, 1–8 (2016)

11. Gurer, D., Ogier, R.G.: An artificial intelligence approach to network fault management. In: 3rd IEEE International Conference on Trends in Electronics and Informatics (ICOEI), pp. 916–920. Tirunelveli, India (1996)

12. Khalid, A., Umer, T., Afzal, M.K., Anjum, S., Sohail, A., Asif, H.M.: Autonomous data driven surveillance and rectification system using in-vehicle sensors for intelligent transportation systems (ITS). Comput. Netw. **139**, 109–118 (2018)

13. Isermann, R.: Fault-Diagnosis Systems. An Introduction from Fault Detection to Fault-Tolerance. Springer-Verlag, Berlin (2006)

14. Koren, I., Krishna, C.M.: Fault-Tolerant Systems, 2nd edn. Morgan Kaufmann, San Francisco (2021). Copyright Elsevier Inc

15. Cioroaica, E., Chren, S., Aktouf, O.E.K., Larsson, A., Chillarege, R., Kuhn, T., Schneider, D., Wolschke, D.: Towards creation of automated prediction systems for trust and dependability evaluation. In: The 28th International Conference on Software, Telecommunications and Computer Networks (SoftCOM 2020) (2020)

16. Yang, L., Mo, T., Li, H.: Research on V2V communication based on peer to peer network. In: 2018 International Conference on Intelligent Autonomous Systems (ICoIAS), pp. 105–110 (2018)

17. Dietzel, S., Gürtler, J., Kargl, F.: A resilient in-network aggregation mechanism for VANETs based on dissemination redundancy. Ad Hoc Netw. **37**, 101–109 (2016). Special Issue on Advances in Vehicular Networks

18. Hiraiwa, M., Ishikawa, H., Aizono, T., Sugiyama, K., Shimura, A., Shinonaga, H.: Implementation and evaluation of autonomous decentralized system based mobile communications platform for ITS services. In: The Sixth International Symposium on Autonomous Decentralized Systems, 2003. ISADS 2003, pp. 277–284 (2003)

19. Saraoglu, M., Hart, F., Morozov, A., Janschek, K.: Fault-tolerant path planning in networked vehicle systems in presence of communication failures. IFAC-PapersOnLine **51**(23), 82–87 (2018). 7th IFAC Workshop on Distributed Estimation and Control in Networked Systems NECSYS 2018

20. Bernsen, J., Manivannan, D.: River: a reliable inter-vehicular routing protocol for vehicular ad hoc networks. Comput. Netw. **56**(17), 3795–3807 (2012)

21. Stetter, R.: Fault-Tolerant Design and Control of Automated Vehicles and Processes. Springer International Publishing, Cham (2020). Insights for the Synthesis of Intelligent Systems

22. Alvi, A.B., Hashmi, M.A., Chuhan, Z.H., Atif, M., Ahmed, I.: Adaptive byzantine fault-tolerance support for agent-oriented systems: the BDARX. Int. J. Adv. Appl. Sci. **6**(2), 57–64 (2019)

23. Rullo, A., Serra, E., Lobo, J.: Redundancy as a Measure of Fault-Tolerance for the Internet of Things: A Review, vol. 11550, pp. 202–226. Springer International Publishing, Cham (2019)

24. Gilbert, S., Lynch, N.A., Newport, C., Pajak, D.: On simple back-off in unreliable radio networks. Theor. Comput. Sci. **806**, 489–508 (2020)

25. Samyoun, S., Rahman, A., Rab, R.: Design and analysis of a fault-tolerant topology control algorithm for wireless multi-hop networks. In: 5th International Conference on Networking, Systems and Security, NSysS 2018, Dhaka, Bangladesh, 18–20 Dec 2018, pp. 37–45. IEEE (2018)

26. Yuan, Y., Li, F., Yu, D., Yu, J., Wu, Y.,. Lv, W., Cheng, X.: Fast fault-tolerant sampling via random walk in dynamic networks. In: 2019 IEEE 39th International Conference on Distributed Computing Systems (ICDCS), pp. 536–544 (2019)

27. Emara, S., Elewa, A., Wasil, O., Moustafa, K., Khalek, N.A., Soliman, A.H., Halawa, H., ElSalamouny, M., Daoud, R., Amer, H., Khattab, A., ElSayed, H., Refaat, T.: Heterogeneous ITS architecture for manned and unmanned cars in suburban areas. In: 2018 IEEE 23rd International Conference on Emerging Technologies and Factory Automation (ETFA), vol. 1, pp. 918–925 (2018)
28. Oubbati, O.S., Lakas, A., Zhou, F., Güneş, M., Lagraa, N., Yagoubi, M.B.: Intelligent UAV-assisted routing protocol for urban VANETs. Comput. Commun. **107**, 93–111 (2017)
29. Addis, T.: Natural and Artificial Reasoning. Springer International Publishing, Berlin (2014)
30. Takahashi, R.: Automobile troubleshooting expert system "ATREX". Future Gener. Comput. Syst. **5**(1), 97–101 (1989)
31. Gelgele, H.L., Wang, K.: An expert system for engine fault diagnosis: development and application. J. Intell. Manuf. **9**(6), 539–545 (1998)
32. Monedero, I., Martín, A., Elena, J., Guerrero, J., Biscarri, F., León, C.: A practical overview of expert systems in telecommunication networks, medicine and power supplies. In: Reiter, A.C., Segura, J.M. (eds.) Expert System Software: Engineering, Advantages and Applications, pp. 178–210. Nova Science Publishers Inc. (Mar 2012)
33. Hinchey, M.: Toward Artificial Intelligence through Knowledge Representation for Awareness, pp. 121–138. Wiley-IEEE Press (2018)
34. Bastani, F.B., Chen, I.-R.: The role of artificial intelligence in fault-tolerant process-control systems. In: Proceedings of the 1st International Conference on Industrial and Engineering Applications of Artificial Intelligence and Expert Systems - Volume 2, IEA/AIE 1988, pp. 1049–1058, New York, NY, USA (1988). Association for Computing Machinery
35. Zadeh, L.A.: Recent Developments in Fuzzy Logic and Fuzzy Sets, vol. 391. Springer International Publishing, Berlin (2020)
36. Solliman, A., Rizzoni, G., Kim, Y.W.: Diagnosis of automotive emission control system using fuzzy inference. IFAC Proc. Vol., **30**(18), 715–720 (1997). IFAC Symposium on Fault Detection, Supervision and Safety for Technical Processes (SAFEPROCESS 97), Kingston upon Hull, UK (Aug 1997)
37. Pearl, J.: Chapter 2 - Bayesian inference. In: Pearl, J. (ed.) Probabilistic Reasoning in Intelligent Systems, pp. 29 – 75. Morgan Kaufmann, San Francisco (1988). http://www.sciencedirect.com/science/article/pii/B9780080514895500084
38. Budiharto, W.: The development of an expert car failure diagnosis system with Bayesian approach. J. Comput. Sci. **9**(10), 1383–1388 (2013)
39. Vargas-Martínez, A., Garza-Castañón, L.: Combining artificial intelligence and advanced techniques in fault-tolerant control. J. Appl. Res. Tech. **9**, 202–226 (2011)
40. Michail, K., Deliparaschos, K.M., Tzafestas, S.G., Zolotas, A.C.: AI-based actuator/sensor fault detection with low computational cost for industrial applications. IEEE Trans. Control Syst. Tech. **24**(1), 293–301 (2016)
41. Garza-Castañón, L., Vargas-Martínez, A.: Artificial intelligence methods in fault-tolerant control. In: Rodic, A. (ed.) Automation Control - Theory and Practice, Chapter 15. Institute Mihajlo Pupin, Serbia (Dec 2009)
42. Russell, S., Norvig, P.: Artificial Intelligence: A Modern Approach, 3rd edn. Prentice Hall, Hoboken (2010)

AI-Based Traffic Queue Detection for IoV Safety Services in 5G Networks

Simone Grilli and Gianmarco Panza

Abstract With the deployment of 5th Generation networks, characterized by a wider bandwidth and an increased computational capability, is now possible to develop more complex services that requires low latency, such as applications for public safety based on artificial intelligence. To this aim, it has been studied a possible use of the so-called Multi-access Edge Computing (MEC) enabled by 5G that reduces the latency thanks to computational resources located closer to the user. This allows to deploy an application that recognizes the formation of traffic queues on the highway through the analysis of video streams, to be used in the context of smart mobility. In order to do so, it is needed to detect the travelling vehicles and to track their movement to understand when a traffic jam is occurring. For the implementation, the Convolutional Neural Network (CNN) paradigm has been leveraged for the detection of the vehicles. Among the several alternatives compared, it has been chosen the third version of You Only Look Once (YOLO) for its trade-off between accuracy and real-time computation. Then, the detections tracked through the Simple Online and Realtime Tracking (SORT) algorithm are exploited to identify the direction of the traffic flows and then to calculate when the vehicles are slowing down or stopping, by either measuring the number of stationary vehicles or the travelling time of the vehicles within a region of interest. The service has been developed with the objective of being employed through the parallel computation offered by 5G MEC servers equipped with modern GPUs, in order to obtain real-time performances.

1 Motivation

During the past few years, there have been a massive increase in the number of devices connected to the Internet, with a subsequent rise of the data volume transmitted that requires higher computation performance and a reduced latency. The 5G

S. Grilli (✉) · G. Panza
Cefriel, Milan, Italy
e-mail: simone.grilli@cefriel.com

G. Panza
e-mail: gianmarco.panza@cefriel.com

© Springer Nature Switzerland AG 2021   61
N. Magaia et al. (eds.), *Intelligent Technologies for Internet of Vehicles*, Internet of Things,
https://doi.org/10.1007/978-3-030-76493-7_3

architecture, thanks to the higher performances provided, allows the development of new use cases such as applications of public safety, Internet of Vehicles (IoVs), Internet of Things (IoT) and autonomous driving. Among the technological innovations introduced by 5G, this research focuses on one in particular: Multi-access Edge Computing (MEC). The purpose of MEC servers is to relocate the computation currently placed in centralized data-centres toward the edge of the network, thus closer to the end-users, in order to reduce the latency due to crossing the whole network. The availability of a high computational capability so close to the user, therefore, paves the way to the development of elaborated algorithms that are in this way mainly limited by the existing hardware. Specifically, the recent improvements of hardware accelerators and hardware architectures in general, make it possible the real-time running of Artificial Intelligence solutions (AI). An example of such an AI solution is the Artificial Neural Networks (ANNs) which represent a model of the biological neural networks widely used in solving complex problems thanks to their ability of being trained on specific tasks. One case is object classification and, to this aim, in the past years one kind of ANN has been studied and improved: the Convolutional Neural Networks (CNN). However, CNNs per-se are not capable to perform the object detection task so they need to be extended, but the complexity of the architecture of these models is a burden that limits the enforceability of AI applications with common general purpose hardware. For this reason, the idea is to exploit MEC computational resources close to the end-user to implement a real-time service for detecting the formation of traffic queues on the road as highways. For this purpose, it has been chosen to leverage on a CNN-based model to develop a multi-layered application that allows to analyse video streams in a real-time fashion. Therefore, several object detection approaches have been considered and their performances compared, both in terms of execution speed as well as accuracy. Then, it has been investigated a way to determine the formation of traffic queues. To this end, an object tracking algorithm has been applied for defining the trajectory of travelling vehicles, necessary to assess when a traffic jam is occurring. On such bases, two solutions have been advanced. The first measures the number of stationary vehicles, whereas the second calculates the average travelling time. These approaches have been then compared keeping in mind the real-time requirements.

Previous works have tried to provide a solution to this problem. In [1] it has been developed an algorithm based on optical flow to estimate the velocity of moving pixels in order to detect variations from normal conditions that can be indicators of an accident, but this approach have resulted far from real-time performances. In [2] edge detection algorithms have been employed to detect the shape of travelling vehicles, then their position is tracked among frames and their speeds estimated. However, the main drawback of these two methods is that they perform only vehicle detection but not classification, that can leave room future improvements through the detection of specific classes of vehicles. The tasks of vehicle detection and classification have been discussed in [3], where it has been proposed a variant of the Faster R-CNN algorithm analysed in Sect. 3.1.4 to detect and classify vehicles travelling on a highway. Nevertheless, in spite of the increase of the performance, the CNN developed runs at 15 FPS. By considering the overhead that would have been introduced by the queue detection algorithm, a faster alternative had to be taken into account. In [4] it has been employed an algorithm based on Adaboost cascade classifiers for detecting vehicles in order to measure the length of a queue. Even though the approach of cascade classifiers has been superseded by the growth of deep neural networks, the authors have considered the greatest computational costs introduced by CNNs, an aspect that brought them to discard the latters because of the employment of equipment with limited capabilities.

Since the aim of this work is to study how the increase of computational power provided by 5G MEC servers in proximity to the traffic sources enables the development of innovative solutions that were not possible with the current hardware (available in cloud data centres), it has been studied an application of complex artificial neural networks to the IoV use case.

In the following section, the concept of CNN as a model of the biological visual cortex will be introduced. In Sect. 3, different artificial intelligence approaches will be analysed and compared, and in Sect. 4 the service developed will be described in details along the results obtained. Lastly, in Sect. 5 considerations about future works will be discussed.

2 From the Biological Neurons to the Artificial Neural Networks

Artificial Neural Networks (ANNs) were born with the aim of emulating biological neural networks. The *feed-forward* neural network is modelled as a series of layers composed by several different neurons. Each neuron is connected to a bunch of neurons from the previous layer, and every link has its own *weight* (biologically, the weights represent the synapses of a neuron). The output of the previous neurons is multiplied by the corresponding weights and the incoming values are then added by the neuron along with a *bias*, that represent the threshold above which the neuron "fires" its action potential. The result of this operation is passed to an activation function which calculates the output that will be propagated to the following neurons.

The purpose of ANNs is to learn a function that maps the input with an output, by starting with a dataset of examples, called *training set*. There are several kinds of ways in which a network can learn:

- **Supervised learning**

  The training set is composed by a series of *labelled* data (i.e. a set of images containing objects whose classes to which they correspond are specified in the labels), in order to let the network learning a function that maps the data to the corresponding labels. In this case, learning means to iteratively change the values of the weights (randomly initialized at the beginning) and the biases to reduce as much as possible the prediction error. A second dataset, called *test set*, is finally exploited to understand if the network is able to generalize on the data that has not seen in the training stage. In case the network has to predict an element of a finite set of values, this is called *classification*. Otherwise, with the *regression* the network has to predict a number.

- **Unsupervised learning**
  As opposed, in the unsupervised learning the training set is not labelled, so the model has the task to discover a structure that subdivides the data. One example is the *clustering* technique, in which the data is grouped in sets based on their similarity.
- **Semi-supervised learning**
  It's a combination of the previous approaches. The training set here is represented by data both labelled as well as unlabelled.
- **Reinforcement learning**
  In this case, the network learns by trying to maximize a cumulative reward, obtained after a sequence of actions taken in a given situation with the purpose of reaching a goal.

In particular, the supervised learning is what interests us the most for the task of image classification. In this type of learning, the training stage takes place in two steps:

- **Forward pass**
  In this phase, the training set is propagated along the network. Each unit of the input layer conveys the value of the input itself, whereas the nodes in the following layers calculate the output through the activation function, as previously explained.
- **Back-propagation**
  After the forward pass, the difference among the vector containing the real estimations and the vector of the predictions is determined. The aim is to minimize this difference through the derivative of the gradient of this error with respect to the weight that is propagated from the output layer back to the input layer. In particular, it is considered the error function E, depending on the all weights of the network. The weights connecting the $i - th$ unit of one layer to the $j - th$ node in the following layer can be represented as w_{ij}, and they are multiplied by the output value of the $i - th$ layer, that is o_i. During the back-propagation stage, the gradient of the error it is calculated as

$$\frac{\partial E}{\partial w_{ij}} \tag{1}$$

for all the weights in the network. Since the error is a function of the output of the previous unit, which is in turn a function of the incoming weights multiplied by the incoming input, then the above formula can be calculated through the chain rule

$$\frac{\partial E}{\partial w_{ij}} = \frac{\partial E}{\partial o_j} \frac{\partial o_j}{\partial w_{ij}} = \frac{\partial E}{\partial o_j} \frac{\partial o_j}{\partial (\sum_i w_{ij} o_i)} \frac{\partial (\sum_i w_{ij} o_i)}{\partial w_{ij}} \tag{2}$$

Each weight is updated by adding the gradient obtained with the Eq. 1 multiplied by a learning rate γ, which determines the speed (or step size) at which the error decreases after each iteration.

$$w_{ij} = w_{ij} - \gamma \frac{\partial E}{\partial w_{ij}} \qquad (3)$$

The negative sign indicates that the gradient converges toward the local minimum, according to the gradient descent optimization.

To evaluate the performance of the model, it is possible to exploit another dataset, the *test set*. In case the network is able to predict the label for the inputs on which it has not been trained, then it means that the model is generalizing well. However, if the neural network is able to generalize with the examples seen in the training set but not in presence of unseen data, this indicates a situation called *overfitting* that is caused by an excess of parameters of the network. It is possible to measure the overfitting during the training by dividing the training set in two distinct subsets, one used to train the network and the other as a *validation set*. The latter is then employed not to train the model by adjusting the weights, but to evaluate the presence of the overfitting on unseen data, in order to be able to adjust the parameters of the network accordingly [5–7].

2.1 Convolutional Neural Network: A Model of the Visual Cortex

As it has already been mentioned, ANNs have risen after studies on biological nervous system. However, in the human body there are distinct sensory receptors that convey the stimuli in specific regions in the brain cortex, such as the Primary Visual Cortex (also called *V1*) which receives the information related to the vision. In the 60s, David Hubel and Torsten Wiesel have discovered a hierarchical structure of the neurons in visual cortex with receptive fields of increasing complexity, namely simple and complex receptive fields. The first, also called simple cells, receive the inputs from the Lateral Geniculate Bodies (LGB), that are the points in which stimuli coming from the retinas of the eyes converge. This kind of cells present distinct excitatory and inhibitory regions, making them responding selectively to different stimuli (i.e. the orientation of an edge or its movement). Thus, a light stimulus hitting the excitatory region will result in the firing on the neuron, whereas the stimulation of the inhibitory region will cause the suppression of the neuron response. These regions are arranged in parallel, sidelong to each other, separated by a straight-line boundary and whose orientation can differ from the one of other simple cells. Therefore, it has been possible to predict how a neuron would have reacted to specific shape and orientation of the stimuli (both stationary as well as in movement) by knowing the layout of the two regions. A group of simple cells conveys in a complex cell, that is a cell with a complex receptive field, which responds to a stimulus with a particular orientation independently on its position on the retina, because of an aggregation of the activations from different simple cells with the same arrangement and orientation of the receptive field. The latter aspect results in a tolerance to variations in the position of the input, however the movement has to happen in a given direction [8]. In further studies [9], Hubel and Wiesel have found another class of cells, the hyper-complex, subdivided in lower order and higher order, receiving inputs from groups of complex cells. In the first case, the excitatory region of the hypercomplex cell is given by one (or more) complex cell, whereas the inhibitory region of the hypercomplex cell is provided by complex cells with receptive field with different orientation. The stimulation of the excitatory receptive field would then result in the activation of the hypercomplex cell, but the additional stimulation of the inhibitory region as well would lead to no response. This means that the hypercomplex cell is selective to the width of the provided stimuli, in addition to their orientation and movement, resulting in a more selective response. Likewise, higher-order hypercomplex cells are able to respond to orthogonal stimuli coming from different lower-order hypercomplex cells, bringing to a greater specialization. Therefore, by progressively aggregating the presented stimuli, the biological brain is able to recognize different complex patterns starting from simple edges. This aspect has been leveraged when the visual cortex has been modelled through the CNNs. The origin of these networks dates back with the Neocognitron [10] developed in 1980 by Kunihiko Fukushima for the pattern recognition of handwritten digits. The Neocognitron was based on the unsupervised learning, but the following CNN models are trained through a supervised approach. In the late 80s, Yann LeCun applied the back-propagation to the CNN model to train the network for the identification of handwritten zip code digits task [11, 12].

During the past years, several re-workings of the CNN model have been devel-oped, notably LeNet [13], AlexNet [14], ZFNet [15], GoogleNet [16], VGGNet [17], ResNet [18] and DenseNet [19].

The CNN architecture is widely used in image recognition, because of its ability to aggregate patterns through the layers of the network, in a manner that resembles the biological visual system. As the biological system, in CNNs each unit has a receptive field that covers only a small portion of the input, that can be the given image or the output of the preceding layer. To extract features such as edges, it is performed the dot product between the values of the receptive field and a set of weights, called filter, that are shared by all the units in order to reduce the number of parameters that the network should learn. The resulting values are then summed along with the bias. Thus, the filter is a window that slides all over the input, in order to detect the same pattern in different positions in the image and to produce an output matrix. This is finally passed to a non-linear activation function to compute the *feature map*. In order to extract different features, multiple filters are used in the same layer. Then, the following layers combine these maps to extract more structured features. The convolutional layers are usually followed by or interleaved with *pooling layers*, also known as sub-sampling layers, that aggregates the values of neighbour pixels by taking, for instance, the maximum, the sum or the average of them. This process allows to summarize the activations resulting in the invariance to small translations and distortions, just as the complex cells in the primary visual cortex. Finally, the last layer of the network comprises a classifier that produces a class probability distribution for any image [6, 13].

3 Object Detection

Convolutional Neural Networks (CNNs) have been created to emulate the biological visual system, that operates through the progressive aggregation of patterns increasingly complex in order to classify objects that appear in images. Nevertheless, this approach per-se fails to localize multiple objects in the same image. For the object detection, several models have been proposed and they can be classified according to whether they operate in two stages or in one stage. In the first case, the classification of the object performed by the convolutional layers is preceded by a region proposal method, which extracts several regions that can contain an object. In the second case, both the localization and classification of the objects in the image are made directly by the convolutional layers. In this section the different approaches are analysed and compared, in order to understand which one could be applied in the use case of the AI-based Traffic Queue Detector.

3.1 Two-Stage Object Detection

Among these approaches, it is possible to find Region-based Convolutional Neural Network (R-CNN), Spatial Pyramid Pooling network (SPP-net), Fast R-CNN and Faster R-CNN.

3.1.1 R-CNN

The R-CNN model [20] considers an external region proposal algorithm, which extracts from an image a set of regions that probably contain an object that should be then classified. An example is represented by Selective Search, in which several segmentation of the image are created at the very beginning by grouping similar pixels. Then, these clusters are recursively aggregated into hierarchically larger groups. In this way, the region proposal module generates 2000 regions to be classified in the second stage. Here, a CNN with five convolutional layers and two fully-connected layers is implemented. The convolutional network takes as input the region proposed with fixed dimensions and produces a feature vector for each of those. Then, the outputs are presented sorted by activation values in decreasing order and a *non-maximum suppression* is performed to eliminate the non optimal regions. Lastly, a series of linear classifier specifically trained for each class carry out the object classification according to the features extracted in the previous layers.

Moreover, R-CNN implements a stage called *Bounding Box Regression*, with the aim of increase the precision in the object localization. During the training, along the region proposed, the algorithm takes as input a set of coordinates representing the ground truth box, that is the box which contains the object in the image of the training set. Then, the weights are modified in order to learn a mapping of the proposed bounding box with the ground-truth box, by iteratively calculating the distances among the coordinates and by updating the weights accordingly.

3.1.2   SPP-net

The main problem of R-CNN and, more in general, of convolutional networks, is the fixed width and height of the input image required by the fully convolutional layers. This fact can lead to distortions, like warping and cropping, that can compromise the accuracy in object detection. SPP-net [21] introduces a layer, namely Spatial Pyramid Pooling (SPP), between the convolutional layers and the fully connected one. This allows to overcome the limitation represented by the fixed dimensionality of the input, by subdividing the image in a fixed number of regions with decreasing size. In other words, grids with increased granularity are applied to the given image. Within each region, the max-pooling operation is then performed. Thus, each cell of the grid sums up the values of neighbouring pixels by taking the maximum of them. This concept recalls the *Spatial Pyramid Matching* introduced by [22] that was a readjustment of the original *Pyramid Matching* used to discover the similarity among two images [23]. In Pyramid Matching, the authors considered a feature space to which they applied grids of increasingly coarse-grained resolutions, with the size of the regions that double between each level to its subsequent. For each granularity level, histograms of the features inside of each region are calculated, and then to measure the similarity of two sets of features it is computed a weighted sum of the number of matchings among them, representing the number of features that fall within the same region. The matchings are posed by the intersection of the histograms of the two input spaces. The coarsest layer of the pyramid represents the *Bag of Visual Words* (BoVW) approach, that consists in a histogram of the features present in the whole image. By comparing two images, if they are similar, their histograms are similar as well. However, the BoVW approach does not allow to identify partial matches, since it compares the images in their entirety. In contrast, Pyramid Matching takes into account the intersections of sets with different cardinalities, assigning a higher weight to matches found in the smaller regions of the image. Then, the differences among the images are measured with respect to the most similar components.

Following this idea, Spatial Pyramid Matching has been introduced to keep the spatial information, by partitioning the image in increasingly small cells similarly to the original Pyramid Matching. For each grid cell, the features are counted and their respective histograms are produced. This process results in a vector of length equal to the number of features for each pyramid layer and with the values of which are given by the dimensions of the histograms. Then, the values contained in the vectors are divided by a weight proportional to the size of the grid cells, in order to give more importance to the histograms obtained from the increasingly smaller regions. Finally, the vectors produced from each layer are combined into an output vector with length equal to the sum of the lengths of the component vectors.

On these bases, SPP-net introduces the SPP layer after the last convolutional layer. To the aim of object detection, there is still need to extract regions that can contain the object. Similarly to R-CNN, an external algorithm (such as Selective Search) can be used to propose the regions. Those are then projected onto the feature maps and they are passed as input to the SPP layers, which subdivide the region proposed in increasingly small cells and the values inside each one are max-pooled. The resulting output is a vector of length given by the number of the cells of each SPP layer. Finally, the output vectors are classified through the fully connected layer. In this way, it is not necessary to crop or warp the proposed regions into fixed sizes and, in addition, the convolutional network is performed just one time for each input image, in contrast to R-CNN that executes the convolutional layers for each region proposal.

3.1.3   Fast R-CNN

Even if SPP-net represents an important improvement in the object detection, it still has some drawbacks that are assessed in [24]. Fast R-CNN has been developed to overcome the limitations of both R-CNN and SPP-net as well. The name of the model recalls the increased speed in training as well as in testing. To understand the benefits introduced, it is appropriate to compare this model with the previous neural networks.

- **Limits of R-CNN**
  The main problem of the R-CNN model is caused by its architecture, composed by the convolutional network, the Support-Vector Machine (SVM) classifier and the bounding box regressor. Each of these components requires training and tuning stages separated from the other components. This causes a waste of computational resources and time, in addition to the higher memory consumption given by the writing on disk of the features extracted from each proposed region in a single image. The resultant performance in object detection stands around 47 s on average for a single image, considering the execution on a GPU Nvidia Tesla K40.
- **Limits of SPP-net**
  Some of the flaws of R-CNN can also be found in SPP-net, since it is formed by separated components such as the convolutional network, the pyramid layers and the final classifier. The drawback of this architecture is that it is not efficient, during the training stage, to update the weights of the convolutional layers through back-propagation, because the receptive field could extends along the whole image. Since the images can have arbitrary sizes, the GPU implementations like *cuda-convnet* and *Caffe* that require fixed-size inputs tend not to perform correctly, and this leads to a lower accuracy in object detection.

Fast R-CNN solves the problems through a heavy modification of the architecture of R-CNN. The structure of Fast R-CNN comprises several convolutional layers, which receive as input the image along with a set of proposed regions and produces feature maps on which it applies the proposed bounding boxes. The features in these regions are then aggregated through a pooling layer that represents a variant of the SPP layer, obtained by employing only one layer. This layer, namely *Region of Interest* (RoI) pooling layer, subdivides the proposed region in a grid of smaller cells, in which the values are aggregated through max pooling in order to produce a feature map of a lower size. Following the RoI pooling layer, there is a sequence of fully connected layers that branch into a pair of parallel output layers. One produces the class probabilities for each object, the other generates a vector containing four values representing the coordinates of the bounding box that are also optimized through the bounding box regression introduced with R-CNN.

3.1.4   Faster R-CNN

Fast R-CNN performs better than SPP-net and the previous R-CNN, but it remains the bottleneck caused by the slowness of the region proposal algorithm. To the aim of solving this issue, it has been proposed a new network model called *Region Proposal Network* (RPN) [25]. In this case, the feature maps generated by the convolutional layers are exploited also by the region proposal task, by sliding a small convolutional network above them. Its output are feature maps of smaller sizes that are passed to two parallel fully connected layers. The first performs the object classification, whereas the second extracts the bounding box coordinates. Thanks to the concept of *anchor box*, different regions are proposed simultaneously. An anchor box is a rectangular (or square) shape, centred on the central point of the sliding window. This approach allows to propose, at each location, regions with different dimensions and aspect ratios that are used to define whether an object is present or not. In particular, a window of size $n \times n$ slides on the feature map with a stride of 16 pixels and, for every position, it extracts 9 anchors with 3 areas of respectively 128^2, 256^2 and 512^2 pixels, and with 3 different aspect ratios (1:1, 1:2 and 2:1). In the training stage, anchor boxes with the highest Intersection over Union (IoU)[1] with respect to the ground truth boxes are labelled as positive examples and the regression is performed, with the purpose of matching as much as possible the coordinates of the predicted bounding box with the real one of the image from the training set. The coordinates of the predicted bounding box are calculated by shifting and/or scaling the ones of the anchor box. This results in regions with different sizes and aspect ratio, for this reason it is implemented a RoI pooling layer to produce a fixed-size output vector finally employed by the classifier to predict the probability distribution for each class.

Despite the progressive improvements of the network, however, Faster R-CNN is still not able to perform in real-time, reaching at its best a frame-rate of only 5 FPS with the most accurate VGG convolutional network that increases to 17 FPS with the ZF network both executed on a GPU Nvidia Tesla K40.

Further improvements of the model, such as Feature Pyramid Networks (FPN) [26] and Mask R-CNN [27] have been developed with the aim of increasing even more the accuracy. However, the overhead introduced leads to a slower execution with respect

---

[1]IoU is an index that measures the closeness between the anchor box and the ground truth box through the division of the area obtained by intersecting the two boxes with the area obtained by the union of them.

to Faster R-CNN. An advancement in terms of execution time is provided by the Region-based Fully Convolutional Networks (R-FCN) model [28]. Notwithstanding, even if it has a high accuracy, it is still slower than the one-stage models that will be discussed in the subsection below.

3.2 One-Stage Object Detection

One-stage algorithms represent faster alternatives introduced to overcome the limitations given by the slowness of the two-stage object detectors. Examples of models that work in a single stage are all the versions of You Only Look Once (YOLO, YOLOv2 and YOLOv3) and the Single Shot Multibox-Detector (SSD).

3.2.1 YOLO

The architecture of You Only Look Once (YOLO) [29] is composed by 24 convolutional layers followed by two fully-connected layers. The first are used to extract the features from a fixed-size input image of 448×448 pixels, whereas the latter computes the probability distributions and the coordinates of the bounding boxes. YOLO proposes several regions simultaneously without the necessity of a sliding window. This results in an execution speed of 45 FPS (obtained with a Nvidia Titan X GPU), making this model useful for the real-time video stream analysis.

YOLO operates by subdividing the image in a grid with $S \times S$ cells, each one responsible for the detection of an object. Every cell predicts a number B of bounding boxes, for which five values are calculated: x and y that represent the coordinates of the centre of the box with respect to the boundaries of the cell in which it falls, w and h that identify respectively the width and the height of the located region and, lastly, the confidence value that equals the IoU between the predicted box and the ground truth box. Moreover, each cell predicts a number C of probabilities that an object belongs to a given class. During the training stage, since several boxes are predicted for each cell, all the predictions that do not have the highest IoU value are dropped in order to keep only one predictor for each cell.

3.2.2 SSD

The elimination of the region proposal stage is the strength of YOLO, since it leads to a simpler architecture which is able to increase the speed. However, compared to Faster R-CNN, YOLO accuracy is lower. Single Shot Multibox-Detector (SSD) [30] introduces improvements to overcome the limitations of YOLO. To this aim, feature maps with different sizes are employed, on which the model performs predictions with ratios commensurate to them.

SSD is based on a CNN, with the last layers (namely *feature layers*) that decrease in dimensionality so as to obtain feature maps with different sizes, through a series of convolutional kernels. The image is subdivided in grids with increasing granularity, and a set of anchor boxes of varying aspect ratios is associated to each cell. These anchor boxes are then exploited to predict the coordinates of the bounding box, estimated as the offsets from the boundaries of the anchor. In other words, the predictions are made for three different sizes, in order to be able to detect both small and large objects. Along with the coordinates, also the class probability scores are calculated. Compared to YOLO, SSD results more accurate still being real-time capable, with a frame-rate of 59 FPS on a GPU Nvidia Titan X.

3.2.3   YOLOv2

In the second version of YOLO [31], several strategies have been introduced to decrease the number of errors in object detection, still maintaining its capability of real-time computation.

- **Batch normalization**
  The problem of training a neural network is given by the variation of probability distributions incoming to each layer, caused by the concurrent update of the weights of the previous layers that alters its output. This leads to an effect called *internal covariate shift*, with which every layer tries constantly to adapt to the new distributions that it receives after each weights update. In case of deep networks, a small change in a layer can be amplified. In order to coordinate the update, it has been proposed a technique called *batch normalization* [32] in which the inputs to the layers are normalized, transforming them into a distribution with zero mean and unit variance. Basically, the cost function is calculated on sub-sets of the training set, namely *mini-batches*. After feed-forwarding a mini-batch in the network, the back-propagation is being performed computing the average and the variance of the sub-set. Then, the normalization factor is determined by dividing the difference between the input and the mean of the mini-batch with the standard deviation of the mini-batch [6, 32]. The batch normalization technique increases the accuracy of YOLOv2 compared to the first version.
- **High-resolution classifier**
  In the first version of YOLO, the classifier is trained with an input of size 224×224 pixels, to increase the resolution to 448×448 for the object detection. In contrast, the second version of YOLO involves the training of the classifier with the image of the largest size, that is 448×448 pixels.
- **Elimination of fully-connected layers**
  The fully connected layers are employed by YOLO to obtain the probability distribution of the objects and to predict the coordinates of the bounding boxes. However, in YOLOv2 these layers have been removed and the predictions of the regions occur through the concept of anchor boxes. In order to predict the confidence and the class probability for each anchor box, a pooling layer is removed

for not reducing the resolution of the feature maps and the input size is limited to 416×416 pixels to obtain a resulting feature map of size 13×13. The reason of having an odd number of pixels in the feature map resides in the necessity of having an odd number of cells. In fact, the major part of the images tends to have objects of large size located to the centre and, thereby, the cell responsible of the detection of the specific object would be the one at the centre of the image.

- **Clustering for anchor boxes generation**
  The anchor boxes employed by the second version of YOLO do not have sizes determined *a priori* but they are computed through a *k-means clustering* algorithm on the bounding boxes of the images from the training set, to generate 5 different shapes.

- **Direct location prediction**
  The concept of the anchor box has been introduced in Faster R-CNN. However, in that case the coordinates of the predicted bounding boxes are calculated by scaling and translating the anchor boxes for a certain offset. The application of this technique to the first version of YOLO resulted in a movement of the box outside of the boundaries of the cells responsible of the prediction. For this reason, in YOLOv2, it has been decided to tie the anchor box to a specific cell. In other words, the coordinates relative to a particular cell are predicted as an offset from its centroid.

In addition, YOLOv2 concatenates feature maps with different grained resolutions and trains the network with images of varying sizes to increase the accuracy further. The architecture of the second version of YOLO, namely Darknet-19, consists of 19 convolutional layers and 5 max-pooling layers. The kernels featured in the convolutional layers are mostly of size 3×3, doubling in number after each pooling layer. The filters with size 1×1 are used to reduce the depth of the outputs by compressing the feature maps.

Finally, YOLOv2 can detect more than 9000 classes through a combination of the two major datasets employed to train neural networks for object detection, that are COCO[2] and ImageNet.[3] However, the accuracy has been compared with the previous models by adopting the PASCAL VOC 2007[4] dataset. In this case, YOLOv2 results more accurate than SSD, still maintaining a real-time execution (67 and 59 FPS with the input resolutions of 416×416 and 480×480 pixels respectively) on a Nvidia GeForce GTX Titan X [31].

3.2.4   YOLOv3

The following update of YOLO, YOLOv3 [33], introduces a new architecture that includes 53 convolutional layers. For this reason, this model is also known as Darknet-53. The convolutional layers alternate kernels of size 3×3 with kernels of size 1×1,

---

[2] *Common Objects in COntext (COCO):* https://cocodataset.org/.

[3] *Imagenet:* http://www.image-net.org/.

[4] *Pascal VOC 2007:* http://host.robots.ox.ac.uk/pascal/VOC/voc2007/.

followed by a *residual layer* that is a concept introduced in the ResNet model [18]. Given a CNN, it has been noticed that with the increasing in depth of the network, during the training stage the accuracy reaches a saturation point before degrading quickly. This situation happens because, for extending a network of small dimension it can be thought to add a (residual) layer that implements an identity function, whereas the remaining layers are a copy of the initial network. In theory, the resulting model should produce a training error similar to the one obtained with the original architecture, but in reality this does not happen. One reason hypothesized is given by the difficulty of the model to learn the linear identity function from a series of non-linear layers. This problem is solved by setting the identity function of the residual layer equal to zero, and by adding a *shortcut connection* that allows to transfer the input of the last convolutional layer before the residual block to the latter, where it is summed to the result of the identity function. Since the identity function is set to zero, the total of this operation is exactly the added input, and this allows to carry on with the convolutions in the following layers. Through the increase in the depth of the network, YOLOv3 obtains better results in terms of accuracy compared to the previous version.

In addition, YOLOv3 predicts bounding boxes with three increasing scales. The last convolutional layer is employed for calculating, for each box at each scale, a 3-dimensional array which represents respectively the coordinates of the bounding box, the confidence score and the class probability distribution. The feature map generated by the second previous layer is doubled in size and concatenated with another feature map engendered by an even higher layer. This lead to the combination of the information of the widest receptive field provided by the up-sampled feature map with the most detailed information supplied by the reduced feature map, allowing a better identification of small size objects. In this way it is possible to obtain a prediction with a size that is the double of the one from the first scale. Lastly, this process is repeated once again to produce the final prediction. For each scale, three reference boxes are chosen through the k-means clustering technique.

Because of the greater complexity of the network architecture compared to Darknet-19, the third version of YOLO is slower. However, it scores a higher accuracy also compared to SSD, against which it achieves a faster execution time. It has to be mentioned that the state of the art comprises several other models, such as RetinaNet [34] and Deconvolutional SSD (DSSD) [35], but they result to be slower than YOLOv3, even though they can be more accurate [33].

3.3   The Final Choice

Based on what it has been examined in the previous sections (and summarized in Table 1), it results that a two-stage approach, in which the models imply a region proposal stage preceding the convolutional layers, is inappropriate for a use in a real-time vehicle detection. For this reason, it has been decided to base the final choice solely on the one-stage detectors, such as each version of YOLO and SSD.

As a result of the analysis among the different alternatives, and coherently with the outcomes of the respective evaluations, it has finally been decided to employ YOLOv3 to develop the AI-based Traffic Queue Detection application.

Because of the high complexity of artificial neural networks, however, it has been required a great computational capability that can be provided by a server. Nevertheless, in order to process video streams produced by traffic monitoring cameras and produce a response in real-time, it is also necessary to support the high data traffic volume generated and reduce the transmission latency as much as possible. This goal can be reach through the employment of the 5th Generation network.

4 Realization of the Traffic Queue Detector

The proposed service has been realized taking in consideration use cases of public safety in highway contexts, by leveraging on 5G along with traffic monitoring infrastructure. An example can be represented by the autonomous driving, in which vehicles must adjust their speeds according to the surrounding traffic without human interaction, and this requires a fast exchange of information between the oncoming vehicles and the infrastructure in order to avoid pile-ups. However, a fully autonomous transport could still require many years to spread on the roads. Thinking about the next few years, a traffic queue detection mechanism can lead to the identification of collisions. This paves the way to another possible application related to the public safety, that is the immediate reporting of traffic queues to emergency bodies that can respond accordingly. The advanced service has also been thought to integrate existing applications, such as satellite navigators and maps, to provide an additional source of information in order to increase the reliability. The latter use case could therefore enable *smart mobility* services, geared to the reduction of the fuel consumption (and consequently of air pollution) and to enhance the comfort of journeys by redirecting traffic flows in relation to the presence of queues.

The infrastructure considered should be composed by traffic monitoring cameras, a MEC server and Road-Side Units (RSUs) that provide an intermediary communication mechanism between the vehicles and the edge platform. Therefore, the development of the application has leveraged the use of a MEC server provided with GPUs to enable the parallel computation of the artificial neural network. In particular, Multi-access Edge Computing represents the solution to the requirements of low latency and high reliability needed to implement IoV use-cases. For the deployment of the infrastructure, it is necessary to keep in consideration how much it costs. Specifically, 5G MEC platform costs vary depending on whether the infrastructure is owned by the telecommunication company or by a cloud provider which offers it as a service. In the first case, the capital expenditure (capex) can range between 65 thousand and 650 thousand euros, with the operating expense (opex) around 25 thousand euros per year and a lifespan up to 5 years. In the second case, the capex ranges from 500 up to 5 thousand euros, with the opex from 10 thousand and 150 thousand euros per year, depending on the dimension of the infrastructure.

Table 1 Summary of the different object detection approaches.

Model	Advantages	Drawbacks
R-CNN	Possibility to detect multiple objects rather than just classify images	Very slow, CNN executed on each region proposed from a single image
SPP-net	It eliminates the need of a fixed-size input. In addition, the convolutional layers are performed once for each image	Inefficiency, both in training as well as in object detection
Fast R-CNN	It addresses the limitations of R-CNN and SPP-net	Region proposal stage is still slow
Faster R-CNN	Introduced the RPN. Anchor boxes extracts regions from the feature maps of the convolutional layers	Faster than the previous models, but still not capable to perform in real-time
YOLO	It eliminates the region proposal stage. Simpler architecture and faster execution speed	Less accurate with respect to Faster R-CNN
SSD	Compared to YOLO it is more accurate, thanks to the use of feature maps of different scales	Surpassed by YOLOv2
YOLOv2	Introduced several improvements to the first version. It results in a higher accuracy than SSD	Still the fastest version of YOLO, however it is less accurate than the third version
YOLOv3	Deeper neural network, it leads to a greater accuracy compared to the other models	The more complex architecture reduces the speed, but on the other side it increases the accuracy

In addition, the cost of cameras should be taken into account as well. This analysis is from the telecommunication operator prospective [36, 37].

Since it was not feasible to test the service on a MEC server, in order to make the implementation as close as possible to the realistic scenario it has been employed a system equipped with an Nvidia Ge-Force RTX2070 GPU and a 9th generation Intel Core i7 processor. In this case, the initial investment is less than 2 thousand euros, with few hundreds euros of opex (as for allocated resources in a MEC server). The edge computing platform deployed by a telco operator is likely provisioned with both off-the-shelf hardware and specific hardware accelerators (e.g. GPUs, FPGAs, SOCs).

To deploy and test the algorithm, two videos found on Internet have been used. These both have the same format but different frame-rates and are listed below[5]:

- **Video 1**

  – Video Format: MP4
  – Length: 93 s
  – Frame-rate: 25 fps
  – Frame size: 640×360 pixels

- **Video 2**

  – Video Format: MP4
  – Length: 306 s
  – Frame-rate: 30 fps
  – Frame size: 640×360 pixels

These videos have been then elaborated to perform both vehicles detection and classification. The model created can be represented as a multi-layered architecture (Fig. 1), where at the bottom we can find YOLOv3 to carry out the object detection task. The vehicles detected are then tracked by the Simple Online and Realtime Tracking (SORT) algorithm [38] that associates a numerical identifier to each vehicle. These identifiers are then used to detect the direction of the traffic flows and to keep track of the motion of the vehicles. The basic idea is to monitor the same vehicles through consecutive frames to understand whether their behaviour has changed. In this specific case, a queue is considered formed when the majority of vehicles are stationary, meaning that their positions from one frame to the following one are the same, or they have slowed down increasing therefore their travelling times through a portion of the carriageway. To this aim, two different approaches have been proposed, namely Stop Detector and Time Detector.

Fig. 1 Architecture of the traffic queue detector.

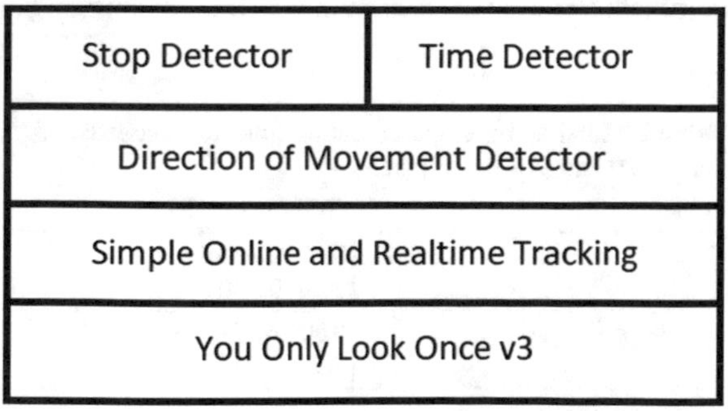

Stop Detector	Time Detector
Direction of Movement Detector	
Simple Online and Realtime Tracking	
You Only Look Once v3	

---

[5]For the sake of simplicity we will refer to them as Video 1 and Video 2, but the original title are: *M6 Motorway Traffic* uploaded by the channel "DriveCamUK" (https://youtu.be/PNCJQkvALVc) for Video 1, and *4K Road traffic video for object detection and tracking* uploaded by Karol Majek (https://youtu.be/MNn9qKG2UFI). Due to copyright reasons, in this section will be shown frames extracted by the second video that have been published under the Creative Commons license.

In the following subsections the model will be explained in detail.

4.1   YOLOv3

In light of the comparison among the different algorithms discussed in the previous section, it has been decided to employ the third version of YOLO given its real-time computation capability. The neural network has been trained on the COCO dataset, which contains about 330 thousand images (more than 200 thousand of them are labelled), for 80 classes representing objects of common use, including vehicles such as cars, trucks, buses and motorbikes. It has been decided to choose the pre-trained network because the gathering of images for the training set and the subsequent labelling would be too much time expensive. In fact, for each image in the training set, YOLO requires the annotation of each object inside of them through the coordinates of the bounding boxes that should be drawn manually.

The images on which YOLOv3 performs the object detection are represented by the single frames of the video streams. The neural network accept inputs of dimension $n \times n$ and the value of n affects both the accuracy as well as the speed. This means that an input with a large size allows to obtain a higher accuracy at the expense of the execution speed which faces an increase in time. Therefore, due to the necessity of processing videos in real-time, it has been chosen to adopt the reference sizes of 320×320 and 416×416 pixels. In case the frames of the video have a lower dimension, the model re-sizes the image and keeps the original aspect ratios with the *zero-padding* technique: since the network requires an image with the same height and width, a pad of zero values is added around the frame in order to match the size needed.

However, the object detection per-sé does not allow to determine the direction of travel of the vehicles. So, for this reason, it is necessary to integrate YOLOv3 with an object tracking algorithm that will be discussed below.

Table 2 Total and average execution time[a], in seconds, of YOLOv3 with inputs respectively of 320×320 and 416×416 pixels.

Video	YOLOv3 - 320	YOLOv3 - 416
1	Tot: 70.8229 s Avg: 0.0302 s	Tot: 90.5160 s Avg: 0.0386 s
2	Tot: 262.2999 s Avg: 0.0286 s	Tot: 335.1132 s Avg: 0.0365 s

[a] For each frame.

4.2   SORT

The main problem faced during the creation of the application in support of the autonomous driving has been to understand how to exploit the detections produced by YOLO (Fig. 2) to locate the traffic flows. This is because the bounding boxes do not provide any clue about the direction of the movement of the vehicles. The solution has been provided by an additional algorithm that identifies the detected vehicles and tracks their movement in the image. The considered model, called SORT [38], operates by analysing the current frame along with the previous frame. The aim is to compare the positions and the sizes of the bounding boxes generated by YOLO, avoiding any additional computation induced by the observation of the object features to keep the model as simple as possible.

To generate the predictions, it implements a Kalman filter [39], which calculates the probability that an object will be in a certain future state on the basis of the current state. Once the possible positions in the current frame are predicted, SORT computes the Intersection over Union (IoU) among the proposed bounding boxes and the present ones. Then, it associates to the target the object with the highest IoU with a numerical identifier through an implementation of the Hungarian method [40]. In Fig. 3 it is shown an example of how SORT works on the input video.

Nonetheless, the introduction of an additional algorithm has increased the global complexity, leading to a rise in the execution times (Tables 2 and 3). In fact, considering the first example video, the frame-rate is approximately 15 fps with YOLO at the resolution of 416×416 pixels (and around 20 fps for the resolution of 320×320 pixels). For the second video the respective frame-rates stand at 20 and 23 fps, depending

Fig. 2 Frame extracted by Video 2. In the picture the bounding box and the predicted labels generated by YOLOv3 are shown.

on the resolutions. However, it remains to be seen how much the migration of the code on a (more) powerful MEC server could bring further benefits.

4.3   Direction of Movement Detector

The algorithms just mentioned are fundamental for the vehicle detection. Nevertheless, these solutions do not result sufficient to detect the formation of traffic queues. For this reason, it has been necessary the development of the two proposed solutions, the Stop Detector and the Time Detector that will be discussed later, which exploits the outputs provided by YOLOv3 and SORT. However, to lay the groundwork for these two approaches it has been required to determine the movement direction of the vehicles by taking advantage of the identifier associated to each vehicle by SORT. During this stage, distinct situations can arise:

Fig. 3 Four consecutive frames extracted by Video 2. SORT has identified and tracked each detection applying a unique identifier.

Table 3 Total and average execution time[a], in seconds, of both YOLOv3 and SORT with inputs respectively of 320 × 320 and 416 × 416 pixels.

Video	YOLOv3 - 320 (with SORT)	YOLOv3 - 416 (with SORT)
1	Tot: 115.7279 s Avg: 0.0494 s	Tot: 155.5433 s Avg: 0.0664 s
2	Tot: 397.1972 s Avg: 0.0432 s	Tot: 508.0500 s Avg: 0.0553 s

[a] For each frame.

- **First occurrence of the identifier**
  This happens when the identifier associated to the vehicle appears in the video for the first time. Hence, the related vehicle has not been paired to any traffic flow yet and its identifier is initially stored in an array (namely, for clarification purposes, *array_start*) along with its *y* coordinate of the upper edge of the bounding box and the current timestamp expressed in milliseconds. The timestamp represent the amount of time elapsed from the so-called *epoch*, that is the point from which the operating system starts to calculate the current time as the number of seconds passed from that date (1 January 1970).

- **Second occurrence of the identifier**
  When the identified vehicle has already been recorded in *array_start* in the previous frame, in this occurrence the actual coordinate is compared with the one stored in the initial array. In this case it is appropriate to consider the direction of movement. If the drive is on the right and the new *y* coordinate of the vehicle within the image is lower than the one related to the previous frame, then the object is located in the traffic flow of the right lanes on the road. That is, the traffic flow that is headed northbound. The traffic flows belonging to the left lanes are identified in the opposite manner. Similarly to what it has been done for the first occurence, in this situation the identifier associated to a vehicle is assigned to the proper array, for instance *right_flow* and *left_flow* along with the timestamp and a flag that indicates if the vehicle is stopped. For the left-hand traffic countries, the flows are inverted. Whether the position from the two frames has not changed and the vehicle does not appear to belong to any flow, the vehicle is kept in *array_start* and the timestamp is updated.

- **Successive occurrences of the identifier**
  Throughout the execution of the video, we proceed in a manner similar to the previous situation. First, it is verified whether the identifier exists in the flow arrays and if it is still moving. Therefore, the new position is compared to the coordinate stored in one of the two array. In case the detected vehicle is still moving, the *y* coordinate in the array is updated, along with the new current timestamp. In case the position is not changed, it means that the vehicle is stopped in its corresponding flow, so the flag in the array is set accordingly and the timestamp is updated.

In the following sections it will be described how the directions of movement, as shown in Fig. 4, can be exploited to detect the formation of traffic queues.

4.4 Traffic Queue Detector

As mentioned previously, the last layer of the application is composed by two approaches. One has been called Stop Detector, whereas the other Time Detector.

Fig. 4 Example of the direction of movement detector applied to Video 2. The image shows two consecutive frames.

4.4.1 Stop Detector

This proposed solution counts the number of stopped vehicles and is based on the previous layers analyzed in depth in the previous sections. Mainly, this algorithm operates in two stages:

- **Counting the stationary vehicles in a given frame**
  In the first stage, the number of stopped vehicles within a flow is counted according to the directions of movement observed for each detected vehicle. More precisely, if the flag for the stationary vehicle in the corresponding array is set, then a counter is incremented. Similarly, the number of moving detections is calculated. If the number of stopped vehicles is greater than the quantity of the travelling ones and the number of stopped vehicles is greater than a given threshold (i.e. 5 vehicles), then the formation of a queue is detected. In addition, a counter keeps track of the number of frames in which the queue has been identified.
- **Counting the number of frames in which the queue has been detected**
  In the previous stage it has been mentioned that a counter is incremented whether a queue is detected in a frame. If this counter is greater than a given threshold (i.e. 15 frames), a signal containing the traffic flow with the congestion is reported.

The threshold mentioned in the first stage has been introduced because it can happen that the bounding box of a travelling vehicle does not move between two consecutive frames due to anomalies in the object detection. In other words, the high frame-rate of the videos, along with the slight movement of the vehicle between two frames, may result in a stationary bounding box, and this effect is emphasized in case the vehicles travel at a slow speed. Therefore, in order to make the estimate more reliable it has been decided to consider the formation of queues with a higher traffic volume. In addition, the counter of frames stated in the second stage has been implemented to avoid effects of intermittency in the signal of the formation of queues.

An example of how the detector individuates the queues is shown in Fig. 5.

Fig. 5 Sample frame of the Position Detector applied to Video 2. The queue has been detected and indicated on the top-left of the image.

4.4.2 Time Detector

The second approach suggested is based on the measurement of the travel time of the vehicles. This time is calculated from the instant in which the vehicles start to transit in a portion of the roadway until the moment they come out. In order to circumscribe the interested area, two parallel lines have been taken as reference points in correspondence to which the current timestamp are computed. To this aim, it is needed to distinguish whether a given vehicle is travelling headed northbound or southbound. This because the coordinates that represents the entry point to the measurement area in one direction of movement constitutes the exit point on the opposite direction.

Similarly to the Stop Detector, the stages of execution will be explained hereafter:

- **The vehicle enters the measurement area**
  To calculate the travelling time, first of all it is necessary to take into account the direction of movement of the detected vehicle. In fact, this allows to record the instant at which the vehicle crosses the boundary to enter the interested region according to the direction of movement. To this aim, two new arrays have been created, each one corresponding to the respective direction of movements.
- **The vehicle travels in the measurement area**
  In this case, the identifier associated to the given vehicle is already stored in the array related to its traffic flow discussed in the previous stage. Therefore, the initial instant of time is not updated.

Fig. 6 Representation of the Time Detection. The figure shows two consecutive frames. The total travelling time is indicated with an integer to the right of the identifier of the vehicle, on top of the bounding box. It is expressed in milliseconds. A value of 0 means that the vehicle travelling time has not been measured yet.

- **The vehicle comes out from the measurement area**

  Similarly to what happened in the first stage, in the last case the time at which the vehicle exits the region is recorded. Then, the total travel time is calculated by subtracting the initial instant of time to the final one, as shown in Fig. 6.

It should be noticed that the measurements are highly variable because of the presence of multiple lanes within each carriageway, that leads to the existence of traffic flows with different speeds. Moreover, within a single lane the vehicles can travel at various velocities. For this reason, it results difficult to understand whether the traffic is slowing down in base of the single vehicles. Therefore, for the implementation it has been considered to employ an array which represents a sliding window of the last 5 total travel times (in this instance the value represents a compromise solution between the reduction of fluctuations and the quick response to change in traffic speed for the scenario with the two example videos, but the size can be chosen according to specific use cases) and on which a moving average is calculated. The resulting average is then added to an additional array that stores the last n values (again, the value of n can be chosen depending on the use case, in this scenario it has been employed a size of 50 averages). On the elements of the latter array a second average is calculated. This is then used to establish a threshold that can correspond, for instance, to the 15% of its value. The need for these two averages has risen to solve the problem of the frequent fluctuations of measurements, that can bring to false detection of queues when the traffic is still flowing. So, if the first average is used to calculate a standard deviation. In case it is greater than the threshold, it means that the traffic is slowing down. The formation of a traffic jam is indicated, similarly to Position Detector, as shown in Fig. 7.

Fig. 7 Sample frame of the Time Detector applied to Video 2. The queue has been detected and indicated on the top-left of the image.

4.5 Comparison Between the Two Solutions

Each of the solutions proposed are able to detect the formation of queues. The main difference, consists in how they work. It is in fact difficult to establish which approach among the two is the best, because their respectively outcomes are heavily influenced by the parameters chosen. High thresholds result in reduced fluctuations of the measurements and, as a consequence, in less false detections (that is, queues detected because the traffic has slowed down for just few frames). However, this leads to a rise in the latency between the beginning of the formation of the queue and its detection. The choice of one way rather than the other resides in how they have been thought. That is, the Stop Detector works by counting the stationary vehicles, whereas the Time Detector measures the average travelling time, therefore one approach does not necessarily excludes the other. The Stop Detector could be used to recognize traffic jams and the Time Detector could be employed for the detection of heavy slowdowns, eventually anticipating the Stop Detector. A combination of both approach could also increase the reliability by filtering false detections in cases only one of them has detected a queue.

Considering the real-time requirements, the Tables 4 and 5 show the execution times in seconds of the two proposed approaches for the two test videos. More precisely, in Table 4 the results of the whole architecture applied to the input of size 320×320 pixels are reported. That is, YOLOv3 runs over videos with frame of height and with each of 320 pixels, performing the detections then tracked by SORT and employed by the queue detection algorithms. Similarly, in Table 5 the performance of the model applied on larger inputs (in this case of size 416×416 pixels) are listed.

Table 4 Comparison between the two algorithms of vehicular traffic queue detection developed, the Stop and the Time Detector. The total and the average time for each frame is expressed in seconds and in this case the size of the input is of 320 × 320 pixels.

YOLOv3 - 320 with (SORT)		
Video	Stop detector	Time detector
1	Tot: 120.1919 s Avg: 0.0513 s	Tot: 116.1275 s Avg: 0.0495 s
2	Tot: 403.1097 s Avg: 0.0439 s	Tot: 402.2968 s Avg: 0.0438 s

Table 5 Comparison between the two algorithms of vehicular traffic queue detection developed, the Stop and the Time Detector. The total and the average time for each frame is expressed in seconds and in this case the size of the input is of 416 × 416 pixels.

YOLOv3 - 416 with (SORT)		
Video	Stop detector	Time detector
1	Tot: 159.366 s Avg: 0.0680 s	Tot: 162.4571 s Avg: 0.0693 s
2	Tot: 507.9745 s Avg: 0.0553 s	Tot: 502.1752 s Avg: 0.0547 s

It should be recalled, to this aim, that the tests have been performed on a Nvidia GeForce RTX 2070 graphics card exploiting the parallel computation. On the basis of what has been seen with the performance of SORT applied to YOLOv3 analysed in the previous sections, an analogous slowdown in the elaboration speeds happens for Stop and Time Detectors and this is given by the increased amount of computation needed by the additional algorithms. Nevertheless, the highest rise is caused by the application of SORT to YOLO network, rather than by the two proposed solutions for the traffic queue detection.

In the light of the outcomes obtained, to reduce the further latency, the code in the stage of frame extraction from videos has been optimized by leveraging on multithreading. In other words, a separated thread has been created to the sole purpose of performing the blocking function that reads the frames from the input video. In fact, usually the main process blocks itself waiting for the read function to complete and return the frame to the process that can continue its execution. So, in order to reduce the computational time and to avoid intervals of deadlock of the process it has been created a thread dedicated to the extraction of the frames that are kept in a queue so that the main process can access it in parallel. In this way, at the time it elaborates the current frame, the others are read and appended to the queue so that the next frame would be ready at the end of the present computation. This optimization allowed to reduce slightly the execution time of the model with respect to the non-optimized algorithms in the majority of the cases, as illustrated in the Tables 6 and 7.

Table 6 Comparison between the execution times of the two algorithms of vehicular traffic queue detection developed, the Stop and the Time Detector, and their optimized versions (indicated with the nomenclature *multithread*) for the two example videos. The total and the average time for each frame is expressed in seconds and in this case the size of the input is of 320 × 320 pixels.

YOLOv3 - 320 with (SORT)				
Video	Stop detector	Stop det. multithread	Time detector	Time det. multithread
1	Tot: 120.1919 s Avg: 0.0513 s	Tot: 115.9606 s Avg: 0.0495 s	Tot: 116.1275 s Avg: 0.0495 s	Tot: 114.8570 s Avg: 0.0490 s
2	Tot: 403.1097 s Avg: 0.0439 s	Tot: 390.5477 s Avg: 0.0425 s	Tot: 402.2968 s Avg: 0.0438 s	Tot: 398.8931 s Avg: 0.0434 s

Table 7 Comparison between the execution times of the two algorithms of vehicular traffic queue detection developed, the Stop and the Time Detector, and their optimized versions (indicated with the nomenclature *multithread*) for the two example videos. The total and the average time for each frame is expressed in seconds and in this case the size of the input is of 416 × 416 pixels.

YOLOv3 - 416 with (SORT)				
Video	Stop detector	Stop det. multithread	Time detector	Time det. multithread
1	Tot: 159.366 s Avg: 0.0680 s	Tot: 154.6966 s Avg: 0.0660 s	Tot: 162.4571 s Avg: 0.0693 s	Tot: 153.7357 s Avg: 0.0656 s
2	Tot: 507.9745 s Avg: 0.0553 s	Tot: 503.4237 s Avg: 0.0548 s	Tot: 502.1752 s Avg: 0.0547 s	Tot: 503.3047 s Avg: 0.0548 s

Besides the minimum temporal differences in the execution of the two solutions deployed for the queue detection, it should be noticed that even it has been chosen the fastest artificial neural network among the alternatives, to maintain the objective of real-time performance has proved to be impossible in this case. The cause has to be attributed mainly, as previously mentioned, to the inevitable increase in the amount of computation introduced by SORT. The further optimization of the code by harnessing the multithreading technique turned out to be a positive strategy, but still not sufficient, in almost all the cases with the exception of the Time Detector employed on the second video. A personal conclusion that has been drawn in order to try to state a reason for this inconsistency can be provided by the fact that YOLOv3-416 works at a frame-rate of 35 fps, and by considering the computational overhead introduced by the frames elaboration and also keeping in mind the high frame-rate of video 2, the application of the multithreading has not provided an actual benefit, since the limit of the frame-rate has already been reached.

A non-trivial aspect to be considered is the hardware employed for the implementation of the model. For demonstrative purposes, Tables 8 and 9 show the difference in terms of execution time of the chosen neural network and of YOLOv3 with the

Table 8 Comparison of the total and average time of execution of YOLOv3 on each frame, with sizes 320×320 and 416×416 respectively, with the employment of two different Nvidia GPUs.

Video	YOLOv3-320		YOLOv3-416	
	GTX 750 Ti	RTX 2070	GTX 750 Ti	RTX 2070
1	Tot: 189.7982 s Avg: 0.0810 s	Tot: 70.8229 s Avg: 0.0302 s	Tot: 261.9133 s Avg: 0.1117 s	Tot: 90.5160 s Avg: 0.0386 s
2	Tot: 731.3657 s Avg: 0.0796 s	Tot: 262.2999 s Avg: 0.0286 s	Tot: 994.9519 s Avg: 0.1090 s	Tot: 335.1132 s Avg: 0.0365 s

Table 9 Comparison of the total and average time of execution of YOLOv3 with SORT on each frame, with sizes 320×320 and 416×416 respectively, with the employment of two different Nvidia GPUs.

Video	YOLOv3-320 (with SORT)		YOLOv3-416 (with SORT)	
	GTX 750 Ti	RTX 2070	GTX 750 Ti	RTX 2070
1	Tot: 252.0297 s Avg: 0.1075 s	Tot: 115.7279 s Avg: 0.0494 s	Tot: 350.0136 s Avg: 0.1493 s	Tot: 155.5433 s Avg: 0.0664 s
2	Tot: 911.4817 s Avg: 0.0992 s	Tot: 397.1972 s Avg: 0.0432 s	Tot: 1222.9473 s Avg: 0.1332 s	Tot: 508.0500 s Avg: 0.0553 s

additional object tracking algorithm deployed through two different Nvidia GPUs. In particular, the GeForce GTX 750 Ti is an obsolescent model that has been super-seded by more modern successors, such as the already mentioned RTX 2070, and has obtained very poor performance, far from the real-time constraints required. This fact emphasizes the importance of having a dedicated hardware, such as a 5G MEC server, whose greater performances lead the way to the execution of complex algo-rithms that otherwise could not be supported by general purpose hardware or systems with limited resources.

4.6 Considerations on the Choice of Parameters

It has already been mentioned how the thresholds employed by the two proposed solutions can be chosen and tuned with respect to the specific scenario. The intro-duction of the thresholds is justified by the presence of false detections. The details for each approach are explained hereafter:

- **Stop Detector**
  With the Stop Detector can happen that the slight movement of the vehicle between two consecutive frames could lead in a stationary bounding box, and this effect is accentuated when the speeds of the travelling vehicles are low. This means that in situations of high traffic volume, a possible slowdown of the vehicles causes multiple stationary bounding boxes for a limited consecutive number of

frames. Therefore, the flow is wrongly depicted as a queue whereas it is still moving, and this causes "intermittent" detections of queues. To reduce this effect, as in Sect. 4.4.2 it has been said, when in a frame the presence of a queue is detected then a counter is incremented. On the opposite, in case of normal traffic this counter is decremented. However, in cases of actual traffic queues, if the vehicles are stationary then they would be detected in more consecutive frames and therefore the counter will exceed the threshold. The queue is then correctly detected, reducing in this way the number of false detections. It should be noted at this point that a high threshold value introduces a latency in the detection, hence a reasonable trade-off of 15 frames has been chosen.

- **Time Detector**
  The same considerations seen for the Stop Detector also applies for the Time Detector. That is, the threshold has been introduced to filter the fluctuations of the measurement. As mentioned in Sect. 4.4.2, a queue is detected if the standard deviation of the average of the last 5 measurements exceeds the threshold. Here, it represents the 15% of the value of the moving average calculated on the last 50 measurements. Even in this case the threshold is a trade-off between the frequent occurrences of false detections and the introduction of high latency with greater threshold value. It has also been noticed that the resolution of the input influences the outcome of the Time Detector. A higher input resolution (in this case, 416×416) results in a more accurate detection and a lower variability of the measurements with respect to the input size of 320×320 pixels.

As a result of these considerations, a combination of both the solutions proposed could further filter out the presence of false detections.

5 Conclusion and Future Perspective

In this research, a possible solution to detect the formation of traffic queues on highways has been developed by leveraging on the AI and 5G technology to support the public safety. More specifically, it has been considered a highway context equipped with an infrastructure suitable for a service based on the employment of artificial neural networks that require a high computational capability, such as the one supplied by a 5G MEC server, for enabling the application of the computer vision intended for the elaboration of video streams generated by traffic monitoring cameras.

To this aim, in Sect. 3 several alternatives of artificial neural networks have been analysed, eventually choosing the benefits both in terms of real-time computation as well as accuracy of the third version of YOLO, belonging to the family of *One-stage* object detectors. In contrast to *Two-stage* object detectors such as R-CNN, SPP-net, Fast R-CNN and Faster R-CNN which operate by extracting a set of region proposal from the input image in the first stage and then classify them through a series of convolutional layers in the second stage, the *One-stage* architecture allows to propose possible bounding boxes and to classify the objects within the regions through the only employment of convolutional layers. This results in a faster execution speed of the network that, if considered along with the quick elaboration and information transmission provided by the Multi-access Edge Computing, enables to promote a context of low-latency communications such as the Intelligent Transportation Systems (ITS) and the autonomous driving.

Therefore, in Sect. 4 it has been shown in details the implementation of the proposed service for the detection of traffic queues. In particular, it harnesses YOLOv3 for the detection of the travelling vehicles and the SORT algorithm to keep track of each object identified. This then allows to understand the directions of movement of each traffic flow, an essential aspect to recognise when they have slowed down or stopped. Nevertheless, YOLO per-se is not enough to realize an application for the traffic queues detection, even with the additional tracking algorithm. For this purpose, two distinct approaches have been developed. The first is based on the recognition of the amount of stationary vehicles in each carriageway, and the latter on the measurement of the average travelling time of a section of the road. The results shown are referred to two videos found on the Internet in which the traffic flows of two highways have been recorded in a perspective analogous to the one provided by traffic monitoring cameras. Moreover, by not being able to deploy the application on a MEC server, it has been chosen to implement the code by leveraging on the parallel computation provided by a local system equipped with a modern graphics unit, also introducing an optimization mechanism based on multithreading video analysis. Both the solutions advanced have proven to detect the formation of traffic queues, but as a future implementation could be conceivable to jointly employ them to detect slowdown of traffic as well as queues in a more robust manner.

Notwithstanding these expedients, the advanced solutions have not been able to satisfy the real-time requirements. This mainly happened because of the additional overhead introduced by the object tracking and by the queue detection. A great challenge that has been faced is exactly the problem of how to exploit the detections of YOLO for understanding when the traffic has formed a queue. To track the vehicles it has been introduced the SORT algorithm, but in the future it would be advisable to research an alternative approach that does not raise the computational time.

However, the progressive and relentless evolution of artificial neural networks could pave the way to even faster and more accurate models, allowing to increase the efficiency of the algorithms developed. Furthermore, a progress in the artificial neural networks could lead to an extension of the service, for example by including the possibility of detecting traffic flows within the different lanes of the same carriageway.

One final important aspect should be addressed. Artificial neural networks are characterized by being able to be trained accordingly to a dataset. This means that, if it is employed an image dataset containing specific kinds of vehicle it is possible to train the network to recognize queues with the presence, for instance, of vehicles used for transportation of hazardous goods or even to determine whether a collision has occurred. This also could allow to instantly alert specialised rescue bodies. In any case, it is necessary to understand how much reliable an object detection model can be by testing it with videos recorded in different meteorological conditions and eventually by integrating it with other sensors and detection systems. A major obstacle with CNNs, in fact, is that the object detection task is largely influenced by the quality of the input. The model has not been tested in different day light conditions or in total absence of illumination and, as a future improvement, the network could be trained with a larger and *tailor-made* dataset.

Nonetheless, the AI undoubtedly represents a powerful and versatile tool in support of ITSs for enhancing safety, comfort and economy on the road.

References

1. Maaloul, B., et al.: Adaptive video-based algorithm for accident detection on highways. In: 2017 12th IEEE International Symposium on Industrial Embedded Systems (SIES). IEEE (2017)
2. Khan, S., et al.: An intelligent monitoring system of vehicles on highway traffic. In: 2018 12th International Conference on Open Source Systems and Technologies (ICOSST). IEEE (2018)
3. Chen, L., et al.: An algorithm for highway vehicle detection based on convolutional neural network. Eurasip J. Image and Video Process. **2018**(1), 109 (2018)
4. Jiang, T., et al.: Fast video-based queue length detection approach for self-organising traffic control. IET Intell. Transp. Syst. **13**(4), 670–676 (2018)
5. Russell, S., Norvig, P.: Artificial intelligence: a modern approach (2002)
6. Goodfellow, I., Bengio, Y., Courville, A.: Deep Learning. MIT Press, Cambridge (2016)
7. Rojas, R.: Neural Networks: A Systematic Introduction. Springer Science & Business Media, Berlin (2013)
8. Hubel, D.H., Wiesel, T.N.: Receptive fields, binocular interaction and functional architecture in the cat's visual cortex. J. Physiol. **160**(1), 106 (1962)
9. Hubel, D.H., Wiesel, T.N.: Receptive fields and functional architecture in two nonstriate visual areas (18 and 19) of the cat. J. Neurophysiol. **28**(2), 229–289 (1965)
10. Fukushima, K., Miyake, S.: Neocognitron: a self-organizing neural network model for a mechanism of visual pattern recognition. In: Competition and cooperation in neural nets, pp. 267–285. Springer, Berlin (1982)
11. LeCun, Y., et al.: Backpropagation applied to handwritten zip code recognition. Neural Comput. **1**(4), 541–551 (1989)
12. LeCun, Y.: Generalization and network design strategies. Connect. Perspect. **19**, 143–155 (1989)
13. LeCun, Y., et al.: Gradient-based learning applied to document recognition. Proc. IEEE **86**(11), 2278–2324 (1998)
14. Krizhevsky, A., Sutskever, I., Hinton, G.E.: ImageNet classification with deep convolutional neural networks. Adv. Neural Inf. Process. Syst. **25**, 1097–1105 (2012)
15. Zeiler, M.D., Fergus, R.: Visualizing and understanding convolutional networks. In: European Conference on Computer Vision. Springer, Cham (2014)
16. Szegedy, C., et al.: Going deeper with convolutions. In: Proceedings of the IEEE Conference on Computer Vision and Pattern Recognition (2015)
17. Simonyan, K., Zisserman, A.: Very deep convolutional networks for large-scale image recognition. arXiv preprint arXiv:1409.1556 (2014)
18. He, K., et al.: Deep residual learning for image recognition. In: Proceedings of the IEEE Conference on Computer Vision and Pattern Recognition (2016)
19. Huang, G., et al.: Densely connected convolutional networks. In: Proceedings of the IEEE Conference on Computer Vision and Pattern Recognition (2017)
20. Girshick, R., et al.: Region-based convolutional networks for accurate object detection and segmentation. IEEE Trans. Pattern Anal. Machine Intell. **38**(1), 142–158 (2015)

21. He, K., et al.: Spatial pyramid pooling in deep convolutional networks for visual recognition. IEEE Trans. Pattern Anal. Mach. Intell. **37**(9), 1904–1916 (2015)
22. Lazebnik, S., Schmid, C., Ponce, J.: Beyond bags of features: spatial pyramid matching for recognizing natural scene categories. In: 2006 IEEE Computer Society Conference on Computer Vision and Pattern Recognition (CVPR'06), vol. 2. IEEE (2006)
23. Grauman, K., Darrell, T.: The pyramid match kernel: discriminative classification with sets of image features. In: Tenth IEEE International Conference on Computer Vision (ICCV'05) Volume 1, vol. 2. IEEE (2005)
24. Girshick, R.: Fast R-CNN. In: Proceedings of the IEEE International Conference on Computer Vision (2015)
25. Ren, S., et al.: Faster R-CNN: towards real-time object detection with region proposal networks. In: Advances in Neural Information Processing Systems (2015)
26. Lin, T.Y., et al.: Feature pyramid networks for object detection. In: Proceedings of the IEEE Conference on Computer Vision and Pattern Recognition (2017)
27. He, K., et al.: Mask R-CNN. In: Proceedings of the IEEE International Conference on Computer Vision (2017)
28. Dai, J., et al.: R-FCN: object detection via region-based fully convolutional networks. In: Advances in Neural Information Processing Systems (2016)
29. Redmon, J., et al.: You only look once: unified, real-time object detection. In: Proceedings of the IEEE Conference on Computer Vision and Pattern Recognition (2016)
30. Liu, W., et al.: SSD: single shot multibox detector. In: European Conference on Computer Vision. Springer, Cham (2016)
31. Redmon, J., Farhadi, A.: YOLO9000: better, faster, stronger. In: Proceedings of the IEEE Conference on Computer Vision and Pattern Recognition (2017)
32. Ioffe, S., Szegedy, C.: Batch normalization: Accelerating deep network training by reducing internal covariate shift. arXiv preprint arXiv:1502.03167 (2015)
33. Redmon, J., Farhadi, A.: YOLOv3: An incremental improvement. arXiv preprint arXiv:1804.02767 (2018)
34. Lin, T.Y., et al.: Focal loss for dense object detection. In: Proceedings of the IEEE International Conference on Computer Vision (2017)
35. Fu, C.Y., et al.: DSSD: Deconvolutional single shot detector. arXiv preprint arXiv:1701.06659 (2017)
36. ICT4CART (2020). WP8 - Evaluation & impact assessment. https://ict4cart.eu/hub/deliverables
37. ICT4CART (2019). WP9 - Communication, Dissemination and Exploitation. https://ict4cart.eu/hub/deliverables
38. Bewley, A., et al.: Simple online and realtime tracking. In: 2016 IEEE International Conference on Image Processing (ICIP). IEEE (2016)
39. Kalman, R.E.: A new approach to linear filtering and prediction problems, pp. 35–45 (1960)
40. Kuhn, H.W.: The Hungarian method for the assignment problem. Nav. Res. Logist. Q. **2**(1–2), 83–97 (1955)

Internet of Vehicles – System of Systems Distributed Intelligence for Mobility Applications

Ovidiu Vermesan, Reiner John, Patrick Pype, Gerardo Daalderop, Meghashyam Ashwathnarayan, Roy Bahr, Tore Karlsen, and Hans-Erik Sand

Abstract This chapter presents the Internet of Vehicles (IoV) concept, technologies and applications used to realise intelligent functions, optimise vehicle performance, control, and decision-making for future electric, connected, autonomous, and shared (ECAS) vehicles mobility scenarios. The concept addresses the convergence of the edge intelligence embedded in the vehicles based on Artificial Intelligence (AI) technologies with the cooperative, collaborative intelligence distributed into the Internet of Things (IoT) devices and edge computing infrastructure federated with the hierarchical cognitive processes and analytics in physical, network, infrastructure, and data spaces. IoV integrates and links the ECAS vehicles' domains with the intra-vehicle networks, vehicle-to-everything (V2X) networks, the processing and cognitive functions provided by federated platforms using the intelligent edge and cloud computing infrastructure. The cognitive transformation of vehicles and the integration of immersive technologies, i.e., virtual reality (VR) and augmented reality (AR) combined with intelligent connectivity, IoT, Distributed Ledger Technologies (DLTs), digital twins (DTs), data, knowledge, security/privacy requirements and learning platforms provide the technological foundation of IoV allowing for entirely new services, applications, and user experiences. The paper advances the latest architectural concepts

O. Vermesan (✉) · R. Bahr
SINTEF AS, Oslo, Norway
e-mail: Ovidiu.Vermesan@sintef.no

R. John
AVL List GmbH, Graz, Austria

P. Pype
NXP Semiconductors Belgium N.V., Leuven, Belgium

G. Daalderop
NXP Semiconductors Netherlands N.V., Eindhoven, Netherlands

M. Ashwathnarayan
Infineon Technologies India Pvt. Ltd., Bangalore, India

T. Karlsen
IoT Proffen AS, Fredrikstad, Norway

H.-E. Sand
Nxtech AS, Fredrikstad, Norway

© Springer Nature Switzerland AG 2021
N. Magaia et al. (eds.), *Intelligent Technologies for Internet of Vehicles*, Internet of Things,
https://doi.org/10.1007/978-3-030-76493-7_4

for ECAS vehicles. It proposes an IoV 3D multi-layered architecture that combines AI, edge computing and connectivity as part of the functional layers while integrating the system properties and trustworthiness properties into the overall architecture to provide efficient new mobility applications and services. The proposed system of systems concept for IoV applications allows for distributed intelligent functions to be embedded into the edge and cloud infrastructure for ECAS vehicles to provide a computing, processing, and intelligent connectivity continuum for IoV applications and services.

1 Introduction

The entire traditional transportation ecosystem is going through a significant transformation with six main trends accelerating the change: new mobility modes and behaviours, the penetration of autonomous driving technologies, the development and use of digital features impacting industry and consumers, the higher need for security, privacy and safety aspects, the electrification of the powertrain, and the introduction of data analytics and AI techniques for implementing intelligent solutions and components.

The mobility evolves towards a safe, green, and connected ecosystem. A push for more ECAS vehicles and mobility solutions is needed to address future mobility challenges.

The incentives and the state of play for the development of each ECAS mobility element are highlighted below.

Electric: Global climate change, limited resources, and increased incentives/regulations, lead to mobility providers embracing electric vehicles and other sustainable mobility forms. Issues related to health and quality of life for the citizens require that local communities limit and ban vehicles with internal combustion engines (ICEs) from city centres. Disruptive automotive original equipment manufacturers (OEMs) and start-ups expand drive ranges and demonstrate competitive technological and cost options for vehicle users. Charging network infrastructures are further expanding, eliminating "range anxiety" for vehicle users and reducing the recharging time. The OEMs support new AI-based electric/electronics (E/E) hardware/software architectures [47] and new energy storage solutions.

Connected: The new vehicles need built-in two-way intelligent connectivity capabilities to address the requirements for the interactions with the environment and infrastructure to provide extra services to the vehicle, exchange real-time data to the edge/cloud computing infrastructure and provide over-the-air software updates. These capabilities will allow vehicles and mobility services to improve over time continuously. Built-in vehicle connectivity will grow significantly as it represents a key element to evolve towards digitising the mobility sector and implementing shared economy services and applications. With increased connectivity, the threat of malicious hackers and compromised cybersecurity also rises, increasing the need for more sophisticated in-vehicle and edge/cloud-based safe, secure, reliable, and un-hackable components, embedded software, and systems.

Autonomous: The automotive industry has defined autonomy levels such as L3-to-L5 for automated and autonomous vehicles. The autonomous driving functions drastically reduce the risk of crashes through alerts and discrete system actuation, alerting drivers of approaching pedestrians and allowing intermittent "hands-off" control for brief periods. Full autonomy in defined geographic spaces will provide greater productivity and less congestion (resulting in less pollution) to engage in secondary tasks and enable new business models by lowering operational human drivers' costs.

Shared: The mobility industry is rapidly deploying new forms of vehicle-sharing, ride-sharing and ride-hailing services to lessen the importance of 1:1 ownership model, smoothen the demand throughout the day, enhancing access to mobility by lowering the capital to participate and increasing utilisation rates of physical assets. Shared mobility services are also taking aggressive and positive steps to fully integrate into public transportation systems and smart cities and offer more seamless multimodal experiences, allowing the consumers to focus on where to go and less on how to combine a patchwork of disparate mobility solutions to get there. Important is that easy access for users' needs to be combined with guaranteed highest privacy standards to avoid potential impacts caused by the General Data Protection Regulation (GDPR) as information about the origin, destination, financial information is being shared. This requires high security and privacy protection at the component, embedded software, and system level.

Electric connected autonomous and shared (ECAS) technologies have transformed the vehicle into a mobile, self-aware, contextual, highly autonomous, software-defined IoT device that is part of the IoV applications. The ECAS vehicles often called "edge computing, connectivity micro servers/centres on wheels", require the highest cognition, security and connectivity performance, with more functional integrated control units, compliance with continuously stricter standards, and rigorously verifiable and proven system-level safety and security for the connected "IoT devices" like cameras, stability sensors, and brake actuators, which puts forward the management of specifications from deep processing technology elements into individual HW/SW components up to the system-of-systems level. The capabilities and benefits of ECAS vehicles are illustrated in Fig. 1.

The AI techniques, methods and functions enable IoV to enhance the perception capabilities, through cognition in intra-vehicle networks, inter-vehicle networks and vehicle-to-everything (V2X) networks. The exchange of information with the road environment, cellular networks, IoT edge nodes, edge computing and cloud infrastructure, provide the data and analytics for implementing new advanced mobility services and applications to assure reduced accident rates, increased mobility efficiency, improved traffic flow, lowered emissions, extended utility, and efficiency for ECAS vehicles. The AI technologies support new mobility applications and services

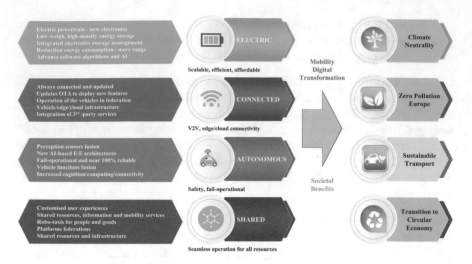

Fig. 1 Electric Connected Autonomous and Shared (ECAS) vehicles features

to achieve more efficient mobility operations, improve vehicle fleet interactions and enhance data management, traffic analytics and efficiency.

As more technology is used to drive the vehicle autonomously, the need to make it completely safe is paramount. Systems must be reliable throughout the time the vehicle is in use, and most of them need to conform to security and functional safety standards. This includes failures due to component breakdown or malfunction due to hacking to maintain device integrity. Furthermore, privacy protection becomes of uttermost importance for autonomous vehicles in a multimodal IoV transport environment.

An overview of the features and capabilities of ECAS vehicles for SAE [20, 30] Level 3, 4 and 5 of driving automation is presented in Fig. 2.

In this article, the vehicles' evolution in the last decade and the further developments and trends for the next decade are mapped into six generations, as illustrated in Fig. 3. The illustration is based on the recent developments and aligned with the work in [28].

The chapter is organised as follows. Section 2 presents the IoV concept and synthesises the relevant related work within IoV field of research. The overview of the ECAS vehicle capabilities, technologies, and architectures are presented in Sect. 3. Section 4 is dedicated to describing the layered IoV architecture, focusing on the functions provided by the AI components and cognitive functions at different architectural layers ranging from ECAS vehicles to network, edge, service, and applications. Section 5 addresses the design issues in implementing the IoV architecture for vehicular edge scenarios at the ECAS vehicle level to advance the security, safety, and performance of the traffic system. In addition, Sect. 5 presents the technologies for implementing the intelligent connectivity for IoV. Section 6 discusses future

	SAE Level 3	SAE Level 4	SAE Level 5
	Awareness for Take Over	No Driver Interaction	No Driver
Ultrasound sensors	8	8	8-10
RADARs (Long-Range - LRR)	2	2	2
RADARs (Mid Range - MRR)	4	4	4
RADARs (Short Range - SRR)	4	4	4
Cameras (Stereo/Trifocal)	2	2-3	2-4
Cameras (Surround)	4	4	4
Cameras (Stereo Vision)	1	1	2
Microbolometer	1	1-2	1-2
LiDARs	1	2-4	4
Dead reckoning	1	1	1
Far Infrared Cameras	0	1	1
Total number perception sensors	24-28	28-34	34-38
Satellites	-	-	x
Mono-Vision Rear	x	x	x
Rear/Surround View	-	x	x
ADAS ECUs	x	-	-
AD DCUs	-	x	x
Real-time High Definition (HD) Map	-	x	x
E-Horizon	-	-	x
E/E architecture	Central gateway	Central cross domain DCUs Zone oriented architecture and vehicle control unit/computer	Zone oriented architecture and vehicle control unit/computer (VCU)
Safety	Safety and Availability	Fail-operational	Fail-operational, Predictive safety
ASIL	ASIL-B, ASIL-C	ASIL-C, ASIL-D	ASIL-D
Software functions/features	>55	>60	>65
AI embedded functions	x	x	
Operating systems	AUTOSAR POSIX Operating System	AUTOSAR + Adaptive AUTOSAR POSIX Operating System	Adaptive AUTOSAR POSIX Operating System
Computing capabilities	>40 000 DMIPS (Dhrystone MIPS)	>250 000 DMIPS	400 000 - 900 000 DMIPS
Hardware	GPUs, CPUs, NNs, AI cores, accelerators	Hybrid GPUs, CPUs, FPGAs, ASICs, on-device NNs, AI cores, accelerators	Embedded High-performance hybrid processing - GPUs, CPUs, FPGAs, ASICs, NNs, AI cores, accelerators
Virtual/Augmented Reality (AR/VR)	-	x	x
Communication capabilities	Wi-Fi, 4G LTE	Ethernet backbone, V2X, Wi-Fi, 5G/6G	Ethernet backbone, V2X, Wi-Fi, 5G/6G
Cognitive capabilities	>100 TOPS	>200 TOPS	>300 TOPS
Memory	8-32 GB RAM	32-256 GB RAM	256-1024 GB RAM
Power consumption	300W	300-600W	600-800W
Driver Monitoring System	-	x	x
Passenger Monitoring System	-	x	x
Edge, cloud platforms	Cloud computing	Edge computing + Cloud computing	Edge computing + Cloud computing platforms federation
Verification, validation, testing	Physical, Virtual	Physical, Virtual	Physical, Virtual, Digital Twin
Standards	ISO26262, ISO 61508, AEC Q100, TS16949	ISO26262, ISO/PAS 21448, ISO 61508, AEC Q100, TS16949	ISO26262, ISO/PAS 21448, ISO 61508, AEC Q100, TS16949
Commercial availability	>2020	>2025	>2030

Fig. 2 Overview of the features and capabilities of ECAS vehicles (SAE Level 3, 4, 5)

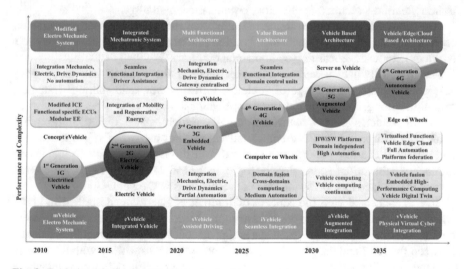

Fig. 3 Evolution of vehicle generations checked

research challenges and raises key open issues related to the implementation of IoV. The conclusions are highlighted in Sect. 7.

2   Concept and Related Work

The Internet of Vehicles mirrors the application of IoT concept to ECAS vehicles addressing resource sharing, real-time information exchange while improving energy-saving emission reduction to provide mobility convenience, intelligence, advanced operational effectiveness, and traffic safety.

IoV reflects the evolution and the convergence of several technologies such as ECAS vehicles (ECAS vehicles as complex, intelligent IoT devices), Internet of Energy (IoE), the mobile Internet, VANETs—Vehicular Ad Hoc Networks (e.g., originated from MANET or Mobile Ad-hoc Network), Ai and edge/cloud computing. IoV applications encompass connectivity and real-time information exchange (e.g., ITS G5, 5G and beyond), environmental perception, energy conservation, security and safety. It enables information sharing and collecting of information on vehicles, roads, and their environments, which bring new functionalities to mobility services and applications while preserving operational performance, safety, security, privacy, and trust.

The IoV is an AI-based IoT system of systems (SoS) that integrates knowledge/information into a collaborative concept providing the aggregation of different domains' architectures. The overall IoV consists of several components that come with their architectures: vehicle, infrastructure, communication [19], data, training/learning, security, application, organisational, strategy.

The components can be structured hierarchically, decentralised, distributed considering the physical components (vehicle, infrastructure), logical components (strategy), knowledge/learning components, etc.

As IoV will include various business models, the concept has to be scalable and adaptable to the type of vehicles, infrastructure, data sharing strategies, training/learning, communication, and applications.

ECAS vehicles access, consume, create, enrich, share digital information enhancing their self-aware and contextual capabilities as part of IoV applications and services. The IoV concept has evolved over the last years with an increasing number of publications addressing the topic. In the following paragraphs, a literature review is provided to address the latest developments in the field.

IoV is connected with the concept of Internet of Energy as described in [1, 2] where IoE is defined as a network infrastructure based on standard and interoperable communication transceivers, gateways and protocols that allows a real time balance between the local and the global generation and storage capability with the energy demand. IoE allows units of energy to be transferred bidirectionally when and where it is needed that is key for the implementation of the IoV applications based on ECAS vehicles.

The IoV is described in [3] as "a distributed transport fabric capable to make its own decisions about driving customers to their destinations", that has "communications, storage, intelligence, and learning capabilities to anticipate the customers' intentions". The article presents the evolution from an intelligent vehicle grid to autonomous, Internet-connected vehicles, and vehicular cloud. The vehicular cloud, the equivalent of Internet cloud for vehicles, is considered the concept that helps the transition to IoV by providing all the services required by the autonomous vehicles.

Implementing IoV applications must consider various security criteria where entire communication can lead to many security and privacy challenges. Blockchains provide the IoV devices with the necessary authentication and security feature for the transfer of data [5] and help eliminate the single source of failure. The AI-Powered Blockchain solution proposed in [5] provides a decentralised approach using auto coding features for smart contracts, speeds up the transaction verification and optimises energy consumption. A collaboration and incentive mechanism for IoV with secured information exchange based on blockchains is presented in [14]. The paper proposes a blockchain framework so that secure information exchange can be handled among participated vehicles in the mobile crowdsensing (MCS) network in IoV.

The ECAS vehicles share a significant amount of information (e.g., internal, environmental, scenery, etc.) in IoV applications. Intelligent decision-making based on AI techniques is more and more used for different applications and services. Driving V2X autonomous vehicles' policies based on reinforcement learning methods are discussed in [6] where the autonomous driving in V2X environment is analysed by proposing a system architecture allowing additional traffic by using V2V technology. The authors study a decision-making method using a reinforcement learning method that could learn in the V2X environment. The proposed model is simulated using OpenAI reinforcement learning framework, and the experimental results demonstrate the effectiveness of the V2X technology for the robust autonomous driving. The concept of the Internet of Autonomous Vehicles (IAV) presented in [4] show how centralised architecture will give way to more distributed architectures for real-time information propagation over the IAV. In this context, the network-centric policies begin to shift to user-centric under more beneficial revenue models by offering network-assisted quality of service (QoS) provisioning. Further expansion of 5G connectivity and the 5G communication model for future implementation of IoV environment in terms of low latency, extremely high bandwidth and reliability is introduced in [17].

The integration of edge computing, as part of the IoV is emphasised in [7]. Edge computing can provide more powerful computing performance, energy efficiency, storage capacity, and mobile performance. The authors apply edge computing and artificial intelligence to the IoV by using an IoV model of intelligent edge computing task offloading, and migration based on SDVN (Software-Defined Vehicular Networks) architecture. The method is called JDE-VCO (Joint Delay and Energy-Vehicle Computational task Offloading), and the IoV model is simulated and can be used for subsequent application of the IoV.

A classified cloud architecture to allow the deployment of edge computing infrastructure and services by combining the internal cloud infrastructure of different stakeholders with multiple highly decentralised edge nodes leased from third-party providers is presented in [10]. The solution can be applied to IoV applications and services.

An IoV framework based on the development of VANETs (vehicular ad hoc networks) and the motivation of IoV is presented in [8]. The paper describes a layered architecture, protocol stack, network model, challenges while giving a qualitative comparison of IoV and VANETs that highlights the benefits of IoV design and development. An IoV authentication protocol for the safety of IoV in data transmission is discussed in [9]. The protocol suffered initially from offline identity guessing attacks, position spoofing attacks, replay attacks and required a long authentication time. After further improvements, the protocol provided improved security and performance.

A communication efficient and privacy-preserving federated learning framework for enhancing the performance of IoV, wherein in-vehicle learning models are trained by exchanging inputs, outputs and their learning parameters locally is described in [11]. The paper presents a systematic approach for IoVs' networking design that transforms the vehicles into mobile data centres, performs federated learning to enhance transmission control protocol (TCP) performance over Wi-Fi and reacts timely to the Internet for the vehicle needs.

The experience readiness level (ERL) concept applied to an autonomous vehicle, and further to IoV is presented in [12]. The ERL allows identifying the overall level of readiness or maturity of a given IoT/IoV technology/system/application as it is related to the experience, usability, and the refinement to be used by end-users/customers. Modelling quality of IoT Experience in autonomous vehicles is presented in [13]. Autonomous services, such as IoVs, are orchestrated by artificial AI, enabling intelligent data processing, reasoning, and decision making. The paper introduces the term Quality of IoT-experience (QoIoT) within the context of autonomous vehicles, where the quality evaluation, besides end users, considers quantifying the perspectives of intelligent machines with objective metrics. In this context, an architecture is proposed that considers Quality of Data (QoD), Quality of Network (QoN), and Quality of Context (QoC) to determine the overall QoIoT in the context of autonomous vehicles.

Trust management and security are critical elements for the development of IoV applications. Latest developments have focused on security and reputation frameworks for identifying denial of traffic service in IoV. The framework proposed in [15] tries to resolve the trustworthiness problem in the application level of the IoV considering that every vehicle communicates with the roadside unit (RSU) directly for traffic event verification, and spreads verified traffic event notification.

Routing and security in IoV are using AI-based and nature-inspired algorithms. (e.g. ant-colony optimisation, grey wolf optimisation, spider-monkey optimisation, swarm optimisation, bees, firefly, genetic algorithms, etc.) An overview of the applicability of nature-inspired algorithms for routing and security in IoV is given in [16]. The paper highlights that for routing, the optimisation aims for optimal and timely delivery of messages between vehicles or between other communicating agents. In contrast, in the case of security, the optimisation seeks to secure the IoV network from different security attacks types.

3   ECAS Vehicle Capabilities, Technologies, and Architectures

The vehicles participating in the IoV as fleets, or individual vehicles for providing services and applications need to have enhanced/augmented human drivers' skills. The vehicles must be able to perceive, locate and interpret their surroundings ("SENSE" and "LOCATE"). Based on the information received about surroundings, the vehicles need to fuse, process the data, and understand the situation ("COMPREHEND"). The information collection and understanding of the situation, need to be followed by the cognition process to plan and decide the vehicles driving strategy "THINK"). The vehicle's decision needs to be exchanged with other vehicles, infrastructure and depends on the information received from infrastructure other vehicles, pedestrians, road signage, etc. ("CONNECT"). When the decision is taken the commands and controls are transferred to the powertrain, steering and braking power to move the wheels so that the planned driving strategy is put into practice ("ACT"). All these operations are recorded, and the vehicles use the data to improve their performance, increase efficiency, enhance the experiences ("LEARN"). The vehicles' overall full capabilities integrated into the IoV services and applications are illustrated in Fig. 4.

The integral digital value chain of IoV systems must integrate vehicles able to SENSE, LOCATE, COMPREHEND, THINK, CONNECT, ACT, COLLABORATE and LEARN.

These "human driver" like enhanced capabilities are pushing the developments of future vehicles and IoV architectures and the requirements and specifications for electronic components and systems. The design approaches are moving towards

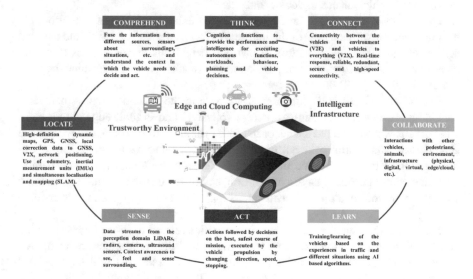

Fig. 4 Vehicles functions for future IoV services and applications

new software, hardware, algorithms, and the use of AI, IoT, DLTs, knowledge, and learning platforms.

The "SENSE" environment perception function combines different sensors such as cameras, radars, light detection and ranging (LiDARs), ultrasound sensors that produce data that are integrated into a single sensor fusion model. Combining different sensor technologies compensates for individual weaknesses under the various environmental conditions and is the only way to come to a vehicle "SENSE" function. The sense functions evolve towards distributed model-based or AI-inference processing at the sense-node and central cognition, supported by edge-intelligence for training, and real-time environment models from the IoVs. The concepts presented in this paper extend the findings and the experience of the authors working with electric, autonomous/automated, and connected vehicles technologies and applications, combined with the research and deployment of IoT/IIoT technologies across various industrial sectors.

The "LOCATE" function detects objects on routes, in-context on 2D or high definition (HD) real-time maps to identify the road network attributes, including lane level accuracy. The function is used to classify the detected objects as vehicles, road lanes, people, traffic signs, objects on drive route or not and is used in combination with the SENSE and CONNECT functions to identify the vehicle's context at any time.

The "CONNECT" function provides the connectivity between the vehicles to the environment (V2E) and everything (V2X). The function implements the communication protocols for connecting the vehicles to all the elements used in providing the IoV applications and services.

The " COMPREHEND" function is used by the vehicle to fuse the information from different sources about surroundings, situations, etc. and understand the context in which the vehicle needs to decide and act.

The "THINK" and "ACT" functions combine the interpretation of (intentions of) the environment with the goal of the journey, to follow a route, and path and detailed actions towards vehicles' actuators. It involves deterministic calculations assessing uncertainty and inaccuracies to minimise the risk of accident and determine the most optimal route.

The "COLLABORATE" function is addressing the activities with other vehicles, animals, environment, infrastructure (physical and digital, edge/cloud, etc.), and humans within a shared space, or in close proximity to produce, create, achieve shared goals, and minimise the risk of accidents and dangerous situations. The collaborate function incorporate one or several features such as natural or non-natural language communication, tactile interaction, safety-rated stop monitoring, teaching by demonstrations/examples, speed and separation monitoring and power/force limiting, etc.

The "LEARN" function addresses the activities related to the vehicles' training and learning based on the experiences in traffic and different situations using AI-based algorithms. The training and learning are processes that will be federated between edge and cloud, and in the future, it will involve the vehicle itself with its extensive computing capabilities used as a training/learning platform.

The IoV concept is created based on vehicles' real-time intelligent collaboration to provide mobility services and applications. The idea must consider integrating legacy mobility systems and vehicles and the management of the interactions between these heterogeneous systems.

3.1 Vehicles Architecture Evolution

The development of ECAS vehicles based on new cognitive-based technologies, the emergence of new mobility business models and the changes in users' mobility behaviours accelerate the automotive industry's disruption that will profoundly impact the future electric and electronic vehicle architecture. The vehicle's transition toward cognitive computing connected intelligent platform on wheels, where perception, sensor fusion, AI, edge computing, connectivity [46], functions virtualisation and resources/information sharing are key elements to implement the higher levels of automation and mobility applications. This transition requires new electric and electronic vehicle architectures [41–43, 45] that must address the existing challenges and next-generation ECAS vehicles' requirements.

The new ECAS vehicle architectures must address the functional complexity with many diverse cross-domain functionalities that need to be handled and distributed, provide flexibility to allow SW functions virtualisation and sharing, enable scalable implementations for vehicle features and functions. The architectures must provide a design paradigm shift by including cognitive capabilities as the core components of the vehicle intelligent mobility platform that converge the functional domains and process/distribute the information with the different functional domains inside and outside the vehicle. The internal and external connectivity concepts should be an integrated part of the architecture with end-to-end (E2E) security and safety embedded at all layers of the architecture to support and enhance the information traffic, eliminate the risks, and provide the needed real-time communication bandwidth for inter-domain and cross-domain communication. The new architectures must provide the optimised computing power and storage for energy-efficient processing of various functions (e.g., functional domains specific or cognitive/decision centric), a fusion of information from different sources, learning/training and real-time upgradability.

The evolution of vehicle architectures is illustrated in Fig. 5. This evolution is correlated with the evolution of vehicles generations, the evolution process of IoV and the development of key technologies that enable these evolutions, including autonomous driving technology, intelligent connectivity (e.g. wired/wireless, cellular 5G/6G, network slicing, etc.), AI, edge/cloud computing, distributing computing, digital twins (DTs), Virtual Reality (VR), augmented reality (AR), Artificial Intelligence of Things (AIoT), E2E security, fail-operational technologies [26], information platforms.

The result of this evolution and steadily increased functional integration is an extensive complex system of electronic control units (ECUs), sensors, actuators, and

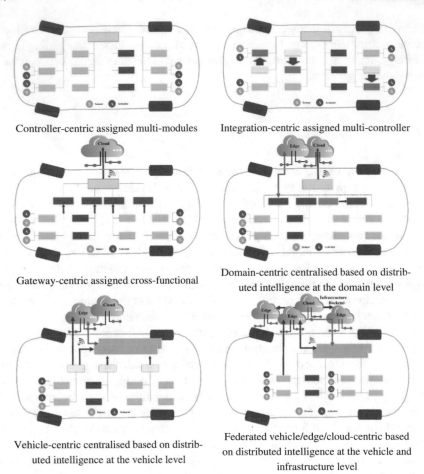

Fig. 5 Evolution of vehicle architectures

wiring. The complexity and the extent of these architectures create new challenges for automotive OEMs and their suppliers.

Today, the traditional vehicle contains between 70–100 ECUs. The electrical-born vehicle architecture must consolidate these into fewer more powerful control units through the use of a centralised architecture with one or few high-performance vehicle platforms managing vehicle functions, a distributed architecture with several high-performance vehicle processing/cognition/connectivity integrated platforms or a federated vehicle/edge/cloud functional platforms to address the requirements for redundancy, safety, security and reliability in-vehicle systems and IoV applications. The developments will lead to more powerful centralised architectures such as the domain control unit (DCU) and multi-domain controller (MDC) appear as an alternative to the distributed ones. In developing new architectures, the DCUs can make systems much more integrated because of the high-performance hardware computing and cognition capabilities and availability of software interfaces, enabling integrating more core functional modules, meaning lower requirements on function perception and execution hardware. Standardised interfaces for data interaction are supporting these components providing modularity and scalability to the whole architecture.

Autonomous vehicles require domain controllers not only to be integrated with versatile capabilities such as multi-sensor fusion, localisation, path planning, decision making and control, V2X and high-speed communication but also to have interfaces for cameras (mono/stereo), multiple radars, LiDAR, inertial measurement unit (IMU), etc. Electronic and components systems producers adapt their product portfolio for the new vehicle architectures by providing new AI-based HW/SW vehicle platforms, connectivity, perception and cognition platforms, AI-based zone ECUs, and virtual ECUs, application-SW, middleware, and algorithms.

The future electric and electronic architectures have to include concepts and modules that address several key features of the next generation ECAS vehicles such as fail-operational by being capable of guaranteeing the full or degraded operation of a function even if a failure occurs, secure, by anticipating, avoiding, detecting, and defending against cyberattacks, updatable, by being capable of over-the-air (OTA) updates of the software, connected, to other vehicles, the infrastructure, Internet, mobile devices, upgradable, in both software and hardware., collaborative, to interact with the other vehicles, infrastructure, services and intelligent transport systems (ITS) applications [60] for optimised seamless transportation, learning/evolving by being artificial-intelligence-enabled.

The HW/SW/AI vehicle platforms will evolve to address these needs from embedded ECUs, to vehicle platforms as integration and fusion platforms to federated vehicle/edge/cloud platforms with distributed intelligence and high-performance real-time processing, cognition, and collaborative capabilities. The high-complexity, scalability, and flexibility of these platforms match the increased automotive requirements on safety, security, reliability, and OTA real-time upgradability/updatability. The move towards vehicle platforms provides a concept for the implementation of vehicle run time environment that requires the virtualisation of vehicle functions by creating software enabling vehicle computing platforms with HW platforms having the lifecycle of up to five years and the software and algorithms are continuously updated to provide new and more performant vehicle functions.

The electric/electronic vehicle architectures will impact energy consumption and CO_2 emissions. Today, with the increased complexity, functionality, higher computing, and communication performance, the HW/SW vehicle platform's energy consumption exceeds 4–6 kW. If the energy consumption needed for AI training/learning functions is added, the total energy consumption is expected to exceed 10 kW.

The development of future vehicle architectures must become more energy-efficient (e.g., HW/SW co-design, local HW-accelerators, AI, etc.) and share the computing, connectivity, learning/training capabilities and use efficient collaborative business models for exchanging real-time traffic, environment, and context information as part of IoV applications and services. The electric and electronics vehicle architectures are continuously evolving. In the next subsections, an overview of past, existing, and future architectures is presented, highlighting the main features and characteristics.

Controller-centric assigned multi-modules architectures are based on independent ECUs (e.g., function-specific control units), addressing isolated functions, with each function having its ECU module providing a one-to-one connection. The topology is based on the controller area network (CAN) with every node in the network sharing the bandwidth. The bandwidth limits the data processing capability of each ECU on the network, so intelligent and high-performance computation units are not integrated into the architecture due to a large amount of data transferred.

Integration-centric assigned multi-controller architectures are based on the collaboration of ECUs within one domain (e.g., body/comfort, chassis, powertrain, infotainment, etc.), using 4–5 independent networks (CAN, FlexRay, LIN, MOST) with limited communication between domains.

Gateway-centric assigned cross-functional architectures are based on robust collaboration between the ECUs via the central gateway using Ethernet as backbone and specific communications networks (CAN, FlexRay, LIN, MOST) for different domains. The gateway provides the cross-functional connection, and the architecture can handle complex functions (e.g., adaptive control).

The architecture can manage up to L3 autonomous driving functions integrating 80–100 ECUs, providing allocated control to the different nodes, and assuring the intercommunication via the common Ethernet backbone gateway. An example of gateway-centric architecture used for the current vehicles up to Level 3 is presented Fig. 6.

An implementation of the service-oriented gateway centric architecture is illustrated in Fig. 7. The service-oriented gateway secure central access to ECAS vehicle

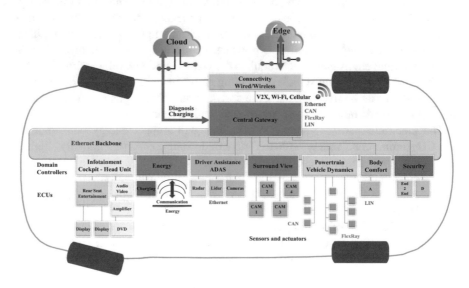

Fig. 6 Example of current complex gateway-centric vehicle architecture

Fig. 7 Example of service-oriented gateway-centric vehicle architecture

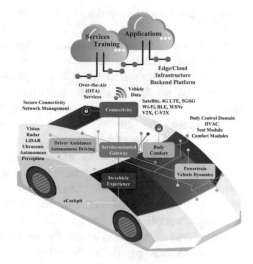

data, lowers the cost to develop, deploy, and integrate software, supports rapid deployment of new services through OTA updates and requires higher performance processing and networking. Edge and cloud infrastructures are used for process data and provide data enrichment, analysis, AI techniques and learning capabilities.

Domain-centric centralised based on distributed intelligence at the domain level architectures are based on a central domain controller (CDC) that can handle several complex functions and provide the consolidation of several functions on the same HW platform. The domain-based topology concept divides the autonomous driving system into several domains, using a domain ECU as the core computation platform for each domain. The vehicle domains are defined according to their functionalities and sensors/actuators shared by different functionalities are grouped as one domain. The DCUs can support complex, intelligent driving functions, as they have high-performance in communication and computation. The architecture allows for virtualisation of functions. The architecture is based on domain-specific functional integration using domain control units (DCUs) that are computing units that can include cognitive features and AI-based techniques and methods. The architecture consists of 4–5 high-performance AI-based DCUs, multiple intelligent sensors/actuators ECUs used for sensor fusion, with CAN bus networks dedicated to the domain zones and 1 Ethernet backbone (including Time-Sensitive Networking) at the vehicle level. The domain ECUs can be connected to sensors in the domain without the issue of sharing bandwidth. The DCUs are used as a computation platform to integrate related control functions into a complex behaviour control function. The distributed computation strategy of domain-based topology has the advantage of being more flexible with the domains being relatively independent of each other by transmitting only the necessary information to other vehicle domains. The data flow within the domain does not utilise the bandwidth and other resources of the backbone. The architecture allows for dedicated domains with consolidated functions at the zone level with the routing between the DCUs being handled by an intelligent gateway.

Multi-Domain-centric centralised based on distributed intelligence at the domain level architectures are based on the domain fusion with several domains integrated using cross-domain control units (CDCU). The architecture provides different HW platforms that are interlinked. Efficient platforms partitioning requires support for functional safety using cross-domain functions. The domain control units control sensors and actuators through common interfaces. By addressing any multi-domain scenario, the architecture facilitates communication over domain borders. The coordination mechanisms influence the whole system performance and availability. In the case of safety–critical functions, mixed integrated on a multi-domain controller with timing constraints, provisions such as appropriate scheduling algorithms must be developed at the software integration phase to guarantee each software component's predicted execution time. The integration of multi-domain architectures requires standardised middleware for the domain controllers to ensure the platform level optimisation, parallelise the software and reduce complexity.

Vehicle-centric centralised based on distributed intelligence at the vehicle level architectures are based on domain-independent units based on vehicle control computation, cognition, connectivity unit (VCU) and specific zone intelligent ECUs. In a vehicle-centric architecture, the primary computation tasks are performed in the VCU rather than in different functional domains. Most of the components are connected to the VCU, which could access all sensors and actuators and perform real-time sensor fusion. The VCU acts as the vehicle's brain to collect and vehicle fusion information, make better decisions, and execute better behaviours.

The vehicle-centric topology requires very high and robust data communication capacity and groups the components into different sub-networks according to their physical placement or the network properties to improve communication efficiency. The controller of the sub-network is the zone controller that connects the zone ECUs. An example of vehicle-centric architecture used for the advanced vehicles implementing Level 3+ functions is presented in Fig. 8. A variant of the architecture integrating zonal gateways is illustrated in Fig. 9.

The vehicle functions are virtualised on the HW platform, and different functions share computing/connectivity resources to provide the various features of virtual functional domains. The specific intelligent hardware is implemented at the zone ECUs and sensor/actuator level. The connectivity is enabled by Ethernet backbone (including Time-Sensitive Networking), allowing the connection between the cluster of embedded high-performance computing units for performing the virtualised vehicle functions. The routing complexity is managed by an integrated intelligent gateway that manages the interactions with the zone intelligent sensor/actuator ECUs that are domain-independent and scalable. The VCU controls sensors and actuators through standardised interfaces. The vehicle-centric architecture will use the zone ECU's and some remaining ECU's for safety–critical applications with strong real-time or latency requirements. (e.g., for braking or airbag control systems). The Zone ECU's bridge the sensors and actuators to the VCU and reduce the wire harness since the modules are placed at different vehicle zones (e.g., front/rear, side, etc.).

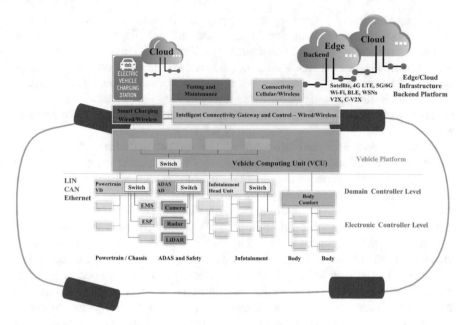

Fig. 8 Example of vehicle-centric architecture

Fig. 9 Example of
vehicle-centric with zonal
gateways architecture

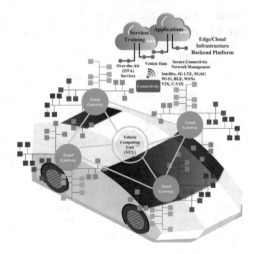

The introduction of domains in the vehicle architecture supports the logical restructure and optimises the software missions facilitating the implementation of autonomous driving functions. The introduction of domains reduces the module count and diversity of ECUs by providing a lower cost, safer vehicle platform to build and maintain that can be used for Level 3+ functions.

The introduction of zones supports the physical restructure to optimise wiring, efficient power distribution implementation and facilitate virtualised functions on the vehicle computing unit and zonal gateways. The safety and security are defined at the vehicle level and brings additional challenges for implementation. The introduction of zones and VCUs allows better update, customisation, reconfiguration of the vehicle functions using hardware and software virtualisation, safe, secure hyperconnectivity [44], and opens the developments to implement service-oriented vehicles.

Federated vehicle/edge/cloud-centric based on distributed intelligence at the vehicle and infrastructure level architectures based on distributed intelligence and shared computing/connectivity/learning/memory resources between a cluster of embedded high-performance computing units in the vehicle, edge, cloud, and infrastructure. The architecture is based on the federation of HW/SW/learning platforms that are physically located in different places and offer common virtualised functions to the vehicle and the services and applications in which the vehicle is part off. These platforms are using the digital twin technologies extensively for modelling, simulating, and executing in real-time the vehicle functions for specific mobility services and applications in particular environments and contexts. The architecture allows virtualisation of functions and integration at a different level: actuators/sensors (e.g. physical devices providing pressure, acceleration, position, image, data to the computing level and the driving modules for the actuators like valves, motors, inverters, etc.), computing/cognition/connectivity/control (e.g. sensor fusion, domain fusion, analytics, planning's and execution of function and services, AI learning, etc.) and edge/cloud back-end for off-board/infrastructure computing, data management, AI training, services, etc.

Automotive-qualified embedded processors do not have enough compute-power to process algorithms like multi-sensor data fusion needed for automated driving. The autonomous driving functions are implemented for L3 using high-performance CPUs and GPUs from the consumer industry, and automotive-qualified master ECUs.

The new ECAS vehicle architectures require automotive-grade embedded AI-based high-performance computing components with larger compute power, increased energy efficiency and scalability to address the Level 5 applications and services. The development needs to accelerate the standardisation of modules AI-based ECU/DCU/VCU, electronic hardware components, interfaces and software/algorithms, and APIs. The vehicle functionality enhancements through virtualisation (e.g., on-demand enabled functions based on standardised set of electronic hardware) and regular OTA updates at the platform levels will use the benefits of separating hardware and software development cycles supporting the mobility stakeholders to speed-up software development.

The advances at the E/E vehicle architecture level are changing the design strategies and concepts of all architecture layers.

The ECU design approaches are changing with the ECUs integrated into domain-specific controllers, vehicle embedded high-performance computing units or virtualised software functions provided by the vehicle or edge.

One trend seen in the next generation architectures is the consolidation of ECUs into 5–6 domain-specific controllers for safety, AI/perception, powertrain/energy comfort/interior, infotainment,

The complex functions to be managed requires the further development of secure real-time robust, energy and computing resource-efficient operating system to manage the multi-tasking, coordinated operations and elimination of redundant components. This will allow the further consolidation of ECUs into an embedded high-performance computing platform that transforms the vehicle into an "edge computing, and connectivity micro servers/centres on wheels".

This further push for federated vehicle/edge/cloud-centric based on distributed intelligence at the vehicle and infrastructure level architectures with a smart-node computing network implemented as a distributed computing platform that comprises smart sensors with dedicated computing power in computing units and virtualisation of processing power using federation edge/cloud/infrastructure.

To manage these new vehicle architectures and provide the reconfigurable capabilities and continuously functions updates/upgrades, advances in the middleware layer's design across domain control units with the ability to access functions of DCUs/VCUs are needed. These functions must support functional safety through fault recognition and tolerance mechanisms, implement plug and play components through standardised interfaces, and provide resource awareness and optimal allocation.

Fail-operational safety and redundancy are achieved by implementing redundant sensors and AI-based algorithms for safety–critical functions. The redundancy is optimised using sensor fusion for autonomous vehicle functions. The research data show that radar and camera will be used in combination to provide safe, secure perception functions, with solid-state LiDAR and camera integration offering optimal solutions for long-term solutions.

The strategies for aggregating the information from intelligent perception sensors are following different directions with solutions where the sensors are connected to several domain controllers as the perception are intelligent and highly integrated (sensing, processing, AI-based algorithms, etc.), and send only critical information to centralised controller to reduce the complexity of the overall vehicle architecture.

Other approaches use centralised control units to perform real-time raw data fusion received from various perception sensors to enable real-time performance and cost reduction of the overall system.

The new vehicle architectures must provide redundant power networks using of x-by-wire functions through fail-safe/fail-operational power supply/energy functionality without additional storage units. New power topologies will be optimised (e.g., star) to ensure the energy paths to safety–critical loads and optimal wiring circuits. The new topologies integrate modular power distribution units that allow load monitoring, fast failure detection, system response and actions on the driving system.

The connectivity is one key element of the ECAS vehicle architectures with redundancy to be implemented for wired and wireless networks. The in-vehicle networks rely on high-data-rate switched Ethernet to ensure reliable inter-domain communication, with topologies (e.g., ring) that offers additional redundancy by duplicating links between switching elements. The switched Ethernet networks will be enhanced by using Ethernet audio–video bridging (AVB) and time-sensitive networking (TSN) [19] for real-time components and computing units.

As presented the evolution of the vehicle architectures for ECAS vehicles, the Ethernet is used as the backbone supporting different bus types, including existing bus systems linked to the central gateway. Several architectures use switches to allow connections to any number of devices and provide flexibility in the architectural design of the ECAS vehicles ensuring interoperability for integration with external networks, high-bandwidth, reduced cost, and less cabling.

End-to-end security is paramount for the ECAS vehicles and IoV applications and access to the safety—and mission-critical components must be restricted and continuously monitored to avoid cyber-attacks and misuse.

In the ECAS vehicles, there are several access points for vehicle information, and various comprehensive information access strategies are developed by the automotive industry. Due to high-economic and liability risks in the future ECAS vehicles development, the OEMs will control the most extensive data sets, and they will provide means to make the information accessible and shared through their backend systems.

The V2X interfaces to access vehicle safety and mission-critical information and messages exchanged with other vehicles, and the OEMs keep the infrastructure with no information access for third parties.

The ECAS vehicle OBD-II port is employed for transmitting information via dongles, and the port is accessible for third party solutions. The port can be restricted due to security risks and the proliferation of secured OTA solutions.

The development of IoV applications and services require sharing of historical and real-time information using the OEMs backend and crating information lakes, spaces, marketplaces where data is available through regulatory decisions. This includes various mobility providers, fleet operators' backend that integrates information relevant for the fleet operator, with information that can be used by licensed fleets.

Information sharing among IoV stakeholders is critical for implementing IoV use cases that rely on data from different data sources with processing done both locally, at the edge and in the cloud, with information flowing in both directions continuously in real-time. The collection, processing, exchange and sharing of information in IoV applications must consider the privacy [44] and the ownership of the data by anonymising and encrypting the data from sensors/actuators, sensitive vehicle, and personal data of the participants in IoV applications/services, etc.

4 IoV 3D Multi-layered Architecture

The IoV architecture is represented in this paper by a 3-dimensional representation of the multi-layered model (Fig. 10), illustrating how to approach the IoV technologies and applications in a structured manner.

The 3D multi-layered model [61] is built along three axes representing the functional layers, the trustworthiness properties, and the system characteristics.

The IoV 3D multi-layered architecture is defined at the high-level abstraction with the IoV functional domains represented on one axis by the physical, communication, control, edge, service, cognition, application, collaborative, and business layers, the IoV trustworthiness properties on the second axis and the IoV system properties on the third axis.

The IoV multi-tier architecture supports the HW/SW partitioning and at lower level enables the interactions of procedures (or methods) within IoV systems designated to perform specific tasks, in distinct conditions and specific context, using the critical, intelligent infrastructure.

The IoV multi-layered architecture approach allows for developing, managing, and maintaining each IoV layer accordingly and adequately integrate the infrastructure development into the overall IoV applications design, development, implementation, and deployment.

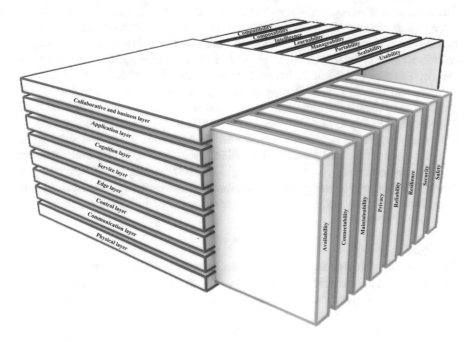

Fig. 10 IoV architecture

4.1 IoV Functional Layers

Physical Layer
The physical layer deals with the physical devices like vehicles, environmental sensors and other objects in the traffic or part of a mobility service, application with the function to sense the information using sensors embedded in the vehicle, in the infrastructure, carried by passengers, pedestrians or other types of vehicles in the traffic and the specific context. The layer integrates the perception and pre-processing of information from multi-source heterogeneous devices providing multidimensional space–time data in physical space from network traffic and resource distribution data in network space. Example of data collected and pre-processed is data from in-vehicle sensors (camera, radars, LiDARs, etc.) route related information, accurate localisation position, charging stations availability, environment conditions, signage status, data from nearby pedestrians, vehicles, mobility operator data, roadside units' information, information on resource occupation by a base station, service request information of users, etc. The physical layer addresses gathering and formatting the data before it is transmitted through different channels/pipelines for processing. Information acquisition is one of the significant IoV challenges in terms of infrastructure requirements and edge heterogeneous vehicles/environment/context.

Communication Layer
The communication layer addresses the vehicle to everything connectivity linked to inter-vehicle communication to perform the information exchange and control functions at networking level using many communication protocols and channels (e.g., wired/wireless, cellular, optical, etc.). This includes transmissions between vehicles, traffic participants and the network, across networks, and between the network vehicles and infrastructure. The layer manages and selects the best network and protocol to exchange the information, based on several selection parameters, functional domains, security requirements, latency, stakeholders, etc. The communication layer performs the virtual network management/coordination for the available heterogeneous network technologies (WAVE/DSRC, 5G/4G/LTE, Wi-Fi, other wireless networks, specific IoT networks, satellites, etc.). The layer implements the transport function and other processing tasks on the information received from heterogeneous networks to create an optimised unified structure with identification capabilities for each type of network.

Control Layer
The control layer deals with the optimisation, coordination and adjusting IoV system performance at the real-time scale needed by the mobility services and applications and matching the traffic requirements and infrastructure capabilities. The control layer performs the orchestration mechanisms for transferring tasks depending on the requirements and infrastructure performance. The layer has the role of an overall control centre that guarantee the optimal partitioning of resources, tasks, and resources among the other architectural layers for different mobility services and applications. The role includes the optimisation of network resources (e.g., type of

network used, type of communication channels, optimisation of parameters for NFV, SDN, SON, network slicing, etc.), processing of resources, learning/training (e.g., vehicle, edge, cloud, data centres, etc.) and storage. Resource optimisation engines are deployed at different cloud/edge locations to increase scalability, stability, and reliability for different IoV scenarios, use cases, and conditions. Resource optimisation engines deployed on edge support delay-sensitive data management considering that the storage, processing, and bandwidth resources available for the edge have to be optimised and a distributed decision-making process is used for orchestrating the federation with cloud and data centres. Based on these processes the V2V cognitive applications responsible for real-time processing of driving data are allocated to the vehicles and edge infrastructure, with the resource in the cloud utilised to effectively integrate the global information of IoV in the decision process.

Edge Layer
The edge addresses the edge computing, data element analysis and transformation, considering that autonomous services are provided through ubiquitous machines in an autonomic way. The layer converts network data flows into information suitable for storage and higher-level processing, control and cognition and provides the ability to process and act upon events created by the edge mobility vehicles and store the data into databases. The requirements for the edge layer are connected to the need for highly real-time, scalable, and complex event processing for high-speed in-memory processing and close to real-time reaction and autonomous actions based on the data and activity of devices and the connected ECAS vehicles and the dynamic systems in IoV mobility services and applications. Edge computing requires processing data at the vehicle's level and using edge computing capabilities to leverage vehicles information to provide end-to-end value streams involving vehicles, participants in the traffic and the infrastructure within digitised processes. The layer functions offer the processing and storing capabilities for the substantial amount of data close to the physical layer to reduces the communication time required for the data to route through the Internet. The management, including control and configuration, is distributed among the edge nodes. The edge nodes can be deployed at a fixed location in the environment along the roads and can be placed in the ECAS vehicles to cover a neighbourhood or traffic zone based on IoV services and applications' coverage requirements. The edge nodes/servers can be integrated into dynamic, collaborative real-time clusters based on the traffic requirements and context. The control layer could orchestrate the dynamic, collaborative real-time clustering to enable sharing resources and facilitate computational load balancing to utilise resources simultaneously.

The edge computing units at the edge layer receives raw or pre-processed information from the communication layer using available communication protocols. The data is further processed, analysed to offer real-time analysis for delay-sensitive applications.

The layer also stores and manages the records of information exchange between different vehicles within the edge server area. The efficient storage and organisation of information and the continuous update of information as it made available through the capturing and processing channels of the vehicles, traffic, environment, infrastructure is integrated into the edge layer and application layer. The information stored is used by the cognition layer to optimise the decisions and by other layers to optimise their processes. Archiving the raw and processed data is addressed by the long-term offline storage of data that is not needed for the IoV system's real-time operations. Centralised storage considers the deployment of storage structures that adapt to the various IoV data types and data capture frequency. The centralised storage is considered at the application and collaborative and business layer.

Service Layer

The service layer for IoV integrates the middleware that is placed on top of networks and IoV streams. It provides data management and data analytics functions in IoV systems where large amounts of vehicles and infrastructure (e.g. traffic, sensors, road, charging stations, weather, etc.) generated information and events must be logged, stored, and processed to create new insights or events on which business decisions can be taken. The service layer creates and manages services for the interoperable consumption of the information through service APIs. The layer includes functional components to implement different mobility service orchestrations to support mashup of various data streams, analytics, and service components. The layer embeds advanced analytics to allow insights from data to be extracted and more complex data processing to be performed and optimised the interaction and exchange between services (e.g., parking, charging, riding, sharing, delivery, fleet management, platooning, etc.). The layer addresses the discovery of mobility services, orchestration of mobility marketplaces and the provision of optimised services for specific mobility scenarios including multi-modality cases. The layer's components create and manage value-added services and expose them using service APIs. The service layer interacts with the control layer to manage the data exchange among the various services. The tasks include handling the information generated by devices, vehicles, roadside infrastructure in the environment, connect them to different mobility services and apply various suitable policies (e.g., traffic management and engineering, packet inspection, etc.).

Cognition Layer

The cognition layer addresses the coordination layer for decision making, combining the information from vehicles, infrastructure and intelligent services provided by different edge and cloud platforms. The layer implements intelligent information management coordination functions that store, process, and analyse the information received from the lower layers and then make decisions based on the different mobility services applications and the environmental conditions and context.

The cognition layer fusions the information and uses information cognitive engine processes to analyse real-time heterogeneous information flows utilising AI methods (machine learning, deep learning, data mining, pattern recognition, analogue neural networks (ANN) etc.). The cognitive engine prepares the information collected (e.g., autonomous vehicle behaviour, traffic patterns, participants in the traffic, fleet use cases, business models, road condition etc.), in the context under control (e.g., time, space, area of rods covered, etc.).

The AI-based methods are applied to process the data and make decisions applied to other layers. The decisions are based on real-time vehicle area network services and non-real-time vehicle area network, and different cognition methods are used for these tasks. The real-time cognition processes are integrated with vehicle area network services closely linked with the control and edge layers.

The cognitive engine can process dynamically information related to computing, storage, and network resources, based on the resource allocation in the edge layer combined with the application layer. For the delay-sensitive tasks, the edge layer needs to assure that the infrastructure and the vehicles involved have sufficient resources to complete the task and decide the partitioning of tasks considering that the functions that are less sensitive to delay are moved to other infrastructure (e.g., cloud, data centres, etc.) that is managed by another layer. In this context, the cognition engines running under the cognition layer are configured in real-time based on different parameters (e.g., services, communication network slices, business models, business types, federation edge-cloud-data centres, application scenarios, infrastructure topology, traffic conditions, network security, etc.)

Applications Layer

The application layer hosts different applications and control and management interface. This layer is responsible for providing IoV applications specific to the stakeholders. The application layer is considering integrating the IoV ecosystem and involves coordination and cooperation among multiple parties, including ECAS vehicles, OEMs, mobile communication operator, networking services provider, manufacturer of intelligent devices, software services provider, road, and IT infrastructure providers, etc. The layer can provide customised applications to reduce safety risks during driving and various intelligent mobility and multi-modal applications.

The application layer offers the platforms that are suited to deliver the critical components for implementing various IoV applications connecting users, partners, vehicles, fleets, platforms, and enterprise systems and providing the needed information interpretation. The components at this layer interact with the service layer components, while the software applications are based on vertical markets. Different types of applications are addressed, including mission-critical, safety–critical applications, enterprise resource planning (ERP), specialised mobility solutions, delivery applications, analytic applications that interpret data for mobility and transportation decisions, etc.

Collaborative and Business Layer

The layer addresses the processes that involve people, organisations that use IoV mobility applications and associated information for their specific needs or apply the business procedures to a range of different purposes utilised to feed the appropriate data, at the right time, to perform the right tasks. End-to-end security is addressed for each layer and as the information is moved across the layers to secure each component or system, provide security for all processes at each level, secure end to end exchange and communication between each layer.

The IoV collaborative and business layer addresses the operational management functions, related to mobility and transportation business aspects, strategies for the development of new mobility business models based on the IoV application usage data and statistical analysis of the data. The analysis tools, including real-time dashboards, graphs, flowcharts, comparison tables, use case diagrams, etc.

The layer provides the mechanisms of collaboration between the mobility stakeholders involved in IoV applications, supporting the optimisation and decision-making processes related to economic investment, usage of resources, pricing, budgets for operation and management, infrastructure development and aggregate data management. The layer processes information using the other layers' data and the available local and remote infrastructures. The decisions are taken based on statistical data analysis and identifying strategies that help apply IoV business models based on the usage of data in IoV applications and the statistical analysis.

4.2   IoV Trustworthiness Properties

IoV systems must be built to be trustable, and the IoV 3D architectural model presented in this paper develops one of the axes on trustworthiness properties such as availability, connectability, maintainability, privacy, reliability, resilience, security, safety. The definitions of these properties are aligned with the product quality model proposed by the ISO/IEC 25010 standards family [22].

Availability
Availability is the ability and the degree to which the IoV system, subsystems, vehicles, or devices are in a specified operable and committable state at the beginning of a drive mission when the drive mission is called for unknown time (e.g., random). The availability is the proportion of time an IoV system is in a functioning condition and represents the probability that the IoV system is up and running and able to deliver the services required as specified. This is defined as a mission capable rate, the degree to which the IoV system connectivity with external systems is operational and accessible when required for use. The availability requirements apply to all the functional layers of the IoV architecture and are essential to select, design and implement the elements and functional platform components in each layer.

Connectability
Availability is the ability and the degree to which the IoV system, subsystems, vehicles, or components are connected securely, anytime, anywhere, to any available network. Most of the IoV applications are safety and mission-critical applications that must have connectivity to perform their mission. ECAS vehicles require IoV systems that are fail-operational and provide redundancy mechanisms at all functional levels, including connectivity. ECAS vehicles must retain full control over

safety–critical functions in certain situations. Should an IoV system connectivity component fail then another takes over so that the vehicle affected can return to a safe state without human intervention and the IoV application and service is not critical disrupted.

Maintainability

Maintainability defines the ability and the degree of effectiveness and efficiency with which the IoV systems and subsystems, vehicles and components can be modified by the intended maintainers and reflects the extent to which the IoV system can be adapted to new services, applications, and business models requirements. Maintainability reflects IoV system's capability to correct defects or their cause, repair or replace faulty or outdated components (vehicles, platforms, HW/SW/algorithms, training sets) and of being restored to serviceable operation. The property definition is aligned with the ISO/IEC 25010 [22] that states that maintainability refers to the degree "of effectiveness and efficiency with which a product or system can be modified to improve it, correct it or adapt it to changes in the environment, and in requirements". The standard specifies the following maintainability sub-characteristics: modularity, reusability, analysability, modifiability, and testability.

Privacy

IoV services and applications inherently must share information, resources, and infrastructure to function optimal, efficiently and offer the required services in time at an affordable cost. Exchanging, and sharing information must protect users' privacy for gaining the trust for using the IoV applications and services and accelerating the adoption of IoV. Privacy protection is becoming more and more critical in a services-driven multi-modal mobility environment. Thereby personally identifiable information (PII) is a significant concern, with many nation-states having strict penalties for companies that fail to protect information entered by the vehicle user.

Total PII may include not only the usernames and passwords we all regularly use to access online services, but it can extend to biometric information employed by vehicle systems to identify the driver or owner. As the system complexity increases within the vehicle, machine learning (ML) techniques [31] may likely be incorporated to detect and prevent user anomalies, penetration techniques, and potential adversarial attacks. Privacy breaches can be put in three categories: illegitimate access to data, unwanted change of data and disappearance of data. This can originate from persons or automated systems and can be due to internal or external operations. Different types of control can be defined, for example, logical access control, encryption, data partitioning, policy handling, anonymisation and specific security measures of computer channels. The main goal is privacy handling and satisfying the GDPR regulations, is to make organizations develop a system of "privacy-by-design" to promote consistent data protection compliance from the start of any initiative. This is a must-do to reduce privacy risks and build trust. A Privacy Impact Assessment (PIA) is an instrument to achieve compliance with the GDPR regulations and introduce innovative privacy [59] enforcing solutions from components to systems. Implementing projects with privacy at the core will help address issues early, making them easier to fix and a lot less costly, increasing privacy and data awareness across an entire organisation, and meeting legal obligations including GDPR. In this context, the end-to-end privacy property defines the ability of the IoV systems and subsystems, vehicles, and component to protect information from disclosure or observation and explicitly listing all information that is or can be disclosed. Privacy reflects a person's right to control access to personal data when using IoV services and applications.

The privacy-related to IoV measures the ability of the IoV system to determine what data in the overall system can be shared with other systems and third parties, in which form and how to anonymise the data, so no personal data is shared without specific consent from the person affected. The privacy property evaluation supports the IoV stakeholders to design and develop IoV systems that anonymise the personal data/information, adapt the functionalities and user interfaces of applications based on environmental conditions or user/applications/context preferences.

Reliability

Reliability specifies the ability and the degree with which the IoV systems and subsystems, vehicles and components deliver and accomplish services specified within given constraints, providing continuity of correct service within given constraints. This is further extended to the IoV system's ability to consistently perform its required function without degradation or failure over time and reflects the capability of the IoV system to keep its level of performance under stated conditions for a stated period of time. The property definition is aligned with the ISO/IEC 25010 [22] that states that reliability refers to the degree "to which a system, product or component performs specified functions under specified conditions for a specified period of time". The sub-characteristics of reliability defined by the standard include maturity, availability, fault tolerance, and recoverability.

Resilience

Resilience is the ability and the degree to which the IoV systems, subsystems, vehicles, or components assure and maintain an acceptable level of service in the case of faults and challenges to normal operation. The property denominates the ability of the IoV system and the degree to which subsystems, vehicles, platforms, or HW/SW components perform specified functions under specified conditions for a specified period of time with the capacity to deal with unexpected (error) situations without the user noticing (best case) or with a defined reduction in IoV service quality. Resilience relates to how an IoV service and application responds to a disturbance or stressor and defines the capability to maintain certain functions, processes, or training sets after experiencing a disturbance, including preserving the continuity of the critical services in the presence of disruptive events (e.g., platforms, vehicle/components failure, SW/HW/algorithms misbehaviour, cyberattacks).

Security

Security describes the end-to-end ability of the IoV systems, subsystems, vehicles, platforms, or components/devices to withstand attack and the degree to which an IoV system protects information and data of protection against danger, loss, and

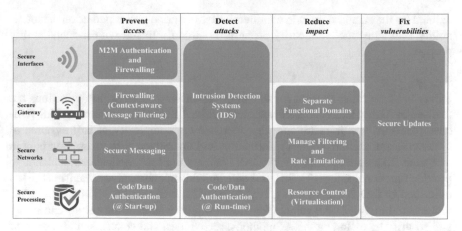

		Prevent *access*	Detect *attacks*	Reduce *impact*	Fix *vulnerabilities*
Secure Interfaces		M2M Authentication and Firewalling			
Secure Gateway		Firewalling (Context-aware Message Filtering)	Intrusion Detection Systems (IDS)	Separate Functional Domains	Secure Updates
Secure Networks		Secure Messaging		Manage Filtering and Rate Limitation	
Secure Processing		Code/Data Authentication (@ Start-up)	Code/Data Authentication (@ Run-time)	Resource Control (Virtualisation)	

Fig. 11 Fundamental security principles in automotive design

criminals so that persons or other services, products or systems have the degree of data access appropriate to their levels and types of authorisation. Security measures the perception of how likely it is that the IoV system and its components can resist accidental or deliberate intrusions and assures that information and IoV subsystem components have not been subject to unauthorised modification or disclosure while listing the category of attacks that the system is certified to resist.

Several core security principles applied to automotive systems design are highlighted in Fig. 11.

The principles equally apply to other electronic systems, such as those used for industrial applications. The key is maintaining a strict separation of the vehicle's external interfaces from the internal domain. Within the vehicle, many electronic control units (ECUs, VCUs, VCUs, etc.) are used for discrete functions such as anti-lock brakes, comfort, drive train, etc.

Another consideration of securing every ECU, DCU, VCU is that they might collectively have 100 million code lines, which presents significant software test and reliability challenges. An in-vehicle gateway is a prudent method of securely and reliably interconnecting all systems across the heterogeneous vehicle networks. Such an approach provides both physical, process, and protocol isolation between all functional domains within a vehicle.

The property definition is aligned with the ISO/IEC 25010 [22] that states that security refers to the degree "to which a product or system protects information and data so that persons or other products or systems have the degree of data access appropriate to their types and levels of authorisation". The sub-characteristics of security defined by the standard include confidentiality, Integrity, non-repudiation, account-ability, and authenticity. Security property is strongly linked with the availability property.

Safety
Safety is an essential element for IoV applications and services. IoV functions are realised by highly interconnected, and networking subsystems based on ECAS vehi-cles, platforms and infrastructure that increase the complexity of IoV systems of systems as automation requires information and interaction with its environment in real-time. To assure the safety for ECAS vehicles and IoV applications demands to provide functional safety [23, 25], fail-safe and fail-operational mechanisms [26] as described in ISO 26262 [23, 24], IEC 61508 [21] and ISO/PAS 21448 [25]. This requires rethinking ECAS vehicle functional safety to address the end-to-end safety and security by including the AI methods, techniques, and concepts at all layers of the IoV architecture to achieve acceptable risk levels.

Automotive Safety Integrity Levels (ASIL) are a core part of the ISO 26262 [24] functional safety standard as it applies to automotive systems. They define the severity, exposure, and controllability of hazards that can be encountered within any system. A "V" model requires that each software component's behaviour is fully specified, verified, and fully traceable. It also covers the potential for enhancements and the requirement for them to conform with and meet the initial specification. When it comes to ML-based systems, for example, conducting inference to determine if the object detected is another vehicle, a human, or a signpost, the software's behaviour is not easy to model since it is incredibly dynamic. There is the belief that autonomous systems need to move beyond the traditionally static nature of functional safety and embrace the concept of behavioural safety. Systems need to learn how to interact with non-automated vehicles and pedestrians, whose behaviour is not necessarily predictable. Being able to anticipate the behaviour of other road users, pedestrians, and different road hazards is crucial to deliver a genuinely autonomous driving experience.

Safety is prescribed as the ability and the degree to which IoV systems, subsys-tems, vehicles, platforms, or components/devices control recognised hazards to achieve an acceptable level of risk, detect a likely dangerous condition resulting in the activation of a protective or corrective mechanism or device to prevent hazardous events appearing or implementing mitigation to reduce the consequence of the hazardous event. IoV services and applications are, in most cases, safety–critical systems whose malfunction/failure could result in loss of life, significant property damage or damage to the environment.

4.3 IoV System Characteristics

Compatibility

Compatibility is characterised as the ability of the IoV systems, subsystems, vehicles, platforms, or components/devices to exist or work together in combination without problems or conflicts. This includes the vehicles' capability to execute a given functions on different types of platforms without modification of the application or the infrastructure. Also, compatibility is the capability of an IoV system that allows the substitution of one subsystem, platform or of one functional unit (e.g., hardware, software), for the originally designated IoV system or functional unit in a relatively transparent manner, without loss of information and without the introduction of errors.

The property definition is aligned with the ISO/IEC 25010 [22] that states that compatibility refers to "a product, system or component can exchange information with other products, systems or components, and/or perform its required functions, while sharing the same hardware or software environment". ISO/IEC 25010 [22] defines co-existence and interoperability as the two sub-characteristics of compatibility.

Interoperability is a crucial characteristic for IoV systems and is defines as the ability of the IoV systems, subsystems, vehicles, platforms, or components/devices and the degree to which two or more decision making IoV systems, platforms, vehicles, or components can exchange information/knowledge and use the information/knowledge that has been exchanged. Implementing interoperability requires that the IoV system interfaces be understood, and they are working with other IoV systems, at any time, in either implementation or access, without any restrictions. Several interoperability levels are defined, such as technical interoperability, syntactic interoperability, semantic interoperability, and organisational interoperability. The different interoperability levels are applied to the various functional domains of the IoV architecture.

Composability

Composability addresses the IoV system design principle that deals with the inter-relationships of components. High composable IoV systems are desirable to provide subsystems, platforms, and components that can be selected and assembled in various combinations to satisfy specific IoV user requirements. Composability is designated as the ability of the IoV systems, subsystems, vehicles, platforms or components/devices to compose subsystems from components (SW/HW/algorithms) providing different features, functions and assemble them (e.g., electrically, wirelessly, functionally) to implement new IoV functions at the new IoV system level. Composability requires features such as generality (e.g. types of entities/components that connect via physical/digital interfaces, types wireless protocols available, kinds of AI algorithms), diversity (e.g. different types of connections—physical/digital/data, different types of wireless protocols, different types of AI algorithms), cardinality (e.g. the total number of physical/digital interfaces available, total connections that can be maintained at any one time) to be implemented.

Intelligence

Intelligence is the characteristic of IoV systems that defines at each architecture level elements such as intelligent components, vehicles, connectivity, interactions, platforms, mobility modes, transportation systems, business models, etc. Intelligence specifies the attributes and the ability of IoV systems, subsystems, vehicles, platforms, or components/devices and the degree of use of AI methods, techniques, algorithms, analytics functions to optimise the IoV system functions and increase the efficiency of the IoV system functions and features. Intelligent functions include autonomous behaviour, perception, propulsion, energy use, search, optimisation, and learning capabilities (e.g., different types of machine learning—supervised, unsupervised and reinforcement learning, etc.), complex automated functions various diagnostic and predictive behaviours.

Learnability

Learnability is considered the ability of the IoV systems, subsystems, vehicles, platforms, or components/devices and the degree to which an AI-based IoV system can be used to achieve specified requirements of learning to use AI system with effectiveness, efficiency, immunity from risk and satisfaction in a specified context of use. For AI-based IoV systems, this includes the ease with which the AI subsystems included in ECAS vehicles, IoV services, and applications are trained, updated, and upgraded to begin effective interaction and achieve maximal performance. The learnability is connected to access and sharing continuously new and more performant data sets collected by multiple ECAS vehicles in various driving scenarios, traffic conditions, environment landscapes/conditions, infrastructures set-ups and contexts.

Manageability

Manageability is characterised as the ability of the IoV systems, subsystems, vehicles, platforms, or components/devices to be adapted to new requirements of the IoV applications, services, and users, to align the functions and the behaviour to changing external environment, or to correct faults. Maintainability can be evaluated by assessing the structuredness of the design, the overall quality of the IoV systems implementation. Insight into different aspects of maintainability is obtained by carrying out virtual validation of different scenarios and simulations using AI-based components and digital twins. Essential characteristics for IoV sustainable maintainability are the updatability and upgradability of IoV systems, subsystems, vehicles, platforms, or components/devices. Updates are considered an enhancement to the current version of the IoV system, HW/SW components, platforms, services, or application, while upgrades are considered the full new version of these elements. In this context, upgradability specifies the ability of the IoV systems, subsystems, vehicles, platforms, or components/devices system to enhance and elevate the functions and features with new patches, new version algorithms/software/operating systems, improved models, increased IoV capabilities/functionalities, and improved IoV services.

Portability

Portability defines the ability of the IoV systems, subsystems, vehicles, platforms, or components/devices and the degree of adequacy and efficiency with which an IoV system using AI-based components can be ported from one HW/SW or other operational or usage environment to another. This includes HW/SW/algorithms' ability to run on numerous platforms, provide data portability, and the ease to be migrated to other IoV systems. The property definition is aligned with the ISO/IEC 25010 [22] that states that portability refers to the degree of "effectiveness and efficiency with which a system, product or component can be transferred from one hardware, software or other operational or usage environment to another". The sub-characteristics for evaluation defined by the standard include adaptability, installability and replaceability.

Scalability

Scalability is the IoV systems' characteristic to handle a growing amount of vehicle, fleets, services, applications, etc. Scalability in this context specifies the ability of the IoV systems, subsystems, vehicles, platforms, or components/devices and the degree to which the IoV system and its ECAS vehicles, platforms, AI-based components can be expanded in size, volume, or the number of interfaces and continue to function correctly. IoV scalability addresses the requirements to meet the demand for stress caused by increased usage and ensures that the IoV system adapts its functionality when scaled more than initially designed.

Usability

Usability is characterised as the ability of IoV systems, subsystems, vehicles, platforms, or components/devices and the degree to which an IoV system can be used by specified IoV services and applications to achieve specified requirements with effectiveness and efficiency in a specified context of use. System usability describes how effectively other systems/users can use, learn, or control the IoV system.

The property definition is aligned with the ISO/IEC 25010 [2] that states that usability refers to the degree "to which a product or system can be used by specified users to achieve specified goals with effectiveness, efficiency and satisfaction in a specified context of use". Key sub-characteristics defined by the standard are appropriateness recognizability, learnability, operability, user error protection, user interface aesthetics, and accessibility.

5   Design Issues in Implementing the IoV Architecture

The vehicle architecture changes will bring changes at the vehicle domains and ECUs level with the introduction of AI techniques and software virtualisation of function on embedded high-performance vehicle computing platforms. An overview of the technology stacks for a selected number of domains is given in Fig. 12.

	Safety		AI/Perception	Powertrain/Energy	Comfort/Interior	Infotainment		
Functions	Airbag control Emergency brake control via external braking mechanism E-call Pedestrian protection Rollover sensing Secondary collision mitigation Pre-crash occupant safety control		Environmental modelling/sensor fusion Perception Image recognition Modelling of decisions Adaptive cruise control Lane departure warning systems	Acceleration Engine braking Battery management Charging optimisation	Supporting systems (e.g., for climate control, electric locking system) Central, door and seat control units Personalised key Smartphone terminal Electric windows and mirrors	Entertainment (e.g., radio) Backup camera system Telematics module Navigation system Vehicle HMI Wi-Fi hub Phone connection / mobile office Apps		
Requirements	Low latency Close link between actuators and sensors Autonomous systems Fail-operational Redundancy		Redundancy Heterogeneous high performance computation Energy efficiency Learnability	Integration of components Range Reliability	Cost-efficiency	Connectivity Usability Third party integration		
HW/SW AI, Connectivity Sensors, Actuators Components			Wired Secured Connectivity Wireless Secures V2X Connectivity			Wired/Wireless (BLE, WSN, Wi-Fi, 4G LTE, 5G) Connectivity		
	Applications	Applications	AI functions and algorithms Learning/training	Power and energy management Energy efficiency/optimisation algorithms	Consumer apps	Apps	Wi-Fi Hub	Smart Mobile Office
			Middleware / OS			Infotainment Middleware / OS		
			Security services			Security		
	HE ECUs	HE ECUs	ECUs, DCU, VCU GPU, CPUs, FPGAs, ASICs, Pre-processing	ECUs DCU	ECUs DCU	ECUs DCU	ECUs DCU	CPUs, GPUs
								Compression, encoding/transcoding
	Sensors	Actuators	Sensors	Actuators	Sensors Radar	LiDAR	Camera	Ultrasound	Sensors	Actuators	Sensors	Actuators	Consumer Circuits
								Audio	Video	Displays

Fig. 12 Overview of ECAS vehicles technology stacks connected for IoV applications

The integration of the ECAS vehicles into IoV applications is illustrated in Fig. 13. The functions needed in the integration requires the use of platforms and federation of platforms that are part of a value network in the IoV ecosystems. The IoV applications and services capabilities include the cooperative and collaborative features that effectively use the vehicular networking and automated driving functions of ECAS vehicles. IoV applications combine different scenarios such as cooperative adaptive cruise control (e.g., platooning) with stringent communication requirements in

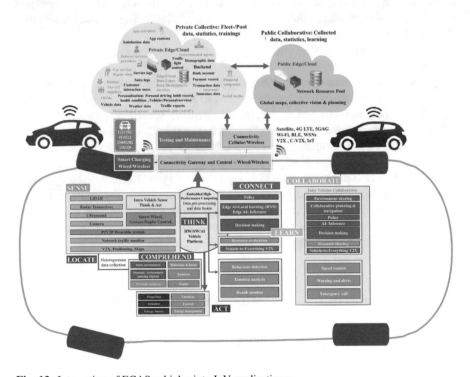

Fig. 13 Integration of ECAS vehicles into IoV applications

terms of high-reliability and low-latency, ensuring the safety distance of less than 4–6 m between the vehicles. These requirements bring new research challenges for implementing cooperative automated driving in terms of scalability, dependability, security, privacy, and user acceptance.

In this context, scalability as a characteristic of the IoV defined in the 3D architecture approach needs to be addressed from the perspective of dynamic resource allocation and density of vehicles operating in a specific area. To provide trustworthiness and full dependability of the IoV systems the underlying communications need to be coordinated by employing different physical communication channels (cellular, wireless, visible light communications, etc.) that should be as uncorrelated as possible. Ensuring availability, connectability, maintainability, privacy, reliability, resilience, security, and safety is essential for implementing IoV architecture.

The fail-operational requirements impose multiple communications links to be used in parallel for redundant information, and for combining data to check for consistency before providing the information to the AI-based algorithms for decision making and propulsion control.

Scalability refers as well to the number of vehicles involved in parallel IoV applications and services, which brings new challenges considering that today many cooperative automated driving scenarios include the collaborative operation of a few of autonomous vehicles, with the vehicles relying on perception provided by local coverage of onboard sensors and single-vehicle control/sensing operations.

ECAS vehicles as part of IoV applications must offer multi-vehicle perception/control and adaptive multimodal features to implement real-time and reliable transfer/share of sensor data and haptic information related to driving trajectories/directions among the vehicles via V2X communications to enable the functionalities of cooperative perception and manoeuvring. The use of digital twin for providing the digital simulation based on the digital/virtual model of ECAS vehicle including different functionalities and features can provide the visualisation, monitoring, and 3D representation for simulating the optimisation of multi-vehicle perception/control and adaptive multimodal features in various use cases and scenarios.

According to [58] "to demonstrate that fully autonomous vehicles have a fatality rate of 1.09 fatalities per 100 million miles (R = 99.9999989%) with a C = 95% confidence level, the vehicles would have to be driven 275 million failure-free miles. With a fleet of 100 autonomous vehicles being test-driven 24 h a day, 365 days a year at an average speed of 25 miles per hour, this would take about 12.5 years".

These challenges require new solutions for virtual validation and testing of ECAS vehicles in different IoV scenarios by combining technologies such as IoT, digital twins, 3D high-definition mapping and AR/VR to test and validate at scale.

The modelling and simulation of digital twins using different scenarios require cross mobility platforms' developments and interoperability with real-time data sharing mechanisms for allowing the federation of these platforms and cross-platform simulation and optimisation. In the virtual simulation model, intersection signals, vehicles, infrastructure, route, weather conditions mirror the physical counterparts.

The new IoV developments require a holistic approach that considers the development of mobility platforms, autopilot functions, computer vision, self-learning algorithms, safety integrity level, high-definition maps modelling and testing, info-tainment systems, and V2X communication. For safety–critical IoV applications, one of the challenges related to connectivity includes achieving ultra-low latency (<1 ms), ultra-high reliability (>99.99%), high data-rate (Gbps to Tbps) and very high backhaul bandwidth.

5.1 AI-Based Design and Technologies

The AI-based design is embedded in implementing of the components included in the different layers of the IoV architecture and in the HW/SW technologies used for implementing the ECAS vehicles architectures.

ECAS vehicles must execute several operations based on sensor information, sensor data fusion, analytics, and optimisation involving AI SW/HW and algorithms. ECAS vehicles collect information, plan the trajectory, and execute the mission requiring new programming approaches, machine learning techniques, and AI-based HW/SW platforms and architectures.

There are many tasks for ECAS vehicles presenting significant challenges that require new resource-efficient approaches to mimic human drivers' cognitive and decision abilities and replace them with AI-based solutions to ensure reliable and safe autonomous driving functions new secure IoV applications and services.

Intelligent processing of the information at all levels in the ECAS vehicles archi-tectures and IoV applications requires the use on AI methods and techniques in the sensor/actuators used in the perception/propulsion domains, in the processing units, ECUs, DCUs, VCUs and in the automated driving functions to provide safer and more deterministic behaviours.

For IoV applications areas where AI techniques are used extensively are the path planning, path execution which are dynamic tasks that must consider many parameters to solve the path optimisation problem while executing the path. The path planning function must use the safest, most convenient, and most economically beneficial routes to complete the mission using the previous driving experiences. The AI techniques, algorithms and agents are deployed for implementing the path planning function. The path execution function uses these AI techniques to support the vehicle to navigate the road conditions by detecting other vehicles, bicycles, objects, pedestrians, traffic lights, and traffic signs to reach the destination defined by the vehicle mission.

AI techniques are applied to advanced driver-assistance systems (ADAS) for ECAS vehicles with Convolutional Neural Networks (CNN), Recurrent Neural Networks (RNN) and Deep Reinforcement Learning (DRL) as the most used deep learning methods applied to automated driving functions and end-to-end solutions.

The amount of data required for training to develop reliable and robust fully automated ECAS vehicles is massive. The rapid developments in implementing deep learning algorithms used for autonomous vehicles require deep learning data sets for different scenarios and environmental conditions. The KITTI benchmark suite

[49], includes multiple data sets for evaluation of stereo vision, optical flow, scene flow, simultaneous localisation and mapping, object detection and tracking, road detection and semantic segmentation [52]. The databases with data sets include the ApolloScape [54], Cityscapes [55], Berkeley DeerDrive [56], Waymo Open dataset [53], etc.

The driving assistance applications and automated driving include road and surrounding scene detection and understanding, with road detection addressing lane identification and road scene understanding with image segmentation that classify an image's pixels into different classes (such as roads and sidewalks, etc.) [50]. Pavement marking identification involves detecting pavement marking positions and recognising their types (e.g., lane markings, road markings, messages, and cross-walks) [51]. The IoV applications have to consider all the road-related topics such as the traffic sign and marking recognition, obstacle detection, detecting parking occupancy, road surface state and road crack recognition [52].

In IoV applications and services environment perception is used to detect other road users and deep learning models, (e.g., CNN), are accurate enough for classification and detection across almost all object types with the main challenges for the pedestrian detection task having a cluttered background and significant occlusions [50]. Deep learning-based methods use deep neural networks to detect positions and geometries of moving obstacles and track their future states based on camera data [51].

For perception task performed by ECAS vehicles, require 3D perception models based on LiDAR, monocular cameras, and stereo cameras. 3D temporal tracking, interactive event recognition and intention prediction are needed in fully automated driving. ML offers potential for predicting the behaviour and intent of vehicles and persons. Vehicle behaviour corresponds to vehicles' actions include braking, steering, lane change, and even moving trajectory. Person behaviour includes motion trajectory and pedestrians' activities (e.g., running, crossing the street, interacting with objects) [50].

5.2   Fail-Operational Design

In the lower levers of autonomy as presented in Fig. 14, most systems are required to be fail-safe, meaning that the driver must be readily alerted to take over should a system detect a fault occurring and safely stop operation. As we progress through to Level 2- and Level 3-based systems, the expectation is that should an error be detected, there is enough capability in the system to continue to operate, in a degraded state. The system will notify the driver to be alerted.

There is a case for redundancy of systems in Level 4 and Level 5 systems, where the emphasis is on failing operationally—a state that alerts the driver—and that the vehicle can promptly bring itself back to a safe condition.

This process may involve hand-off to the driver, as in Level 0 to Level 3 systems. Concerning handing off the vehicle's control back to the driver, a fair amount of

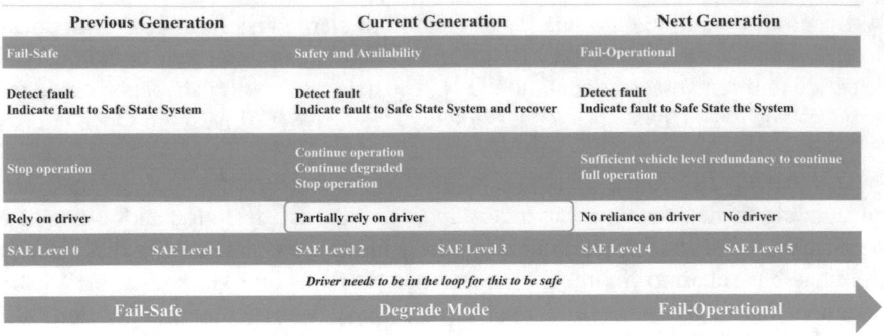

Previous Generation	Current Generation	Next Generation
Fail-Safe	Safety and Availability	Fail-Operational
Detect fault Indicate fault to Safe State System	Detect fault Indicate fault to Safe State System and recover	Detect fault Indicate fault to Safe State the System
Stop operation	Continue operation Continue degraded Stop operation	Sufficient vehicle level redundancy to continue full operation
Rely on driver	Partially rely on driver	No reliance on driver No driver
SAE Level 0 SAE Level 1	SAE Level 2 SAE Level 3	SAE Level 4 SAE Level 5

Driver needs to be in the loop for this to be safe

| Fail-Safe | Degrade Mode | Fail-Operational |

Fig. 14 Industry approach to the safety concept evolution

research has gone into the time it takes for the human to take back control. A notable study presented in [48] found that the completion time for humans to grasp the situation and actively respond adequately can take anything from 2 to 26 s.

Bearing in mind that, at speed on a highway with the best-case reaction time, the vehicle could travel 50–60 m. In the worst-case scenario, nearly 500–800 m would be covered. The likelihood of accidents occurring during either of these scenarios is exceptionally high. The requirements of Level 3 and above should consider that the vehicle fails operationally, and the vehicle is brought to a safe stop as presented in Fig. 15.

The concept that any safe autonomous vehicle always follows the rules is a tricky one. The societal norms apply in several situations while driving, and these norms often are contrary to the rules used by autonomous systems. Under certain conditions, drivers crossed into the opposite oncoming lane of traffic to overtake a stopped or broken-down vehicle.

These are deviations from the rules that drivers learned to deal with to prevent further accidents. From a systems perspective, ISO/PAS 21448 SOTIF is highly relevant to such driving scenarios. Safety and security go together as the foundations

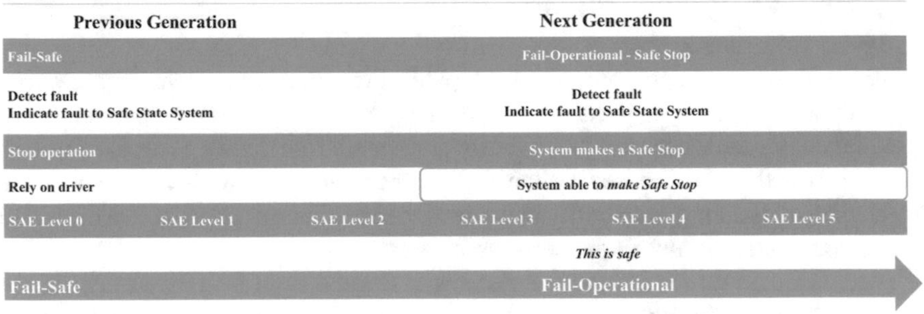

Previous Generation	Next Generation
Fail-Safe	Fail-Operational - Safe Stop
Detect fault Indicate fault to Safe State System	Detect fault Indicate fault to Safe State System
Stop operation	System makes a Safe Stop
Rely on driver	System able to *make Safe Stop*
SAE Level 0 SAE Level 1 SAE Level 2	SAE Level 3 SAE Level 4 SAE Level 5

This is safe

| Fail-Safe | Fail-Operational |

Fig. 15 An automotive approach to the safety concept evolution

for an autonomous system. Bringing together, ISO 26262 and ISO/PAS 21448 SOTIF will advance the design, test, and deployment of safe autonomous systems.

5.3   End-to-End Security

The end-to-end security is vital for adopting IoV services and applications and must be addressed at all layers of the IoV architecture. The new 5G architectures to provide the communication layer's functions and components are designed to close security gaps from previous cellular networks. However, the pervasive nature of 5G introduces new security challenges outside the traditional space. Security considerations need to examine various aspects of the software, virtualisation, automation, orchestration, and Radio Access Network (RAN) considerations. Zero-Trust security, and several other techniques are discussed to mitigate the threats, and various recommendations are proposed for security in [33]. The paper considers that the new architectures that allow 5G to progress can also expose new vulnerabilities. Securing 5G must be designed-in and patched afterwards. In this context, a careful approach to these unique aspects of cloud-native services, opensource software, APIs, SDN and NFV can improve their security. Taking a Zero-Trust approach, combined with the advanced techniques of cyber threat intelligence, and network slicing further can enhance 5G's security.

The Zero-Trust model is critical to mitigate security risks, and in 5G, each component should utilise a robust code signing stack at both the silicon and software layers. At the silicon layer, a secure implementation of code signing should be in place.

At the software layer, each element's code must be verified before it can be successfully loaded into the software stack (e.g., the firmware, operating system/hypervisor, and network function layers must be validated sequentially before the software stack layer can be fully enabled to serve network traffic and host data).

End-to-end security implementation is a stack of signed elements, so all the layers involved can be trusted as illustrated in Fig. 16. The end-to-end security is aligned with the IoV architecture by addressing the security over the continuum of vehicles, platforms, edge, cloud, data centre by using concepts such as Security-as-a-Service for intelligent connectivity for IoV.

The new elements like cyber threat intelligence must be considered, as threat intelligence is evolving from static configuration to reactive and predictive protection based on AI techniques. In this context, it is vital to address the 5G security dimensions [34] including networks access (e.g., UE authentication and access for 3GPP, non-3GPP), networks (e.g., control plane security, user plane security) and devices (e.g., device security capabilities, USIM-Universal Subscriber Identity Module).

The IoV applications' security (e.g. the protection of application workloads at the edge and central locations in the application domain), the service-based architecture security implementation (e.g. authentication and transport security protection between network functions, authorisation framework, etc.), and finally visibility

Fig. 16 Zero-Trust
Validation checks required at
both the hardware and
software layers [33]

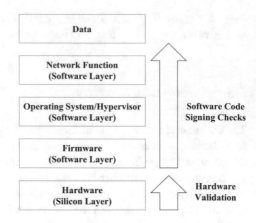

and security monitoring (e.g., full visibility can support achieving end-to-end security for different IoV applications) need to be addressed in the implementation and deployment phases.

Security aspects, threats and attacks, network and threat models related to the IoV environment are addressed and discussed in [18]. The paper presents a taxonomy of security protocols for IoV, focusing on different authentication protocols needed for mutual authentication among the involved entities in the IoV environment for secure communication and summarises the future challenges for IoV security protocols. The authors provide a comparative analysis among different authentication protocols proposed in the related IoV environment to exemplify their effectiveness and functionality features.

5.4 Intelligent Connectivity for IoV

Intelligent connectivity is achieved by combining the wired and wireless communication technologies with edge computing, IoT, AI, automation, telepresence, augmented reality (AR) and virtual reality (VR) to provide new experiences, capabilities, services and applications in transport and mobility.

Intelligent connectivity is critical for IoV applications to continuously link humans with vehicles, vehicles with all digitally connected environments by providing ultra-reliable, real-time, scalable, dependable, robust, context, content-aware and cost-effective connectivity.

Haptic applications for ECAS vehicles provide the remote driving support of vehicle (e.g., passenger vehicles, trucks, shuttles, and delivery vehicles) in the area where it is difficult to maintain or serve, and where remote driving requires immediate feedback to make reliable decisions in safety- and mission-critical use cases and scenarios.

Connectivity technologies are an integrated part of the communication/network layer of the IoV architecture, including several communication technologies based on the convergence of IoT, ITS, and VANET technologies.

The domains of interactions between the vehicle and the environment through communication and sensing capabilities, use several channels and different communication protocols [29] and technologies, as presented in Table 1 and illustrated in Fig. 17.

V2X communication is one crucial element of IoV applications and services with several communication technologies used to implement the functions and components of the communication/network layer components in the IoV architecture. The development of the new communication technologies, the infrastructure development, co-existence of existing V2X with future cellular (C-V2X), end-to-end security for IoV communications, the overall cost, the interaction between multi stakeholders' ecosystems, and the balance between the intelligence provided by the vehicle perception systems and V2X are only a few challenges for the future IoV research and innovation.

The V2X communication for road infrastructure services (e.g., physical traffic guidance systems, traffic lights. parking management, dynamic traffic signage, etc.), are operated by regional/national road and local transportation authorities to monitor traffic and control the flow. The services provided include multiple use cases, with various requirements across the different services integrated with complex IoV applications. The development of cellular (e.g., 5G, 6G and beyond) and other wireless technologies (e.g., Wi-Fi 6, Wi-Fi 7, wireless sensor networks (WSNs), etc.) provide cost-efficient, robust, and redundant solutions for developing horizontal multi-service networks for IoV applications and services.

In the next years, 5G technology is instrumental in supporting maximising the safety, efficiency, and sustainability of road mobility. In this context, an overview of the automotive and road transport that require cellular connectivity is presented in Fig. 18 [35]. Regulated Cooperative-Intelligent Transport Systems (C-ITS) provide government-regulated services for road safety and traffic efficiency with mixed latency requirements. Regulated C-ITS services may also use (depending on the region) dedicated ITS spectrum (e.g., for direct short-range communication using 3GPP PC5 or IEEE 802.11p technologies) [35].

There are two key technologies considered today for intelligent transportation systems, namely ITS-G5 and Cellular-V2X (C-V2X) [27, 32, 37]. ITS-G5 and C-V2X are rooted in totally different design principles, leading to fundamentally different radio interfaces, although they have many commonalities at higher layers and can mainly share the same protocol stack above the PHY/MAC radio layers.

ITS-G5 is specified by ETSI, whereas 3GPP specifies C-V2X. Its radio air-interface is based on Wi-Fi-like IEEE 802.11p (known as DSRC in the US), explicitly designed to deal with high-speed mobility and frequent network topology changes of the vehicular networks. The current realisation of CV2X is LTE-V2X for short-range and long-range communications, while 5G NR-V2X is the future implementation of C-V2X [36, 39].

Table 1 Vehicle to environment domains of interactions and the communication technologies

Domain	Description	Application	Features	Technology
V2B	Wire/wireless exchange of information with the internal body components	Vehicle-to-Body Internal information transmission	Real-time, higher reliability	Ethernet, CAN, LIN, MOST, FlexRay
V2C	Wireless transfer of information between vehicles and the edge or cloud infrastructure and data centres is used for tracking and usage-based insurance	Vehicle-to-Cloud/Edge	Medium distance, low/high-speed movement	4G, 5G, Wi-Fi 6
V2D	Wired and/or wireless exchange of information between the vehicle and devices (e.g., IoT devices) either inside or outside the vehicle	Vehicle-to-Device Short-range wireless communication	Short distance	Bluetooth, BLE, WSNs
V2G	Wired and/or wireless transfer of information between EVs and the charging station/power grid, such as battery status, correct charging, energy storage type and power grid load/peak balancing	Vehicle-to-Grid	Short distance	Different protocols
V2H	Wireless transfer of information between vehicles and a fixed or temporarily home used for real-time routing, charging and maintenance	Vehicle-to-Home	Short distance	Wi-Fi 6, different other protocols

(continued)

Table 1 (continued)

Domain	Description	Application	Features	Technology
V2I	Wireless transfer of information between vehicles and roadside infrastructure (e.g., traffic lights, traffic signs, road and weather condition alerts, traffic control, upcoming traffic lights information, or parking lot information)	Vehicle-to-Infrastructure Vehicle and external communication device Vehicle and external traffic facilities	Long distance, high-speed movement Short distance and high-speed movement	GSM, GPRS, 3G, 4G, 5G, GPS WSNs Microwave, infrared, V2I (DSRC/ITS G5, C-V2I), Wi-Fi 6
V2M	Wireless transfer of information between the vehicle and the vehicle condition responsible automotive OEM or maintenance shop), along with vehicle condition monitoring, predictive maintenance notification or alerts	Vehicle-to-Maintenance	Medium distance, low/high-speed movement	4G, 5G, Wi-Fi 6
V2N	Wireless transfer of information between vehicles and cellular networks, used for value-added services like real-time routing, traffic jam information and availability of charging stations for EVs)	Vehicle-to-Network	Medium distance, low/high-speed movement	4G, 5G

(continued)

Table 1 (continued)

Domain	Description	Application	Features	Technology
V2O	Wireless transfer of information between vehicles and their owners. Use cases may be vehicle rental, fleet management, freight tracking, etc	Vehicle-to-Owner	Short distance, low/high-speed movement	4G, 5G, Wi-Fi 6, BLE, NFC, RFID
V2P	Wireless exchange of information between vehicles and vulnerable road users (VRUs) for safety-related services	Vehicle-to-Pedestrian	Short distance, low/high-speed movement	4G, 5G, Wi-Fi 6, BLE
V2U	Wired/wireless exchange of information between the vehicle and its current user(s) including situational information	Vehicle-to-Users	Short distance, low/high-speed movement	4G, 5G, Wi-Fi 6, BLE
V2V	Wireless exchange of information between vehicles about speed, position of surrounding vehicles and their actions (brake, change lane, change direction, stop, etc.)	Vehicle-to-Vehicle Transmission between mobile vehicles	Security, ultra-reliability, real-time	Microwave, infrared, V2V (DSRC/ITS G5, C-V2V)

ITS-G5 and LTE-V2X short-range mode (PC5) supports communications between vehicles (V2V), between vehicles and roadside infrastructure (V2I), and between vehicles and pedestrians (V2P) or other VRUs.

Where ITS-G5 and C-V2X were initially intended for the delivery of massages/warnings for driver assistance functionalities (SAE level 1 and 2) rather than for fully autonomous (self-driving) vehicles; with an appropriate safe system design (e.g., utilizing redundancy), these short-range communication technologies are now also used for specific highly automated driving operational contexts,

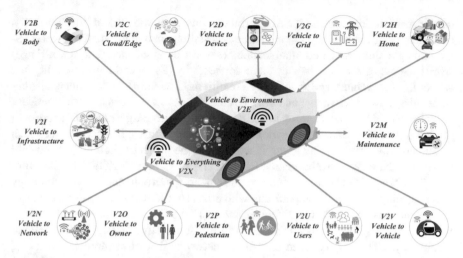

Fig. 17 Vehicle to everything communication

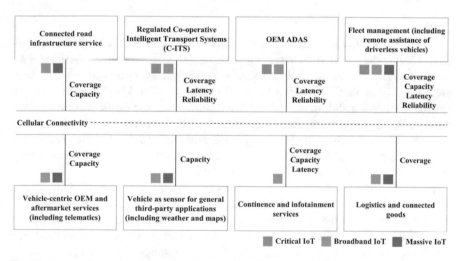

Fig. 18 Automotive and road transport that require cellular connectivity [35]

such as Platooning. V2X communication via ITS-G5 or C-V2X extends the line-of-sight-limited operation of sensors such as cameras, radars, and LIDARs.

ITS-G5

ITS-G5 is designed for short-range radio communications between vehicles (V2V) and between vehicles and roadside infrastructure (V2I) and is based on the Wi-Fi-like 802.11p specifications standardised by IEEE. An ITS-G5 signal uses OFDM and is designed to occupy a 10 MHz channel in the license-exempt 5.9 GHz band (harmonised for ITS in Europe). The MAC layer of ITS-G5 is based on a Wi-Fi-like carrier sense multiple access (CSMA) protocol with a random back off to allow statistical sharing of the medium among numerous stations in a distributed manner. ITS-G5 supports short-range communications (several hundred meters to a km-range) whilst achieving very low-latency radio-connectivity (2 ms) and high-reliability and works in high vehicle speed mobility conditions. ITS-G5 operates independently of cellular network coverage and does not require a SIM or license from a network operator to be deployed. A group in IEEE has begun to investigate the 802.11bd technology (referred to as new generation V2X) for future more advanced services such as autonomous driving. The group is working in the project called P802.11bd [38] to develop the Project Authorization Request (PAR), Criteria for Standard Development (CSD) and TGbd definitions and requirements for an enhanced V2X technology while keeping backward compatibility with 802.11p. ITS-G5 is a more mature technology that has been tested in different countries but still has few co-existence issues with LTE-V2X (PC5). Today, ITS-G5 is more cost-effective than LTE-V2X cellular technology. LTE-V2X is based on cellular networks that offer more services and, therefore, more complex, and expensive. For future IoV applications, the ITS-G5 and the C-V2X technologies combination could be the right solution.

C-V2X

LTE-V2X is today's realisation of C-V2X and was standardised in 3GPP Release 14. LTE-V2X supports both short-range and long-range communications. The LTE-V2X short-range mode signal uses OFDMA and occupies a 10 MHz channel in the license-exempt 5.9 GHz band (harmonised for ITS in Europe). The MAC layer is based on semi-persistent scheduling and allows deterministic sharing of the medium among multiple stations in a distributed manner. LTE-V2X short-range used in its mode 4 operates independently of (and does not require the availability of) cellular networks and cellular base-stations. In mode 3 cellular base stations are needed orchestrating the planning of sending and receiving LTE-V packets.

LTE-V2X long-range mode (Uu) supports communications between vehicles and the base stations of a cellular LTE network (V2N). LTE-V2X long-range mode is intended for operation in bands licensed for mobile cellular networks and has already been used to deliver services such as traffic jam warnings, weather condition warnings, road hazard warnings, etc. LTE-V2X long-range mode benefits from the investments made in LTE cellular mobile networks.

5G-V2X is the future realisation of C-V2X and will provide more services, supporting the safety requirements for highly automated driving. Standardisation of 5G-V2X is on-going in 3GPP with a first step completed with Release 15, and the second step in Release 16. C-V2X will be supported in 5G NR-V2X, providing lower latency, ultra-reliable communication, and high data rate for autonomous driving.

Different communication protocols and technologies are used to support ADAS built-in ECAS systems and exchange information (e.g., vehicle sensor/perception information, collaborative functions, etc.) across vehicle brands. As part of IoV applications, fleet management services are using different communication technologies to monitor vehicle locations and the vehicle/passengers, the status for operations monitoring and remote assistance, which can include full remote driving.

Wi-Fi 6 and 5G have been designed to work together and complement each other. 5G will continue to rely on Wi-Fi for data offloading as 5G and LTE will be governed by mobile providers charging for services and the 5G coverage will be limited, especially in rural areas.

Vehicle OEMs are continuously connected to the vehicle to monitor the vehicle's health and offer maintenance, predictive maintenance, and real-time services addressing vehicle performance and usage. The OEMs are collecting vehicle diagnostics data that enables to monitor and adjust/optimise the vehicle to the mission, and the real-time conditions (e.g., traffic, weather, context, etc.) and advise on the usage of the vehicles in a fleet or in an area to improve driving efficiency.

In IoV applications, the intelligent vehicle is acting as an intelligent sensor/actuator and processing device connected to other vehicles and infrastructure to provide/share information acquired by its one sensors/perception units. This information can be used to provide anonymised data to other parties (e.g., to maintain maps, monitor road infrastructure, road status, give accurate real-time weather and environment information).

The different communication technologies are used for IoV services in logistics for tracking transported objects (commodities, merchandise goods, cargo, etc.) during the production and transport cycle and delivery of persons and goods.

6 Future Research Challenges

The future ECAS vehicles require developments on AI-based computing, control, cognition, connectivity platforms for fusion at the domain and vehicle level the functions of sensing, detection, perception, processing, decision functions so that the vehicle sees/senses the surroundings, perceives obstacles and act safely based on its perceptions. The transition from fail-safe, fail aware to fail-operational, requires more research dealing with security-safety issues, updatability of the AI-based platform functions, connectability, upgradeability, collaborative function (V2X), decision function/processes, learning/self-learning and acting.

The next research challenges are related to the further development of the IoV multi-layered architecture that needs to be aligned with the in-vehicle architecture and the back-end federated edge, cloud architecture and infrastructure. The architectural approaches become services/functions oriented, with new layers that include cognition/intelligence components for addressing the autonomous capabilities, including sensor- and domain-fusion algorithms as a complement to hardware. New research is needed to address the move to the deep edge with software and AI algorithms moving further down the stack to hardware and sensors (intelligent deep edge). In this context, more research is needed for energy and processing efficient platforms for in-vehicle, at the edge and in the cloud AI learning/training.

The technology stacks are becoming horizontally integrated with AI functions shared across the vehicle domains that require research on the hardware/software platforms' and tools development with common and standardized interfaces.

Further work is needed to provide safety–critical, interoperable, and custom "plug and play" capabilities to enhance the autonomous driving functions and features in ECAS vehicles. New ECAS vehicle architectures are needed for embedded high-performance edge computing "vehicle brain" capabilities to integrate and fuse the information from the vehicle domains/zones that allow the federation with other edge computing and cloud infrastructure for updates off-vehicle computations. Innovations are needed to develop energy-efficient AI-based electronic components for safety–critical ECAS vehicles functions with substantial real-time or latency requirements allowing to bridge the sensors and actuators to the vehicle processing/computing units and help to reduce the wire harness as they are placed at different zones of the vehicle (e.g., front/rear, side) [40].

Integration of immersive technologies, i.e., VR/AR combined with intelligent connectivity, IoT, and digital twins, for providing efficient validation tools and environments for simulating ECAS vehicles in different IoV scenarios.

The work on integrating end-to-end security and safety by design features based on "fail-operational" behaviour, integrating closely controlled add-on apps and modules due to safety considerations. The use of different communication protocols requires further research to address communication security in heterogeneous environments. Attention will have to be paid to privacy-preserving technologies, a must-have in the services-oriented IoT based mobility network. Further research is needed on blockchain and DLTs as the concepts to enhance privacy in IoV to protect several dimensions of privacy. The use of decentralised solutions allows removing the need for any trusted third party and opens new research opportunities for blockchain-based authentication mechanisms for securing data in the IoV environment.

The intelligent connectivity and convergence of communication technologies to improve autonomous driving capabilities need further development to integrate solutions providing secured redundancy and co-existence for fail-operation with more than one communication technology providing the safety- mission-critical functions (e.g., V2X).

More research is needed to solve the communication co-existence issues to support both technologies ITS-G5 and C-V2X by implementing a hybrid solution. Proposals to allocate separate 10 MHz channels to both technologies to avoid any interference are supported by different stakeholders. The use of software define networks (SDN) for vehicles is a venue to be further investigated to provide an architecture that can enable the ITS-G5 and C-V2X co-existence allowing the management and the centralised control of the heterogeneous 5G.

Next-generation 6G communication technologies need to be developed to build reconfigurable communication and radio stations with combined perception solutions to create more performant and secure fail-safe operations of the mobility ecosystem. 6G is expected to undergo an unprecedented transformation that will substantially differ from the previous generations of wireless cellular systems. 6G may go beyond mobile Internet and will be required to support ubiquitous AI services from the core to the network's end devices for IoV applications and services.

New control software and AI-based algorithms are required to take full advantage of new solid-state batteries for an extended life and driving range in vehicles.

Lifetime is key for batteries used in the mobility system. Tools for accelerated lifetime testing, diagnostic methods, and control systems extending the lifetime and limiting degradation are essential for the success of electrified green mobility. New power-electronics based on SiC and GaN are needed to ensure energy-efficient operation. AI and model predictive control algorithms supported by high-performance multi-core real-time operating systems must provide necessary intelligence based on ultra-low-power/high-performance control units.

The increase in AI-based vehicle applications requires further research on optimising embedded software/hardware co-design, abstract applications from hardware, energy-efficient AI algorithms and new energy-optimised real-time AI learning/training. To improve vehicles' power efficiency and reduce the amount of data to be transmitted via the wireless network, the concept of AI at the edge (or edge AI) needs to be extended. The idea is to process the data provided by the perception sensors locally using in-vehicle computing capabilities and data fusion. Moreover, processing data locally reduces streaming and storing a large amount of data to the cloud, which could create some vulnerabilities from a data privacy perspective. This requires developing new distributed intelligent functions and mechanisms to be embedded into the edge and cloud infrastructure for ECAS vehicles to support the efficient real-time computing, processing, and intelligent connectivity continuum for IoV applications and services.

Research on methods and techniques to share resources and combine in-vehicle data with environmental data, analyse data for real-time decisions and autonomous driving to learn and improve the vehicle driving capabilities and enhance the performance and efficiency of the IoV services and applications in which the vehicle is involved. A particular focus is on the verification, validation and certification of embedded AI-based systems and the required training data for the respective machine learning algorithms used in the ECAS vehicles. Ecosystems for the creation and maintenance of reliable labelled data are envisioned. To integrate with various legacy systems, ecosystems supporting open platforms are required.

Reliable simulation models for environmental sensors, vehicles, drivers, traffic participants, and traffic are required. The development of these models, as well as the corresponding test systems, are essential. To test safety–critical scenarios using real vehicles in a safe environment requires creating stimulators for the different environmental sensors under different weather, traffic, and road conditions.

The verification, validation and certification of vehicles will be done in a combination of virtual test environments using model/software-in-the-loop (MIL/SIL), mixed virtual /real environments (vehicle/hardware-the-loop VIL/HIL) as well as proving ground or real-world public road testing. Road testing will result in amounts of data larger than 20 TB per hour generated by the vehicle. Therefore, adequate data acquisition, management (edge or cloud processing), and evaluation systems capable of handling the sensors' specific data types are necessary but do not exist yet. Additionally, OTA data collection from in-use operations is required to continuously collect unknown scenarios, which can be fed back into development to improve the systems' quality.

An important future research area is related to edge and cloud federation and the integration of different IoT and IoV interoperable platforms to unlock vehicle data's full potential and enable new services and business models.

7 Conclusions

With the introduction of ECAS vehicles and IoV services and applications, the automotive and mobility landscape changes. Developing system-of-systems distributed intelligence for mobility applications requires to address the complete vehicles and IoV architectural layers to provide AI-based domain knowledge and optimise performance/cost at the system level, module level, component level and technology level.

The future IoV applications call for integrating system knowledge and design capabilities with competencies across AI and edge computing, electronics hardware, and software, security and privacy frameworks and system of system architectures. In this new context, the roles of the OEMs, Tier 1 and Tier 2 stakeholders are dynamically changing. The Tier 1 automotive electronics stakeholders are under pressure by up- and downstream players with new solutions provided by innovative mobility companies and enhanced capabilities offered by OEMs.

Therefore, the automotive industry is evolving from a classical value chain to a flexible and adaptable value network. The automotive electronics components and systems must pass the rigorous and extensive qualification, and the manufacturers of these components must respect stringent quality and reliability processes, validated by supplier audits. The technology development, the innovation in the field, the introduction of AI-based methods and virtual validation and testing allow for solutions to accelerate the qualification processes.

These developments will support the penetration of new suppliers. As the automotive electronics volumes are relative "small-scale" compared to other consumer electronics market segments, it will keep the cost of these intelligent embedded HW/SW/AI components relatively high.

The automotive electronic components and systems manufacturers will further reduce these components' costs by producing efficient platforms around optimised E/E architectures with standardised HW/SW components and AI-based algorithms to continuously support the vehicle functions' updates/upgrades to enhance the performances and the lifetime of the vehicles.

This is even more stringent as the development of future vehicles and domain controller computing units will include a combination of functions offered by CPUs, GPUs, FPGAs, ASICs, neuromorphic components, accelerators that will be implemented in advanced below 5 nm technologies and implemented in 300 mm wafers by few semiconductor foundries or in very specialised and customisable mixed-signal More-than-Moore foundries above 5 nm technologies.

The ECAS vehicles must provide fail-operational functions that require very high ruggedness and reliability to withstand harsh environments, extreme temperatures, and humidity, and coexist without interference.

The IoV services and applications will push towards 24/7 use of the vehicles and a different lifecycle for automotive electronics components that can span fewer years and be updated and upgraded much more often using on-chip health lifetime monitoring capabilities and OTA techniques.

The development of the new ECAS E2E architectures requires that the ECUs/DCUs/VCUs are standardised and the vehicle will have few domain controllers performing the functions and fusion the information from sensors/actuators.

The trend will result in new architectural layers with integrated HW/SW/AI components connected across the existing functional domains. The new embedded software and algorithms are optimised and virtualised on the hardware platform to provide full-safe efficient vehicle functions and support the co-design and integration of hardware and embedded software.

The multi-layered ECAS vehicles architecture is extending middleware functions to abstract applications from the hardware as the middleware layer supports the abstraction and virtualisation on top of the vehicle embedded high-performance computing platform accelerates the use IoV multi-layered architecture, and distributed computing.

To ensure functional safety, the vehicle's perception domain's functions will integrate enhanced sensor fusion capabilities and intelligent solutions to optimise the number and efficiency of sensors/costs. This will drive innovative smart sensors' design with the intelligence functions migrating into the deep vehicle edge. The innovations allow sensors to pre-process data, trigger actuators directly, and inform ECUs/DCUs/VSUs or even the edge computing infrastructure about the actions.

The fail-operational functions for ECAS vehicles require redundancy of power/energy, connectivity, and information networks.

The IoV applications based on ECAS vehicles implement most of the safety- and mission-critical functions and must deploy fully redundant solutions for connectivity, information transmission and power/energy supply, to support the vehicle highly automated driving features.

The increased data rates between the intelligent vehicle units (ECUs, DCUs, VCUs), between vehicles and the external infrastructure and the redundancy requirements, make automotive Ethernet a key enabler for the next vehicles architectures' central data bus.

Data connectivity for functional safety and autonomous driving functions is controlled by OEMs using their computing infrastructure.

The vehicles are designed with the central connectivity gateways transmitting/receiving safety–critical information connect to the OEM back-end which 3rd parties and the service providers connected via firewalls to access data for implementing the necessary services and applications.

Privacy protection will also be an essential requirement to make the next generation ECAS vehicles societally acceptable.

The infotainment domain will continue to be implemented using several open interfaces that allow for content deployment according to the OEMs' standards and IoV applications' requirements.

The evolution towards federated vehicle/edge/cloud-centric based on distributed intelligence at the vehicle and infrastructure level architectures provides an increased use of edge/cloud infrastructure to combine in-vehicle data with environmental and infrastructure information.

Implementing these types of architectures requires that the vehicle HW/SW/AI components and platforms are updateable and communicate bi-directionally.

The validation and test systems in the vehicle allow for function and integration tests of updates, and virtual validation/testing integrating learning and training functions as the basis for the lifecycle management and feature unlocking/enhancement.

Acknowledgements This work was supported by the European Commission within the European Union's Horizon 2020 research and innovation programme funding, project AUTOPILOT under Grant Agreement No. 731993, ECSEL Joint Undertaking project AutoDrive under Grant Agreement No. 737469, ECSEL Joint Undertaking project SECREDAS under Grant Agreement No 783119, and ECSEL Joint Undertaking project ArchitectECA2030 under Grant Agreement No 877539.

References

1. Vermesan, O., et al.: Internet of energy – connecting energy anywhere anytime. In: Meyer, G., Valldorf, J. (eds.) Advanced Microsystems for Automotive Applications 2011. VDI-Buch. Springer, Heidelberg (2011). https://doi.org/10.1007/978-3-642-21381-6_4
2. Vermesan, O., Friess, P.: Internet of Things: Converging Technologies for Smart Environments and Integrated Ecosystems. River Publishers, Gistrup (2013). ISBN 978-87-92982-96-4
3. Gerla, M., Lee, E., Pau, G., Lee, U.: Internet of vehicles: from intelligent grid to autonomous cars and vehicular clouds. In: 2014 IEEE World Forum on Internet of Things (WF-IoT), Seoul, pp. 241–246 (2014). https://doi.org/10.1109/WF-IoT.2014.6803166
4. Qazi, S., Sabir, F., Khawaja, B.A., Atif, S.M., Mustaqim, M.: Why is Internet of autonomous vehicles not as plug and play as we think? Lessons to be learnt from present internet and future directions. IEEE Access **8**, 133015–133033 (2020). https://doi.org/10.1109/ACCESS.2020.3009336
5. Raja, G., Manaswini, Y., Vivekanandan, G.D., Sampath, H., Dev, K., Bashir, A.K.: AI-powered blockchain - a decentralized secure multiparty computation protocol for IoV. In: IEEE INFOCOM 2020 - IEEE Conference on Computer Communications Workshops (INFOCOM WKSHPS), Toronto, ON, Canada, pp. 865–870 (2020). https://doi.org/10.1109/INFOCOMWKSHPS50562.2020.9162866
6. Wu, Z., Qiu, K., Gao, H.: Driving policies of V2X autonomous vehicles based on reinforcement learning methods. In: IET Intelligent Transport Systems, vol. 14, no. 5, pp. 331–337, May 2020. https://doi.org/10.1049/iet-its.2019.0457
7. Lv, Z., Chen, D., Wang, Q.: Diversified technologies in internet of vehicles under intelligent edge computing. IEEE Trans. Intell. Transp. Syst. (2020). https://doi.org/10.1109/TITS.2020.3019756
8. Kaiwartya, O., et al.: (2016) Internet of vehicles: motivation, layered architecture, network model, challenges, and future aspects. IEEE Access **4**, 5356–5373 (2016). https://doi.org/10.1109/ACCESS.2016.2603219

9. Chen, C., Xiang, B., Liu, Y., Wang, K.: (2019) A secure authentication protocol for Internet of vehicles. IEEE Access **7**, 12047–12057 (2019). https://doi.org/10.1109/ACCESS.2019.289 1105

10. Moreno-Vozmediano, R., Huedo, E., Montero, R.S., Llorente, I.M., Pallis, G.: A disaggregated cloud architecture for edge computing. IEEE Internet Comput. **23**(3), 31–36 (2019). https://doi.org/10.1109/MIC.2019.2918079

11. Pokhrel, S.R., Choi, J.: Improving TCP performance over WiFi for internet of vehicles: a federated learning approach. IEEE Trans. Veh. Technol. **69**(6), 6798–6802 (2020). https://doi.org/10.1109/TVT.2020.2984369

12. Vermesan, O., Bahr, R., Crouch, J., Stratford, A.: ERL Policy memo and associated communication activities, H2020 (2019). https://european-iot-pilots.eu/wp-content/uploads/2020/06/D03_06_WP03_H2020_CREATE-IoT_Final.pdf

13. Minovski, D., Åhlund, C., Mitra, K.: Modeling quality of IoT experience in autonomous vehicles. IEEE Internet Things J. **7**(5), 3833–3849 (2020). https://doi.org/10.1109/JIOT.2020.297 5418

14. Yin, B., Wu, Y., Hu, T., Dong, J., Jiang, Z.: An efficient collaboration and incentive mechanism for Internet of Vehicles (IoV) with secured information exchange based on blockchains. IEEE Internet Things J. **7**(3), 1582–1593 (2020). https://doi.org/10.1109/JIOT.2019.2949088

15. Tian, Z., Gao, X., Su, S., Qiu, J.: Vcash: a novel reputation framework for identifying denial of traffic service in internet of connected vehicles. IEEE Internet Things J. **7**(5), 3901–3909 (2020). https://doi.org/10.1109/JIOT.2019.2951620

16. Sharma, S., Kaushik, B.: A comprehensive review of nature-inspired algorithms for Internet of Vehicles. In: 2020 International Conference on Emerging Smart Computing and Informatics (ESCI), Pune, India, pp. 336–340 (2020). https://doi.org/10.1109/ESCI48226.2020.9167513

17. Kombate, D., Wanglina: The Internet of Vehicles based on 5G communications. In: 2016 IEEE International Conference on Internet of Things (iThings) and IEEE Green Computing and Communications (GreenCom) and IEEE Cyber, Physical and Social Computing (CPSCom) and IEEE Smart Data (SmartData), Chengdu, pp. 445–448 (2016). https://doi.org/10.1109/iThings-GreenCom-CPSCom-SmartData.2016.105

18. Bagga, P., Das, A.K., Wazid, M., Rodrigues, J., Park, Y.: Authentication protocols in internet of vehicles: taxonomy, analysis, and challenges. IEEE Access **8**, 54314–54344 (2020). https://doi.org/10.1109/ACCESS.2020.2981397

19. Brunner, S., Roder, J., Kucera, M., Waas, T.: Automotive E/E-architecture enhancements by usage of Ethernet TSN. In: 2017 13th Workshop on Intelligent Solutions in Embedded Systems (WISES), Hamburg, pp. 9–13 (2017). https://doi.org/10.1109/WISES.2017.7986925

20. SAE J3016. Taxonomy and Definitions for Terms Related to Driving Automation Systems for On-Road Motor Vehicles J3016_201806 (2018). https://www.sae.org/standards/content/j3016_201806/

21. IEC 61508 Edition 2.0 - Functional safety of electrical/electronic/programmable electronic safety-related systems IEC 61508 - Part 1 - Part 7 (2010). https://www.iec.ch/functionalsafety/standards/page2.htm

22. ISO/IEC Std. 25010. Systems and software engineering – Systems and software Quality Requirements and Evaluation (SQuaRE) – System and software quality models (2011). https://www.iso.org/standard/35733.html

23. ISO 26262-1:2011 Road vehicles — Functional safety — Part 1: Vocabulary. https://www.iso.org/standard/43464.html

24. ISO TC22/SC32/WG08 Publications. Road Vehicles - Functional Safety Parts 1–12. https://www.din.de/en/getting-involved/standards-committees/naautomobil/international-committees/68242/wdc-grem:din21:227765085!search-grem-details?masking=true

25. ISO/PAS 21448:2019 Road vehicles — Safety of the intended functionality. https://www.iso.org/standard/70939.html

26. AutoDrive project (Advancing fail-aware, fail-safe, and fail-operational electronic components, systems, and architectures for fully automated driving to make future mobility safer, affordable, and end-user acceptable). https://autodrive-project.eu/

27. AUTOPILOT project (Automated Driving Progressed by Internet of Things). https://autopilot-project.eu/
28. Vermesan, O., et al.: Advanced electronic architecture design for next electric vehicle generation. In: Müller, B., Meyer, G. (eds.) Electric Vehicle Systems Architecture and Standardization Needs. Lecture Notes in Mobility. Springer, Cham (2015). https://doi.org/10.1007/978-3-319-13656-1_8
29. Vermesan, O., et al.: IoT technologies for connected and automated driving applications. In: Vermesan, O., Bacquet, J. (eds.) Internet of Things – The Call of the Edge Everything Intelligent Everywhere. River Publishers, Gistrup (2020). https://www.riverpublishers.com/downloadc hapter.php?file=RP_9788770221955C6.pdf
30. SAE International Releases Updated Visual Chart for Its "Levels of Driving Automation" Standard for Self-Driving Vehicles (2018). https://www.sae.org/news/press-room/2018/12/sae-international-releases-updated-visual-chart-for-its-%E2%80%9Clevels-of-driving-aut omation%E2%80%9D-standard-for-self-driving-vehicles
31. Mohseni, S., Pitale, M., Singh, V., Wang, Z.: Practical solutions for machine learning safety in autonomous vehicles. In: Accepted at AAAI's Workshop on Artificial Intelligence Safety (Safe AI) (2020). https://arxiv.org/abs/1912.09630v1
32. Naik, G., Choudhur, B., Park, J.-M.: IEEE 802.11bd & 5G NR V2X: evolution of radio access technologies for V2X communications. IEEE Access 7, 70169–70184 (2019). https://ieeexp lore.ieee.org/document/8723326
33. 5G Americas white paper. Security Considerations for the 5G Era (2020). https://www.5ga mericas.org/wp-content/uploads/2020/07/Security-Considerations-for-the-5G-Era-2020-WP-Lossless.pdf
34. 3GPP. TS 33.501, Security architecture and procedures for 5G System (2019)
35. Ericsson Mobility Report (2019). https://www.ericsson.com/4acd7e/assets/local/mobility-rep ort/documents/2019/emr-november-2019.pdf?_ga=2.258613962.1969153473.1594892941-1856930902.1594892941&_gac=1.212672800.1594892967.EAIaIQobChMIkb6xoL_R6g IVS9OyCh3EWA8tEAAYASAAEgJNYPD_BwE
36. 3GPP TS 22.261 v16.5.0. 3rd Generation Partnership Project; Technical Specification Group Services and System Aspects; Service requirements for the 5G system; Stage 1 (Release 16). 5G and 3GPP, September 2018 (2018). https://www.3gpp.org/ftp/specs/archive/22_series/22. 261/22261-g50.zip
37. IoT Relation and Impact on 5G Release 2.0 (2019). https://aioti.eu/aioti-report-on-iot-relation-and-impact-on-5g/
38. IEEE Standards Association, P802.11bd - Standard for Information technology - Telecommunications and information exchange between systems Local and metropolitan area networks - Specific requirements - Part 11: Wireless LAN Medium Access Control (MAC) and Physical Layer (PHY) Specifications Amendment: Enhancements for Next Generation V2X. https://sta ndards.ieee.org/project/802_11bd.html
39. GSMA Spectrum, 5G Spectrum: Public Policy Position (2019). https://www.gsma.com/spe ctrum/wp-content/uploads/2019/09/5G-Spectrum-Positions.pdf
40. Shankar, A.: Future automotive E/E architecture. IEEE India Info. 14(3), 68–73 (2019). https://site.ieee.org/indiacouncil/files/2019/10/p68-p73.pdf
41. McKinsey & Company: Automotive software and electronics 2030. Mapping the sector's future landscape (2019). https://www.mckinsey.com/~/media/mckinsey/industries/automo tive%20and%20assembly/our%20insights/mapping%20the%20automotive%20software% 20and%20electronics%20landscape%20through%202030/automotive-software-and-electr onics-2030-final.pdf
42. Buechel, M., et al.: An automated electric vehicle prototype showing new trends in automotive architectures. In: 2015 IEEE 18th International Conference on Intelligent Transportation Systems, Las Palmas, pp. 1274–1279 (2015). https://doi.org/10.1109/ITSC.2015.209
43. ALTRAN: Software defined vehicles: The path to autonomy? (2016). https://ignition.altran. com/wp-content/uploads/2016/09/software-defined-vehicles_position_paper_20160608.pdf

44. Karnouskos, S., Kerschbaum, F.: Privacy and integrity considerations in hyperconnected autonomous vehicles. In: Proceedings of the IEEE, vol. 106, no. 1, pp. 160–170 (2018). https://doi.org/10.1109/JPROC.2017.2725339

45. Sommer, S., et al.: RACE: a centralized platform computer based architecture for automotive applications. In: 2013 IEEE International Electric Vehicle Conference (IEVC), Santa Clara, CA, pp. 1–6 (2013). https://doi.org/10.1109/IEVC.2013.6681152

46. Hassan, N., Yau, K.A., Wu, C.: Edge computing in 5G: a review. IEEE Access 7, 127276–127289 (2019). https://doi.org/10.1109/ACCESS.2019.2938534

47. Kugele, S., et al.: Research challenges for a future-proof E/E architecture - a project statement. In: Eibl, M., Gaedke, M. (Hrsg.) INFORMATIK 2017. Gesellschaft für Informatik, Bonn, pp. 1463–1474 (2017). https://doi.org/10.18420/in2017_146

48. Eriksson, A., Stanton, N.: Takeover time in highly automated vehicles: noncritical transitions to and from manual control. human factors. J. Hum. Factors Ergon. Soc. 59, 689–705 (2017). https://doi.org/10.1177/0018720816685832

49. The KITTI Vision Benchmark Suite. https://www.cvlibs.net/datasets/kitti/

50. Yuan, T., Borba, W., Obraczka, K., Barakat, C., Turletti, T., Esteve Rothenberg, C., da Rocha Neto, W.: Harnessing machine learning for next-generation intelligent transportation systems: a survey (2019). https://doi.org/10.13140/RG.2.2.14242.79043

51. Badue, C., Guidolini, R., Carneiro, R.V., Azevedo, P., Cardoso, V.B., Forechi, A., Jesus, L.F.R., Berriel, R.F., Paixão, T.M., Mutz, F., Oliveira-Santos, T., Souza, A.F.D.: Self-driving cars: a survey, pp. 1–31. https://arxiv.org/abs/1901.04407(2019)

52. Kuutti, S., Bowden, R., Jin, Y., Barber, P., Fallah, S.: A survey of deep learning applications to autonomous vehicle control. https://arxiv.org/abs/1912.10773(2019)

53. Waymo Open Dataset. https://waymo.com/open

54. Apollo Scape. https://apolloscape.auto/index.html

55. CityScapes - Semantic Understanding of Urban Street Scenes. https://www.cityscapes-dataset.com/

56. Berkeley DeerDrive. https://bdd-data.berkeley.edu/

57. Reich, T., Budka, M., Robbins, D., Hulbert, D.: Survey of ETA prediction methods in public transport networks. https://arxiv.org/abs/1904.05037(2019)

58. Karla, N., Paddock, S.M.: Driving to Safety. RAND Corporation (2016). https://www.rand.org/content/dam/rand/pubs/research_reports/RR1400/RR1478/RAND_RR1478.pdf

59. Pype, P., Daalderop, G., Schulz-Kamm, E., Walters, E., von Grafenstein, M.: Privacy and security in autonomous vehicles. In: Watzenig, D., Horn, M. (eds.) Automated Driving. Springer, Cham (2017). https://doi.org/10.1007/978-3-319-31895-0_2

60. Pype, P., Daalderop, G., Schulz-Kamm, E., Walters, E., Blom, G., Westermann, S.: Intelligent transport systems: the trials making smart mobility a reality. In: Watzenig, D., Horn, M. (eds.) Automated Driving. Springer, Cham (2017). https://doi.org/10.1007/978-3-319-31895-0_30

61. Vermesan, O., Bacquet, J.: Next Generation Internet of Things - Distributed Intelligence at the Edge and Human Machine-to-Machine Cooperation. River Publishers, Gistrup (2018). ISBN 978-87-7022-008-8

Software-Defined Networking/Network Function Virtualization

Cross Network Slicing in Vehicular Networks

Amani Ibraheem ⓘ

Abstract Internet of things (IoTs) has been emerging significantly in recent years and has its impact on different industries, one of such is the automotive industry. In automotive industry, Internet of Vehicles (IoVs) gain considerable attention as it is one of the important constituents of IoTs. There are different scenarios with different network requirements to be considered in IoVs. For instance, safety messages require low latency network, while on the other hand, infotainment services demand high bandwidth network. For such various requirements, the network should be able to allocate appropriate resources to accomplish the desired service. One of the promising technologies that is leveraged to fulfil such goal is Network Slicing. With network slicing, the underlying infrastructure is divided into multiple slices each equipped with required resources to meet specific need. To slice the network, the infrastructure should be controllable in a way that allow a central unit to guide the slicing process. For this purpose, Software-defined Networking (SDN) is utilised to decouple the control plane from the data plane, allowing a separate unit to take control. In this chapter, we discuss vehicular SDN slicing and how to boost it with intelligent capabilities using recent advances in machine and deep learning.

1 Introduction to Network Slicing and SDN

The next generation networks, namely 5G (Fifth Generation) is one of the significant technologies that lead to a new different era of communications and connected devices as in what so-called the Internet of Things (IoTs). 5G networks [1] promised to connect more devices with different user requirements that may require high bandwidth, low latency, reliable connection and so forth. As a result of connecting more devices under different requirements, diverse and huge amount of data is generated from these devices and according to [2], the data rate is expected to increase by a factor of 1000 with the emergence of Internet of Things (IoTs) and 5G. In addition,

A. Ibraheem (✉)
School of Informatics and Engineering, University of Sussex, 3A04 Richmond,
University of Sussex, Brighton BN1 9RH, UK
e-mail: A.Ibraheem@sussex.ac.uk

© Springer Nature Switzerland AG 2021
N. Magaia et al. (eds.), *Intelligent Technologies for Internet of Vehicles*, Internet of Things,
https://doi.org/10.1007/978-3-030-76493-7_5

different requirements lead to heterogeneity in technologies and resources used to achieve the required service. Moreover, the problem with the existing networks such as 4G LTE is that they were designed as one-size-fits-all architecture where the same network is used for different purposes. This is not going to be ideal with the growing demand of connecting more devices with different performance requirements. Therefore, new mechanisms have to be introduced to efficiently integrate heterogeneous technologies to enable interoperability and to optimise the network differently based on specific service requirements in order to deliver the appropriate service to the end user with a dedicated network tailored to meet such requirements. A solution to this is Network Slicing.

Internet of Vehicles (IoVs) is one of the important branches of IoTs [3], it deals with vehicles that require the communications to be ultra reliable given the sensitivity of the situation with vehicles carrying humans, thus, ensuring the safety of passengers is of top priority. Therefore, networking mechanisms in vehicular communications have to be carefully designed and implemented with the focus on guaranteeing security and safety. Additionally, vehicular networks include diverse communications' techniques ranging from wired networks such as Ethernet network in in-vehicle network, to wireless networks such as Bluetooth, WiFi, and cellular 4G/5G. Therefore, having these heterogeneous networks in one domain making the field of IoVs an excellent candidate to implement and test network slicing for different IoVs applications. In the following sections, we will define the term of network slicing from both general and technical aspects. We will also discuss how network slicing can be used in different approaches and domains. Moreover, Software-Define Networking (SDN), a technology used to achieve network slicing, will be demonstrated. Additionally, network slicing in vehicular communications will be discussed with possible applications and examples. Then, intelligent methods including machine and deep learning approaches will be illustrated for the use in network slicing, existing algorithms and applications will further be stated. Finally, challenges and open issues in network slicing will be discussed.

1.1   What Is Network Slicing?

Network slicing is one of the promising technologies enabled in 5G networks and it is currently attracting the researchers both in industry and academia. The term network slicing in the current literature differs in several aspects due to the fact that it is still under ongoing research in addition to the possible uses of the term in different cases. For example, slicing the network can be considered in terms of connectivity or in terms of routing. Therefore, there is no consent on the definition of network slicing. However, the fundamental concept that most parties agree on is that network slicing enable the logical partitioning of the underlying network resources in order to create virtualised subnetworks that share the physical infrastructure, where each subnetwork (or slice), is customized to serve specific requirement to achieve certain QoS [4–11].

A single network slice is defined by FCC Technological Advisory Council 5G IoT Working Group as [12]:

> *"A Network Slice is a logical (virtual) network customized to serve a defined business purpose or customer, consisting of an end-to-end composition of all the varied network resources required to satisfy the specific performance and economic needs of that particular service class or customer application."*

And according to 3GPP (3rd Generation Partnership Project), a network slice is referred to as network slice instance. When 3GPP studied network slicing, they defined three main types of slices: (1) eMBB (enhanced Mobile BroadBand), (2) URLLC (Ultra-Reliable Low Latency Communications), and (3) mMTC (massive Machine Type Communications) or sometimes referred to as mIoT (massive Internet of Things). Figure 1 shows all three cases with their requirements.

In eMBB, the network provides high bandwidth with moderate latency and can be used in scenarios that require high data rates such as streaming a high-resolution video. On the other hand, URLLC provides communication for reliability and latency sensitive applications like factory automation, remote surgery, and autonomous driving. In IoTs, numerous devices are connected to each other and for such scenarios, mMTC provides support of communicating massive number of devices. These three slices are standardised slices that support roaming between operators; however, other types of non-standardised slices can be introduced and customised based on other requirements and for these non-standardised slices, roaming with other operators is not supported. Moreover, the standardised slices are assigned with slice/service type

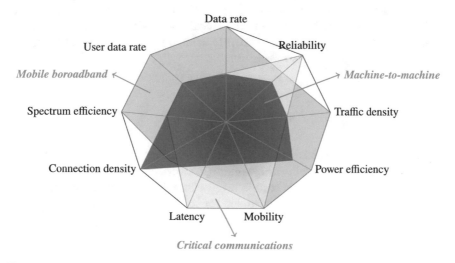

Fig. 1 Key 5G usecases and their requirements [13]

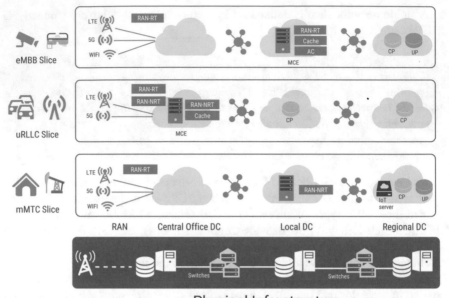

Fig. 2 5G use case slices

(SST) value as eMBB, URLLC, and mMTC are assigned with values 1, 2, and 3, respectively. Network slices can be provided as on-demand service, where the slice will be created, deployed and, once consumed, removed dynamically. It is worth mentioning that the concept of network slicing considers slicing the end-to-end network. End-to-end network slicing means slicing the core network, transport network and access network, and in terms of 5G networks, this means consider slicing both the 5G Core Network (CN) and 5G Radio Access Network (RAN). Figure 2 shows end-to-end slicing of the network to provide tailored services for different applications.

1.2   Network Slicing Enabling Technologies

Network slicing in 5G is enabled by a number of existing technologies. These technologies include software-defined networking, network function virtualisation, cloud and edge computing [14]. This section briefly discuss each one of these technologies, where more details will be illustrated for software-defined networking in later sections since this technology is the key enabler for network slicing.

1.2.1   Software Defined Networking SDN

Software Defined Networking (SDN) is a network solution that decouples the control plane (CP) from the data plane (DP) to allow for more solid security, lower OPEX, and reduced CAPEX. SDN allow for network softwarisation and programmability by enabling network operators to program, manage and control the network from a centralised entity that is not necessary to be coupled with the hardware device. This centralised entity is called *SDN controller* and it can be a physical controlling entity or simply a logical piece of software. Currently, there are a number of existing SDN controllers available such as *Floodlight* [15], *NOX* [16], *RYU* [17], *OpenDayLight* [18], and many others. Later, we will explain SDN in more details and show how it is used in network slicing.

1.2.2   Network Function Virtualisation NFV

As the name suggest, network function virtualisation (NFV) is about converting the network nodes functions into virtual network functions (VNFs) which can be chained together to provide different network services [19]. Traditionally, network functions, such as firewall, were implemented and managed in hardware, while with NVF, network functions can be instantiated virtually with virtual appliances without the need to install hardware. NFV still, however, need to run these virtual functions from commodity hardware that enable virtualisation as shown in Fig. 3 [20]. For network slicing, NFV facilitates the management of VNF, service chanining and latency oriented VNF embedding [21]. SDN and NFV are related technologies in which each one requires the other.

1.2.3   Cloud Computing

Cloud computing is an online pool of shared resources that users can access on-demand. These shared resources include network, computing and storage resources. Three delivery models are defined in cloud computing, see Fig. 4:

- **Infrastructure as a Service (IaaS):** this model provides the required infrastructure in virtualised form of resources that are accessible via the Internet. Examples of IaaS include *Google Compute Engine, Amazon EC2, DigitalOcean*, and so on.
- **Platform as a Service (PaaS):** delivers the platform in which the developers use to implement software applications and services. Examples of PaaS are *Windows Azure, Amazon Web Services (AWS), Engine Yard*, and others.
- **Software as a Service (SaaS):** this model is to deliver the end-user application on-demand with no need to install the application. Examples are *Gmail, Dropbox, Google drive* and many others.

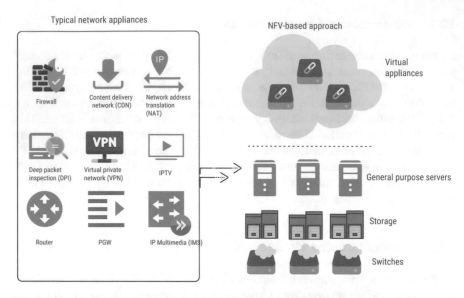

Fig. 3 Instantiating virtual network functions (VNFs) with NFV [20]

Fig. 4 Cloud computing
service models

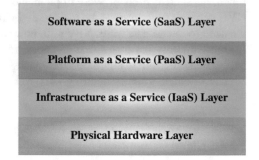

Furthermore, NIST defines five main deployment models for cloud computing [22]:

1. Public cloud where the cloud resources are owned and provided to the public by a third party called service provider. This type of cloud has the advantage of low cost and more reliability, but on the other hand, has disadvantage of security and privacy issues.
2. Private cloud is unlike public cloud, this model owned by a single organisation in which the users data will be protected and more secured.
3. Virtual private cloud allows the user data to be segregated with private IP address, storage and services. This enhance the security even more than with private cloud.
4. Community cloud allows multiple organisations with similar interests or purposes to share the same infrastructure.

5. Hybrid cloud can combine two or more models in order to achieve the best of each model. An example can be to use public and private clouds where the private cloud will store the users data to be protected and the public cloud is used to advertise a company services.

1.2.4 Edge Computing

Even though cloud computing provides benefits of resource pooling, scalability and elasticity, it has the limitation of high latency when resources are located on the cloud (data) center. To solve this issue, edge computing has come into play in minimising the latency. Different terms are used for edge computing in the literature; cloudlets [23], fog computing [24], and mobile edge computing [25]. Edge computing pushes the computing resources closer to the edge of the end user so that accessing such resources will be faster and not suffer from high delays.

The use of cloud and edge computing can provide great benefits when implementing network slicing where cloud comouting can be used in slices that do require more computational process and storage while using edge computing for the slices demanding low latency.

1.3 Network Slicing as a Solution

Network slicing can solve a number of network problems that includes resource allocation, isolation, management and control. In this section, we will discuss different networking issues where network slicing can be used to address.

Network slicing provides efficient utilisation of resources and helps reducing wasted resources by slicing the network resources. For example, consider allocating resources among two different use cases; where one requires high bandwidth and tolerable latency while the other case requires very low latency and moderate bandwidth. For such scenario, network slicing can efficiently allocate the network resources between the two cases by instantiating two slices; equipped with resources supplying high bandwidth for the first slice and resources that are enabled with low latency capabilities for the second slice and then assign each slice to the allotted use case.

Moreover, in vertical industries, network slicing can be used to overcome shortage in resources such as in [26] where the authors defined a Vertical Slicer (VS) that manage and regulate service requests and manage the resources in case a Service Orchestrator (SO) informs about resource shortage. In addition, even though, network slicing raises concerns of security issues, as described in challenges section, it can be thought of as a solution to some security problems, such as isolating part of the network through concept of slicing, hence, protecting this part (or slice) of the network that may require high level of security from the rest of the network.

Another way of utilising network slicing for security purposes is to assign different security levels for different scenarios based on the sensitivity of the use case. For example, consider a scenario of automotive applications with one application for driving assistant and another application for navigation system. In the prior application, it is of high risk to tamper with the system as it may cause major damages and pose a great risk to passengers' lives. While for the second application, the risk is less harmful. Therefore, security level should be higher than with the second application. And with network slicing, this is can be achieved by instantiating and assigning two slices with different security levels. A study in [27] uses a similar concept based on network slicing called Micro-Segmentation that allow for more fine grained isolation. Moreover, in factory applications, more specifically industry 4.0, network slicing methods are used for industrial communication protocols to facilitate the management of such complex networks [8].

1.4 Network Slicing from Business Perspective

From a business point of view, in terms of infrastructure and network service providers, network slicing solutions can provide significant enhancements to the overall infrastructure, hence, reducing the CAPEX through efficiently utilising the resources. For instance, the authors in [28] provided a proof-of-concept (PoC) where network slices can be created and deployed dynamically and on-demand over several domains for real time contents in a Stadium. The architecture offered functions for brokering, federation, and orchestration. It exploited the benefit of edge resources via the resource federation concept by allowing a network operator to federate and orchestrate third party resources that are placed at the edge of the operator infrastructure through the use of registration portal and application programming interfaces (APIs). In addition, and with the ability of slice sharing, network slicing can facilitate the idea of multi-tenancy solutions in which the infrastructure provider can serve multiple mobile network operators [29–31]. In [29], the physical RAN (Radio Access Network) infrastructure is shared among multiple mobile virtual network operators (MVNOs), where MVNO is considered as a tenant while the physical infrastructure is owned by an infrastructure provider (InP) and all the physical resources were virtualised into shared slices between the MVNOs.

Blockchain concept was introduced for network slicing in [30] in which a network operator can run the blockchain based slice ledger with known tenants and for new tenants and MVNOs they could be added to a participant list as required. Another model of cost optimal deployment of network slices is studied in [31], enabling a mobile network operator to assign the underlying resources based on its users' requirements. The work in this paper introduced a new architecture to instantiate a fined-grained Network Slicing (NS). Using a Mixed Integer Linear Programming (MILP) model to attain cross-domain network slicing deployment without considering the underlying layers. In addition, a heuristic algorithm was designed to address the problem of exponential runtime, hence allowing a faster decision-making.

Beside Internet and mobile network providers, new business entrants shall benefit from network slicing. In ITS (Intelligent Transport System) for instance, applying network slicing in vehicular networks can allow verticals other than traditional Internet providers to play roles in providing the tenants with the required services. For example, road municipality can provide a V2V-based data exchange and a vehicle manufacturer can provide services related to vehicle diagnostics [32].

1.5   Network Slicing Architecture

A number of ongoing studies have been setting the architecture for network slicing in 5G networks. A generic 5G framework studied by [13] is illustrated in Fig. 5. It consists of three main layers: infrastructure layer, network function layer, and service layer. We will briefly define each layer of the framework:

Infrastructure Layer. The infrastructure layer is basically the physical hardware layer that involves the CN and the RAN as well as the control and allocation of slices. From cloud computing concept, the infrastructure layer can benefit from IaaS (Infrastructure as a Service), where the infrastructure elements can be leased specifically to meet the needs of different slices. Moreover, in addition to the central cloud infrastructure, edge cloud computing infrastructures are needed to accommodate the low latency requirements since the resources are located at the edge closer to the access network.

Network Function Layer. The second layer, network function layer, is composed of control plane and user plane functions and this layer is responsible for the configuration and life cycle management of such functions. In network slicing, each slice is a composition of a number of network functions.

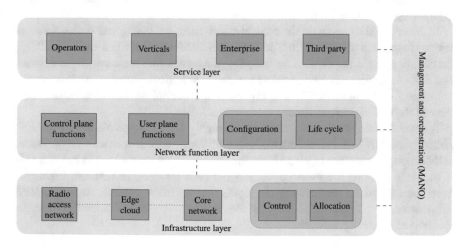

Fig. 5   Generic 5G framework for network slicing [13]

Service Layer. This layer includes different service providers that produce the service (in the form of specified slice) to the end user. It is responsible for the service description and the SLA requirements.

Management and Orchestration (MANO). MANO is responsible for managing and allocating tasks in the layered architecture. In particular, MANO performs mappings of services to network components. It supports two kinds of mapping:

1. Service level agreement (SLA) mapping of the slice requirements to network functions and types of infrastructure.
2. Mapping of network functions and infrastructure types to vendor specification.

1.6 Software Defined Networking (SDN)

Network slicing is achieved by network softwarization that requires high level of programmability and in traditional networks this level of programmability is very limited, and it is even absent in most networks. Therefore, new mechanisms had to be introduced to facilitate network softwarization. This mechanism is referred to as *Software-defined Networking* or *SDN*.

In traditional networks, the network devices hold both the data and control planes. The data plane is the underlying network devices that may include: switches, routers, hub, firewall devices, and so forth, with their different ports used for receiving and transmitting packets. On the other hand, the control plane is responsible for setting the routing rules of packets to be transmitted and received in the data plane. In traditional networks, the control plane is often responsible for handling control protocols which sometimes have effects on the forwarding table of the data plane, while the majority of the forwarding rules is hard-coded within the network device making it a single entity that has the capability of handling most of the traffic without any intervention from the control plane [33].

In contrast, SDN separate the control plane from the data plane where network devices in the data plane are dumb devices with no forwarding rules, the controller in the control plane is responsible for inserting the forwarding logic into these devices. The SDN controller can be a physical hardware or logical software and it is usually centralised to facilitate controlling process from central unit that maintain global view of the underlying network.

It is worth noting that SDN technology is highly dependant on *Network Function Virtualisation (NFV)*. In fact both SDN and NFV are dependant on each other. Where SDN decouples the control plane from the data plane as we will see in more details in the next sections, NFV decouples the network services, or network functions, from the data plane (or network devices), like switches, routers, firewalls, VPN terminators, etc. Therefore, NFV provide network services without the need for specialized network hardware.

1.6.1 SDN Architecture

The SDN architecture consists of three main layers as depicted in Fig. 6. The first layer from bottom is the infrastructure layer which mainly composed of network devices in the data plane. The second layer is the control layer where the SDN controller resides and it communicates with the underlying data plane layer using standard interface called *southbound interface (SBI)*. The upper layer is the application layer that allow network applications development for various purposes such as security applications. The application layer communicates with the control layer through another interface referred to as *northbound interface (NBI)*. In the following section we will elaborate more on the interfaces used in SDN to facilitate the communication between the three layers.

1.6.2 SDN Workflow

The network communication in SDN starts with message transmission between the data plane and control plane in which the SDN controller makes use of the information provided by the data plane such as network topology and features. This type of communication between control plane and data plane uses southbound interfaces such as OpenFlow protocol which is the most well-known protocol used for SBI and it is standardised by Open Network Foundation (ONF). Other protocols are also available such as Forwarding and Control Element Separation (ForCES) which is standardised by Internet Engineering Task Force (IETF) [34].

After the SDN controller establish a global view of the network, the application layer can use this information to provide the controller with the suitable actions to take over the network such as directing the traffic to specific path based on certain criteria by setting proactive or reactive flows on the network devices. An example of network application can be a firewall software application. The application layer communicate with the control layer using the northound interface. There are no standard NBIs available as with the SBIs, however existing APIs such as REST API, Python API, or Java API can be used as NBI between the two layers.

There are two types of flows installation in SDN:

- *Proactive flows:* where the SDN application proactively set the flows when it starts up. This type usually referred to as *staic flows*.
- *Reactive flows:* the flow will be installed upon the arrival of packets.

Figure 7 shows the operation of SDN where each device in the underlying layer holds a data plane and a forwarding plane. Initially, the forwarding plane will be empty, and after the application starts and the controller provide it with the required information from its global view of the network, the flows will be installed into the devices, this is in case of proactive flows. Otherwise the devices will wait for incoming packets to forward to the controller and in turn the controller will communicate with the application layer that instructs the controller on how to handle the packets and then install the flows for future use.

Fig. 6 Three Layer SDN
Architecture [34]

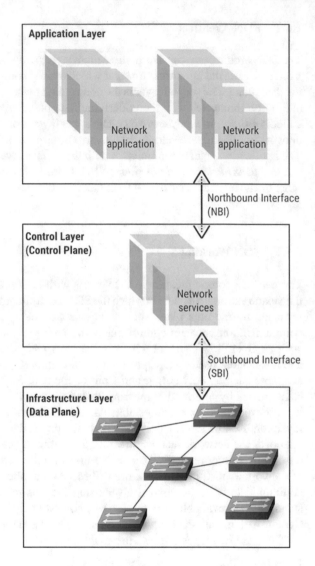

1.6.3   Benefits of SDN

The main idea in SDN is to separate the control plane from the data plane. This
separation has the advantage of making network monitoring and management more
flexible. For instance, in case of a network failure in one link, the SDN controller can
recover the network by changing set of forwarding rules to avoid passing through the
failed link and hence provide another path. This way, with SDN, the effect of failure
can be mitigated. Other benefits can also be gained with SDN. The following are the
most paramount:

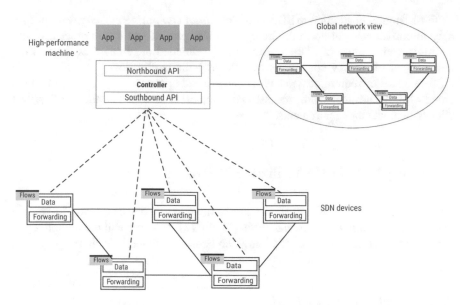

Fig. 7 SDN operation [33]

- Centralised control with SDN controller maintaining overall view of the network.
- Network automation with programmable network functions that enable policy-driven networks as well as efficiently implementing quality of service (QoS) for different applications.
- Less hardware expenses by which the network devices in SDN do not have to be complex as in traditional networks. Hence, reducing the cost of such devices which basically considered as white-box switches in SDN that can easily be made.

1.6.4 Network Slicing with SDN

As we saw how SDN operates, we can induce that network slicing require the functionalities provided by SDN in order to construct slices and direct the traffic to the appropriate slice based on its requirements. This can be achieved by leveraging the advantage of global network view that the SDN controller is capable of handling. In addition to one single centralised controller, multiple distributed SDN controllers are often needed in network slicing where the physical network infrastructure is shared among multiple controllers with each controller responsible for managing specific set of slices.

Furthermore, in addition to SDN concept, *cloud computing* and *edge computing* play important roles in network slicing for management and traffic processing. In addition, with the use of SDN, traffic can be offloaded or rerouted as to meet the requirement of latency constraints by allowing the applications that have low to moderate latency requirement to have their processing computation offloaded on the cloud while on the other hand, processing the request at the edge for other applications with critical latency requirements.

2 Network Slicing in Vehicular Networks

Network slicing in vehicular networks can be divided into two categories: *in-vehicle network* and *vehicular ad hoc networks (VANETs)*. Next we will define both categories and describe how network slicing can be implemented in each category.

2.1 In-Vehicle Network Slicing

By in-vehicle network, we mean the network used inside the vehicle to communicate different parts or electronic control units (ECUs) with each other. Most existing in-vehicle networks consist of CAN (Controller Area Network), LIN (Local Interconnected Network), FlexRay, and MOST (Media Oriented System Transport). However, these networks have several limitations such as low bandwidth and high cost [35].

2.1.1 Ethernet-Based In-Vehicle Network

As future autonomous vehicles will require high demand of bandwidth to operate various safety and infotainment applications [36], the current in-vehicle networks such as CAN and LIN will not be able to meet such demands. Accordingly, contemporary vehicles should be equipped with more capable technologies. Hence, the automotive industries are gradually shifting from CAN-/LIN-based network to Ethernet-based network solutions where higher bandwidth can be supported. In addition, with Ethernet enabled in in-vehicle network, the in-vehicle SDN notion can grow more rapidly and flexibly in in-vehicle networks. Figure 8 shows an in-car backbone network with three Ethernet switches connecting different ECUs and gateways.

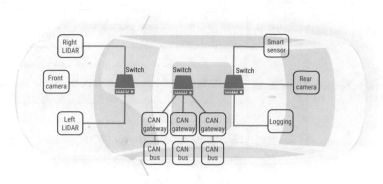

Fig. 8 Ethernet-based In-vehicle Network

2.1.2 In-Vehicle Software-Defined Network

Since SDN is an important key enabler for network slicing to come to life, implementing network slicing in in-vehicle network requires integrating the SDN framework into the in-vehicle network. To this extent, designing the in-vehicle network might be different in order to incorporate the control plane within the architecture. A limited number of studies have been conducted to applying SDN concept for in-vehicle networks, such in [37–40].

Network slicing can be employed in in-vehicle network to enhance passengers safety and provide better security mechanisms within the network. On this regard, G. Parisis and P. Fussey in [37] have promoted an in-vehicle SDN architecture as a promising network technology for Connected and Autonomous Vehicles (CAVs). They have emphasised the importance and benefits of such architecture in terms of security; by enabling firewalling and flow isolation, and safety; by allowing a failover mechanism to spot faults and dynamically reconfigure the network to avoid service discontinuity.

The security benefits of applying SDN in in-vehicle network has also been studied in [38], in which the authors proposed a MACsec extension over the SDN. The proposed extension has stretched the security scope of MACsec from point-to-point to end-to-end where the frames do not have to be encrypted/decrypted from within the intermediate nodes, rather, the SDN controller is delegated to authenticate ECUs (Electronic Control Units) and switches in the vehicle.

On the other side, heterogeneous data that is generated from various ECU sources is addressed by K. Halba and C. Mahmoudi in [39] where the SDN approach is integrated within the vehicle network that itself uses a Time Triggered Ethernet (TTEthernet) technology and form an SDIVN (Software Defined In-Vehicle Network). Data interoperability for in-vehicle network is enabled by this approach by which it allows the interaction between different ECUs irrespective of technology or protocol used in each ECU. The design is composed of three planes (see Fig. 9),

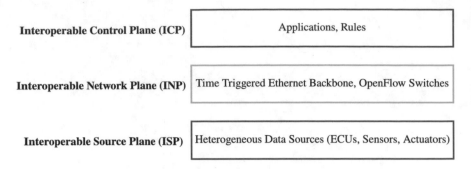

Fig. 9 Planes composing the network interoperability design [39]

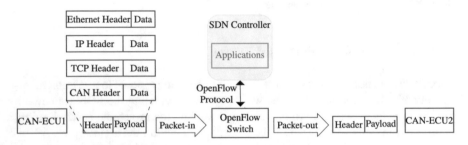

Fig. 10 ECU CAN message in SDN-enabled Network [39]

namely: interoperable sources plane (ISP), interoperable network plane (INP), and interoperable control plane (ICP).

In addition to the interoperability design, they have also introduced a universal adapter to enable interoperability between different data sources with different protocols. Two adapter elements have been used; IPtoLIVN adapter and LIVNtoIP adapter to allow the interaction between data messages generated by the legacy ECU and the TTEthernet frames. Figure 10 depict how an ECU message is handled by the OpenFlow switch and the SDN controller.

Same authors in [40] offered, in addition to the SDIVN [39], a fast failover mechanism to support self-healing in autonomous vehicles in the event of a link failure. Moreover, with the SDN concept, they adopted the use of unicast communication instead of broadcasts in which the unicast has an advantage of protecting the vehicle components from potential danger of snooping by external parties. In the proposed design, the ECUs are connected to a TTE (Time Triggered Ethernet) backbone using OpenFlow switches which in turns connected to an SDN controller that host a number of applications. The failover application run on the controller inserts rules into SDN switches' flow tables that help repairing the network in case of a failure by enabling a backup path. Results showed that the recovering process in SDIVN is relatively fast as opposed to the legacy network that failed to recover from a link failure because it does not support a failover mechanism.

2.2 VANETs Slicing

Vehicular ad hoc networks (VANETs) are usually deemed as part of mobile ad hoc networks (MANETs). There are two main categories of VANETs referred to as V2I (Vehicle-to-Infrastructure) and V2V (Vehicle-to-Vehicle).

2.2.1 Vehicle-to-Vehicle (V2V)

In V2V, the communication is established between vehicles to exchange information. This type of communication typically uses DSRC (Dedicated Short Range Communications). It is a wireless communication technology based on the IEEE 802.11p standard. DSRC provides multiple benefits to VANETs such as designated licensed bandwidth, high reliability, and priority for safety applications [41].

2.2.2 Vehicle-to-Infrastructure (V2I)

In V2I, the vehicle essentially communicates with the infrastructure such as the Road Side Unit (RSU). DSRC can also be used in this type of communication in addition to others like cellular networks. Vehicles can also communicate with other types of infrastructure such as base stations, it is referred to as V2N (Vehicle-to-Network) in this case. Vehicle-to-Pedestrians (V2P) is another V2I communication where the vehicle can communicate with the pedestrian's device e.g. smart phone.

2.2.3 Software-Defined Networks in VANETs

Network slicing has been widely studied in vehicular networks especially in VANETs since the resiliency and programmability of SDN makes it a qualified technology to be employed in highly dynamic scenarios such as vehicular and mobile networks.

There are numerous studies that consider the use of SDN concept in ITS domain. One of the studies that applied SDN into VANETs, is [42] where the authors proposed an SDN-based VANET architecture and compared the simulation results with the traditional VANET routing protocols. The architecture is composed of three main components: SDN controller, SDN wireless nodes, and SDN RSU (see Fig. 11). The SDN controller is the central intelligence part of the system which controls the network actions; hence, it is deemed as the control plane element. On the hand, the SDN wireless nodes and RSUs are the data plane elements, where the vehicles are the mobile wireless nodes and the RSUs are the stationary infrastructure nodes. Different operational modes have also been proposed: central control mode, distributed control mode, and hybrid control mode. The results of this approach have better packet delivery ratio than other traditional Ad hoc routing protocols such as AODV (Ad

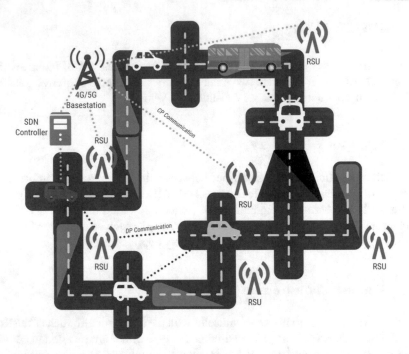

Fig. 11 Software-defined VANET communications [42]

hoc On-demand Distance Vector) and GPSR (Greedy Perimeter Stateless Routing) protocols.

VANETs SDN and network slicing can be utilised to tackle number of networks' issues such as:

Heterogeneous Networks. Network heterogeneity issue in vehicular communications has been addressed using SDN concept [43]. Authors in [43] first shed the light on the differences between SDN and SDNV (Software-defined Vehicular Network). These differences include:

- The data plane elements in SDN are stationary, while in SDVN, they can be stationary (e.g. RSU) or mobile (e.g. moving vehicle).
- With regard to the controller overhead, that is in SDVN, switch status is maintained by direct status collection and estimation as well, where in SDN, the status is collected from switches.
- In terms of heterogeneity, SDN data plane components have the same hardware interface (i.e. Ethernet), while SDVN data plane components have heterogeneous wireless interfaces (e.g. WiFi, DSRC, cellular network, etc.).

The SDVN incorporates vehicle-to-vehicle, vehicle-to-infrastructure, and vehicle-to-cloud communications. In addition, multiple tenants' isolation has been introduced for vehicles by applying the notion of network slicing. Particularly, slicing the network is performed based on the vehicle direction in order to avoid broadcast storming.

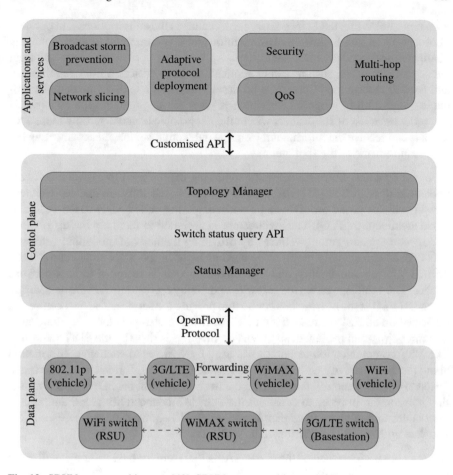

Fig. 12 SDVN system architecture [43]. SDVN system architecture [43]

So the vehicles moving on the opposite direction will not receive irrelevant notifications. The authors in this study have also extended the OpenFlow protocol to be capable of handling the dynamic vehicular mobility behaviours. They compared the results of the simulation with two routing protocols: GPSR and OLSR (Optimised Link State Routing). They have also evaluated the multi tenant's isolation in terms of network slicing and found that with network slicing, the likelihood of packet collision can be reduced substantially, and the bandwidth utilisation will be improved. Figure 12 illustrate SDVN system architecture.

Information Dissemination. Information can be delivered to vehicles with the support of SDN approach where vehicles can subscribe to receive information based on their geographical locations. On this regard, an SDN-based Pub/Sub middleware has been proposed [44] which consists of two layers: the SDN control layer and data plane layer. The latter basically includes the Pub/Sub vehicles along with the infrastructure. While the SDN control layer manages the subscriptions and publications as well as location-aware dissemination and it is composed of geographically distributed independent SDN controllers. As vehicular applications have different latency requirements, such as safety applications have critical latency requirement and need to receive the information on time. Therefore, a QoS-aware data dissemination algorithm is proposed, in which it takes the application type as an input and produces a list of subscribers within the geographic area where only these subscribers can receive the information from the publishing vehicle, hence, meeting the QoS requirements of the application.

Software Updates. Software-defined networking can be used to expedite software updates as described in [45] where the authors proposed software updates architecture in vehicles based on SDN and cloud computing called SDN-based vehicular cloud architecture (SVC). The architecture is designed to facilitate the distribution of software updates among vehicles by providing methods of modelling vehicular networks as connectivity graph to be used as input in the SDN architecture. They utilised V2V communication and the software update is assumed to be a network service. The vehicle receives the software updates by replicating the vehicle's virtual machine (VM) at the nearest data centre that provide the service. The data centre can be hosted on an RSU or base station (BS) which can also host the SDN controller. The migrations and management of VM replications is done by the SDN controller. The V2V communication technology is used by vehicles to exchange information about their connectivity with other vehicles and then this information will be transferred to the SDN controller using LTE connection. In turn, the SDN controller uses this information in order to construct the vehicle's connectivity graph and update flow tables and allocate frequency bands. Figure 13 shows different cases on how software updates will be distributed in VANETs:

(a) after the vehicle received the update (A|B|C|D) from the base station, it will share it with the neighbor vehicle,
(b) half of the updates will be received by the vehicles located at the two ends of the road and share it with vehicles in proximity.
(c) the vehicle receive the updates from the base station and split it to be delivered to neighbors.
(d) the vehicle receive the updates from the base station and split it to be delivered to neighbors that in turn share it to their neighbors.

We can observe from above cases that the vehicles in close proximity participate to receive the full update. This is done by the instructions given by the SDN controller. This has advantages of reducing both the usage of cellular bandwidth as well as the delivery delay.

Handover Between Vehicles and RSUs. To tackle frequently handover between vehicles and RSUs, authors in [46] have proposed fog cells approach in addition to a vehicular network architecture supported by 5G communication technologies and SDN along with cloud computing. The architecture is composed of three main planes: application plane, control plane, and data plane. The control plane is responsible of

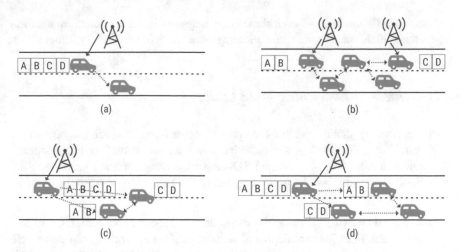

Fig. 13 Content distribution in VANETs [45]

determining control instructions, while the data plane is responsible of collecting data, and the application plane is in charge of setting up policies and strategies. The topology structure used in this study includes cloud computing centres, SDN controllers (SDNCs), RSU centres (RSUCs), BSs, fog computing clusters, vehicles and users. RSUCs and SDNC are considered to be in the control plane, while the others are in the data plane. For the communication between these entities, different links are used, such as infrastructure-to-infrastructure (I2I) links, vehicle-to-infrastructure (V2I) links, and vehicle-to-vehicle (V2V) links. The task of fog computing clusters is to control information sharing between vehicles and users. Moreover, fog computing clusters are configured in the edge of 5G software defined vehicular networks in order to provide fast responses from vehicles and users. Once the data is processed in the edge by the fog computing cluster, the SDNC gathers state information about the cluster and transfer this information into the cloud computing centre. The fog computing approach adopted in this study uses a multi-hop relay network which can minimise the frequent handover through establishing fog cells at the edge of 5G software defined vehicular networks. The simulation results of this approach yield minimum transmission delay of the network, taking into account different vehicle densities. Furthermore, the throughput measurement shows better result compared with the conventional transportation management systems.

3 Applications of Vehicular Network Slicing

In the previous section, we showed how network slicing can be employed in both in-vehicle network and in VANETs between vehicles and the infrastructure. In this section, we will show how network slicing can be applied in vehicular communication where the vehicle can communicate with anything i.e. Vehicle-to-everything (V2X).

3.1 Vehicle-to-Everything (V2X) Slices

Vehicular network slicing with SDN can provide several services such as autonomous driving, tele-operated driving, vehicular infotainment, and vehicle remote diagnostic services. Authors in [47] studied 5G network slicing for such services, V2X (vehicle-to-everything) as shown in Fig. 14. In the following, we briefly describe these services:

- **Autonomous driving slice:** autonomous driving requires ultra reliable and low latency communication to be deployed at the edge of the network to help the vehicle build 3D map of the surrounding environment and guarantee response time. This slice might be provided by the road authority.
- **Tele-operated driving slice:** as in autonomous driving, tele-operated driving requires ultra reliable and low latency communication to ensure end-to-end connectivity between the operator, located remotely, and the controlled vehicle. This kind of service is limited to certain number of vehicles and only activated when needed. Emergency department is usually using this slice.
- **Vehicular infotainment slice:** infotainment services such as video streaming require high bandwidth communication, hence it is expected to use multiple RATs (Radio Access Technology) to provide high throughput and the contents can be hosted at the cloud or cached at the edge closer to the user.
- **Vehicular remote diagnostic slice:** this slice can be employed by car manufacturer to support information exchange with vehicles. The user plane (UP) is responsible for handling the interaction while the control plane (CP) is instantiated correspondingly.

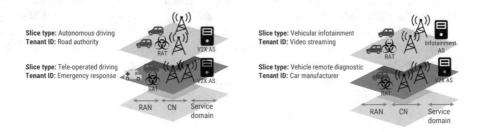

Fig. 14 V2X service slices [47]

3.2   Real-World Scenario with V2X Network Slicing

As mentioned before, 5G network is driving the applicability of network slicing, however, other communication technologies can be used in network slicing such as 4G LTE and LTE-Advanced. Moreover, other than cellular communications, wireless communications can be part of sliced networks. Generally, in V2X, these are the main two access technologies; WLAN-based such as DSRC [48] and cellular-based with 4G and 5G [49]. A typical real world application of implementing network slicing is presented in [50] that uses different communication technologies including 5G and 4G. The scenario is shown in Fig. 15 where an emergency vehicle needs to pass the road to reach the incident location. When the emergency vehicle is approaching the traffic light, it will send signals so that the traffic light will update its status and change the light for the emergency vehicle lane into green and for other crosstraffic lanes into red. Additionally, to make space for the emergency vehicle to pass, it will also send signals to the vehicles in front of it, the yellow car in Fig. 15. When the yellow car receive the signal, it will check if free space is available in the other lanes to move to or if the other lanes are not free, it will calculate which lane is more suitable to move to by measuring the distances from other cars in the other lanes. The car then will adapt its speed and for safety reasons it will send another signal to the cars in the right lane and then it will pass through to the right lane. Moreover, there is a passenger in one of the cars back seat streaming a HD video and there is also a pedestrian watching video on YouTube on his smartphone with intention to cross the road. The emergency vehicle will also warn the pedestrian by sending signal to his smartphone so that he will not cross the road.

The above scenario has three different V2X communications as:

- Vehicle-to-infrastructure (V2I) where the emergency vehicle is communicating with the traffic light to change status.
- Vehicle-to-vehicle (V2V) where the yellow car is communicating with other vehicles in proximity.

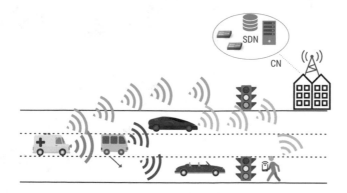

Fig. 15 Scenario of different communications in V2X [50]

- Vehicle-to-pedestrian (V2P) where the emergency vehicle is communicating with the pedestrian through his smartphone.

Furthermore, each one of the above communication require different technical needs, thus, network slicing is a great choice to be implemented in such scenario to provide the required QoS and at the same time preserve the network resources and allocate them in an efficient way. Here, there are four logical communications:

- **Communication A**, shown in blue color in Fig. 15, is when the emergency vehicle is sending signals to cars in front of it (V2V) in which the pedestrian crossing the road will also recieve the warning (V2P).
- **Communication B**, shown in red color, for V2V when the cars communicate to avoid collisions.
- **Communication C**, shown in green color, is when the emergency vehicle sends signals to the traffic light to change status (V2I).
- **Communication D**, shown in yellow color, is for the passenger in the car streaming HD video and for the pedestrian whose watching a YouTube video on his smartphone.

Therefore, for the above scenario, network slicing can be used to create different slices, each providing the necessary services as depicted in Fig. 16 and demonstrated in the following:

- **Slice 1**: for communications A, B and C. These communications require ultra-low latency and low bandwidth in addition to privacy.
- **Slice 2**: for communication D where it requires low latency and high bandwidth.

We can see from the previous example that different service applications ranging from enhancing real-time road safety and efficiency to streaming of HD videos can be supported with the network requirements that each application needs.

Fig. 16 Two slices created from scenario in Fig. 15

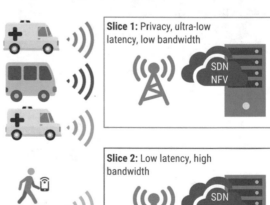

4 Intelligent Network Slicing

Slicing the network can be done in smart manners by leveraging the emerging developments in big data and machine learning in addition to SDN. In SDN, the network intelligence resides in the SDN controller that has the ability to make decisions and take actions based on the network's state and information. Several aspects of network slicing can be addressed by machine and deep learning solutions. And since vehicular networks produce vast amount of traffic data and because machines can process this amount of data faster than humans with more accurate results, the demand is shifting to high-volume accurate road and traffic data that allow for frequent updates as changes occur rapidly. Therefore, machine learning plays important role to efficiently process such big data and thus enable taking decisions dynamically whenever needed without desiring any external or human interventions.

In this section, we will discuss applying machine learning methods including deep learning to simplify the implementation of network slicing. We will start by discussing the existing network issues that machine learning methods can tackle and then we will draw a picture on how to achieve smart slicing in vehicular communications.

4.1 Machine Learning in Network Slicing

Machine and deep learning methods can be applied to address variety of issues in networking such as resource management, traffic identification and forecasts, among others. In network slicing for instance, network resources can efficiently be managed with machine learning techniques. Resource management in network slicing are categorised into two classes:

- **Slice admission control** to either accept or decline tenants' requests for slices.
- **Cross-slice resource allocation** to allow sharing of network resources among slices.

Policy-based decision is an approach used to make decisions on tenants selections or resource allocations [51]. Policy-based decisions are typically Markov decision processes (MDPs) in which the policy matches every system state with a corresponding action and then produce a reward. The issue in resource management in network slicing is that the reward function is non-convex over a large policy space [52]. Therefore, Reinforcement Learning (RL) can be used here since it has high capability in solving Markov decision problems.

A study by [53] used RL to improve network slicing policy-based decisions. Their Q-learning approach can provide approximation of the optimal slice admission policy. Moreover, it support online learning with acceptable computation cost.

On the other hand, cross-slice resource allocation can also be improved with RL. Such as cross-slice congestion control introduced by [54]. The proposed framework

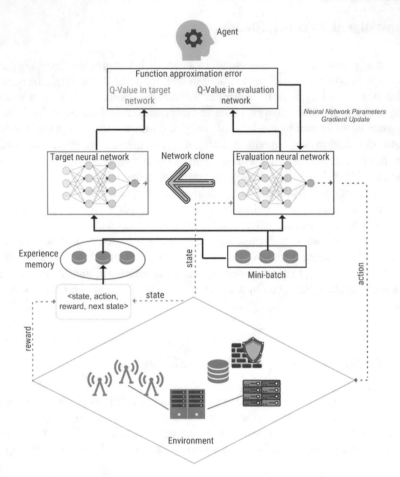

Fig. 17 Deep Q-learning [55]

achieved real-time slice elasticity with admission-control-like mechanism using Q-learning.

In addition to Q-learning, deep Q-learning (DQL) was used in [55] to allocate resources to slices by matching users' activity demand with resource allocation. Figure 17 illustrate the use of deep Q-learning for resource allocation in network slicing.

Other machine learning approaches were considered in the realm of networking that can be used in conjunction with network slicing, such as [56] that uses neural networks to classify traffic based on the application. The traffic classification process is performed over SDN by making use of the information provided by the packet header received from the OF switches and statistics from the controller. The proposed approach involves two phases: online and offline phases as shown in Fig. 18.

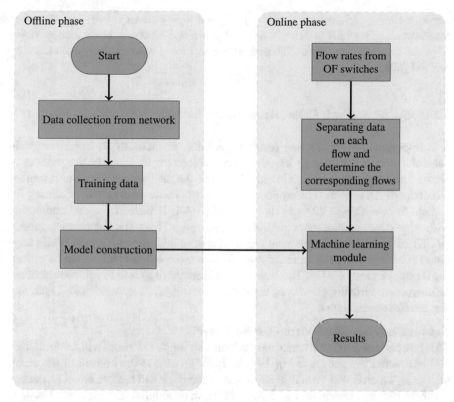

Fig. 18 Online and offline phases of traffic classification [56]

By correctly classifying the traffic, the application requirements can be identified and hence provide the suitable slice that serve such requirements.

Traffic clustering and forecasting solutions highly rely on machine learning methods. For example, [57] used machine learning algorithms to predict traffic. Traffic forecasts can be used in multiple network applications such as congestion avoidance and control applications, and energy saving applications. In [57], traffic forecasting is accomplished by investigating linear and non-linear machine learning algorithms, such as Auto-Regressive (AR) and Neural Networks (NNs). For traffic clustering, the well-known K-mean clustering algorithm is used. It divides cells into clusters by ensuring that the traffic behaviors in each cluster are similar.

4.2   Smart Vehicular Network Slicing

With the above machine learning-based solutions, intelligent network slicing can be leveraged in vehicular communications. In this section we will describe a smart V2X network slicing architecture given by [58], and then a smart scheduling algorithm for network slices [59] is discussed. Finally, we will shade light on a cooperative edge and cloud slicing approach based on deep learning with reinforcement learning method [60].

4.2.1 Smart Network Slicing Architecture for V2X

It is important for considering machine learning methods to be implemented in vehicular network slicing to have a basic architecture to simplify the process of network slicing with intelligent methods where the involved entities and resources are defined. An architecture proposed by [58] to enable smart network slicing for vehicle-to-everything (V2X) is shown in Fig. 19. It is based on cloud computing with edge cloud and remote cloud. Services required by vehicular user equipment (VUE) will be hosted on the edge cloud to ensure end-to-end latency. While edge cloud cannot provide the necessary services, remote cloud can be used to offload the service request of VUE. The architecture is categorised into four layers: network infrastructure virtualisation layer, intelligent control layer, network slice layer, and service customised layer.

Network Infrastructure Virtualisation Layer
This layer is based on NFV where network infrastructure resources will be virtualised and abstracted. As shown in Fig. 19, this layer located at the bottom of the architecture in which it will virtualise and abstract the edge and remote cloud resources. Furthermore, this abstraction with NFV helps in decoupling the network functions from the physical hardware. The process in this layer starts with the infrastructure resource virtualisation that is responsible for mapping the physical resources in each network domain to the multi-dimension resource pool. Then these resources are further divided into three main dimensions; communication, storage, and computation. Because VNFs are normally implemented in software and to facilitate their programmability and make them available, a VNF pool is used to combine all VNFs from each network domain. The appropriate set of VNFs available in VNF pool will then be selected by the control layer to create a slice.

Intelligent Control Layer
The managing and deploying of network slices are accomplished by this layer. In particular, this layer is responsible for managing and allocating the proper VNFs for each slice. Further, to benefit from machine learning algorithms, the intelligent control layer collects data produced from vehicular networks and saves it in a database. The collected data will be used by machine learning algorithms to extract and find patterns of vehicular networks that help in performing self-configuration of network slices.

Because machine learning methods require large amount of data, they take longer time to be processed. As a result, real-time V2X service delivery will not be guaranteed if the machine learning algorithm is to be processed by the control layer solely. For this reason, a hierarchical control layer is designed to quickly and efficiently

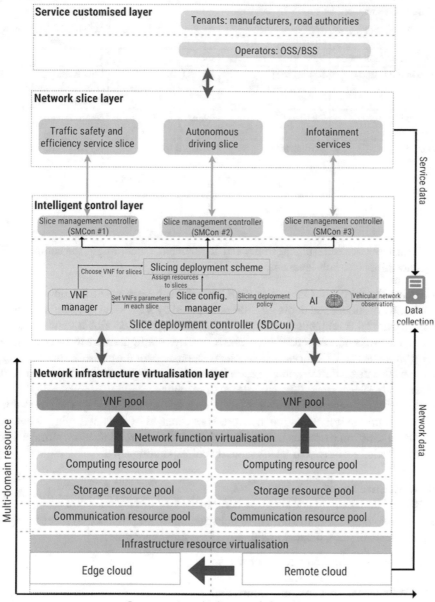

Fig. 19 Intelligent network slicing architecture for V2X [58]

Table 1 QoS requirements for V2X services

Service category	Traffic safety efficiency	Autonomous driving	Vehicular Internet and infotainment
Communication mode	V2V, V2P, V2I	V2V, V2I, V2N	V2N
Maximum latency	100 ms [61]	10 ms [62]	Low latency is not critical for media streaming [63]
Reliability requirement	99% [63]	99.999% [62]	Not a concern
Data rate	1 Mbps [63]	10 Mbps [62]	0.5 Mbps for web browsing, up to 15 Mbps for HD videos [63]

perform network slicing management and deployment. Accordingly, the control layer is divided into two sub-layers as:

- **Slicing Deployment Controller (SDCon):** this sub-layer performs global control of network slicing and deploy the slices required by V2X services. Further, it is an AI (Artificial Intelligence) learning agent that can provide global view of vehicular networks and therefore adapt network slices configurations to meet the QoS requirements of V2X services. SDCon is responsible for the network slice lifecycle from slice instantiation and orchestration to slice configuration and management, and lastly slice termination and release of resources.
- **Slicing Management Controller (SMCon):** this second sub-layer is a real-time local control unit that controls network slices created by the underlying layer, i.e. SDCon. SMCon performs managing tasks, allocate communication, storage, and computation resources to VUEs from its corresponding slice.

Network Slice Layer

This layer defines network slices as combination of multi-dimension resources in the multiple network domains. Three network slices are proposed to meet the QoS requirements of V2X defined in Table 1. These three slices are:

1. Traffic safety and efficiency service slice
2. Autonomous driving slice
3. Vehicular Internet and infotainment slice

Service Customised Layer

This is the last layer in the architecture that resides at the top of other layers and it is responsible for the SLA by which it captures the service requirements so that the service provider can supply on-demand network slices for each V2X service according to its requirements.

4.2.2 Intelligent Slice Schedule with Vehicular Fog-RAN

Usually, vehicular networks require cooperation from different entities such as road infrastructure which provide resources to vehicles in vicinity. And because vehicles are dynamic in nature, the density of vehicles in one road is different than the density in other roads, leading to unbalanced distribution of vehicles. This unbalance distribution cause infrastructure resources to be underutilised in roads with low traffic density while on other roads with higher density this cause the infrastructure resources to drain. Consequently, high investment should be put in place to provide sufficient number of infrastructure resources to accommodate for the increasing demand of resources in high traffic area, this in turn will increase both CAPEX and OPEX. One promising solution to reduce CAPEX is to exploit resources available in the vehicles to assist along with the infrastructure resources.

Exploiting both edge computing and vehicular fog computing, authors in [59] proposed V-FRAN (Vehicular Fog-RAN) framework that uses, jointly, edge fog resources (e.g. RSU) and vehicular fog resources. In addition, they developed smart slice scheduling algorithm with Cross-Entropy (CE) [64] based Monte Carlo Tree Search-Rapid Action Value Estimation (MCTS-RAVE) [65] that do not require any prior knowledge about the network traffic. Moreover, coordination between V-FRAN and road traffic management can be achieved by employing reinforcement learning. The framework is composed of three algorithms that form a close-loop:

- **Fog formation algorithm:** this algorithm selects the appropriate vehicular resources to form the vehicular fog.
- **Slice allocation algorithm:** with CE based MCTS-RAVE algorithm, slice allocation between different fogs is performed.
- **Road traffic management algorithm:** it provides a solution for adaptive control of the average speed of road traffic.

4.2.3 Adaptive Network Slicing for Intelligent Vehicular System

The above framework works well when the service delivery cannot tolerate any delay. Although edge and fog computing can solve the latency issue of cloud computing, fog nodes are equipped with limited storage and computation resources. Therefore, for applications that demand intensive computational processes and tolerable delay, cloud computing is the one to use. However, some applications are both latency-critical and computational-intensive, such as real-time video streaming of traffic management in ITS.

To utilise capabilities of both edge/fog RAN and cloud RAN, and to efficiently assign the limited edge resources to the vehicles, [60] proposed an intelligent solution based on network slicing to adaptively allocate edge resources in F-RAN to dynamic IoV and smart city applications with varying latency requirements. The model is presented in Fig. 20. The model consists of two logical slices: cloud slice and edge slice.

The edge slice is connected to the cloud slice via high-capacity fronthaul links, represented by green dashed lines. Solid arrows between the environment and the edge slice represent services presented by the edge slice to meet QoS, while the dashed blue arrows represent process delegation to the cloud slice in order to save the edge slice resources for more latency-critical tasks. The contents of each slice is described as follows:

- **Edge slice:** is composed of multiple fog nodes (FNs), as shown in Fig. 20, and each FN provides services to entities in its coverage area. Adjacent FNs are connected via extremely fast and reliable optical links. Moreover, FNs are equipped with storage and computing abilities to autonomously supply network services at the edge of the network.
- **Cloud slice:** includes cloud controller (CC) with powerful and full computing capabilities, pool of enormous storage capacity, centralised baseband units (BBUs), and operations and maintenance center (OMC) that is responsible for monitoring key performance indicators (KPIs) and producing network reports.

Fig. 20 Network slicing model with cooperative edge slice and cloud slice [60]

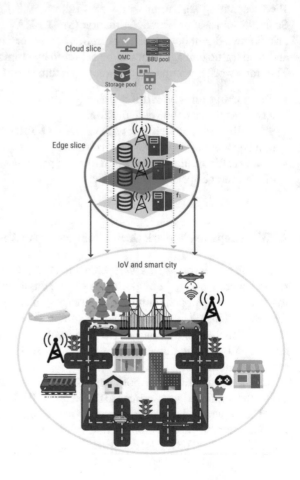

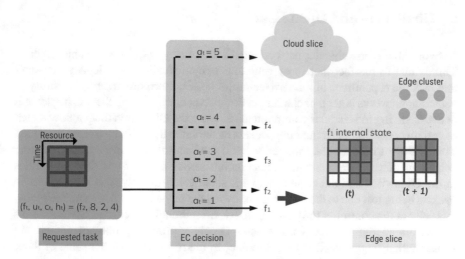

Fig. 21 EC decision process for a request from f_2. The learning agent i.e. EC is f_3

As mentioned earlier, vehicular networks are considered very dynamic with rapid changing of densities and hence, edge resources can go under-utilised with low vehicle density and low delay-sensitive application requests. A rule-based network slicing policy cannot guarantee efficient utilisation of edge resources in such dynamic circumference. In contrast, a statistical based learning policy can adjust its decisions to match the changes caused by the dynamic behaviour of vehicular networks. In fact, such policy can learn to prioritise requests based on the delay requirement of each application request. The idea here is to use a Markov decision process (MDP) to enable considering expected rewards for all potential actions in deciding the network slice to use. This is achieved by employing a reinforcement learning (RL) approach that can learn the optimal policy for the MDP problem. The RL agent here is an edge controller (EC), it is basically one of the FNs in edge slice. It is usually chosen based on its central location, e.g. f_1 or f_3 in Fig. 20.

The EC learns to maximise the rewards by trial and error. As shown in Fig. 21, if a service request received by the edge cluster from IoV application at time t with utility u_t, the number of available resource blocks c_t is calculated by the primary FN. In addition, FN will also calculate the holding time h_t required by the task to be served locally at the edge. After that, the FN will share the tuple of (u_t, c_t, h_t) with the learning agent i.e. EC. In turn, EC keeps track of the available resources at all FNs. If the required resource blocks c_t are neither available at the primary FN nor at any other FN in the cluster for the duration of h_t, then the EC declines serving the task locally at the edge and therefore, it will refer the task to the cloud slice.

5 Challenges and Open Issues

Network slicing in vehicular networks is in its infancy with limited achievements. This is due to the fact that most vehicular communications use legacy networks that lack the capabilities of modern technologies. Furthermore, the high mobility of vehicular networks is another challenge. For instance, when using SDN to implement network slicing in vehicular communication, once the OF switch finds a flow match of certain packet, all the remaining packets of the same flow will be treated the same. This is done to reduce the interaction with the controller. However, the dynamic behaviour of vehicular networks cause inconsistency between the flow table rules in the OF switch and the current network state. Consequently, this may results in packets being lost due to the mismatch between the physical topology and the global topology at the controller [66]. Additionally, even though the current advancements in network slicing in other types of networks show promising outcomes, there are some challenges that need to be overcome. In the following, we list some of these challenges:

- **Network security and privacy:** Although network slicing can improve security by isolating traffic in different slices, security concerns still exist [10, 67]. Since network slicing based on partitioning the underlying network infrastructure, security and privacy are major concerns due to the fact that some network resources will be shared among different slices that might be owned by external entities or third parties. This is one of the barriers in adopting multi-tenancy approaches in 5G network slicing. To mitigate the risk of data breaches and cyber attacks, access to slice should be restricted to authorised users. Moreover, the open APIs that support network programmability in SDN bring more concerns about potential attacks. This calls for more effort to effectively design and implement security mechanisms for network slicing.
- **End-to-end slice management and orchestration:** Implementing network slicing involves number of operations that need to be carefully managed. These operations include slice creation, slice activation, slice maintenance, and slice deactivation at the service level. Managing the life cycle of network slices is challenging, especially in multi-tenant scenarios. Additionally, managing network resources, monitoring, adjust load balancing, and charging policies at the network level cause extra burden on management and orchestration tasks [68].
- **Dynamic network topology:** In mobile as well as vehicular networks, the network topology is not stable due to the high mobility which causes rapid changes in the physical network topology [10, 67]. This is very challenging when assigning network slices to moving objects such as vehicles. Thus, new mechanisms should be introduced to address such issue. One technique that can be used is fog computing at the network edge. However, the current advances in fog computing are not sufficient to handle such high dynamic behaviour. More effective solutions can be achieved by predicting vehicle's future directions with e.g. machine learning tools.

- **Single point of failures:** When utilising single centralised SDN controller, the possibility of launching attacks on the controller is high due to single point of failure induced by the central control [69]. Such attacks may include DDoS (Distributed Denial of Service) attacks that force the whole network to go down. Therefore, fail-over mechanisms should be put in place in case of problems or attacks occur in the control plane.
- **Inconsistent data:** For implementing AI and especially when using machine learning algorithms, the algorithms need vast amount of data to learn and converge to deploy the final model. However, with IoVs, the data generated by different sources of vehicles and even by different entities in the ITS such as RSUs, is usually inconsistent and most of the time has redundancy. Therefore, mobile data collected from heterogeneous sources need to be pre-processed [58].
- **Cost of implementation:** Since network slicing is driven by 5G networks, and as 5G uses new frequencies and millimeter waves, the size of the cell is minimised which result in the demand of increasing the number of base stations to improve the network coverage [50]. Such increase in base stations, in turn, increases the cost of implementation and installation.

6 Conclusion

In this chapter, we discussed network slicing in vehicular networks. We first defined the term of network slicing and how it is represented in 5G networks by three main use cases: ultra reliable and low latency communication (uRLLC) use case, massive machine type communication (mMTC) use case, and enhanced mobile broadband (eMBB) use case. While these three cases have their own network requirements, other use cases might need different requirements, so one can include both uRLLC and eMBB in one use case. In addition, using network slicing to solve various networking issues is described such as using network slicing for flow isolation to improve security. We also shed a light on network slicing from a business perspective and showed how it can reduce capital expenditures (CAPEX). Moreover, we identified SDN, software define networking that facilitate network slicing by providing network programmability, where SDN can decouple control plane from data plane unlike in conventional networks that have both planes in the network hardware. Then, we thoroughly discussed network slicing in vehicular networks. We categorised vehicular network slicing into two main categories: in-vehicle network slicing and vehicular ad hoc networks (VANETs) slicing. In both categories, we described the network communication and how network slicing can be performed by investigating the current studies in this field. In addition, we described, with real-world example, the application of V2X for different slices to support various demands. Further, we emphasised on the importance of improving network slicing by utilising the emerging advances of AI and machine learning to achieve intelligent network slicing. Finally, we stated the potential challenges and discussed some open issues in applying network slicing such as the high dynamic behaviour of vehicular networks.

References

1. Larsson, C.: 5G Networks: Planning. Academic Press, Design and Optimization (2018)
2. Matthiesen, B., Aydin, O., Jorswieck, E.A.: Throughput and energy-efficient network slicing. In: WSA 2018; 22nd International ITG Workshop on Smart Antennas, pp. 1–6, VDE (2018)
3. Paul, A., Chilamkurti, N., Daniel, A., Rho, S.: Intelligent vehicular networks and communications: fundamentals, architectures and solutions. Elsevier (2016)
4. Abbas, M.T., Khan, T.A., Mahmood, A., Rivera, J.J.D., Song, W.-C.: Introducing network slice management inside m-cord-based-5G framework. In: NOMS 2018-2018 IEEE/IFIP Network Operations and Management Symposium, pp. 1–2. IEEE (2018)
5. Nakao, A., Du, P., Kiriha, Y., Granelli, F., Gebremariam, A.A., Taleb, T., Bagaa, M.: End-to-end network slicing for 5G mobile networks. J. Inf. Process. **25**, 153–163 (2017)
6. Wei, H., Zhang, Z., Fan, B.: Network slice access selection scheme in 5G. In: 2017 IEEE 2nd Information Technology, Networking, Electronic and Automation Control Conference (ITNEC), pp. 352–356. IEEE (2017)
7. Yoo, T.: Network slicing architecture for 5G network. In: 2016 International Conference on Information and Communication Technology Convergence (ICTC), pp. 1010–1014. IEEE (2016)
8. Kalør, A.E., Guillaume, R., Nielsen, J.J., Mueller, A., Popovski, P.: Network slicing in industry 4.0 applications: Abstraction methods and end-to-end analysis. IEEE Trans. Industr. Inf. **14**(12), 5419–5427 (2018)
9. Khan, H., Luoto, P., Bennis, M., Latva-aho, M.: On the application of network slicing for 5G-v2x. In: European Wireless 2018; 24th European Wireless Conference, pp. 1–6, VDE (2018)
10. Li, X., Samaka, M., Chan, H.A., Bhamare, D., Gupta, L., Guo, C., Jain, R.: Network slicing for 5g: Challenges and opportunities. IEEE Internet Comput. **21**(5), 20–27 (2017)
11. Lee, M.K., Hong, C.S.: Efficient slice allocation for novel 5G services. In: 2018 Tenth International Conference on Ubiquitous and Future Networks (ICUFN), pp. 625–629. IEEE (2018)
12. Sparks, K., Sirbu, M., Nasielski, J., Merrill, L., Leddy, K., Krishnaswamy, P., Johnston, W., Gyurek, R., Daly, B.: 5G network slicing whitepaper - FCC technological advisory council - 5G IoT working group (2016)
13. Foukas, X., Patounas, G., Elmokashfi, A., Marina, M.K.: Network slicing in 5G: survey and challenges. IEEE Commun. Mag. **55**(5), 94–100 (2017)
14. Kazmi, S.A., Khan, L.U., Tran, N.H., Hong, C.S.: Network Slicing for 5G and Beyond Networks. Springer, Cham (2019)
15. Floodlight openflow controller. https://floodlight.atlassian.net/wiki/spaces/floodlightcontroller/overview
16. Nox openflow controller. https://github.com/noxrepo/nox
17. Ryu openflow controller. https://ryu-sdn.org
18. Opendaylight (odl) controller. https://www.opendaylight.org
19. Farrel, A.: Recent developments in service function chaining (SFC) and network slicing in backhaul and metro networks in support of 5G. In: 2018 20th International Conference on Transparent Optical Networks (ICTON), pp. 1–4. IEEE (2018)
20. Han, B., Gopalakrishnan, V., Ji, L., Lee, S.: Network function virtualization: challenges and opportunities for innovations. IEEE Commun. Mag. **53**(2), 90–97 (2015)
21. Afolabi, I., Taleb, T., Samdanis, K., Ksentini, A., Flinck, H.: Network slicing and softwarization: a survey on principles, enabling technologies, and solutions. IEEE Commun. Surv. Tutorials **20**(3), 2429–2453 (2018)
22. Singh, S., Jeong, Y.-S., Park, J.H.: A survey on cloud computing security: issues, threats, and solutions. J. Netw. Comput. Appl. **75**, 200–222 (2016)
23. Satyanarayanan, M., Bahl, P., Caceres, R., Davies, N.: The case for VM-based cloudlets in mobile computing. IEEE Pervasive Comput. **8**(4), 14–23 (2009)
24. Bonomi, F., Milito, R., Zhu, J., Addepalli, S.: Fog computing and its role in the internet of things. In: Proceedings of the first edition of the MCC Workshop on Mobile Cloud Computing, pp. 13–16 (2012)

25. Mach, P., Becvar, Z.: Mobile edge computing: a survey on architecture and computation offload-
    ing. IEEE Commun. Surv. Tutorials **19**(3), 1628–1656 (2017)
26. Casetti, C., Chiasserini, C.F., Deiß, T., Frangoudis, P.A., Ksentini, A., Landi, G., Li, X., Molner,
    N., Mangues, J.: Network slices for vertical industries. In: 2018 IEEE Wireless Communications
    and Networking Conference Workshops (WCNCW), pp. 254–259. IEEE (2018)
27. Mämmelä, O., Hiltunen, J., Suomalainen, J., Ahola, K., Mannersalo, P., Vehkaperä, J.: Towards
    micro-segmentation in 5g network security. In: European Conference on Networks and Com-
    munications (EuCNC 2016) Workshop on Network Management, Quality of Service and Secu-
    rity for 5G Networks (2016)
28. Boubendir, A., Guillemin, F., Le Toquin, C., Alberi-Morel, M.-L., Faucheux, F., Kerboeuf, S.,
    Lafragette, J.-L., Orlandi, B.: 5G edge resource federation: dynamic and cross-domain network
    slice deployment. In: 2018 4th IEEE Conference on Network Softwarization and Workshops
    (NetSoft), pp. 338–340. IEEE (2018)
29. Oladejo, S.O., Falowo, O.E.: 5G network slicing: a multi-tenancy scenario. In: 2017 Global
    Wireless Summit (GWS), pp. 88–92. IEEE (2017)
30. Backman, J., Yrjölä, S., Valtanen, K., Mämmelä, O.: Blockchain network slice broker in 5G:
    slice leasing in factory of the future use case. In: 2017 Internet of Things Business Models,
    Users, and Networks, pp. 1–8. IEEE (2017)
31. Addad, R.A., Bagaa, M., Taleb, T., Dutra, D.L.C., Flinck, H.: Optimization model for cross-
    domain network slices in 5G networks. IEEE Trans. Mob. Comput. **19**(5), 1156–1169 (2019)
32. Campolo, C., Molinaro, A., Iera, A., Fontes, R.R., Rothenberg, C.E.: Towards 5G network
    slicing for the v2x ecosystem. In: 2018 4th IEEE Conference on Network Softwarization and
    Workshops (NetSoft), pp. 400–405. IEEE (2018)
33. Goransson, P., Black, C., Culver, T.: Software Defined Networks: A Comprehensive Approach.
    Morgan Kaufmann, New York (2016)
34. Braun, W., Menth, M.: Software-defined networking using openflow: protocols, applications
    and architectural design choices. Future Internet **6**(2), 302–336 (2014)
35. Huang, J., Zhao, M., Zhou, Y., Xing, C.-C.: In-vehicle networking: protocols, challenges, and
    solutions. IEEE Network **33**(1), 92–98 (2018)
36. Steinbach, T., Müller, K., Korf, F., Röllig, R.: Real-time ethernet in-car backbones: first
    insights into an automotive prototype. In: 2014 IEEE Vehicular Networking Conference (VNC),
    pp. 133–134. IEEE (2014)
37. Fussey, P., Parisis, G.: Poster: an in-vehicle software defined network architecture for connected
    and automated vehicles. In: Proceedings of the 2nd ACM International Workshop on Smart,
    Autonomous, and Connected Vehicular Systems and Services, pp. 73–74 (2017)
38. Choi, J.-H., Min, S.-G., Han, Y.-H.: Macsec extension over software-defined networks for in-
    vehicle secure communication. In: 2018 Tenth International Conference on Ubiquitous and
    Future Networks (ICUFN), pp. 180–185. IEEE (2018)
39. Halba, K., Mahmoudi, C.: In-vehicle software defined networking: an enabler for data inter-
    operability. In: Proceedings of the 2nd International Conference on Information System and
    Data Mining, pp. 93–97 (2018)
40. Halba, K., Mahmoudi, C., Griffor, E.: Robust safety for autonomous vehicles through recon-
    figurable networking, arXiv preprint arXiv:1804.08407 (2018)
41. Peng, H., Liang, L., Shen, X., Li, G.Y.: Vehicular communications: a network layer perspective.
    IEEE Trans. Veh. Technol. **68**(2), 1064–1078 (2018)
42. Ku, I., Lu, Y., Gerla, M., Gomes, R.L., Ongaro, F., Cerqueira, E.: Towards software-defined
    vanet: architecture and services. In: 2014 13th Annual Mediterranean ad Hoc Networking
    Workshop (MED-HOC-NET), pp. 103–110. IEEE (2014)
43. He, Z., Cao, J., Liu, X.: SDVN: enabling rapid network innovation for heterogeneous vehicular
    communication. IEEE Network **30**(4), 10–15 (2016)
44. Mendiboure, L., Chalouf, M.A., Krief, F.: A SDN-based pub/sub middleware for geographic
    content dissemination in internet of vehicles. In: 2019 IEEE 90th Vehicular Technology Con-
    ference (VTC2019-Fall), pp. 1–6. IEEE (2019)

45. Azizian, M., Cherkaoui, S., Hafid, A.S.: Vehicle software updates distribution with SDN and cloud computing. IEEE Commun. Mag. **55**(8), 74–79 (2017)
46. Ge, X., Li, Z., Li, S.: 5G software defined vehicular networks. IEEE Commun. Mag. **55**(7), 87–93 (2017)
47. Campolo, C., Molinaro, A., Iera, A., Menichella, F.: 5G network slicing for vehicle-to-everything services. IEEE Wirel. Commun. **24**(6), 38–45 (2017)
48. Jiang, D., Delgrossi, L.: Ieee 802.11 p: towards an international standard for wireless access in vehicular environments. In: VTC Spring 2008-IEEE Vehicular Technology Conference, pp. 2036–2040. IEEE (2008)
49. Zheng, K., Zheng, Q., Chatzimisios, P., Xiang, W., Zhou, Y.: Heterogeneous vehicular networking: a survey on architecture, challenges, and solutions. IEEE Commun. Surv. Tutorials **17**(4), 2377–2396 (2015)
50. Šeremet, I., Čaušević, S.: Benefits of using 5G network slicing to implement vehicle-to-everything (v2x) technology. In: 2019 18th International Symposium INFOTEH-JAHORINA (INFOTEH), pp. 1–6. IEEE (2019)
51. Han, B., Schotten, H.D.: Machine learning for network slicing resource management: a comprehensive survey. arXiv preprint arXiv:2001.07974 (2020)
52. Han, B., Feng, D., Schotten, H.D.: A Markov model of slice admission control. IEEE Network. Lett. **1**(1), 2–5 (2018)
53. Bega, D., Gramaglia, M., Banchs, A., Sciancalepore, V., Samdanis, K., Costa-Perez, X.: Optimising 5G infrastructure markets: the business of network slicing. In: IEEE INFOCOM 2017-IEEE Conference on Computer Communications, pp. 1–9. IEEE (2017)
54. Han, B., DeDomenico, A., Dandachi, G., Drosou, A., Tzovaras, D., Querio, R., Moggio, F., Bulakci, O., Schotten, H.D.: Admission and congestion control for 5g network slicing. In: 2018 IEEE Conference on Standards for Communications and Networking (CSCN), pp. 1–6. IEEE (2018)
55. Li, R., Zhao, Z., Sun, Q., Chih-Lin, I., Yang, C., Chen, X., Zhao, M., Zhang, H.: Deep reinforcement learning for resource management in network slicing. IEEE Access **6**, 74429–74441 (2018)
56. Parsaei, M.R., Sobouti, M.J., Khayami, S.R., Javidan, R.: Network traffic classification using machine learning techniques over software defined networks. Int. J. Adv. Comput. Sci. Appl. **8**(7), 220–225 (2017)
57. Le, L.-V., Sinh, D., Lin, B.-S.P., Tung, L.-P.: Applying big data, machine learning, and SDN/NFV to 5G traffic clustering, forecasting, and management. In: 2018 4th IEEE Conference on Network Softwarization and Workshops (NetSoft), pp. 168–176. IEEE (2018)
58. Mei, J., Wang, X., Zheng, K.: Intelligent network slicing for v2x services toward 5G. IEEE Network **33**(6), 196–204 (2019)
59. Xiong, K., Leng, S., Hu, J., Chen, X., Yang, K.: Smart network slicing for vehicular fog-RANs. IEEE Trans. Veh. Technol. **68**(4), 3075–3085 (2019)
60. Nassar, A., Yilmaz, Y.: Deep reinforcement learning for adaptive network slicing in 5G for intelligent vehicular systems and smart cities,arXiv preprint arXiv:2010.09916 (2020)
61. Zheng, K., Meng, H., Chatzimisios, P., Lei, L., Shen, X.: An SMDP-based resource allocation in vehicular cloud computing systems. IEEE Trans. Industr. Electron. **62**(12), 7920–7928 (2015)
62. 5G-Infrastructure-Association et al.: 5G automotive vision, 20 October 2015
63. Meredith, P.M.J.: Study on lte-based v2x services (release 14). Tech. rep., technical specification. Technical report (2016)
64. Wang, Y., Guo, C., Wu, Q.: A cross-entropy-based three-stage sequential importance sampling for composite power system short-term reliability evaluation. IEEE Trans. Power Syst. **28**(4), 4254–4263 (2013)
65. Gelly, S., Silver, D.: Monte-carlo tree search and rapid action value estimation in computer go. Artif. Intell. **175**(11), 1856–1875 (2011)
66. Jaballah, W., Conti, M., Lal, C.: A survey on software-defined vanets: benefits, challenges, and future directions. arxiv 2019 arXiv preprint arXiv:1904.04577

67. Mahmood, A., Zhang, W.E., Sheng, Q.Z.: Software-defined heterogeneous vehicular networking: the architectural design and open challenges. Future Internet **11**(3), 70 (2019)
68. Zhang, H., Liu, N., Chu, X., Long, K., Aghvami, A.-H., Leung, V.C.: Network slicing based 5G and future mobile networks: mobility, resource management, and challenges. IEEE Commun. Mag. **55**(8), 138–145 (2017)
69. Hackel, T., Meyer, P., Korf, F., Schmidt, T.C.: Software-defined networks supporting time-sensitive in-vehicular communication. In: 2019 IEEE 89th Vehicular Technology Conference (VTC2019-Spring), pp. 1–5. IEEE (2019)

Towards Artificial Intelligence Assisted Software Defined Networking for Internet of Vehicles

Sachin Sharma

Abstract In the Internet of Vehicles (IoV), the Internet of Things (IoT) is integrated with Vehicular Ad hoc NETworks (VANET). This enables gathering, processing and sharing of lots of information (regarding vehicles, roads and their surroundings) through the Internet and hence, helps in making intelligent decisions. On the other hand, Software Defined Networking (SDN) has the capability of designing a flexible programmable IoV network that can foster innovation and reduce complexity. Applying SDN in IoV will be useful, as SDN enabled IoV devices can be controlled seamlessly from an external server (called a controller) which can be located in the cloud and may have computational resources to run resource-intensive algorithms, making intelligent decisions. This chapter provides an introduction about SDN, describes the benefits of integrating SDN in IoV and reports the recent advances. It also presents an Artificial Intelligence (AI) based architecture and open challenges. Finally, the chapter presents an automatic configuration method with which SDN can be deployed automatically in IoV without any manual configuration. The experiments are performed on a publicly available European testbed using an emulator for wireless SDN networks. Experiments are conducted for automatic configuration of SDN in IoV network's topologies and for data collection in SDN enabled IoV. The results show the effectiveness of the proposed automatic configuration method. Furthermore, AI-assisted intelligent decisions supported by SDN enabled IoV are introduced. The challenges and solutions presented in this chapter may have a huge impact on the speed at which IoV infrastructure can efficiently evolve with market evolution.

1 Introduction

Wireless Mobile Ad hoc NETworks (MANET) are infrastructure-less networks that do not depend on a pre-existing infrastructure (e.g., access points or base stations) for communication. In a MANET, wireless mobile devices can either directly

S. Sharma (✉)
School of Electrical and Electronic Engineering, Technological University Dublin (TU Dublin), Grangegorman Campus, Dublin, Ireland
e-mail: Sachin.Sharma@tudublin.ie

© Springer Nature Switzerland AG 2021   191
N. Magaia et al. (eds.), *Intelligent Technologies for Internet of Vehicles*, Internet of Things,
https://doi.org/10.1007/978-3-030-76493-7_6

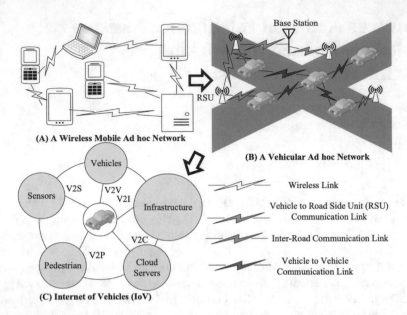

(A) A Wireless Mobile Ad hoc Network

(B) A Vehicular Ad hoc Network

(C) Internet of Vehicles (IoV)

Fig. 1 From a wireless Mobile Ad hoc Network (MANET) to a Vehicular Ad hoc Network (VANET) and from the VANET to the Internet of Vehicles. Here, RSU is the Road Side Unit, V2V is a Vehicle-to-Vehicle interface, V2I is a Vehicle-to-Infrastructure interface, V2C is a Vehicle-to-Cloud server interface, V2P is a Vehicle-to-Pedestrian interface and V2S is a Vehicle-to-Sensor interface

communicate or can communicate through other wireless devices in the network (see Fig. 1A). Vehicular Ad hoc NETworks (VANET) are originated from the MANETs where wireless devices are vehicles, Road Side Units (RSU) and/or base stations (see Fig. 1B). The main objective of VANETs is to provide safety to vehicles. The Internet of Vehicles (IoV) is the evolution of VANETs where the Internet of Things (IoT) is integrated with VANETs. IoV has additional capability over VANETs in terms of communications, storage, intelligence and decision making [1]. In Fig. 1C, IoV has the capability to communicate with vehicles, sensors, pedestrians, cloud servers and infrastructures such as signs, base stations and RSUs.

IoV enables sharing of lots of data of the vehicles and their surroundings, and thereby, intelligent decisions could be taken for vehicles to improve vehicle experience [2]. This chapter is about the use of Software Defined Networking (SDN) in IoV to increase the flexibility and programmability in IoV. SDN started with the decoupling of control from network devices with the main goal of enabling (fast) independent innovation of networking hardware and software, as well as reduction of CAPital EXpenditure (CAPEX) and OPerational EXpenditure (OPEX). SDN works on three principles: (1) decoupling of control from traffic forwarding, (2) logically centralized control, and (3) programmability of network services. The first two principles enable IoV networks to control their vehicular networks from a centralized server called a controller. The third principle supports the specific requirements of

IoV applications, including traffic isolation, multiple Quality of Service (QoS) levels and customized network behavior [3].

There are several SDN protocols such as OpenFlow, OpenFlow Configuration protocol (OF-Config) and Open vSwitch Database (OVS-DB) management protocol. Using the OpenFlow protocol, the data plane of network devices can be programmed from the controller. Using OF-Config and OVS-DB management protocol, a network can be configured from a server called OF-Config server or OVS-DB server. All these protocols are useful in controlling and configuration of the network behaviour.

This chapter presents solutions (architecture, automatic configuration, etc.) to deploy SDN in IoV. The contributions of this chapter are as follows:

1. Background and Benefits
   This chapter first provides the background necessary to understand SDN enabled IoV (Sect. 2) and then presents the benefits of SDN for IoV (Sect. 3).
2. SDN enabled IoV Architectures
   The chapter also proposes an architecture for SDN enabled IoV. The proposed architecture is the enhanced version of architecture proposed for SDN by Open Networking Foundation (ONF).[1] The architecture includes an Artificial Intelligence (AI) based application on top of SDN architecture. An AI-based application is needed to add up additional intelligence in IoV (Sect. 4).
3. The State of the Art, Limitations and Solutions for SDN enabled IoV
   The state of the art related to SDN enabled IoV is presented in Sect. 5. Section 6 discusses the limitations of current SDN software to deploy it in IoV. It also presents solutions to overcome those limitations.
4. Automatic Configuration of SDN
   Later, in Sect. 7, an automatic configuration framework for SDN enabled IoV is presented. Using this framework, SDN can be configured in IoV without any manual configuration. This proposal might revolutionise the way IoV can be managed and will pave the path for scalable, low-cost, self-configurable, and programmable future IoV networks supporting a plethora of use cases such as self-driving cars and safety applications. The challenge is that IoV or VANETs are very dynamic networks where vehicles may move at a different speed and direction. The speed and direction may also change at any time. Therefore, without deploying SDN automatically, it may be difficult to deploy VANETs or IoV with SDN. The framework is recently proposed for wireless ad hoc networks to configure SDN in wireless mobile ad hoc networks [8, 9]. This chapter explores the use of this algorithm in IoV. The approaches proposed for the automatic configuration of SDN in IoV in this chapter may have a huge impact on the speed at which IoV infrastructure can efficiently evolve with market evolution.
5. Data Collection and AI Framework
   Data collection and AI-assisted framework inspired by SDN is provided in Sect. 8.

---

[1] https://www.opennetworking.org/.

6. Experiments
   Experimental scenarios and results are then presented in Sect. 9. Experimentation scenarios and results are related to the automatic configuration of SDN and data collection in SDN enabled IoV. Experiments include movement scenarios in IoV.
7. Open Challenges
   Open challenges about using SDN in the IoV context are demonstrated in Sect. 10.
8. Conclusions
   Finally, Sect. 11 concludes the chapter and provides future work.

2 Background

This section provides the background information necessary to understand the topic discussed in this chapter.

2.1 Internet of Vehicles

Internet of Vehicles (IoV) is a new field of study for connecting vehicles to the Internet. It integrates the Internet of Things (IoT) with VANETs. This will tackle traffic congestion and driving issues, and thus, will lead to the safety of users, simplification of the driving experience and improvement of user experience. In IoV, the devices are vehicles, sensors, personal devices, cloud servers and infrastructure devices such as RSU and many more. In the literature [24], several different types of communications are discussed for IoV. The following illustrates the five of them (see Fig. 1C):

1. Vehicle-to-Vehicle (V2V) Communication
   It consists of communication between vehicles regarding speed, direction, location, braking, and loss of stability, etc. This prevents accidents, reduces congestion, improves fuel efficiency and optimizes routes.
2. Vehicle-to-Infrastructure (V2I) Communication
   It consists of an exchange of data between vehicles and roadside infrastructure. The roadside infrastructure includes lane markings, road signs, traffic lights and RSUs. With the data captured and shared through the V2I interface, several applications can be built to enhance safety, mobility, and environmental benefits.
3. Vehicle-to-Cloud (V2C) servers Communication:
   It consists of an exchange of information between vehicles and cloud servers. It supports several applications such as security, firmware, updates and entertainment.
4. Vehicle-to-Pedestrian (V2P) Communication
   It includes communication between vehicles and pedestrians. It prevents possible accidents by enabling communication between vehicles and pedestrians through applications on personal devices such as smartphones and wearable.

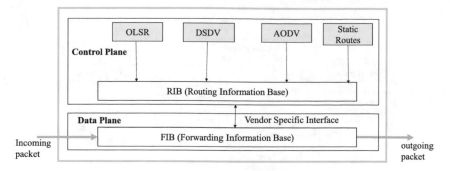

Fig. 2 Standard Network Router Design. OLSR is Optimized Link State Routing, DSDV is Destination-Sequenced Distance Vector Routing and AODV is Ad hoc On Demand Routing

5. Vehicle-to-Sensor (V2S) Communication

It provides communication between sensors and vehicles. This is important for cases such as collision protection by detecting objects and people in front of vehicles, and therefore, can provide safety against accidents.

The above vehicular communications open up the implementation of several applications to enhance user experience. The applications are self-driving vehicles, emergency rescue systems, intelligent traffic management, on-board social networks and intelligent navigation system, etc., [10]. All these applications require lots of data from IoV devices that may assist in making intelligent decisions.

2.2 Standard Traditional Vehicular Router Design and Networking

The basic design of routers such as wireless ad hoc or vehicular ad hoc router is presented in Fig. 2. It consists of a data plane, control plane and vendor-specific interface between them. The data plane is in charge of forwarding data traffic and is implemented in the Forwarding Information Base (FIB) [11]. On the other hand, the control plane runs routing protocols such as OLSR (Optimized Link State Routing), DSDV (Destination-Sequenced Distance-Vector Routing) and AODV (Ad hoc On-Demand) to make forwarding decisions. These forwarding decisions are first fed into the RIB (Routing Information Base) [11] and then fed into the FIB through the vendor-specific interface. In comparison to the data plane, the control plane is mostly run on a low-end Central Processing Unit (CPU). Hence, the processing of packets/traffic is slower in the control plane than in the data plane. Therefore, routes from the RIB are inserted into the FIB for fast forwarding of packets.

In traditional networks, each router takes its own control plane decisions to forward packets. A tradition network is shown in Fig. 3. These networks have become too costly to build, too complicated to manage or control, too vulnerable to vendor-

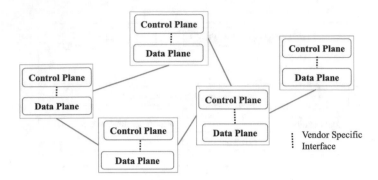

Fig. 3 Traditional networking

locking, and too inflexible to fulfill the needs of new requirements [7]. This is because the control plane and data plane are located in the same device and the interface between them is vendor-specific.

Software Defined Networking (SDN) has emerged to address the above problems of traditional networks by making networks more programmable. One of the major advantages of SDN is its ability to make networks simple. It simplifies the traditional networks by decoupling the control plane (complex software) from network devices.

2.3 Software Defined Networking (SDN)

SDN is an emerging approach that allows a network to be centrally controlled and programmed using software applications located in an external server called controller [12]. OpenFlow is currently the de-facto standard protocol of SDN [13]. OpenFlow proposes a standard protocol for communication between the data plane and the control plane of network devices. By opening up the traditionally closed vendor-specific interface between the control and data plane and implementing a common control plane using OpenFlow, the entire network can be managed/controlled seamlessly regardless of the complexity of the underlying network technology [13]. There are other SDN protocols that are also popular such as OpenFlow Configuration protocol (OF-Config) [14] and Open vSwitch Database Management protocol (OVS-DB) [15]. These protocols can be applied together with OpenFlow. The following section describes all the above SDN protocols:

2.3.1 OpenFlow

OpenFlow is an SDN protocol that allows one or more entities (called controllers) in a network to communicate with the data plane of network devices and to make adjustments so that it can be adapted to meet the changing requirements [13]. OpenFlow

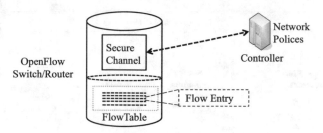

Fig. 4 Openflow switch/router

is developed in a clean-slate future internet program (the future Internet re-design program at Stanford University) [13]. OpenFlow has been released in the form of specifications. The first two versions of the specifications (i.e., v1.0 and v1.1) were released by Stanford University in 2009 and 2011 respectively. In 2011, several Industrial players such as Deutsche Telekom, Google, Microsoft and Yahoo formed the Open Networking Foundation (ONF)[2] to standardize the OpenFlow protocol, meeting industry requirements. Hence, since the third version (v1.2), ONF has been releasing the next versions of OpenFlow. The current major OpenFlow version is 1.5.

Figure 4 shows the switch or router design, originally proposed for OpenFlow. It consists of two components (see Fig. 4): (1) an OpenFlow switch or router, which contains the FlowTable (FIB of the router) and a secure channel (i.e., to establish a secure session with the controller) and (2) the controller, which contains a set of policies to add in the OpenFlow switches/routers through the secure channel. A Flow Entry in the FlowTable consists of two parts: (1) the matching part and (2) the action part. When a packet arrives at an OpenFlow switch/router, its header is compared against the matching part of Flow Entries in the FlowTable. If a match is found, the action of that entry is performed. If no match is found, the packet is sent to the controller and the controller, thereafter, can insert a new entry in the switch to forward packets. Here, the controller is responsible for decisions on how to forward packets.

Figure 5 shows an OpenFlow network in which a controller controls the entire OpenFlow network. Practically, more than one controller may be needed to control the entire OpenFlow network. Figure 5 shows that the whole control plane is moved to the controller. However, in the current market, most OpenFlow switches or routers are hybrid in nature. This means that today's OpenFlow switches/routers support traditional protocols as well as the OpenFlow protocol. This means that the whole control plane is not moved to the controller. Traditional protocols can still be run through hybrid OpenFlow switches/routers without the controller. However, new protocols or functionalities can be implemented using the controller. The benefit of these hybrid switches/routers is that operators can choose both traditional networking protocols and OpenFlow protocols in their networks. In addition, they have the

---

[2] www.opennetworking.org.

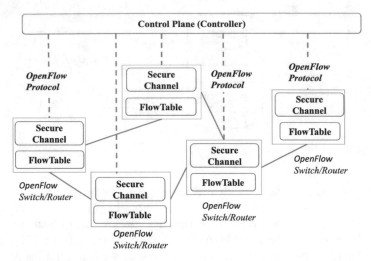

Fig. 5 An OpenFlow switch/router network

flexibility to choose either of them depending on the current network's requirements. Figure 6 shows a hybrid OpenFlow switch/router network where part of control plane functionality is located external to the network and implemented in an external server called the controller.

There are several Open-Source OpenFlow switch/router software such as Open-Flow Reference switch,[3] Open vSwitch[4] and OpenFlow CPqD switch.[5] Among the switches, Open vSwitch is a production quality switch software. In addition, there are several open-source OpenFlow controller software such as NOX,[6] POX,[7] Floodlight,[8] OpenDayLight[9] and ONOS.[10] Currently, OpenDayLight and ONOS support both OpenFlow and configuration protocols, and have the capability to support resiliency and scalability requirements of carrier-grade networks.

2.3.2   OpenFlow Configuration Protocol (OF-Config)

In parallel with the evolution of OpenFlow, the ONF configuration and management working group (WG) focused on configuration and monitoring aspects of the Open-

---

[3]https://www.opennetworking.org/reference-designs/.

[4]http://openvswitch.org/.

[5]https://cpqd.github.io/ofsoftswitch13/.

[6]https://github.com/noxrepo/nox.

[7]https://github.com/noxrepo/pox.

[8]https://floodlight.atlassian.net/wiki/spaces/floodlightcontroller/overview.

[9]https://www.opendaylight.org/.

[10]https://www.opennetworking.org/onos/.

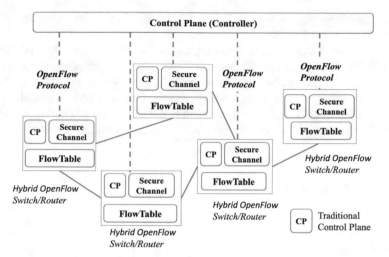

Fig. 6 An hybrid OpenFlow switch/router network

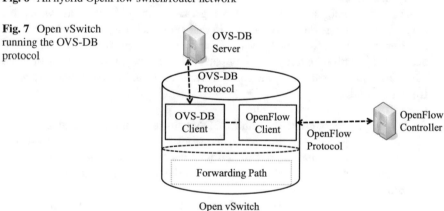

Fig. 7 Open vSwitch running the OVS-DB protocol

Flow switch/router. Current versions of OpenFlow do not cover management and configuration features sufficiently (e.g., addresses have to be configured manually for control channel establishment). This WG thus specified OF-Config, a mechanism for configuring OpenFlow capable devices. OF-Config 1.0 was released in January 2012, and is based on NETCONF (RFC 4741), a transnational protocol that makes use of Remote Procedure Calls (RPC) on top of a secure transport layer protocol to configure remote devices. The current version is OF-Config 1.2.

2.3.3 Open vSwitch Database (OVS-DB) Management Protocol

The OVS-DB management protocol is an OpenFlow configuration protocol designed for the configuration of Open vSwitch (one of the production quality OpenFlow switch implementations). It uses JSON (RFC 4627) and is based on JSON RPC (Remote Procedure Calls).

Figure 7 shows Open vSwitch running the OVS-DB and OpenFlow protocol. In the OVS-DB management protocol, the OVS-DB client is located at the vSwitch and the OVS-DB server could be located external to the vSwitch. Both the OpenFlow and OVS-DB protocol create the session on top of the transport layer protocol. As both OVS-DB and OpenFlow clients are located at the same vSwitch node, both should use a different transport layer port number for communication.

3   Benefits of SDN

The followings are the foreseen benefits of SDN for IoV:

1. Programmability
   Using the OpenFlow protocol, SDN enables the programmability of the data plane from an external server. This has a great advantage for IoV, as it provides the ability to change device behaviour in real-time according to current needs or requirements. Therefore, the Quality of Service (QoS) can be maintained in IoV networks by setting up a high-quality path. All these low-level decisions can be taken from an external server that can be located external to the IoV network (e.g., from the cloud).
2. Centralized management or control
   As shown in Fig. 5, one of the benefits of SDN is that the centralized controller has a view of the whole topology and hence, decisions could be made that are consistent and predictable. In addition, new functionalities can be deployed seamlessly, as the centralized controller has full control over the underlying network.
3. Flexibility
   Currently, traditional networks including IoV networks take a long time to offer new services, which require modification in available protocols or algorithms of network or vehicular devices. These services must wait for vendors (and standards bodies such as IETF) to approve and incorporate new solutions in corresponding devices. Currently, the standardization process for a new solution (or protocol) takes a long time. Using SDN, new solutions can be implemented in the controller without requiring new hardware from the network or vehicular device's side. Therefore, it reduces the time to add new solutions or services in the network.
4. Simplicity
   SDN also adds simplicity from the network side, as the complex part (part thereof) of the devices i.e., control plane could be moved somewhere else in the network (i.e., external to the network).
5. Vendor Interoperability
   As a vendor-specific interface (part thereof in Fig. 6 and the complete interface in Fig. 5) between the control plane and data plane is replaced by an SDN protocol such as OpenFlow protocol, SDN makes the network vendor agnostic.

6. Slicing on Demand

   Network slicing is an innovative architectural solution in 5G including vehicular ad hoc networks that enables isolated slices being created on the same physical infrastructure to cater to the new requirements. Adding, moving, and changing resources is one of the advantages of SDN. Therefore, it is easier to create a new slice using SDN at any time in all networks.

7. Automatic configuration

   There are several SDN protocols such as OF-Config, OVS-DB protocols which are designed for automatic configuration of SDN devices without any manual configurations. This is one of the advantages of SDN for IoV.

8. Variety of data collection and use of machine learning or AI techniques

   As the controller can have control over all the network devices using SDN protocols, it can collect different types of data including low-level data from IoV devices such as buffer capacity, configured MAC layer protocol, the number of successfully delivered messages, hop count, CPU usage and memory usage. Therefore, different machine learning and deep learning techniques (random decision tree, support vector machine, K-Nearest Neighbour, Neural networks) can be applied to predict the future best configurations (such as the best data traffic path). In addition to best configuration, machine learning could be used to decide the important parameters of the system such as: (1) which data (buffer capacity, hop count, processing speed, etc.) should be collected, (2) from where (e.g., which device) it should be collected and (3) the frequency of data collection. SDN gives the flexibility to collect lots of data. This now gives the flexibility to run different machine learning techniques to predict the best configurations.

4   Towards SDN Architecture for IoV

This section first introduces the present architecture of SDN, given by Stanford University and ONF, and then proposes SDN architecture mapping to IoV using pure and hybrid SDN devices.

4.1   Present Architecture for SDN

Figure 8A shows the architecture of today's network in which control and data plane are present in the same device. Figure 8B depicts the network architecture proposed by Stanford University for future networks in which the control plane is decoupled from the data plane. In this design, the control plane is located in an external server (the controller) and then the external server communicates with the data plane through the OpenFlow protocol. In this architecture, applications (which may be network or service-related or AI-based) are the pieces of software that are coupled in the

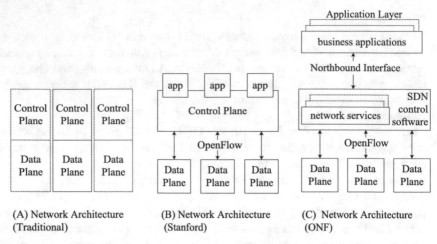

Fig. 8 Traditional network architecture, network architecture proposed by stanford university and network architecture proposed by ONF (Open Networking Foundation)

controller. These applications can introduce new features in the network such as routing and security-related features.

Figure 8C presents the network architecture proposed by ONF. It extends the architecture proposed by Stanford given in Fig. 8B by placing applications in the application layer (a separate entity) and the application layer then communicates with the controller through the northbound interface (see Fig. 8C). Several open-source projects and groups are currently dedicated to developing the northbound interface standard between the applications and the controller. For example, the Linux Open API Initiative has been focusing on programmable APIs (Application Program Interfaces) that can be used across multiple controller software.

4.2 SDN Architecture Mapping to IoV

Researchers are currently applying mapping of SDN architecture (presented by ONF and given in Fig. 8C) in IoV. One of such mappings is presented in Fig. 9. Similar mapping is presented in [4, 5] and [6]. In this mapping, IoV applications such as automatic driving, safety driving and infotainment applications are deployed as business applications and they communicate with the SDN control plane (controllers) through the northbound interfaces. All the IoV devices such as vehicles, RSU, Pedestrians and Road Side Infrastructure just run the SDN data plane and communicate with the controller using the OpenFlow protocol.

The following part explains each component depicted in Fig. 9:

1. Data Plane (Vehicles, RSU, Pedestrians and Sensors)
   As shown in Fig. 10, the data plane consists of the forwarding plane and ports/inter-faces. Ports/interfaces are responsible for sending/receiving messages and could

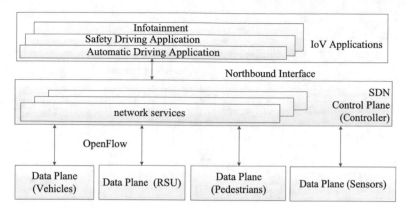

Fig. 9 IoV SDN architecture

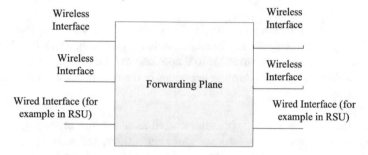

Fig. 10 IoV data plane

be a wireless or wired interface. Some IoV devices such as vehicles and sensors may just contain wireless interfaces (e.g., WAVE, long-range WiFi, 4G/LTE). The other IoV devices such as RSU may also contain wired interfaces. In order to enable SDN, no changes are required from the physical and MAC layer side of the wireless devices.

The forwarding plane of the data plane contains the forwarding entries and is responsible for forwarding traffic to a particular port (wireless/wired). The forwarding entries in the forwarding plane are inserted by the controller. To enable SDN, IoV devices need to run SDN software (such as OpenFlow) and should have an in-band or out-of-band connection to communicate with the controller [17].

2. Controller (SDN Control Plane)

The followings are important tasks of the controller: (1) to make control plane decisions and incorporate them in the data plane through an SDN protocol (e.g., the OpenFlow protocol) and (2) to provide network services to IoV applications. The services that the controller can provide are: (1) data collection service, (2) traffic engineering, e.g., QoS and (3) mapping service to data planes. The data collection service may be responsible for collecting different types of data from

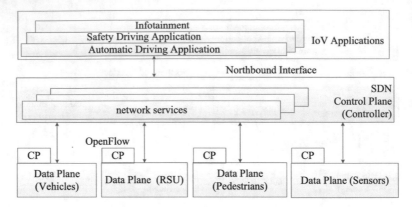

Fig. 11 Modified SDN architecture for hybrid IoV devices. Here, CP is the partial control plane

the data plane and making it available to IoV applications when it is required. The traffic engineering service may be responsible for providing a specific QoS to the traffic sent/received by a particular IoV application. The mapping service may be responsible for adding forwarding decisions in the data plane, as specified by the application.

3. IoV Applications

   There can be several IoV applications such as automatic driving applications, safety driving applications, infotainment applications, and security applications. Some or all of them could be AI-based applications where AI/machine learning is used to make intelligent decisions. The IoV applications will be able to communicate with the controller through a northbound API (see Fig. 9).

4.3 Modified SDN Architecture Mapping to IoV

Figure 11 shows the modified SDN architecture mapping where part of control plane functionality is present in the IoV devices. This is applicable to hybrid SDN IoV devices, specified in OpenFlow Specification 1.1 and later versions of OpenFlow. These hybrid devices are common for many manufacturers such as Brocade, Juniper and Cisco.

The architecture presented in Fig. 11 is general to cover all hybrid switches functionalities, as CP in Fig. 11 is the partial control plane functionality which can be traditional control plane or any other new control plane functionality implemented in IoV devices together with the data plane of IoV devices. The difference between the modified SDN architecture for hybrid IoV devices (Fig. 11) and IoV SDN architecture (Fig. 9) for pure SDN IoV devices is that in modified architecture, CP is available in SDN IoV devices together with the data plane and in IoV SDN architecture, CP is not available in SDN IoV devices together with the data plane. All the other details of the data plane, controllers and IoV applications are the same as given in the previous subsection.

Several researchers have already used hybrid SDN switches for the implementation of new functionalities in IoV. For example, a multi-criteria routing protocol based on a hybrid SDN IoV switch/router is proposed in [16] in order to increase failure resiliency while reducing resolution costs and optimizing delay.

5 The State of the Art for SDN Enabled IoV

In recent years, the idea of SDN in IoV has drawn significant attention in research communities. Due to the increased flexibility and programmability, several solutions are currently proposed to meet QoS and scalability challenges of the highly dynamic networks of IoV using SDN [24]. In [25], SDN is used in IoV to maximize the utilization of the Internet of Vehicles by reducing the transfer of tasks between clouds, fog and IoV. In [26], load balancing strategies are proposed for SDN enabling IoV to balance the load between IoV devices and cloud servers. In [27], SDN is used to implement different strategies for routing in IoV.

Using SDN, IoV can be controlled through a centralized controller. Therefore, there can be new security attacks in SDN enabled IoV due to vulnerabilities of a centralized system [29]. To keep this in mind, the impact of DDoS (Distributed Denial-of-service) attacks on SDN enabled IoV is studied in [28]. To overcome such impacts, several solutions such as a hybrid approach using probabilistic data structures [30] and a machine learning-based DDoS [31] solution are provided for SDN based IoV.

In [32] blockchain is used in SDN enabled IoV in the 5G. The blockchain is integrated in this research to improve trust among vehicles as the data received from a vehicle should be trusted.

Moreover, different AI-based/machine learning-based applications using supervised, unsupervised and deep learning are also proposed for IoV [33] for channel estimation, traffic flow prediction, vehicle trajectory predictor, network congestion control and resource management. As lots of data can be available through SDN, these approaches will benefit from SDN.

As given above, currently several approaches to use SDN in IoV have been given in the literature. Recently, we proposed a method for the automatic configuration of SDN in wireless ad-hoc networks, which can be applied in IoV. This chapter introduces the work on the automatic configuration of SDN in IoV.

6 Limitations and Possible Solutions

The limitation of current SDN software such as Open vSwitch, reference switch and CPqD switch is that they follow wired network's MAC layer standards (such as

Fig. 12 MAC addresses in IEEE 802.11

the IEEE 802.3 standard) for forwarding frames. However, wireless devices such as vehicular devices follow wireless standards such as IEEE 802.11 and IEEE 802.11p. This problem is already discussed by M. Rademacher et al. in [18]. The result is that frames sent by SDN software are not correctly received or sent by the wireless interface of vehicular or roadside units of IoV devices.

There are the following three different ways to address the above problem, but not all the ways are applicable for all the devices (see below):

1. Enable 4-address mode
   The IEEE 802.11 standards define five different address types (see Fig. 12): (1) the transmitter address (TA), the MAC address of a wireless device that actually transmits the frame, (2) the receiver address (RA), the MAC address of a wireless device which receives the frame, (3) the source address (SA), the address of a device which is the source of the frame, (4) the destination address (DA), the address of a device to which the frame destined to and (5) Basic Service Set Identifier (BSSID), the address of a set of stations (or wireless devices) that are associated successfully.
   When sending a frame from a wireless interface, a maximum of four addresses can be used. In the 4-address format, TA, RA, SA and DA addresses are used as the frame Layer 2 addresses. This is also called WDS (Wireless Distribution System).
   Compared to the above IEEE 802.11 standard, an IEEE 802.3 frame uses two addresses - source and destination address - in its frame. If a device needs to send an IEEE 802.3 frame from a wireless interface, the 4-address mode (or WDS) mode can be enabled in the device. Otherwise, the IEEE 802.3 frame cannot be correctly transmitted over a wireless interface [18].
   The 4-address mode enables IEE 802.3 frames to be transmitted successfully over a wireless interface. The problem is that not all the devices support enabling this mode due to security reasons [34].
2. Implement a protocol translator
   In this case, an extra MAC replacing a component is needed for packet header translation from IEEE 802.3 to IEEE 802.11 and vice versa. The problem is that this is not currently supported in vehicular devices.
3. Enable Tunneling
   Several devices such as vehicular devices support creating tunnels such as GRE (Generic Routing Encapsulation) or (Vxlans) between devices. When these tunnels are created, a GRE (or Vxlan) header is added into an IEEE 802.3 frame, transmitted from OpenFlow software such as OpenvSwitch and sent through a tunnel. The MAC addressing scheme of these GRE frames is the MAC addressing scheme of the underlying transmission medium (in our case it will be IEEE

802.11 or IEEE 802.11p). Vehicular devices already support the enabling of these tunnels. The only problem with these tunnels is that an extra header is added to each original frame and hence, these increase the bandwidth requirement of the network.

The 4-address mode is the simplest way for forwarding IEEE 802.3 frames generated from OpenFlow software through a wireless interface. However, as not all vehicular ad hoc devices support this mode, the proposed automatic configuration method (presented in the next section) uses the tunneling, presented above, for forwarding IEEE 802.3 frames through a wireless interface.

7 Automatic Configuration Framework for SDN Enabled IoV

As stated in Sect. 3, SDN can provide automatic configuration of several functionalities such as QoS, routing and security in IoV data plane networks. However, before that, IoV needs to automatically configure SDN in its network. Automatic configuration means that the IoV devices (such as vehicles, pedestrians and RSUs) have to make an SDN session with the controller. In the following section, an automatic configuration method of one of the protocols (i.e., OpenFlow) in IoV is presented. This method is proposed for hybrid IoV devices and applicable for the architecture proposed for IoV in Sect. 4.3. The automatic configuration method is proposed for a network where not all the devices have direct connectivity with the controller. Therefore, the controller has to make sessions with the IoV devices through other devices in the network.

Previously, an automatic configuration method was applied to configure Open-Flow in wireless mobile ad hoc networks [8, 9]. This section discusses the use of the same method in IoV. The method is different from the automatic configuration method of OpenFlow for wired networks, presented in [17], as devices in IoV are wireless devices and they may also move at different speed and direction. Furthermore, the standards of wireless and wired networks such as MAC layer standards are different.

7.1 Assumptions

The followings are the assumptions of the proposed automatic configuration method:

1. The SDN controller is located external to IoV networks. For example, it may be located at cloud servers, as shown in Fig. 13.
2. SDN IoV devices (such as vehicles and RSU) should implement hybrid functionality, as presented in Sect. 4.3. The hybrid functionality of IoV devices should

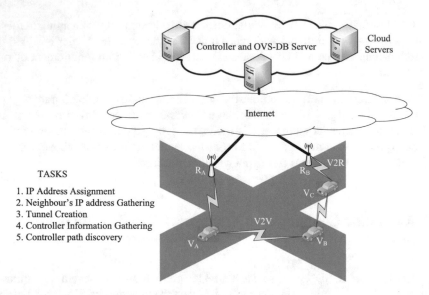

TASKS

1. IP Address Assignment
2. Neighbour's IP address Gathering
3. Tunnel Creation
4. Controller Information Gathering
5. Controller path discovery

Fig. 13 Automatic configuration of SDN in IoV

contain at least traditional routing protocols (such as OLSR) together with the OpenFlow protocol.

3. Each SDN IoV device should have a path to connect to the controller. The path can be through other IoV devices in the network (see Fig. 13).

7.2 Requirements

As the OpenFlow session is built on top of the transport layer session, the following requirements are designed for a hybrid IoV SDN device to implement the proposed automatic configuration method:

1. Unique IP address
   The IoV device should have a unique IP address to establish a transport layer session (and then an OpenFlow session) with the controller.
2. Knowledge of neighbor's IP address
   The IoV device needs to know the IP address of each of its neighboring OpenFlow IoV devices (i.e., to establish GRE or Vxlan tunnels between the device and a neighbour IoV device).
3. Tunneling Requirement
   Each IoV OpenFlow device needs to establish GRE (or VXlan) tunnels with its direct neighbours (as presented in the previous subsection).
4. Knowledge of the controller

Each IoV OpenFlow device needs to know the IP address and OpenFlow session parameters (including transport layer parameters) of the controller.

5. Knowledge of a path to the controller

Each IoV device also requires to discover a path to reach the controller (the path may be through other vehicles in the network, see Fig. 13). This is needed to establish an OpenFlow session with the controller.

7.3 Deployed Solutions

The possible solutions of each of the above requirements for the proposed automatic configuration method are given below:

1. IP address assignment

The Dynamic Host Configuration Protocol (DHCP) can be used to configure the unique IP address for vehicles (the first requirement stated above). DHCP is used extensively in computer networks to get an IP address. There are several other approaches proposed for vehicular networks to get a unique IP address as given in [19, 20]. Any of the approaches can be used. The motive is just to get a unique IP address for an IoV device.

2. Knowledge of neighbor's IP address

To address this requirement, each IoV device runs a traditional routing protocol (e.g., OLSR) in the network. Each device running traditional routing protocols such as OLSR sends hello messages to its direct neighbors after a regular time interval. A hello message contains the IP address of the device. Once a device receives a Hello message, it knows the neighbor IP address.

3. Tunneling Requirement

To address this requirement, the proposed automatic configuration method runs an agent in an IoV device. This agent creates a GRE or Vxlan tunnel with a neighbouring IoV device when the traditional routing protocol such as OLSR discovers a neighbor. The agent also removes a tunnel with a device when the routing protocol detects that the device is no longer a neighbor, as it stopped receiving hello messages from the device.

4. Knowledge of the controller

To get the controller information, the proposed automatic configuration method deploys the OVS-DB server in a network external to the IoV network and reachable to each IoV device. In Fig. 13, this OVS-DB server is deployed in a cloud server and runs the routing protocol to reach each IoV device. This OVS-DB server has the information about the controller e.g., IP address, transport layer and OpenFlow session parameters. This information is fed already to the OVS-DB server by an administrator. Further, each IoV device runs an OVS-DB client. The OVS-DB server configures the controller information in an IoV device through its OVS-DB client, when the routing protocol running in the server detects a new IoV device in the IoV network.

5. Knowledge of a path to the controller
   To get a path to the controller, the controller also runs the routing protocol. As all
   the IoV devices also run the routing protocol, they all know a path to reach each
   other. The path is decided by the routing protocol.

7.4 Proposed Automatic Configuration Method

This chapter explains the proposed automatic configuration using four hybrid Open-
Flow IoV devices (two car vehicles, V_A and V_B, and two RSUs, R_A and R_B) as shown
in Fig. 13. The RSUs are connected to the controller through the Internet. Figure 13
is an example topology to explain the proposed configuration method. However, the
proposed configuration method can be applied to any topology, following the three
assumptions mentioned in Sect. 7.1.

Each IoV device (V_A, V_B, R_A, R_B) runs the following process stack: (1) an automatic IP configuration protocol such as DHCP or address automatic configuration
protocol [20]: This is to get a unique IP address, (2) a routing protocol (e.g., OLSR):
This is to get neighbor's information and to find a path to the controller and OVS-DB
server, (3) the tunnel agent: This is to create tunnels with neighbors, (4) the OVSDB
client: This is to configure the controller information in an IoV device, (5) a transport layer protocol: This is to build a transport layer session with the controller and
(6) the OpenFlow protocol: This is to enable the OpenFlow protocol in the device.
Furthermore, the OVS-DB server and the controller (see Fig. 13) also run the routing
protocol to reach IoV devices.

The following two automatic configuration scenarios - (1) Initial phase setup and
(2) Movement phase setup - are described in the following part of the section. In the
initial phase setup, an IoV device is in a bootstrapping phase where the IoV device
joins the IoV networks and establishes an OpenFlow session with the controller automatically. In the movement phase setup, the configured path between the controller
and an IoV device changes (e.g., due to movement) and a new path between the
controller and the IoV device is setup between the controller and IoV devices.

7.4.1 Initial Phase Setup Scenario

Each IoV device (V_A, V_B, R_A, R_B) gets an IP address using any of the IP configuration protocols such as DHCP or address automatic configuration protocol, described
in the previous subsection. Each IoV device also runs a routing protocol. Therefore,
each IoV device gets the neighbor information (e.g., neighbor's IP address), once it
receives a hello message from a neighbor. On the discovery of a new neighbor, the
tunnel agent deployed on the IoV device creates a tunnel between the IoV device
and a discovered neighbor.

In the OVS-DB server, the controller information is already fed by the administrator and the routing protocol is also run on it. On discovery of a new IoV device

by the OVS-DB server through the routing protocol, the OVS-DB server configures the controller information in the new IoV device through the OVS-DB management protocol. The path between the OVS-DB server and the IoV device is decided by the routing protocol.

Once the controller information is configured into an IoV device (V_A, V_B, R_A, R_B), the IoV device tries to create an OpenFlow session by first establishing the transport layer session. As the path between the controller and the IoV device is decided by the routing protocol. The IoV device establishes the transport layer session along the path to the controller and then the OpenFlow session is established on top of the transport layer session.

7.4.2 Movement or Failure Phase Setup Scenario

As some of IoV devices such as car vehicles and pedestrian (V_A, V_B) may move from one location to another (or IoV devices may fail to function properly), IoV devices may become unreachable with their neighbour vehicles or RSUs (discovered previously by the routing protocol). The routing protocol detects this when it stops receiving hello messages from a neighboring IoV device. When the path between the controller (or the OVS-DB server) and an IoV device contains an unreachable neighbour, the communication between the controller (or the OVS-DB server) and the IoV device does not work until a new valid working path (discovered by the routing protocol) is established.

The following two cases can happen when a new path is established: (1) the path is established before the transport or OpenFlow session detects a failure and (2) the path is established after the transport layer or OpenFlow session detects a failure. In the former case, the previously established OpenFlow session with the IoV device does not disconnect and the communication between the controller and the IoV device begins to work after the new path is established. However, in the latter case, the previously established OpenFlow session between the IoV device and the controller disconnects, and the IoV device tries to establish a new session with the controller. Once the new path is established, the IoV device is able to establish a new session with the controller.

Further to the above changes, when the routing protocol running on IoV device (V_A, V_B, R_A, R_B) detects an unreachable neighbour, the tunnel agent (running on the same device) removes the tunnel with the unreachable neighbour. Moreover, when the routing protocol detects a new neighbour, the tunnel agent establishes a tunnel with that neighbour.

Moreover, the IP address of an IoV device may change when moving from one location to another. In this case, the old OpenFlow session with the controller is disconnected and a new OpenFlow session is established after the IoV device does not receive any message from the controller containing the new IP address of the device. There can be a case when more than one controller (not shown in Fig. 13) are used to control IoV devices. In this case, it is the responsibility of the OVS-DB server to add the correct controller information in the case when the controller information changes (maybe due to a change in the location of an IoV device). If the controller information changes, the IoV device disconnects the old OpenFlow session and establishes a new session. Note that the algorithm that decides, which controller will control which IoV device, is out of scope for this chapter. However, several works exist in the literature regarding multiple controller architectures or algorithms [21, 22].

8 AI Assisted SDN Enabled IoV

8.1 Data Collection

After configuring SDN in IoV (as discussed in the previous subsection), different types of data can be collected, as using SDN the controller has low-level access to IoV devices. Low-level data may be: (1) the number of packets or bytes of data successfully received or sent by an IoV device, (2) the number of packets or bytes received incorrectly, (3) buffer space left, (4) processor usage, (5) location of the device and (6) temperature of the vehicle. There are many other different types of data that can be collected. The information of data that can be collected is available in ONF OpenFlow specifications and in the OVS-DB protocol specification.

The data can be collected after a regular interval or on-demand. It depends on a particular application that needs the data. More important is that SDN can help in collecting data and an AI application can use the data to solve a problem.

8.2 AI Management in IoV

AI will be a key enabling technology in the management of future networks (also pointed by several researchers [35, 36]), as it can provide learning capability and better decision making. First and foremost, to effectively apply AI/machine learning to any problem, data availability is crucial. As stated before, SDN can help in collecting data. Some AI-related applications in IoV, which can make management simpler, are given below:

1. AI-assisted IoV applications
   AI can be used to enable the cars to move together with other road vehicles and make real-time decisions (such as security and safety decisions). Using AI and other IoT sensors, such as cameras, it will be easier to ensure safe driving. The other IoV applications where AI could be applied are: (1) traffic guidance system, (2) intelligent vehicle control, (3) safety applications, (4) infotainment applications, and many more.

2. Anomaly detection in IoV

   Using machine learning/AI, one can make decisions on whether a sudden burst of traffic in the network is due to some anomalies or attacks like Distributed Denial of Service (DDoS) attacks or due to a normal surge in network traffic.
3. Identifying a faulty/vulnerable component

   Similar to anomaly detection, a faulty component can also be identified using AI/machine learning.
4. Prediction of future conditions

   AI/machine learning can help in predicting future conditions such as which component will fail in the future, which is a safe path at a particular time and so on.

8.3 AI Assisted Framework for SDN Enabled IoV

An AI framework for IoV can be described using the architecture figure given in Sect. 4.3. In the AI/machine learning framework, the applications such as Automatic driving, safety driving and infotainment, given in Fig. 11, can be AI applications. Here, AI applications mean that applications use AI/machine learning to solve a particular problem.

9 Experimental Scenario and Results

In this section, experimental scenarios and results are provided for the automatic configuration of SDN and data collection in IoV networks. The emulation scenarios are similar to the wireless mobile ad hoc network scenario tested in [9].

9.1 Emulation Scenarios

Emulations are performed on a node of the virtual wall testbed facility at the Fed4Fire testbed facility in Europe.[11] A personal computer generation 4 (pcgen04) node of this testbed is used for the emulation of IoV. The following configuration of the node is available in this testbed: 8 CPU cores of Intel E5-2650v2 processor with 2.6 GHz speed, 48 GB RAM and $1 \times 250GB$ hard disk. A single physical CPU core with hyper-threading enabled appears as four logical CPUs in this node. Therefore, emulations are performed using 32 logical CPUs.

For IoV emulations, Ubuntu 16.04 LTS image is deployed on the above pcgen04 node (n064-10) available through the Fed4Fire interface and a real-time emulator,

---

[11] https://www.fed4fire.eu/testbeds/virtual-wall/.

Mininet-WiFi[12] is deployed. Mininet-WiFi is a fork of the Mininet SDN network emulator[13] and enhances it by adding virtualized wireless stations and access points based on the standard Linux wireless drivers. Open vSwitch[14] version 2.5.5 and POX controller[15] (version 0.5.0) are deployed to perform SDN emulations in IoV. Open vSwitch comes with the OVS-DB protocol implemented in its software. In emulations, one controller is used and the OVS-DB server is also located at the controller node. The controller is deployed on a separate node of the Fed4Fire testbed.

OpenFlow uses ECHO_REQUEST and ECHO_REPLY messages to check the aliveness of its sessions. If an OpenFlow switch or the controller does not receive a reply to its ECHO_REQUEST message, a failure is declared and the session is broken. In the performed experiments, this failure is declared in about 15 s.

In the emulations, Transmission Control Protocol (TCP) is used as a transport layer protocol for the communication between the IoV devices and the controller. In addition, OLSR version 0.6.8[16] is used for running a routing protocol. The hello interval of OLSR is kept as 2 s and the neighbor hold time is 40 s. Hence, OLSR detects the failure in communication in 40 s.

OLSR is used as a routing protocol in our emulation. However, any other routing protocol such as AODV (Ad hoc on Demand), DSR (Dynamic Source Routing) and GRP (Geographical Routing Protocol) could be used in our automatic configuration method. The decision on which routing protocol to choose depends on the delay, throughput, routing overhead requirements. The detail on the performance comparison of different routing protocols can be found at [38].

A separate CPUs are dedicated to each emulated IoV device. Three types of IoV topologies are emulated: (1) linear, (2) sparse and (3) dense network. In the linear network, 20 ad hoc car vehicles are connected with each other in a linear fashion and connected with the controller through an RSU access point. The distance between a car vehicle and its neighbour is 40 m and the radio range of each car vehicle is 74 m. In the sparse network, 20 car ad hoc vehicles are deployed over 250×250 meter square area. However, in the dense ad hoc network, 20 car vehicles are deployed over 125×125 m square area. All other parameters like radio range are the same as the linear network. Like the linear IoV network, car ad hoc vehicles connect with the controller through 1 RSU access point.

The OpenFlow session time (i.e., the OpenFlow automatic configuration time) is calculated to show the effectiveness of the proposed automatic configuration method. In addition, movement experiments are performed. In these experiments, car vehicles are moved to a different location and the OpenFlow session re-connection time is calculated.

---

[12]https://github.com/intrig-unicamp/mininet-wifi.

[13]http://mininet.org/.

[14]https://www.openvswitch.org/.

[15]https://github.com/noxrepo/pox.

[16]http://www.olsr.org/?q=download.

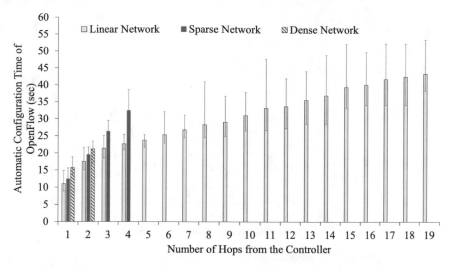

Fig. 14 Automatic configuration time (seconds)

Furthermore, to show that low-level data can be collected from IoV devices, the number of packets successfully matched with Flow Entries is collected from IoV devices through the controller and the time of data collection is calculated.

9.2 Emulation Results

Figure 14 shows the automatic configuration time of an IoV device which is located at a certain hop (hop 1 to 19) away from the controller. This time is calculated at the initial setup time. This time includes: (1) the time to run OpenFlow software (Open vSwitch and POX controller), (2) the OLSR configuration time and running time, and (3) the time to run the proposed automatic configuration method. The automatic configuration time is calculated for all the considered topologies: linear vehicular, sparse vehicular and dense vehicular networks. If there is no device at a certain hop from the controller in a particular topology, no result is shown in Fig. 14 for that topology of the vehicular network topology. For example, there is no result in the sparse vehicular network in Fig. 14 after hop 4 and there is no result in the dense vehicular network after hop 2. All the results are calculated 50 times and the minimum, average and maximum values of the automatic configuration time are shown in Fig. 14.

Figure 14 shows that the first-hop IoV device takes a significantly long time (from 10 to 15 s) to configure OpenFlow. This is because it also includes the time run appropriate software such as Open vSwitch, OLSR, POX, etc. After the first hop, the automatic configuration time increases either linearly or non-linearly as shown in Fig. 14. The automatic configuration time in the sparse vehicular network is shorter

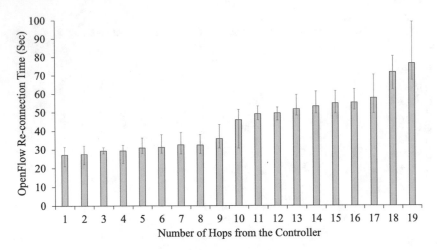

Fig. 15 The OpenFlow re-connection time

than in the dense vehicular network. This is because in the dense network, a larger number of nodes is available at a certain hop to configure (i.e., a large number of OpenFlow session paths needs to be established for the certain hop). Furthermore, the automatic configuration time is shorter in the linear vehicular network than in the sparse vehicular network.

Figure 15 shows the OpenFlow re-connection time when the path between the controller and an IoV device changes and the controller disconnects from the IoV device. This happened, as IoV devices moved from one location to another. OpenFlow re-connection time is calculated for a linear vehicular network, as described in the previous subsection.

In the mobility scenarios, all the nodes move from one location to another. Therefore, all the devices along a path to the controller have to reestablish an OpenFlow session with the controller. If an IoV device arrives at a location that is a few hops away from the controller, an OpenFlow path establishment takes a short time.

Figure 16 shows the data collection time. In this experiment, packet statistics such as the number of packets successfully matched with Flow Entries of an IoV device are requested from the controller. It shows the data collection time from hop 1 to 18. As the number of hops increases, the data collection time increases, as the data has to travel a long distance

Future experiments will focus on running a machine learning application on top of the controller and use the data collected from SDN enabled IoV to predict future conditions. One of such experiments will be focused on detecting anomalies in IoV according to the method given in [23].

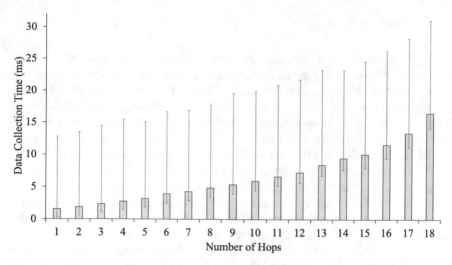

Fig. 16 Data collection time

10 Open Challenges for SDN in IoV

The followings are important open challenges for SDN in IoV:

10.1 Transition from Non-SDN IoV to SDN IoV

As presented in Sect. 3, SDN has a number of advantages to IoV. Therefore, several researchers are currently finding ways to deploy SDN into IoV [5, 6, 16]. The transition from a non-SDN IoV to a completely SDN IoV may take a long time. Hence, during this time, SDN IoV devices may need to communicate with non-SDN devices in a network. Using hybrid SDN IoV devices, it is now possible to communicate between non-SDN devices and SDN devices. The current open challenges in this respect are:

1. How many non-SDN IoV devices (e.g., traditional RSU and vehicles) need to be upgraded to use SDN to get the benefits of SDN?
2. How important is the actual placement of the IoV devices that have to be upgraded to SDN?

The answers to the above questions will have a huge impact on the transition from non-SDN to SDN IoV.

10.2 Performance

Using SDN, control plane decisions are moved to external servers, i.e., to controllers and hence, this may delay the decisions to be mapped in IoV devices. Furthermore, as the number of IoV devices increases, more events/requests will be sent to the controller and therefore, it may not be able to handle all the requests. There are the following two options to solve the above problems: (1) inserting proactive forwarding entries and (2) deploying multiple controllers [7]. The challenge with the above options is that additional resources are required to deploy the above solutions in IoV.

Another open challenge is how many devices of IoV can be controlled through a single controller, meeting all requirements (e.g., latency, throughput, and reliability requirements). For a small IoV network, a single controller may be enough to control all the devices of a network. However, as the network size increases, a single controller may not be enough to control all the devices.

The next challenge, which can affect the performance, is that where to place the controller? Currently, much of the research work on the side of the controller placement problem is available as an optimization problem [37]. The problem here is that these optimization algorithms take a very long time to provide a solution.

10.3 Mobility

Compared to MANETs, in IoV/VANETs, mobility is limited, as vehicles have to move through the roads only. However, the high speed of vehicles is a challenge for IoV/VANETs in SDN. Due to high speed, vehicles can move from one location to another in a short time. It can cause rapid changes in IoV topology and fluctuations in radio channels. The frequent such changes can also prevent the real-time collection of data. In addition, it may be that IoV devices have to re-establish a session with a new controller, as different controllers could be deployed in different locations. Therefore, there is a need for a seamless transition from one controller to another. Furthermore, a large amount of data may be needed to be exchanged between one controller to another. The challenge is also where to store that data such that the transition from one controller to another takes place smoothly.

Moreover, with high speed, there are more chances that conditions of the network (e.g., location, neighboring devices) may also change in a short time. Therefore, every time the condition changes, data is needed to be recollected (e.g., by the controller to make AI decisions) to reflect new conditions of the network. Hence, there are the following open challenges for AI assisted SDN enabled IoV: (1) which data is needed to be collected, (2) when the data should be collected and (3) where the data should be stored.

10.4 Security

Non-SDN IoVs have natural protection against security threats. This is because of the closed (proprietary) nature of IoV devices, their fairly static design, and the decentralized nature of the control plane. A common SDN standard (e.g., OpenFlow) among vendors increases the risk of threats, by the possible introduction of security attacks in SDN enabled IoV.

OpenFlow defines a transport layer security (TLS) session between OpenFlow devices and the controller. The same TLS session could be used to collect different types of data in OpenFlow. However, the problem is that OpenFlow does not provide any details of TLS operations. This could lead to interoperability issues [17]. Furthermore, the OVS-DB SDN protocol can also be used on top of the TLS protocol. However, as lots of data could be collected using SDN, there is still an open challenge for SDN on where and how to store the data such that an attacker should not be able to attack it.

Furthermore, there are several security concerns related to SDN enabled VANETs/ IoV from the data plane, controller and the application layer side. In [39], several such concerns related to DoS, DDoS, forwarding rule conflicts, privacy attacks and forgery attacks are presented. Future work could be focused on mitigating such attacks.

11 Conclusions

This chapter is focused on enabling Software Defined Networking (SDN) in the Internet of Vehicles (IoV). More particularly, the chapter is focused on presenting the benefits of using SDN in IoV, providing SDN enabled IoV architecture and applying a method to automatically configure SDN in IoV. The automatic configuration method is applicable for the networks where some IoV devices such as vehicles and pedestrian personal devices do not have Internet connectivity to reach the controller. In addition, experimental results on a European testbed are provided using a wireless SDN emulator. The results showed that OpenFlow can be automatically configured in IoV in a short time (i.e., in 60 s for a network of 20 vehicles). Furthermore, low-level data can be collected from IoV devices and it may be useful to predict future conditions using an AI application. Furthermore, AI-based architecture and open challenges of SDN enabled IoV related to performance, security, mobility and transition from non-SDN IoV to SDN IoV are provided.

Currently, there is also significant interest in applying Network Function Virtualization (NFV) in IoV [40]. Network Function Virtualization (NFV) refers to the process of running network functions (e.g., router, firewall, load-balancer, etc.) in virtualized IT infrastructures as softwarized Virtual Network Functions (VNFs). SDN and NFV are related terms, they are not really dependent on each other. Both the technologies help in achieving flexibility, automated network management and efficient network resource orchestration, and are envisioned as a key enabler to future IoV.

AI-assisted solutions will be useful in SDN as well as in NFV. Currently, AI-assisted solutions provide a black-box approach to solve a particular problem [41]. Currently, researchers are also working towards making AI explainable to users [42]. This will be useful in IoV, as decisions taken in IoV can be very critical and explanations of a solution given by an AI application would be useful in selecting the solution for a particular decision.

References

1. Hamid, U.Z.A., Zamzuri, H., Limbu, D.K.: Internet of vehicle (IoV) applications in expediting the implementation of smart highway of autonomous vehicle: a survey. In: EAI/Springer Innovations in Communication and Computing book series (EAISICC) (2018)
2. Sharma, S., Kaushik, B.: A survey on internet of vehicles: applications, security issues & solutions. Veh. Commun. **20** (2019)
3. Jiacheng, C., Haibo, Z., Ning, Z., Peng, Y., Lin, G., Xuemin, S.: Software defined Internet of vehicles: architecture, challenges and solutions. J. Commun. Inf. Netw. **1**(1), 14–26 (2016)
4. Tuyisenge, L., Ayaida, M., Tohme, S., Afilal, L.: Network architectures in Internet of vehicles (IoV): review, p. 11253. Lecture Notes in Computer Science book series, vol, Protocols Analysis, Challenges and Issues (2018)
5. Zhuang, W., Ye, Q., Lyu, F., Cheng, N., Ren, J.: SDN/NFV-empowered future IoV with enhanced communication. Proc. IEEE **108**(2), 274–291 (2020)
6. Mahmood, A., Zhang, W.E., Sheng, Q.Z.: Software-defined heterogeneous vehicular networking: the architectural design and open challenges. Future Internet, **11**(17), 70 (2019)
7. Sharma, S.: Towards High Quality and Flexible Future Internet Architectures. Ghent University, Faculty of Engineering and Architecture, Ghent, Belgium (2016)
8. Sharma, S., Nekovee, M.: Demo abstract: a demonstration of automatic configuration of openflow in wireless ad hoc networks. In: IEEE INFOCOM 2019 - IEEE Conference on Computer Communications, pp. 953–954 (2019)
9. Sharma, S., Nag, A., Stynes, P., Nekovee M.: Automatic configuration of OpenFlow in wireless mobile ad hoc networks. In: The 17th IEEE International Conference on High Performance Computing & Simulation, pp. 367–373 (2019)
10. Fantian Z., Chunxiao L., Anran Z., Xuelong H.: Review of the key technologies and applications in internet of vehicle. In: 13th IEEE International Conference on Electronic Measurement & Instruments (ICEMI), Yangzhou, pp. 228–232 (2019)
11. Kurose, J.F., Ross, K.W.: Computer Networking: A Top-Down Approach Featuring the Internet. Addison-Wesley, Boston (2001)
12. Casado, M., Koponen, T., Shenker, S., Tootoonchian, A.: Fabric: a retrospective on evolving SDN. In: The First Workshop on Hot Topics in Software Defined Networks (HotSDN 2012) (2012)
13. McKeown, N., Anderson, T., Balakrishnan, H., Parulkar, G., Peterson, L., Rexford, J., Shenker, S., Turner, J.: OpenFlow: enabling innovation in campus networks. SIGCOMM Comput. Commun. Rev. **38**(2), 69–74 (2008)
14. OpenFlow Management and Configuration Protocol (OF-Config 1.1.1). ONF Documents (2013)
15. Pfaff, B., Davie, B.: The Open vSwitch Database Management Protocol. RFC 7047 (2013)
16. Alouache, L., Nguyen, N., Aliouat, M. Chelouah, R.: Toward a hybrid SDN architecture for V2V communication in IoV environment. In: Fifth International Conference on Software Defined Systems (SDS), pp. 93–99 (2018)
17. Sharma, S., Staessens, D., Colle, D., Pickavet, M., Demeester, P.: In-band control, queuing, and failure recovery functionalities for openflow. IEEE Netw. **30**(1), 106–112 (2016)

18. Rademacher, M., Siebertz, F., Schlebusch, M., Jonas, K.: Experiments with OpenFlow and IEEE802.11 point-to-point links in a WMN. In: Twelfth International Conference on Wireless and Mobile Communications, pp. 100–105 (2016)
19. Tuan, C., Wu, Y.: IP address exchanging scheme for vehicle ad hoc networks. Adv. Intell. Syst. Appl. **1**, 409–418 (2013)
20. Munjal A.: Address Auto-Configuration Protocols and their message complexity in Mobile Adhoc Networks. PhD Dissertation, IIT Kanpur, India (2015)
21. Blial, O., Mamoun, M., Benaini, R.: An Overview on SDN Architectures with Multiple Controllers. J. Comput. Netw. Commun. 8 (2016)
22. Hu, T., Guo, Z., Yi, P., Baker, T., Lan, J.: Multi-controller based software-defined networking: a survey. IEEE Access **6**, 15980–15996 (2018)
23. Abdelkefi, A., Jiang, Y., Sharma, S.: SENATUS: an approach to joint traffic anomaly detection and root cause analysis. In: 2nd Cyber Security in Networking Conference (CSNet), Paris, pp. 1–8 (2018)
24. Smida, K., Tounsi, H., Frikha, M., Song, Y.: Software defined internet of vehicles: a survey from QoS and scalability perspectives. In: 15th International Wireless Communications & Mobile Computing Conference (IWCMC) (2019)
25. Kadhim, A.J., Hosseini, S.A.: Maximizing the utilization of fog computing in Internet of vehicle using SDN. IEEE Commun. Lett. **23**(1), 140–143 (2019)
26. He, X., Ren, Z., Shi, C., Fang, J.: A novel load balancing strategy of software-defined cloud/fog networking in the Internet of vehicles. China Commun. **13**(Supplement 2), 140–149 (2016)
27. Kadhim, A.J., Seno, A.A.H., Shihab, R.A.: Routing strategy for Internet of vehicles based on hierarchical SDN and fog computing. JUBPAS **26**(10), 309–319 (2018)
28. Siddiqui, A.J., Boukerche, A.: On the impact of DDoS attacks on software-defined internet-of-vehicles control plane. In: 14th International Wireless Communications & Mobile Computing Conference (IWCMC), pp. 1284–1289 (2018)
29. Zio, E.: Critical infrastructures vulnerability and risk analysis. Eur. J. Secur. Res. **1**, 97–114 (2016)
30. Garg, S., Singh, A., Aujla, G.S., Kaur, S., Batra, S., Kumar, N.: A probabilistic data structures-based anomalydetection scheme for software-defined internet of vehicles. IEEE Trans. Intell. Transp. Syst. (2020)
31. Singh, P.K., Jha, S.K., Nandi, S.K., Nandi, S.: ML-based approach to detect DDoS attack in V2I communication under SDN architecture. In: TENCON IEEE Region 10 Conference, pp. 0144–0149 (2018)
32. Gao, J., Agyekum, K.O., Sifah, E.B., Acheampong, K.N., Xia, Q.: A Blockchain-SDN-enabled internet of vehicles environment for fog computing and 5G networks. IEEE Internet of Things J. **7**(5), 4278–4291 (2020)
33. Liang, L., Ye, H., Li, G.V.: Toward intelligent vehicular networks: a machine learning framework. IEEE Internet Things J. **6**(1), 124–135 (2019)
34. Chen S., Dong D., Nai W., Zheng W.: Research on wireless access in vehicular environment based on IEEE 802.11p. In: The Twelfth COTA International Conference of Transportation Professionals (2012)
35. Bock, H., Morais, R.M., Pedro, J., Krombholz, B.S., Sadasivarao, A., Syed, S., Paraschis, L., Kandappan, P.: Coming of age of AI-assisted network management & control. OSA Advanced Photonics Congress (2020)
36. Majd, L., Levent, T.: Artificial intelligence enabled software-defined networking: a comprehensive overview. IET Netw. **8**(2), 79–99 (2019)
37. Lu, J., Zhang, Z., Hu, T., Yi, P., Lan, J.: A Survey of Controller Placement Problem in Software-Defined Networking. IEEE Access, **7**, pp. 24290–24307 (2019)
38. Rivoirard, L., Wahl, M., Sondi, P., Berbineau, M., Gruyer, D.: Performance evaluation of AODV, DSR, GRP and OLSR for VANET with real-world trajectories. In: 2017 15th International Conference on ITS Telecommunications (ITST), Warsaw, pp. 1–7 (2017)
39. Arif, M., Wang, G., Geman, O., Balas, V.E., Tao, P., Brezulianu, A., Chen, J.: SDN-based VANETs, security attacks, applications, and challenges. Appl. Sci. **10**, 3217 (2020)

40. Zhuang, W., Ye, Q., Lyu, F., Cheng, N., Ren, J.: SDN/NFV-empowered future IoV with enhanced communication, computing, and caching. Proc. IEEE **108**(2), 274–291 (2020)
41. Nekovee, M., Sharma, S., Uniyal, N., Nag, A., Nejabati, R., Simeonidou, D.: Towards AI-enabled microservice architecture for network function virtualization. In: 2020 IEEE Eighth International Conference on Communications and Networking (ComNet), pp. 1–8 (2020)
42. Sharma, S., Nag, A., Cordeiro, L., Ayoub, O., Tornatore, M., Nekovee, M.: Towards explainable artificial intelligence for network function virtualization. In: Proceedings of the 16th International Conference on Emerging Networking EXperiments and Technologies (CoNEXT 2020), pp. 558–559. Association for Computing Machinery, New York (2020)

IoV with ML/DL Technologies

Machine Learning Technologies in Internet of Vehicles

Elmustafa Sayed Ali, Mona Bakri Hassan, and Rashid A. Saeed

Abstract Recently, there was much interest in Technology which has emerged greatly to the development of smart cars. Internet of Vehicle (IoV) enables vehicles to communicate with public networks and interact with surrounding environment. It also enables vehicles to exchange information in addition to collect information about other vehicles and roads. However, actual applications of smart IoV systems face many challenges. These challenges are related to different problematic issues like big data connection with IoV, cloud network, data processing, and efficient communication between a large amount of different vehicles types, in addition to optimum decision data processing on or off board. Intelligence of the huge amount of data that can be processed to reduce road congestion and improve traffic management, as well as ensuring road safety is an important issue in future IoV trends.

Artificial Intelligence (AI) technology with Machine Learning (ML) mechanisms offers smart solutions that can improve IoV network efficiency. For example, decision for data processing at various layers i.e. on-board units (OBUs), Fog level or cloud level are one of the problems which need ML algorithms. Other critical issues that can be resolved by ML mechanisms are time, energy, rapid topology of IoV, optimization quality of experience (QoE) and channel modeling. These issues need to be optimized. This chapter provides theoretical fundamentals for ML models, algorithms in IoV applications and future directions.

Keywords IoV · Machine learning · V2E · Deep learning · Optimization QoE · Autonomous driving · Smart transportation

E. S. Ali (✉) · M. B. Hassan · R. A. Saeed
Department of Electronics Engineering, Sudan University of Science and Technology, Khartoum, Sudan

E. S. Ali
Department of Electrical and Electronics Engineering, Red Sea University, Port Sudan, Sudan

R. A. Saeed
Department Computer Engineering, Taif University, Al-Taif, Kingdom of Saudi Arabia

© Springer Nature Switzerland AG 2021
N. Magaia et al. (eds.), *Intelligent Technologies for Internet of Vehicles*, Internet of Things, https://doi.org/10.1007/978-3-030-76493-7_7

1 Introduction

The rapid development of intelligent transportation systems (ITS) and computational systems opened new scientific researches in intelligent traffic safety. These systems provide comfort and efficiency solutions for smart vehicles applications. Artificial intelligence (AI) has been widely used for optimizing traditional data-driven approaches in different smart vehicles research areas [1]. AI based vehicle to everything (V2X) system obtains information from various sources i.e. cars, trains, buses, and etc. It enables to increase realization of drivers and forecast to avoid accidents. This progression has directed the opportunity of understanding smart driving which built on idea of copying real driving comportment. The real driving behavior artificial learning helps to avoid human mistakes and bring comfortable, safety to drivers. Many related services have been invented to issues of crowd and light road traffic. These services enable to adapting traffic, legacy from self-based vehicles systems to Internet of vehicles (IoV) [2]. IoVs is addressed to change an interaction between vehicles, roadside stations, on-board stations and environments. It enables vehicles to communicate data and multimedia between various networks.

Machine learning (ML) is responsible for wide range of AI applications. There are many ML techniques used as methods for different AI solutions, such as unsupervised, supervised and reinforced learning. In unsupervised ML scheme, training depends on untagged data. It tries to find an effective representation of unlabeled data. While in the supervised learning, it learns from a group of labeled data. In supervised learning, regression and classification schemes are used to training the discrete and continuous data for prediction and decision making. The reinforcement learning (RL) studies the activities of learning agent from consistent reward in order to capitalize the notion of cumulative rewards. Markov decision process (MDP) is sample of RL [2, 3]. MDP is a perfect technique for taking many issues research problems in vehicular networks. It presents different solutions such as collaborative optimization of oil consumption for a specific area, optimum path forecasting of electric vehicles and minimize in traffic congestions. Moreover, the use of ML in IoV promises to solve different related issues such as, traffic prediction based path, data routing, vehicular block chain, congestion, load balancing, cyber-physical attack mitigation, resource management based energy efficiency. Other considerations of ML based IoV applications are related to critical challenge for IoV during multimedia communication. In critical applications i.e. healthcare, quality of experience (QoE) optimization requires to manage mobility of wireless channel between vehicles. ML based approaches have entirely changed the landscape of IoVs. It enables portable devices for transmitting multimedia content in IoV system, to end users in their respective fields.

Given the importance of artificial intelligence (AI) use in IoV, and according to ML powerful benefits, ML provides smart models in most of IoV applications. This chapter contributes a brief concept about machine learning methods and possibility of its use in several specific aspects related to IoV. The contributions of this chapter is organized as follows; Sect. 2 presents chapter background, in addition to show the motivation behind conducting review of using ML in internet of vehicles. The chapter reviews a brief concept about AI in IoV in Sect. 3, considering use of AI in multimedia, IoV edge based and vehicles to every thing's internet communications. In Sect. 4, chapter provides a clear concept about contribution of AI to enabling QoS and QoE optimization. Section 5 provides a detailed description of using machine learning algorithms with IoV in different aspects. Most common use cases of ML in IoV applications are presented in Sect. 6. Section 7 gives a brief review about possible future research directions and tentative solutions related to ML in IoV. Finally, chapter conclusion is given in Sect. 8.

2 Background and Motivation

Due to significant research and technology development in wireless communication, traditional intelligent transport system (ITS) has care about vehicular communication field. Recently, numbers of vehicles are increased due to transport huge number of people from region to region. This increment in number of vehicles would create issues such as crowding and accidents on the roads. These issues could be considered as one of the main problems in daily life. Most of general form of vehicular networking is known as vehicular ad hoc network (VANET) [4]. VANET consist of vehicle to vehicle (V2V) and vehicle to roadside (V2R) communications to transfer information between vehicles. The VANETs communication depends on road side unit (RSUs) to support wireless access in vehicular environments (WAVE). The road side units (RSUs) along the road work as wireless access points to support communication to vehicles inside its coverage area [5]. The hybrid vehicular network architecture interacted to cellular communication architectures will perform operation to cellular communication services i.e. voice in collaborations way. Due to current trend to connect vehicular networks to information centers, and need to exchange data, IoV allows to enable internet access among on road vehicles. One of the most important IoV applications is to improve the features of VANETs in order to reduce various issues in urban traffic and accident environment [6]. IoV enables vehicular road networks to interconnection with different wireless network technologies i.e. Wi-Fi and 4G/LTE for V2I, IEEE WAVE for V2V and V2R, MOST/Wi-Fi for V2S. This progress occurred by intelligent transportation system (ITS) to enhance traffic monitoring system environment, and to reduce accident for improving travel roads [7].

In recent years, need arises to introduce artificial intelligence technologies in IoV applications to face some challenges. These challenges are related to questions of how to make special decisions and forecast for different aspects related to IoV such as, traffic monitoring and management, big data processing, energy and resource management, intelligent interaction with users to provide high quality services [6, 7]. Several studies have been done on how to use artificial intelligence techniques in smart vehicle applications. Machine learning used to develop solutions to most of these challenges [8]. Due to current developments in field of AI, especially in machine learning techniques, intelligent decisions in IoV applications can be optimized for different related issues and aspects. It is useful to provide a comprehensive presentation to study some concepts about use of ML in IoV, in addition to explain areas that could contribute to development of these networks. This is one of the most important motivations for which this chapter was written.

3 AI in IoV Network

AI technology is more related to the layer that responsible for presentation and functionalities in IoV layered architecture. This layer can be described by a term of virtual cloud infrastructure, and responsible for storing, processing, analyzing information received from IoV network, and for decision making based on analyzed information. In IoV, computation and analysis are provided by big data analysis (BDA) and vehicular cloud computing (VCC) systems which they are used as an information management center [9]. According to IoV applications, many services can be provided by IoV cloud environment which is requires an intelligent service management. The smart cloud computing servers provide many smart services i.e. safety traffic administration, entertaining and subscriptions, which are the foundations of elegance in IoV. The cloud servers based on AI enable to procedure and develop AI in real time (RT) huge data traffic. AI based cloud servers provide a smart decision for intelligent customer services. Vehicular cyber physical system (VCPS) is considered as vehicular network model that concern dissemination of information using next generation Internet [10]. The VCPS is depends on AI technology to provide smart processing in huge data traffic utilizing fog and cloud computing for civilian and safety applications respectively.

In IoV networks, the edge computing and caching problems are most considered challenges which are need an intelligent optimization method. Edge computing and caching challenges are related to many factors, i.e. channel condition, dynamic communication topology and resource allocation management. AI in IoV provides an intelligent approach to solve most of these challenges. The use of ML provides a means of interaction to IoV environment, and enable to create an agent learns challenges factors to optimize overall IoV network utilization [11]. Q-learning and deep neural networks are ML algorithms, developed to make decisions according to learned IoV resource actions. In IoV network architecture, presentation of artificial intelligence in a separated layer is responsible for virtual cloud infrastructure. AI layer act as an information management brain [10, 11]. AI layer in IoV architecture consists of big data analysis, cloud computing, and expert systems. Its play an important task to store processes and analyzes information received from coordination layer and takes decisions according to network status.

3.1 AI for IoV Multimedia Communication

The deployment of IoV in multimedia communications requires a mobile device that allows data exchange and communication with other devices. In addition, implementation of IoV in multimedia communications needs interaction between personal area networks, IoT technologies, sensors, drives, and connectors in a wireless sensor network (WSN). Scalable and flexible in data transfer is quite important for IoV by integrating sensors, vehicles, humans, actuators, machines, etc. The sensor in intelligent IoV help to enhance safety of vehicle and traffic systems, while harmonized traffic data transfer in IoV system network enhanced vehicular system efficiency. Due to consumption of energy, capacity and buffer as well as heavy data exchange, it probable to make great compromise risk of quality as in health media data transfers [12]. AI based on self-driven vehicles encourages deploying several types of applications with many benefits of intelligence. AI enables to avoid low quality risks, especially when amount of data and complexity increases, as machine learning algorithms help to provide accurate and highly effective solutions for future trends. As growing of high traffic of information in IoV, it required a smart utility, tendencies and follows to efficiently monitor and manage demand of smart IoV technologies [13]. With rapid revolution in digital technologies, IoVs multimedia development relies on a mobile device to collect a massive amount of information to assist and guide the specific trend. In addition, development on IoVs multimedia depends to possibility of analyzing the transportation industry through IoT based platforms. As an example, the structure of multimedia communication can be deployed through sensor nodes in IoV system (see Fig. 1). The structure consists of three main parts for IoV data and information network techniques and models.

In the given example, data and information network techniques and models are developed with main server. The inter and intra-vehicle network connections amongst various sections is executed by transferring urgent data and sensitive throughout vehicle via adaptive and smart wireless communication. The vehicles client enables QoS monitoring [12, 13]. In this structure, IoV traffic can be arranged based on to category contain sensitive/normal, pre-stored, real time, or high definition resolution, respectively. For accomplishing real time and jitter tolerant data and information exchange with low buffer storage and scarce power supply, it should be fortified to tolerate the raw unprocessed data and information into regular and synchronized format with good and clean visibility.

3.2 Intelligent IoV Edge Based Algorithm

Mobile edge computing (MEC) is a new technology which enables communications with cloud computing to deliver cloud services directly from network edge and support delay critical mobile applications. It could be achieved by placing computer servers at radio access points or base stations. AI in edge caching and computing

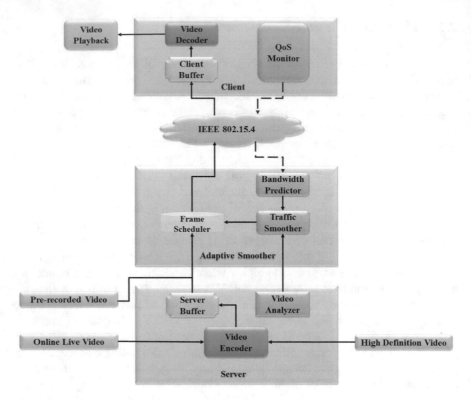

Fig. 1 IoV multimedia communication models

platforms trains and deploys powerful ML models at the edge servers and mobile devices. Edge AI techniques change the structure of semiconductor industry [14]. In IoV edge information system (EIS) plays a vital and unique role. It is able to help the key functionalities of intelligent vehicles, from data acquisition, data processing, to actuation. Data processing in network edge can satisfy the low latency requirement for mission critical tasks and save an amount of communication bandwidth. The AI edge based IoV serve typically high spatial locality for road conditions, map information, and temporal locality for traffic conditions. On the other hand, with big sensing data, intelligent vehicles are facing tremendous computation burdens [12–14].

Offloading computation and load balancing are most important factors determine the maximize system utility in IoV. Cooperative edge caching and edge computing can serve to improve the performance of these factors. But indeed, the edge computing and caching policies are of limited in dynamic systems applications such in IoV networks. AI cognitive capability help to develop edge cognitive computing architecture which enable to provide dynamic computing service [15]. AI cognitive will improve energy efficiency and user experience, since it able to interacts with other IoV components to perform efficient resource management (see Fig. 2). IoV architecture-

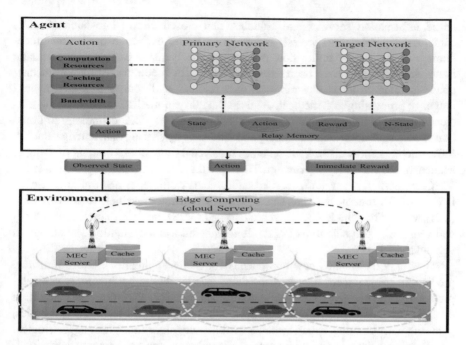

Fig. 2 IoV edge based AI architecture

based AI algorithms enable to perceive real time behavior of vehicular environment information, by interacting to environment according to current state related to offloading, cooperative caching and edge computing [14, 15].

The IoV edge-based AI architecture able to intelligently drive the edge computing resources in a manner depends on cooperative caching to efficiently managing the edge computing policies. Such edge-based AI architecture can use deep ML algorithm for efficient IoV resource management. Other considerations related to system utility are IoV network mobility and vehicles/RSUs handover mechanisms. These considerations are major factors that greatly affect temporary storage resources [14]. Therefore, it is necessary to trade-off between accuracy of prediction, temporary storage of content on move, and handoff implementation. AI enables to prediction of handover and intelligently share allocated bandwidth and edge caching.

3.3 AI for Vehicle to Everything

AI in vehicular applications enables to execute tasks intelligently, such as enhanced plug-in electric vehicle (PEV) charge, minimize fuel consumption, enhancing location based on services as well as traffic congestion rectification. The traffic flow information can be obtained from multiple sources as induction loops, crowd souring based information services and vehicles, and closed circuit television (CCTV) cameras [16]. Modeling precise and accurate traffic exchange prediction procedures utilizing legacy traffic flow predication mechanisms is a vital problematic issue. AI techniques have been extensively used for modeling estimation mechanisms in research areas like robotics, data science, and computer vision, natural language processing and medical. AI used data driven method facilitate more efficiency to tackle little and multimedia data. The aims of V2X technology to transportation system are to enhance safety and efficiency by sharing data among vehicles, infrastructures and walker. V2X schemes received a tremendous amount of using in academia, industry and government. There are three fundamental aspects related to V2X communication system, i.e. road safety, energy efficiency and traffic efficiency [17]. The V2X scheme based on sharing information among vehicle to infrastructure (V2I), vehicle to vehicle (V2V), vehicle to self (V2S), vehicle to pedestrian (V2P) and vehicle to road side units (V2R). It considered as an evaluation technology to vehicular networks.

AI with V2X enables new approaches for many applications like traffic flow prediction, management for real time data, location based applications, vehicular platoons, data storage in vehicles, autonomous transport facilities and congestion control. The most widely utilized AI techniques are heuristic techniques, robotics, game theoretic learning, expert systems, evolutionary algorithms, turning test, logical AI, planning, schedule and optimization, natural language processing, swarm intelligence, inference, fuzzy logic, machine learning and etc. [18]. One of ML based V2X applications is autonomous driving, where AI used to enable basic features of human driving. ML in V2X play an important role to enhance both safety and efficiency in vehicle networks [17]. It has been widely applied by modern machines for applications such as competing at highest level in strategic games, autonomous vehicles, understanding human speech and intelligent network routing in content delivery networks.

4   AI enabled QoE/QoS Optimization

In internet of vehicles, quality of experience (QoE) provides measurement deals with network performance and perception, in addition to IoV application experience. QoE considers IoV experience to ensure high quality of data transmission provided through continuously measure QoE of network and update. Low connection quality is considered as one of the major challenges in IoV applications, especially for IoV end users [19]. In addition, flexible and scalable connection between integrated components of IoV system i.e. vehicles, sensors, actuators, humans, machines are very vital for IoV, which must fit with requirement of user perception enhancement and decrease power consumption. Moreover, to improve safety of transportation system and traffic data exchanging in vehicular network, power aware and buffer aware QoE/QoS via IoV will optimize transportation system especially for high risk of quality compromise during sensitive IoV applications like in medical field. Quality of service (QoS) in IoV is related to routing paths quality, impact of velocity and position of vehicles in addition to network topology. These aspects mainly effect on IoV energy efficiency [20]. The QoS optimization with energy efficiency regarding to IoV network efficiency is regard to develop a solution to multi-attribute decision making, and able to optimize many IoV network operations.

AI techniques have changed landscape of IoV through multimedia communication. It helps to improve overall IoV network by optimize route selection in an efficient way to obtains a stable transmitting multimedia content in IoV system [20, 21]. AI also helps to develop energy and buffer aware optimization mechanisms to optimize QoE and QoS during multimedia communication in IoV system. Machine Learning (ML) techniques can provide a framework to analysis QoE services with high level of optimization. Its help in assessing and examining faults and quality degrading factors prospected from important collected information by IoV systems to enhance IoV user's satisfaction. In addition, ML helps to evaluate QoS by considering several impacts to IoV network related to communication, energy and resource management operations [21].

4.1   Buffer Aware QoE/QoS Optimization

Due to high demand of video traffic in IoV networks, a development of intelligent solutions must fulfill expectations and ensure maximum quality of experience (QoE). The optimizations of QoE during multimedia communication in IoV system can be obtained by deploy a novel algorithm based on buffer allocation mechanism. This mechanism enables to control high peak variable rate of multimedia by allocating proper buffer size in IoV. The buffer aware QoE optimization must consider requirements related to energy and video rate adaptation. In IoV applications based on video transmission, dynamic adapting coding rate of requested videos enable to ensure optimizing in QoE by encoding rate depends on video content itself. ML algorithms provide an automatic video processing with additional complexity given by temporal dimension of the data [22]. Different video processing schemes can be achieved by ML in pixel level or higher-level representations obtained after additional pre-processing of raw images. The ML schemes enable to optimize process of buffer allocation and dynamic video rate adaption.

In IoV networks, it is difficult to achieve QoS and efficiency for multimedia streaming specially in high mobility feature. Buffer aware streaming approach allows users to play multimedia streaming over IoV network. AI based buffer aware QoS adopted for vehicle streaming services will evaluate multimedia content that may be preloaded by IoV servers according to the user's mobility information. The use of buffer aware QoS streaming approach over IoV network will help to provide various priority levels of streaming service [23]. ML will evaluate direction and speed of vehicles mobility, strength of IoV signals, in addition to size of media content stored in buffer to optimize the quality of streaming service on IoV network.

4.2   Energy Aware QoE/QoS Optimization

Energy management in IoV systems is considered one of the most main challenges facing IoV applications. It's very important to effectively manage power resources during communication in IoV system [24]. In most of IoV applications, electric vehicles (EVs) charging and discharging time have critical impact on quality of experience (QoE). Power aware QoE optimization in vehicle to grid (V2G) networks; express degree of satisfaction with state of charge (SOC) and charging cost of using an EV. In charging scheduling, service of enough CSs is important QoE metric, especially in the peak charging hours [25]. AI based charging scheduling schemes must consider QoE optimization. The QoE of vehicles in IoV network with a higher vehicle's mobility and limited coverage area of road side unit (RSU) can be degraded and significantly affect in communication quality by decreasing percentage of flow satisfied. In addition, due to limited IEEE 802.11p based vehicular communication bandwidth, and because of needs to provide a fair share of network resources among vehicles, flow management will face a crucial quality problem [26]. Moreover, any growth of energy consumption in RSU leads to inefficient IoV network management. AI based energy and flow management schemes provide an intelligence decisions controller to overcome complexity of energy and traffic operation by providing efficient solutions.

5   ML Algorithms in IoV Network

Machine learning has different models, classification and training which have been widely used for prediction problems and intelligent managing. In IoV applications, reinforcement learning (RL) will provide a guidance behavior to promote resilience and scalability. It can provide path selection or route optimization in IoV networks. The uses of ML with software defined network (SDN) in IoV ensure delay minimization and throughput maximization as the operation and maintenance strategy. ML and SDN together will improve IoV network performance with stable and superior routing services [27]. They ensure optimal routing policy adaptation according to IoV environment sensing and learning to achieve better utilization. Figure 3 shows ML functions can be deployed in IoV networks. In the domain of IoV network security, ML with SDN brings some unique advantages to deployments of security solutions. For security issue, centralized control on software layer with API access will a convenient to develop ML software interact with SDN data plane to provide statistical reports to application layer upon vehicles requests [28].

In cognitive internet of vehicles (CIoV) applications such as automatic driving, the automation and connectivity are very important in self-driving aspects which they should be sufficient of intelligence to reduce road accidents. ML can take control of vehicles to enable error free driving. CIoV enables to deploy ML based cloud into transportation system for security risks and privacy issues [29]. In CIoV cognition and control layer, ML provides strategic services for different function levels i.e. driving

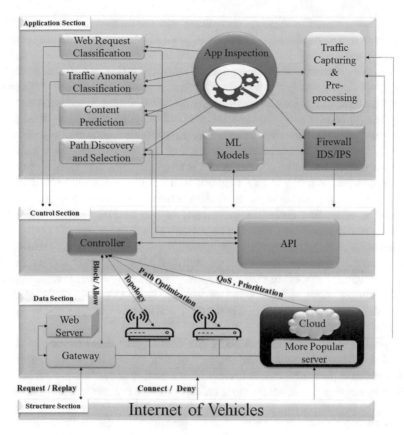

Fig. 3 ML functions in IoV network

behaviors and health monitoring, pattern and emotion analysis, in addition to network resource allocation and optimization. To improve driving safety and efficiency in IoV transportation system, deep learning (DL) schemes provide an intelligent decision-making to evaluate the important influential collisions probability factors and risk of possible accidents in IoV [30]. There are different DL techniques can be used for collision prediction and accident forecasting i.e. generic algorithms (GA), neural networks (NN), fuzzy logic and support vector machine (SVM).

5.1 ML Based Edge Caching Mechanisms for IoV

The operational excellence and cost efficiency in IoV are depending on caching and computing design. To efficiently improve QoS in IoV applications, edge caching placements and computing offloading at vehicles and road side units (RSUs) can ensure to guarantee efficient QoS. Machine learning provides schemes to tackle problems encountered in caching, computing and communications for IoV. Many ML schemes can be used for edge caching in IoV [31]. Supervised learning provides relatively good caching decisions, in addition to IoV traffic levels classification, prediction and content demand. Unsupervised learning can be applied to edge caching design by clustering numbers vehicles into different groups according to their behavioral and data request history information [32]. The ML based clustering scheme can predict data demand depends on interests or social relations of entire vehicles group.

The reinforcement learning scheme like Q-learning technique, will enable to distribute cache replacement strategy according to content popularity. Moreover, it can estimate the unknown popularity of caching contents. Integrated mobile edge computing (MEC) servers in IoV network will help to reduce workload at roadside stations. It also enable to make vehicle requesting content, perform data and computation offloading during its movement as like mobility-aware caching and computational scenario (see Fig. 4).The use of deep Q-learning will optimize the parameters of caching, computing for resource allocation. Deep Q learning will determine optimal actions from collected status of MEC and RSU servers, in addition to each vehicle's mobility, channel information, caching contents and computing [31, 32]. These actions are forwarded to vehicles. Deep Q learning will select best set of caching action for RSU, MEC and vehicles to serve requesting and to compute offloading tasks for IoV.

Integration of ML with edge caching has challenges related to data processing and analysis. The diffusion and high density of data are challenges for learning and training process. In addition, insufficient of computing resources in to manipulate the high dimensional information that cannot provide precise buffering decisions. To strongly cooperative ML at IoV network edge to enhance smart duties of edge, it requires an effective learning approaches for massive high dimensional information that be established in order to offer precise estimation of buffered information at IoV network edge. Moreover, ML schemes deployment in IoV applications will extract much sensitive and critical information, and if any leakage of information, it can

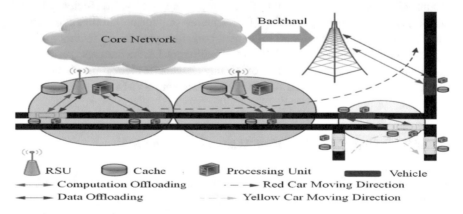

Fig. 4 IoV mobility-aware caching and computational scenario

cause serious confidentiality, security, and privacy concerns [32]. For these concerns, edge caching system must be secured by security and privacy preserving schemes and should be developed in different system levels, i.e. transmission/collection, data processing, data access, and storage levels for both edge network and vehicles.

5.2 Deep Reinforcement Learning Based Offloading Algorithm

The execution of computing-intensive applications on resource-constrained vehicles still faces a challenge related to offloading IoV system. Deep reinforcement learning (RL) will provide an intelligent offloading system for vehicular edge computing. The integration of deep RL with vehicular edge computing help to schedule offloading requests, in addition to allocate IoV network resources. Deep RL optimize scheduling and resource allocation in IoV to maximize QoE. In IoV, vehicles calculate utility values related to their available RSUs, and passed offloading requests to roadside stations. The stations perform task scheduling and resource allocation and inform the RSUs. RSUs enable to receive all vehicles offloading tasks to perform computation offloading. The deep RL algorithms help to optimize offloading decision by intelligent task scheduling. Figure 5 shows deep reinforcement learning based task offloading framework. Offloading requests scheduled in the task according to action-value function Q. RSU is selected by vehicles from the available accessing list with

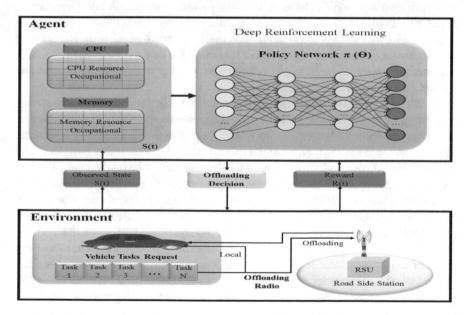

Fig. 5 Offloading decision optimization based deep reinforcement learning

probability ε and largest Q-value of the current action-value function [33]. The use
of deep RL in IoV offloading optimization will guaranteeing their venue of network
operators by ensuring cooperative offloading in IoV network, which will maximize
QoE of vehicles.

In IoV offloading computing, optimization parameters are related to offloading
ratio Γi for each task. The vehicles utility values constraints are related to limitations
of vehicles CPU and memory resource. Offloading optimization is depending on
how to minimize function task latency and energy costs [34]. Cost function can be
calculated as follows,

$$\text{Function} = \sum_{i=1}^{N} (L_i + w_i E_i) \tag{1}$$

Where, N represents number of tasks, L_i denoted for latency cost, and E_i for
energy cost is w_i denoted for weight ratio between latency and energy cost. Each
offloading decision is depending on resource unit time slot. This means, flow of
scheduling tasks is a sequence in time. For sequence of the N tasks arrives during a
limited observing time L_{obs}, cost function F can be calculated depending on reward
function R(t) by,

$$\text{Function} = \sum_{t=1}^{L_{obs}} \left(\gamma^{t-1} R(t) \right) \tag{2}$$

Where, $R(t)$ is reward of time slot t, and γ denoted for reward discount ratio
to describe affection of rewards of the future time slot on overall cost function.
This total cost function can be used to make policy network training. Deep neural
network will provide a policy for mapping from perceived states of IoV environment
to probabilities of actions to be taken. The policy network based deep RL training
will achieve an optimize computing via reward value of each time slot [34]. This
will minimize weighted sum of offloading latency and power consumption cost and
ensures offloading decision optimization.

In IoV based edge computing, vehicles act like clients connecting over edge
computing node in roadside without accessing to remote cloud. In this scenario,
offloading decision for heterogeneous resources is considered a complex operation.
This because, environment of vehicular edge computing is changing each time and
requires that offloading decision should be re-computed, which will make delay in
services providing. In addition, for vehicular service, task execution progress cannot
guarantee fairness offloading queuing. Deep RL provides a unique decision algorithm
to achieve intelligent vehicular controlled services based on edge computing model
[35]. It helps to learn service offloading knowledge, and observation functions related
to environment data of vehicular mobility and edge computing nodes. The offloading
decision model is trained at powerful edge computing nodes and distributes decision information to vehicles for services offloading. During decision model training,
vehicles transmit parameters to roadside station edge computing node for updating
basic offloading decision periodically.

5.3 ML for Dynamic and High Mobility IoV

IoV networks may have dynamics features in many aspects i.e. topology, traffic and wireless propagation channels due to its mobility. An efficient learn and dynamic prediction are required to provide a degree of optimization in routing, traffic load, and for assist the channel estimation module to track channel variations [36]. Machine learning (ML) methods lead to better results for modeling the dynamic changes of vehicular channels in addition to optimize vehicle routing and traffic flow. ML system integrated into RSUs help to estimate traffic patterns by collecting information about vehicles. ML can provide intelligent IoV routing protocol with critical information for highly dynamic environment. It able to predict network capability of paths to optimize vehicles route selection based on vehicles mobility and transmission capacity. In dynamic IoV, RSUs based ML can predict the vehicle moves and direction among them [37]. The prediction is depending on information provided by vehicle when it moves from RSU to another which will help RSUs to enable estimate traffic flows.

5.4 ML Based Decision Making in IoV

In recent years, autonomous vehicle (AV) growth generates a novel tendency to implement several intelligent approaches and methods to enhance efficient and quality of adaptive decision-making. The combination of AI, ML, RI and IoV offers high efficient control systems that can exploited in various applications to accommodate more adaptive, automatic and robust embedded systems [38]. Decision making in IoV networks requires intelligent algorithms to handle processes related to driving environment perception, path planning, strategy network control, and resource management. For intelligent driving vehicles system, a module integrates path, behavior, and motion planning, is needed to operate in highly optimized decision-making algorithm. In addition, decision making algorithm must take into account operations of vehicles control. It must able to predict and learn information related to vehicle platform faults, trajectory and energy [39]. These considerations are deal with vehicles platform as shown in Fig. 6. For cognitive driving decision making, localization, semantic understanding, and sensor fusion are more contribute in decision-making process.

Furthermore, intelligent vehicles and IoV systems applications face the decision-making challenges associated with collecting and distributing IoV big data to vehicles and interested users with the aim of enhancing road intelligence experience. In addition, making decisions related to traffic managing, road congestion and safety. Huge volumes of big data require a more powerful and intelligent mechanism in decision-making procedures to reduce road congestion, and improve traffic operations. Moreover, intelligent decision-making enables to get ride the challenges related to effective communication link between different types of vehicles and smart devices, security and privacy problems [40]. Many machine learning methods can used to

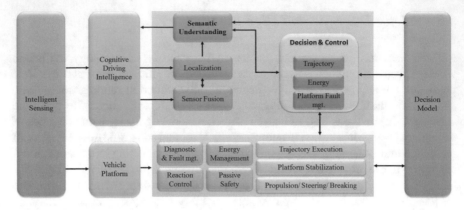

Fig. 6 Intelligent driving vehicles decision making framework

contribute in solving the above challenges. Where, these methods enable to model channels in different IoV network scenarios. In addition, it provides intelligent solutions to avoid road accidents by smart learning and analysis of driving environment using data collected from sensors. Since IoV networks are interested in exchanging messages everywhere and sharing content between smart vehicles [41]. ML based smart resource management for IoV networks has become extremely important to make decisions on policy of connection method of power control, selection, and resource allocation and assignment.

5.4.1 Network Control

Higher IoV network performance demands efficient solutions for network operation and optimization. ML in network domain will leverage the powerful of ML abilities for new network management in IoV applications. The capabilities of ML will provide an efficient way for intrusion detection and performance prediction. In addition, ML enables IoV network to make intelligent decisions for network scheduling and adaptation depends on network characteristic and environment [42]. ML algorithms will facilitate IoV network to classify and predict of traffic patterns and network states. In general, use of ML in communication networks promise many solutions for different networking aspects, i.e. data collection and analysis, clustering decision making and prediction, model construction a validation, in addition for network deployment and interference as shown in Fig. 7 [43]. Because of IoV characteristics depends on internet, data and traffic prediction, analysis and classification are most important aspects related to IoV network control.

A. **Traffic Prediction**

Data collection and analysis are related to the steps of collecting a large amount of representative network data, and ability of characterize the network factors. Based on IoV application, data collection can be gathered from different

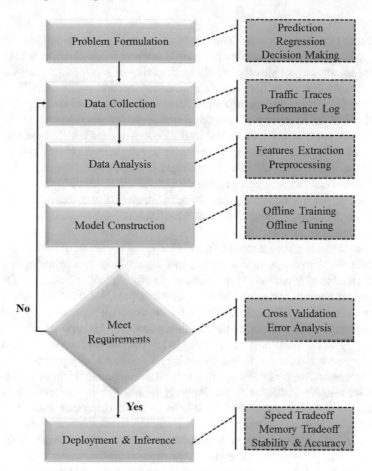

Fig. 7 ML for IoV network control cycle

network layers. According to IoV network state, offline data collection with high quality is required for data analysis, while online data collection will enable to learn network performance and adaptation [44]. For IoV critical applications, data analysis needs to find a proper network feature i.e. to predict the best network traffic performance by analyzing historical data. For data collection and analysis, it's important to prepare network data by normalization, discretization, and missing value completion. ML is a good choice to help extract the network features. For IoV networks, ML plays an important role in traffic prediction and network management [45]. Accuracy in traffic volume estimation in IoV networks is considered one of the main factors impact the performance analysis of network operations, i.e. resource allocation, network routing, congestion and data streaming control. Many studies try to reduce the cost of traffic measurement by using ML algorithms [44, 45].

B.   Traffic Classification

Traffic classification represents the need of IoV network applications to be matched with internet traffic flow. In IoV, internet traffic classification is important aspects to provide efficient network quality of service and quality of experience. Moreover, in network edge, accurate internet traffic classification is a critical challenge and an important component of network security domain. In this case, the important of network traffic classification is to recognize vehicle network applications and control traffic flow as needed with in balancing value or in priority over each other. In security issue, traffic classifications provide a means of intrusions and malicious attacks detection [45]. The use of ML based on statistical features will provides a classification scenario to more realistic situation for IoV network traffic for network control and security. Moreover, it achieves efficiency, adaptability, and performance enhancement.

C.   Traffic Management

Other considerations rerated to network control are network traffic monitoring and management. In IoV network, to ensure efficient network optimization, ML enable to adapt the dynamic internet traffic in IoV to maximize QoS/QoE without compromising end user experiences [45]. ML can help to overcome the shortcoming of classical TCP congestion control algorithms by classifying a packet loss due to congestion or link errors. By ML approaches, it will be easy to customized best suited congestion control scheme that able to adapt to network unique requirements. ML can systematically prospect important information from data held by vehicles and automatically identify very complex links, allowing vehicles to intelligently monitor their environment, and use data for training purposes [45, 46]. ML enable to predict and adapt to the evolution of environmental features, including wireless channel dynamics and traffic and mobility patterns, in addition to configure the network, which gives great possibility to control and manage network traffic.

Other ML solutions relate to developing accurate channel models in different environments and reducing path loss. These solutions lie in predicting IoV topology, and treating severe interference from other IoVs using navigation data and vehicles connectivity. In IoV applications, Internet traffic may be impacted by the weakness of wireless communications [44]. ML technologies are able to assess wireless conditions without the need for a large amount of data sets and by using ANNs methods, an RSS prediction can be performed in an IoV environment.

5.4.2   Location Prediction

Automation is an important advantage of IoV network. The vehicles contain perception system to be able object detection and prediction. In most IoV applications, the behavior of vehicles depends on sensory data, and ability of classify objects in the surrounding environment. These factors help to develop autonomous vehicle applications by using an efficient vehicles behavior prediction and decision making [47].

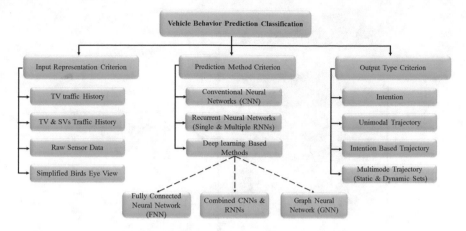

Fig. 8 IoV behavior prediction models

Intelligent prediction will help to optimize the decision making to vehicles trajectories to avoid any risks. Self-driving and autonomous IoV are depend on the location prediction. The prediction requires information about position of vehicle itself and behaviors of the surrounding vehicles, in addition to road geometry and traffic rules. Different vehicle behavior prediction models are developed i.e. intention trajectory, maneuver-based, and interaction aware models [48]. These kinds of models are categorizing as input representation and output types criterion as shown in Fig. 8. In recent years, researchers try to use ML prediction methods to optimize precision of location prediction.

ML uses recorded vehicles historical mobility patterns to predict the next location prediction according to mining trajectory patterns. This strategy is depending on the availability of enough historical trajectory data. To have gain accurate prediction, ML provides an efficient method to get ride the problem of suffering from data sparsely and little historical trajectory, in addition to impact of unknown dynamic contexts traffic flows, weather. ML enables to incorporate this contextual information into vehicle movement prediction. ML helps to model the contextual information characteristics between trajectories and builds learning model by integrates for example neural network with long short-term memory (LSTM) to predict next location as shown in Fig. 9 [49]. The LSTM can easily incorporate heterogeneous features by integrate trajectory variables to effectively predict the next location.

5.4.3 Intelligent Resource Management

Since IoV applications depend on IoT, it found that resource management is in this technology facing many challenges especially in large-scale IoT networks. These challenges are related to massive channel access, power allocation and interference management, energy management, and coexistence between V2V or V2I and IoT

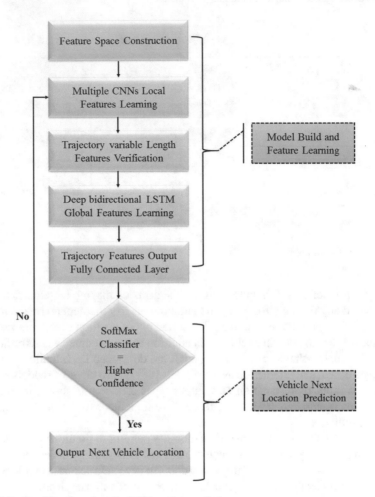

Fig. 9 ML algorithm flowchart for IoV location prediction

traffic. Massive channel accessing causes overloads to networks and congestion [50]. For resource management, there is a need to develop proper load balancing and access management techniques. The crowded vehicles traveling over roads make interference problems become which requires an efficient power allocation and interference management techniques. In IoV, the nature of IoT is characterized by continuous data traffic, leads to high energy consumption. Moreover, harmonious coexistence between V2V or V2I existing networks and IoT traffic, require an intelligent resource management [50, 51]. ML algorithms play an important role in addressing the mentioned challenges related to resource management.

For intelligent resource management, ML able to make classification, regression and density estimation to exploit the traffic of data and develop automated solutions for IoV services. ML provides intelligent prediction for unknown IoV system parameters system behavior, i.e. reinforcement learning (RL) can enable to control the decisions of system actions from unknown monitored system behavior during network activities. Moreover, ML provides suitable solutions for enabling careful channel and power allocation and extracts network parameters to make decisions for CSI, traffic characteristics, and demands of vehicle's users [51]. Deep learning promising smart solutions to characterize inherent relationships between IoV system input and output to develop traffic control system in order to optimize network management routing and scheduling adaption [42]. This will help to optimize IoV network QoE.

Other consideration related to ML use in resource management is to maximize overall network capacity and guarantee best QoS. Q learning can attain a great regulation and strategy by utilizing network learning policy to accomplish smart resource control, assignment and management with continuous-valued activities. It can be employed to obtain an optimal policy for resource allocation in V2V communications to maximize long term expected accumulated discounted rewards, where Q function is approximated by a deep neural network [42]. The optimal policy with Q-values can be found by the following equation.

$$Q_{new}(s_t, a_t) = Q_{old}(s_t, a_t) + \alpha \left[r_{t+1} + \gamma max_{s \in S} \cdot Q_{old}(s, a_t) - Q_{old}(s_t, a_t) \right] \tag{3}$$

The observed state represents by $s \in S$, where S represents the state space. t denoted for time. s_t is an agent state and a_t represents action. The Q learning can be deployed by what is known as actor-critic learning algorithm (AC) which is discussed in [42] by (Mowei Wang, et al., 2017). The frame of AC learning consists of actor and critic parts where are responsible for control strategy adoption with action selections based on the tested network status and the entered policy of environment parameter reward function respectively as shown in Fig. 10. This mechanism enables IoV vehicles to make decisions itself based on its learned policy strategy [52]. Each IoV communication link will observe the current network state i.e. resource block allocation, channel quality and the requirements of QoS to enable selection of actions related to resource block assignment and power level according to the policy strategy and provide new IoV network state.

6   Machine Learning Applications in IoV

ML contributes in many IoV applications related to emergent message transmission for road safety and dangerous activities. In addition, ML provides new smart solutions for IoV services and entertainment. In order to minimize the overall energy consumption of computational facilities and vehicles while satisfying the delay constraint for traffic offloading. The use of ML technology in data mining, pattern recognition, processing, cognitive computing, is an alternative for decision making, which

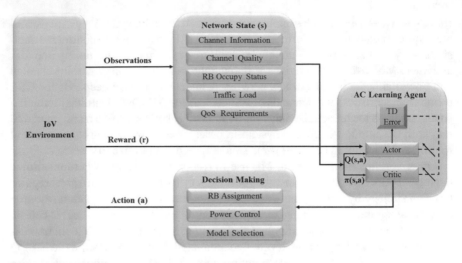

Fig. 10 AC learning algorithm for IoV resource management

will open new opportunities for intelligent IoV networks i.e. in driver safety, smart transportation and autonomous driving applications.

6.1 Intelligent Autonomous Driving

Machine learning plays a vital role in vehicles intelligent driving application, which is make vehicles able to perception and estimation to efficiently manage vehicle driving system. ML makes vehicles self-automated which will improve society by reducing road accidents. In general, self-driving vehicles are very closely associated with IoV. The combination of IoT with ML and smart computing, will provide and intelligent driving system. Machine learning algorithms in self-driving enable IoV to predicts the possible changes in the surrounding driving environment and provide different tasks i.e. object detection and identification, in addition to prediction of another vehicle's localization and movement [53]. Many ML algorithms can be used to provide the mentioned tasks. Regression algorithms provide a localization schemes to develop a prediction and feature selection models for self-driving vehicles. Clustering algorithms provide a way to modeling approaches such as centroid based and hierarchical for intelligent localization [53, 54]. For intelligent decision making, decision matrix algorithms will help to identifying, analyzing, and rating the performance of relationships between sets of values and information.

To enable self-driving vehicles, intelligent decision making is required to process streams of observations coming from different vehicle devices, like cameras, radars, LiDAR's, ultrasonic sensors, GPS units and sensors. The information gathered by these sources, help the vehicle ML based computer to make driving decisions, studied by (Sorin Grigorescu et al., 2020) [50]. Decision making can take place by modular perception-planning-action or by End2End learning fashion. The modular perception-planning-action uses AI and deep learning methodologies to make various permutations of learning and non-learning based components. End2End learning is based on deep learning which perform direct mapping from sensory data to control commands. End2End learning can also be formulated as a back propagation algorithm scaled up to complex models [55]. Such deep learning-based algorithms will able to find a route between vehicle start position and a desired location, which represents path planning. It's able to consider all possible obstacles that are present in the surrounding environment, in addition to find out a trajectory with free of collision route.

6.2   Deep Learning for Driver Safety and Assistance

Due to the increasing accidents and urgent need to reduce road accidents and improve traffic safety, modern vehicles are equipped with a set of sensors, and connected to high-speed mobile communication networks. The vehicle sensors allow collection of a large amount of data that is used in vehicle safety analysis procedures. The data analyzed in real time by AI algorithms in autonomous driving systems applications to reach a high level of safety through several designs to designate road safety index and its prediction of parameters such as street engineering, human behavior and traffic flow. The description of road safety by deep learning will predict real-time road safety index based on deep dense neural network. Moreover, ML helps to learning association between visual entities and city characteristics to estimate road safety based on image processing [56]. The extraction of associations between captured pictures and estimated road safety with multiple cross-domain factors can achieve high prediction accuracy of road safety index (SI). The Real-time road safety index estimation will enhance vehicle safety. The road safety index (SI) can be defined as a number used to inform public about safety of area as published by (Zhe Peng et al., 2018) [57]. Safety index can be calculated based on traffic accident rate per 100,000 inhabitants Ra as follows,

$$SI = (1 - Ra) * 100 \qquad (4)$$

Driver assistance systems (ADAS) are quickly being established for self-directed vehicles which are considered as one of driver safety and assistance methods. To develop an advanced driver assistance systems, ML and embedded computing are considered as main driving factors enabling the development [57]. ML will enable driver assistance systems to perceive obstacles, objects, lanes, pedestrians, and other cars. Also it allows predicting obstacle trajectories and targets. ML helps to detect and tacking obstacle to avoid collision and for path planning. Vehicle camera based deep learning improves quality enhancement and cost reduction of blind spot predication rather than radar [58] as studied by (John E. Ball and Bo Tang, 2019). It uses a lightweight and computationally effective neural network (NN).

6.3 ML in Smart Transportation

Intelligent transport is one of the most important applications of vehicle internet, as it covers a number of applications including improving track, parking lots, avoiding and detecting accidents and other applications related to infrastructure. ML technologies serve to develop advanced models of smart transport systems (ITS). In general, traffic congestion is one of the most important problems facing transportation systems in urban areas, especially with cities that contain high vehicle density [59]. The use of ML with smart transport systems provides an optimization for traffic network configuration. Other smart transportation application in modern cities is parking. Smart cities try to find out an intelligent method for parking to provide reservation services, and selecting parking for vehicles. IoT and ML technologies enable free parking methods. ML helps manage parking for different drivers. It can classify parking according to the requirements of drivers, i.e. regular drivers or those with special needs. The IoT help to exchange the mapping of parking information to the vehicles or for mobile users through cloud servers [60]. Moreover, IoT will improve traffic monitoring, live location streaming, and vehicle performance monitoring.

Since IoV network consists of multiple types of smart vehicles. Transport data processing of these multiple vehicles in real time requires an intelligent schedule and data processing mechanism [60]. Distributed systems provide an efficient and fast method for such a situation. These needs to efficiently deal with big transportation data collected from heterogeneous sources within an efficient database solution. ML based SQL database enables smart database queries and flow data processing. ML will enable balance between accuracy of the algorithms and the size of data, and determination of circumstances in which it becomes useful to implement distributed systems. For purpose of transportation route optimization, ML provides reliable predictions to make routing decisions [61]. ML enables a clear understanding of available route options, their associated energy and environmental costs in real-time. ML provides predictability of changes which can help convoy operators choose vehicles and methods that save fuel costs while maximizing efficiency [62].

7 Future Directions and Tentative Solutions

It is well known that artificial intelligence plays a great role in most IoT applications that depend on perception and predictions of events. As one of these applications, IoV networks require a development of smart algorithms to manage intelligent technology, such as example in the case of self-driving cars. Self-driving cars are a high-risk test for machine learning authorities, as well as a test case for social learning in technology management [63]. In IoV applications, the convergence between machine learning and IoT promises future progress in efficiency, accuracy and improved resource management. The use of machine learning with IoV provides high performance in field of communication and computing in order to achieve efficient control, management and decision-making processes [63, 64]. ML allows extraction of big sensory data to get better insights into range of problems associated with IoV and surrounding environment, as well as ability to make critical operational decisions. It also promises in near future to upgrade performance of vehicle networks and make them more interactive with other things Internet applications. In addition, using ML in IoV enables interaction between cyber and physical components together and can significantly improve efficiency and reliability of processes and systems [64]. Moreover, machine learning offers smart solutions to improve decision-making in event of cyber-attacks.

ML provides solutions for many ITS applications, especially in 2D level realization and forecasting. However, it can develop AI techniques that have the ability to develop collaborative mobility applications based on the description of realistic 3D objects and 4D perceptions for autonomous driving [65]. For different IoV applications, like driving managements, route and localization prediction, a smart ITS camera devices can create holograms to provide 3D object visualization. Due to hybrid ITS context, the combination of data from different resources to improve 3D visualization accuracy is an exciting tentative solution and critical future research direction. The fifth generation (5G) of IoV network is expected to provide some AI technologies to provide network management in a completely smart and provide innovative services, but the sixth generation (6G) is expected to pack machine learning techniques an important role in its operation through self-reconfiguration on demand to ensure a doubling in network performance and service types [66]. ML techniques can provide 6G network model that have ability to rapidly respond to IoV management processes by learning in real time state of network.

8 Conclusion

Machine Learning (ML) one of the most powerful tools for intelligent forecasting and decision making that can assist in analyzing big data in IoV networks. Many potential ML-based IoV applications have been addressed to improve the performance of IoV networks. ML technologies offer solutions that are extremely useful in addressing congestion problem in high density and rapid topology change IoV networks in order to ensure quality of services and experience. Moreover, the scope of employing machine learning technology in network management and control, data flow, site forecasting and resource tools across different layers of communication networks were discussed. In general, we found that in most automated learning applications, performance depends on the amounts of data available where more data available better performance would be achieved. Recently, parallel computing capabilities and machine learning methods have been developed to build smart and integrated IoV networks and systems. Various operations associated with IoV such as multi-dimensional signal/image processing, and wireless communications are need extensive computing and data processing. Energy consuming efficiency one of the problematic issues that need ML solution.

References

1. Tong, W., et al.: Artificial intelligence for vehicle-to-everything: a survey, vol. 4. IEEE (2018)
2. Yang, H., et al.: Artificial intelligence-enabled intelligent 6G networks. arXiv (2019)
3. Eltahir, A.A., Saeed, R.A., Mukherjee, A., Hasan, M.K.: Evaluation and analysis of an enhanced hybrid wireless mesh protocol for vehicular ad-hoc network. EURASIP J. Wirel. Commun. Netw. **2016**(1), 1–11 (2016)
4. Dai, Y., et al.: Artificial intelligence empowered edge computing and caching for internet of vehicles. IEEE Wirel. Commun. **26**, 12–18 (2019)
5. Ji, H., et al.: Artificial intelligence-empowered edge of vehicles: architecture, enabling technologies, and applications. IEEE Access **8**, 61020–61034 (2020)
6. Sodhro, A.H., et al.: Artificial Intelligence based QoS optimization for multimedia communication in IoV systems. Fut. Gener. Comput. Syst. **95**, 667–680 (2018)
7. Hassan, M.B., Ali, E.S., Mokhtar, R.A., Saeed, R.A., Chaudhari, B.S.: NB-IoT: concepts, applications, and deployment challenges, chap. 6. In: Chaudhari, B.S., Zennaro, M. (eds.) LPWAN Technologies for IoT and M2M Applications. Elsevier, March 2020. ISBN 9780128188804
8. Dai, Y., et al.: Artificial intelligence empowered edge computing and caching for internet of vehicles. IEEE Wirel. Commun. **26**(3), 12–18 (2019)
9. Ahmed, E.S.A., Saeed, R.A.: A survey of big data cloud computing security. Int. J. Comput. Sci. Softw. Eng. (IJCSSE) **3**(1), 78–85 (2014)
10. Mohammed, Z.K.A., Ahmed, E.S.A.: Internet of things applications, challenges and related future technologies. WSN **67**(2), 126–148 (2017)
11. He, W., et al.: Developing vehicular data cloud services in the IoT environment. IEEE Trans. Ind. Inf. **10**(2), 1587–1595 (2014)
12. Ahmed, Z.E., Saeed, R.A., Ghopade, S.N., Mukherjee, A.: Energy optimization in LPWANs by using heuristic techniques, chap. 11. In: Chaudhari, B.S., Zennaro, M. (eds.) LPWAN Technologies for IoT and M2M Applications. Elsevier, March 2020. ISBN 9780128188804
13. Borcoci, E., et al.: Internet of vehicles functional architectures - comparative critical study. In: The 9th International Conference on Advances in Future Internet, AFIN 2017 (2017)
14. Mao, Y., et al.: A survey on mobile edge computing: the communication perspective. IEEE Commun. Surv. Tutor. **19**, 2322–2358 (2017)
15. Xu, J., et al.: Joint service caching and task offloading for mobile edge computing in dense networks. arXiv (2018)
16. Cao, Y., et al.: An EV charging management system concerning drivers' trip duration and mobility uncertainty. IEEE Trans. Syst. Man Cybern. Syst. **48**, 596–607 (2016)
17. Nguyen, H., et al.: Deep learning methods in transportation domain: a review. IET Intell. Transp. Syst. **12**(9), 998–1004·(2018)
18. Alhilal, A., et al.: Distributed vehicular computing at the dawn of 5G: a survey. arXiv (2020)
19. Jagadessan, J., et al.: A Machine Learning Algorithm for Jitter reduction and Video Quality Enhancement in IoT Environment. Int. J. Eng. Adv. Technol. (IJEAT) **8**(4), 667–672 (2019). ISSN 2249-8958
20. Hu, S., et al.: A fuzzy QoS optimization method with energy efficiency for the internet of vehicles. Adv. Netw. **4**, 34–44 (2016)
21 Sodhro, A.H., et al.: Artificial Intelligence based QoS optimization for multimedia communication in IoV systems. Fut. Gener. Comput. Syst. **95**, 667–680 (2019)

22. De Filippo, M., Grazia, De., Zucchetto, D., Testolin, A., Zanella, A., Zorzi, M., Zorzi, M.: QoE multi-stage machine learning for dynamic video streaming. IEEE Trans. Cogn. Commun. Netw. **4**(1), 146–161 (2018)
23. Lai, C.-F., e al.: A buffer-aware QoS streaming approach for SDN-enabled 5G vehicular networks. IEEE Commun. Mag. **55**, 68–73 (2017)
24. Saeed, R.A., Mokhtar, R., Khatun, S.: Spectrum sensing and sharing for cognitive radio and advanced spectrum management. ICGST Int. J. Comput. Netw. Internet Res. (CNIR) **9**(2), 87–97 (2009)
25. Zeng, M., et al.: QoE-aware power management in vehicle-to-grid networks: a matching-theoretic approach. IEEE Trans. Smart Grid **9**, 2468–2477 (2016)
26. Bozkaya, E., et al.: Software-defined management model for energy-aware vehicular networks. EAI Endorsed Trans. Wirel. Spectr. **3**, e5 (2017)
27. Zhao, Y., et al.: A survey of networking applications applying the software defined networking concept based on machine learning. IEEE Access **7**, 95397–95417 (2019)
28. Nguyen, T.N.: The challenges in ML-based security for SDN. In: 2nd Cyber Security in Networking Conference (CSNet) (2018)
29. Hasan, K.F., et al.: Cognitive internet of vehicles: motivation, layered architecture and security issues. In: International Conference on Sustainable Technologies for Industry 4.0 (STI) (2019)
30. Chen, C., et al.: A rear-end collision prediction scheme based on deep learning in the Internet of Vehicles. J. Parallel Distrib. Comput. **117**, 192–204 (2017)
31 Tan, L.T., et al.: Mobility-aware edge caching and computing in vehicle networks: a deep reinforcement learning. IEEE Trans. Veh. Technol. **67**(11), 10190–10203 (2018)
32 Chang, Z., et al.: Learn to cache: machine learning for network edge caching in the Big Dta era. IEEE Wirel. Commun. **25**(3), 28–35 (2018)
33 Ning, Z., Dong, P., Wang, X., Rodrigues, J.J.P.C., Xia, F.: Deep reinforcement learning for vehicular edge computing: an intelligent offloading system. ACM Trans. Intell. Syst. Technol. **10**(6), 1–24 (2019)
34. Zhang, H., et al.: Deep reinforcement learning-based offloading decision optimization in mobile edge computing. In: IEEE Wireless Communications and Networking Conference (WCNC) (2019)
35. Li, M., Gao, J., Zhang, N., Zhao, L., Shen, X.: Collaborative computing in vehicular networks: a deep reinforcement learning approach. ICC 2020 - IEEE International Conference on Communications (ICC). pp. 1–6 (2020)
36. Ye, H., et al.: Machine learning for vehicular networks: recent advances and application examples. IEEE Veh. Technol. Mag. **13**(2), 94–101 (2018)
37. Lai, W.K., et al.: A machine learning system for routing decision-making in urban vehicular ad hoc networks. Int. J. Distrib. Sens. Netw. **2015**, 1–3 (2015)
38. Khayyam, H., et al.: Artificial intelligence and internet of things for autonomous vehicles. In: Nonlinear Approaches in Engineering Applications. Springer, Heidelberg (2020)
39. Li, J., et al.: Survey on artificial intelligence for vehicles. Automot. Innov. **1**, 2–4 (2018)
40. Fan, C.-Y., et al.: using machine learning to forecast patent quality – take "vehicle networking" industry for example. In: Volume 5: Trans disciplinary Engineering: A Paradigm Shift (2017)
41. Gu, J., et al.: Introduction to the special section on machine learning-based internet of vehicles: theory, methodology, and applications. IEEE Trans. Veh. Technol. **68**(5), 4105–4109 (2019)
42. Wang, M., et al.: Machine learning for networking: workflow, advances and opportunities. IEEE Netw. **32**, 925–999 (2017)
43. de Hoog, J., et al; Improving machine learning-based decision-making through inclusion of data quality. In: BNAIC/BENELEARN Computer Science (2019)
44. Zerilli, J., et al.: Algorithmic decision-making and the control problem. Minds Mach. **29**, 555–578 (2019)
45. Usama, M., et al.: Unsupervised machine learning for networking: techniques, applications and research challenges. In: ACCESS 2019 (2019)
46 Bithas, P.S., et al.: A survey on machine-learning techniques for UAV-based communications. Sensors **19**, 5170 (2019)

47. Mozaffari, S., et al.: Deep learning-based vehicle behavior prediction for autonomous driving applications: a review. arXiv (2019)
48. Fan, X., et al.: A deep learning approach for net location prediction. In: Proceedings of the 2018 IEEE 22nd International Conference on Computer Supported Cooperative Work in Design (2018)
49. Jiang, H., et al.: Trajectory prediction of vehicles based on deep learning. In: The 4th International Conference on Intelligent Transportation Engineering (2019)
50. Hussain, F., et al.: Machine learning for resource management in cellular and IoT networks: potentials, current solutions, and open challenges. arXiv (2019)
51. Chen, M., et al.: Artificial neural networks-based machine learning for wireless networks: a tutorial. arXiv (2019)
52. Yang, H., et al.: Intelligent resource management based on reinforcement learning for ultra-reliable and low-latency IoV communication networks. IEEE Trans. Veh. Technol. **68**, 4157–4169 (2019)
53. Abduljabbar, R., et al.: Applications of artificial intelligence in transport: an overview. Sustainability **11**, 189 (2019)
54. Xing, Y., et al.: Driver activity recognition for intelligent vehicles: a deep learning approach. IEEE Trans. Veh. Technol. **68**, 5379–5390 (2019)
55. Grigorescu, S., et al.: A survey of deep learning techniques for autonomous driving. arXiv (V)
56. Peng, Z., et al.: Vehicle safety improvement through deep learning and mobile sensing. IEEE Netw. **32**, 28–33 (2018)
57. Moujahid, A., et al: Machine learning techniques in ADAS: a review. In: International Conference on Advances in Computing and Communication Engineering, ICACCE-2018, Paris (2018)
58. Ball, J.E., Tang, B.: Machine learning and embedded computing in advanced driver assistance systems (ADAS). Electronics **8**, 748 (2019)
59. Zantalis, F., et al.: A review of machine learning and IoT in smart transportation. Fut. Internet **11**, 94 (2019)
60. Veres, M., et al.: Deep learning for intelligent transportation systems: a survey of emerging trends. IEEE Trans. Intell. Transp. Syst. **21**, 3152–3168 (2019)
61. Howard, A.J., et al.: Distributed data analytics framework for smart transportation. In: IEEE 20th International Conference on High Performance Computing and Communications (2018)
62. Lana, I., et al.: From data to actions in intelligent transportation systems: a prescription of functional requirements for model actionability. arXiv (2020)
63. Stilgoe, J.: Machine learning, social learning and the governance of self-driving cars. Soc. Stud. Sci. **48**(1), 25–56 (2018)
64. Adi, E., et al.: Machine learning and data analytics for the IoT. Neural Comput. Appl. **33**, 16205–16233 (2020)
65. Yuan, T., et al.: Harnessing machine learning for next-generation intelligent transportation systems: a survey. HAL archive-ouverter.fr (2019)
66. Nawaz, S.J., et al.: Quantum machine learning for 6G communication networks: state-of-the-art and vision for the future. IEEE Access **7**, 46317–46350 (2019)

Deep Learning Approaches for IoV Applications and Services

Lina Elmoiz Alatabani, Elmustafa Sayed Ali, and Rashid A. Saeed

Abstract Internet of vehicles (IoV) has become an important revolution of intelligent transportation system (ITS). It became an emerging research area as the need for it has increased tremendously. With a great number of applications available, in addition to the intention to improve the quality of life and quality of services, the application of artificial intelligence (AI) techniques would dramatically enhance the performance of the IoV overall system. This chapter will discuss deep learning networks as a type of machine learning use in IoV with influence of Neural Networks (NN), where great amounts of unlabeled data are processed, classified and clustered. Deep learning network approaches i.e., Convolutional Neural Networks (CNN), Recurrent Neural Networks (RNN), Deep Reinforcement Learning (DRL), classification, clustering, and predictive analysis (regression) will briefly discussed in this chapter, in addition to review its ability to obtain better performing IoV applications.

Keywords IoT · AI · IoV · Deep learning · Neural networks · CNN · RNN · Reinforcement learning · Classification · Clustering · Regression

1 Introduction

Deep learning (DL) refers to the intelligent machine learning concept, which mimic the functions of a human brain in processing data and creating patterns for decision making which is also known as deep neural network [2]. The depth comes from the multi hidden layered approach. For example, having a number of hidden layers between the input and output layers, this contains a number of connected processing nodes or neurons. Deep learning has evolved significantly during the last decade,

L. E. Alatabani · E. S. Ali (✉) · R. A. Saeed
Department of Electronics Engineering, Sudan University of Science and Technology, Khartoum, Sudan

E. S. Ali
Department of Electrical and Electronics Engineering, Red Sea University, Port Sudan, Sudan

R. A. Saeed
Department of Computer Engineering, Taif University, Al-Taif, Saudi Arabia

© Springer Nature Switzerland AG 2021
N. Magaia et al. (eds.), *Intelligent Technologies for Internet of Vehicles*, Internet of Things,
https://doi.org/10.1007/978-3-030-76493-7_8

helping in many advances in the IoT and its applications [2]. Having a tremendous impact in improving lives and a simplified experience that dealing with daily routines. Such routines like driving cars, items delivery, and smart governments. DL was first introduced in 1943 and has been developed expeditiously since then, to cope with the growing need of Artificial Intelligence (AI) applications and approaches. It improves the quality of service (QoS) by using IoT applications in the medical field, transportation, defense, and smart cities among many other applications. In addition, AI will improve the QoS for many IoV applications includes; smart cities, safe and autonomous driving, convenience service, traffic, traffic flow monitoring and crash response, traffic guidance system, safe navigation, intelligent vehicle control, accident and crash prevention, electronic toll collection and vehicle autonomy.

Recently, many new applications introduced rapidly to the AI and IoV fields according to the needs and users' expectations. For the importance of artificial intelligence in IoV applications, this chapter provides details on the concept of deep learning in the IoV applications and services. The chapter is organized as follows; first, the chapter provides an introduction to the history and evolution of deep learning concept, briefly introduces the ML mechanism, fast optimization algorithms in IoV applications and services. Then chapter discuss in details the deep learning approaches such as convolutional neural networks (CNNs), recurrent neural networks (RNNs), and deep reinforcement learning (DRL). Subsequently, the chapter discusses the IoV network architecture and data analysis with the implications of clustering algorithms and network control. Followed by a brief discussion on the regression problem and how deep learning provides a practical solution such problem. Finally, the chapter reviews the consideration related to the deep learning application to IoV applications and Services.

2   The Evolution of Deep Learning in IoT

The concept of using deep learning approaches introduced by creating a computer model based on the neural network of the human brain. The study uses the threshold logic which is a combination of algorithms and mathematics. The year 1965 was a milestone in the development of DL, where the earliest efforts in developing DL algorithms were released. It contributes to developing group method of data handling and cybernetics and forecasting techniques respectively [1]. Another turning milestone was in 1979 when convolutional neural networks (CNN) were first introduced, and reviews the designed neural networks with multiple pooling and convolutional layers. These new neural networks were called neocognitron using a hierarchical multi-layered design. It allows computers to learn visual patterns recognition [3]. In 1989 the first practical demonstration of back-propagation (BP) was introduced. In 1999 the deep learning evolved significantly when the speed of computers became higher and graphical processing units (GPUs) where developed. By 2009, an AI research staff at Stanford University launched a project known as Image Net, an assembled database containing labeled images, which are used to train the neural networks [49].

As for today's growing need of fast data processing for the purpose of analysis and prediction with accurate and precise output, the use of deep learning has lifted the computational structure of neural networks [2, 3]. DL is one of the most remarkably complex fields. The breaking through research activities will lead to much more unparalleled trends in the future. Thus, DL tools can be used for making the programming structure easier, and make great development in the embedded systems interact with. DL contributed to improve many IoT applications as a result of evolving throughout the years, such applications are, speech recognition, image recognition, time-series signal analysis, and machine translation and natural language processing [3]. Figure 1 shows the DL model which is designed to extract patterns from input data, so as to learn and predict new patterns of unseen data. These models learn to improve their own DL by mimics the function of a human brain, which has pattern recognition modules that main function is to recognize individual patterns.

Current IoT applications depend on more user-friendly services on mobile, embedded devices, and sensors. Thus, applying DL networks to IoT devices could bring a generation of applications capable of performing complex sensing and recognition tasks to support the interaction between humans and their surroundings [6]. Recent researches describe an effective deep learning compression algorithm called deep IoT, which it can directly compress the structure of commonly used deep neural networks (DNN).

Challenges can be addressed by using a general DL framework. Deep sense is a recent framework demonstrating a feasible solution, where it integrates convolutional neural networks (CNN) and recurrent neural networks (RNN). Sensory data in the

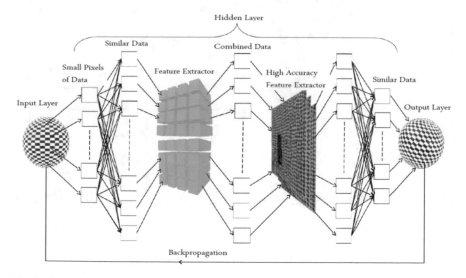

Fig. 1 Typical deep leaning neural network structure

input layer is aligned and split into times intervals for processing. This framework solves the general problem of learning multi-sensor blending tasks for the purpose of estimation or classification from time-series data [5, 7]. One constraint against deploying DL model based IoT is the resource, there for it requires inquire the possibility of compressing deep neural networks (DNN) to overcome the resource constraint. Figure 2 shows a compression framework introduced with the name deep IoT. The deep IoT takes the idea of dropping hidden elements from a widely used deep learning regularization method called dropout. The concept is to give each hidden element a dropout probability during the process. Hidden elements can be cut based on their dropout probabilities leaving a slim network structure for the purpose of preserving the accuracy of sensing applications while decreasing the consumption of resources [3, 6].

In IoV applications, deep learning promises to develop new intelligent technologies for vehicles systems and services. For self-driving and automotive vehicles, the use of DL will enhance the process of self-learning mechanism and optimize the analysis calculations for the data gathered from different surrounding vehicles objects in the driving environment [8]. Moreover, the DL will provide a solution to optimize the QoS for data streaming between vehicles and other resources.

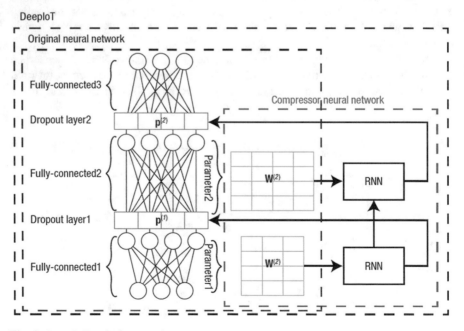

Fig. 2 Deep IoT main framework

3   Distributed Machine Learning Systems

The demand for machine learning has an increasing need and tremendously grown over the last decade. Continuous growth of data collection has triggered rapid techno-logical development of machine learning (ML) algorithms to cope with the growing need to analyse datasets and build decision making systems. Examples include autonomous driving, speech recognition, consumer behaviour prediction [5–7]. The long runtime of training has guided solution designers towards using distributed systems for enhanced parallelization and total amount to I/O bandwidth. In other cases, when data is too big to be stored in one machine, a centralized solution is not feasible. To make these datasets available as training data for ML, algorithms that allow parallel computation, data distribution, and flexibility to failure are developed [8, 9].

3.1   Distributed Machine Learning Architecture

Every algorithm has its unique communication pattern. Thus, designing an inclusive system that allows efficient distribution of machine learning can be challenging [8]. Generally, machine learning issues can be divided into training phase and prediction phase as shown in Fig. 3. The training phase includes feeding a large amount of

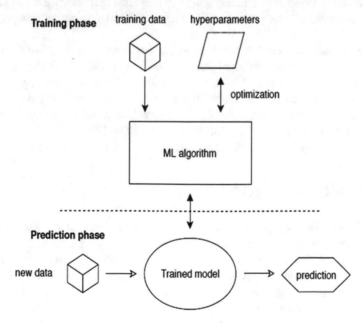

Fig. 3   General machine learning overview

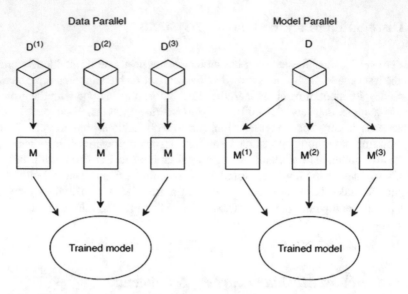

Fig. 4 Parallelism in distributed machine learning environment

training data to allow the machine learning model to train using ML algorithms. A suitable algorithm is used depending on the purpose or application of the system, along with a specific set of hyper-parameters. The outcome of the training phase is a trained model. The prediction phase is used to position the trained model in practice. The trained model receives new data as an input and gives an output as a prediction result.

There are two approaches when applying parallelism in distributed machine learning, data parallel or the model parallel which are illustrated in Fig. 4. The two methods can be applied at the same time [10]. First, in the data-parallel approach, the data is partitioned the same amount of the working nodes in the system as applying the same algorithm to different datasets [8]. Second, in the model parallel approach, a replica of the entire datasets is processed by the nodes [4].

3.2 Parallelization Algorithms

AI calculations can be classified dependent on the capacity of the model. It very well may be isolated into administered and solo techniques. Directed arrangements with marked information sources or datasets, while unaided worked on unlabelled datasets [11]. The calculations utilized, for example, slope plunge, relapse, and K-implies bunching. Conveyed AI frameworks can be separated into three principal classes; they are, information base, general, and reason.

A.   Database Management Systems

Ordinary database management systems (DBMS) in light of SQL don't convey significant parts of AI calculations. In this way, researchers have concocted frameworks that permit clients to execute AI inside the (DBMS). Bismarck executes a reflection layer that gives a mixed design and spotlights on inclination plunge. Another framework known as MADlib gives extra augmentations to SQL that permits clients to execute worked in AI, for example, strategic relapse in a current information base [12].

B.   General Systems

The primary intention is to allow clients to compose custom information preparing work processes utilizing a bunch of Programming interface administrators inside a host language. Message passive interface (MPI) is a low-level structure intended for superior conveyed calculation. MPI gives numerous natives, for example, send, get, broadcast, dissipate, and so on It allows clients to execute a wide scope of uses, for example, ML. Be that as it may, for its low-level nature, it very well may be defenceless to mistake and work comprehensively [11]. Well-known framework inactivity is called Hadoop, it is an open-source framework intended to execute work processes on huge groups of ware machines. Hadoop permits administrations, for example, programmed adaptation to internal failure, and basic programming, permitting clients to examine the enormous size of information across countless machines. Notwithstanding, Hadoop can't uphold or give iterative work processes, requiring accommodation of a solitary occupation for each application [12].

C.   Built Systems

These frameworks give one of two choices, area explicit dialects for AI or calculation explicit enhancement that are not for the most part material [4]. Framework ML gives a significant level of language. This language gives R-like punctuation a programming language utilized in information examination and permits worked-in administrators to perform framework activities. Work processes are changed into map decrease occupations and revised to stay away from numerous ignores input information. Opti-ML is a direct polynomial math-based language that is scale-inserted and area explicit [7]. It incorporates vector, lattice, and chart information types alongside sub-information types that permit extra enhancement [11]. This is to sum up the accessible frameworks and not expected to incorporate all frameworks. Circulated AI is proposed to permit clients to finish up their ideal outcomes from monstrous datasets within a perceptibly short measure of time with the objective of upgrading the assets [12].

3.3 Fast Optimization Algorithms

Deep Learning is an iterative process involving the continuous training of models to reach the final training model. Optimization algorithms are the way to train the network for a given purpose or result in order to find the optimum result. Attention should be given to the performance, i.e., time taken to train the model and the cost constraint [6]. There are a set of optimization algorithms that allow us to achieve the purpose of deep learning within the given constraints. Assume an object function needs to be minimized $f(x)$, and $x \in \mathbb{R}n$. The gradient is $\Delta f(x)$, step side for iteration k is tk. Generally, optimization refers to minimizing of cost/loss function $f(x)$ with the parameter $x \in \mathbb{R}n$. Gradient descent is the most popular and commonly used optimization algorithm in deep learning and is a first-order optimization algorithm. Meaning it takes the first derivative when performing the updates on parameters into account [8]. After updating the parameter of iterations, a learning rate α determines the step size to take on iterations. Batch gradient descent and stochastic gradient descent are not exactly suitable for deep learning, as deep learning requires the use of a huge amount of training data because of their slow performance and decreased efficiency [12].

A. Mini-Gradient Descent

It integrates both the advantages of batch and stochastic gradient descent, working on updating the parameters after getting the gradient of a mini-batch of samples given by the following equation.

$$x_{k+1} = x_k - t_k \Delta f(x_k)^{(i:i+m)} \tag{1}$$

Where m is the Mini-Batch size., $f(x)$ is the objective function which needs to be minimized, $\Delta f(x)$ is the corresponding gradient, and k is the step size of iteration K. Mini-batch gradient descent relay on choosing examples of the large training data, thus taking samples of the large datasets leads to faster performance. The random selection of samples to redundancy avoidance [12, 54].

B. Gradient Descent with Momentum

Navigating ravines is a common issue in stochastic gradient descent. For example, areas of steeping dimensions, in one dimension have a sharper angel than the other. SGD oscillates over the angels of the ravine, slowly making a tentative progress from the bottom moving towards the local optimum [13, 56]. It can be given by the following equations.

$$u_k = mu_{k-1} + t_k \Delta f(x_k) \tag{2}$$

$$x_{k+1} = x_k - u_k \tag{3}$$

Where, $m \in (0, 1)$ determining how many of the previous iterations of the preceding gradients are included in the current update. t_k is the step size, $\Delta f(x)$ is the corresponding gradient, $f(x)$ is the objective function which needs to be minimized, and u_k is the step sequence. Mainly m values is set to 0.5 and then increased up to 0.9 after the initial learning is stabilized [13].

C. Nesterov Accelerated Gradient

Generally, this optimization method improves the algorithms and makes them smarter and more efficient, in addition to making the algorithm more capable of predicting the next position of the parameter. Nesterov accelerated gradient (NAG) will improve the momentum and make it anticipate the suitable direction [13, 55, 56].

$$u_k = mu_{k-1} + t_k \Delta f(x_k - mu_{k-1}) \tag{4}$$

$$x_{k+1} = x_k - u_k \tag{5}$$

D. Adagard

Is a gradient-based algorithm mainly adapting the learning rate to the parameter, executing larger updates. It divides the current gradient in update rule by the sum of all the preceding gradients by scaling the step size for each parameter based on the gradient history for that parameter.

$$G_k = G_{k-1} + \Delta f(x_k)^2 \tag{6}$$

$$x_{k+1} = x_k - \frac{t}{\sqrt{G_k + \varepsilon}} \Delta f(x_k) \tag{7}$$

Where G = Accumulated Gradient history. ε = Smoothing term avoids division by zero, $f(x)$ is the objective function which needs to be minimized, $\Delta f(x)$ is the corresponding gradient, and k is the step size of iteration K. Note that we need to manually tone the step size t [14, 56].

E. Adadelta

Is an improved algorithm based on adagard that tones down the aggregation of adagard. It has two main problems; first problem is reduction in training data. The second problem related to manual selection of the global learning rate. Adadelta combines the advantages of momentum and Adagard to augment the efficiency of the algorithm. It scales the step size using only the last update instead of using the whole history as Adagard [15, 56].

$$\mathbb{E}\left[\Delta f(x)^2\right]_k = \rho \mathbb{E}\left[f(x)^2\right]_{k-1} 1 + (1 - \rho) \Delta f(x_k)^2 \tag{8}$$

$$\hat{x}_k = -\frac{\sqrt{\mathbb{E}\left[\hat{x}^2\right]_{k-1} + \varepsilon}}{\mathbb{E}\left[\Delta f(x)^2\right]_k + \varepsilon} \tag{9}$$

$$\mathbb{E}[\hat{x}^2]_k = \rho\mathbb{E}[\hat{x}^2]_{k-1} + (1 - \rho)\hat{x}_k^2 \tag{10}$$

$$x_{k+1} = x_k + \hat{x}_k \tag{11}$$

Where ρ is decay constant, usually at 0.95. ε is small value to stabilize the equation for the purpose of numerical stability, x is the parameter and E is an accumulation variable.

F. RMSprop

RMSprop proposed to solve adagard problem of step size disappearing. It is an unpublished adaptive learning technique. Where ρ is a decay constant, ε is small value to stabilize the equation for the purpose of numerical stability, x is the parameter and E is an accumulation variable, $f(x)$ is the objective function which needs to be minimized, $\Delta f(x)$ is the corresponding gradient, k is the step size of iteration K, and x is the parameter [13].

$$\mathbb{E}\left[\Delta f(x)^2\right]_k = \rho\mathbb{E}\left[f(x)^2\right]_{k-1}1 + (1 - \rho)\Delta f(x_k)^2 \tag{12}$$

$$x_{k+1} = x_k - \frac{t}{\sqrt{\mathbb{E}\left[\Delta f(x)^2\right]_k + \varepsilon}}\Delta f(x_k) \tag{13}$$

G. AdamAdam

Computes adaptive step size for each parameter, it uses the decay advantage of training data history along with their squared values. Adam also stores an exponentially decaying average of past gradients.

$$m_k = \beta_1 m_{k-1} + 1(1 - \beta_1)\Delta f(x_k) \tag{14}$$

$$u_k = \beta_2 u_{k-1} + (1 - \beta_2)\Delta f(x_k)^2 \tag{15}$$

$$\widehat{m}_k = \frac{m_k}{1 - \beta_1^k}, \hat{u}_k = \frac{u_k}{1 - \beta_2^k} \tag{16}$$

$$x_{k+1} = x_k - \frac{t}{\sqrt{\hat{u}_k + \varepsilon}}\widehat{m}_k \tag{17}$$

Where, $\beta_1 = 0.9$, $\beta_2 = 0.999$, $and\ \varepsilon = 1e - 8$. m is a momentum vector, u is a second momentum vector, $\beta_1\ and\ \beta_2$ are Exponential decay rates for the moment estimates, and k is the initial step size. There are many more advances in the optimization algorithms as it is an emerging research area to serve the improvement of deep learning as it is a fast-growing field [16].

4 Internet of Vehicles DL Applications and Services

IoV has as of late stood out as a piece of intelligent transport systems (ITS) and being critical assistance in our day by day lives in numerous applications [16]. The sending of IoV is convoluted and requires uncommon contemplations for their extraordinary qualities in which the high versatility and dynamic change in geography are unequivocally introduced. Consequently, the European Telecommunications Standards Institute (ETSI) design normalization introduced a correspondence standard for the IoV that appeared in Fig. 5.

In the ETSI design, the offices' layer is answerable for dealing with the VANET related applications, for example, cooperative awareness message (CAM), decentralized notification message (DENM), and local dynamic map (LDM) just as the correspondence cycle. Organization and transport layers are consolidated in one

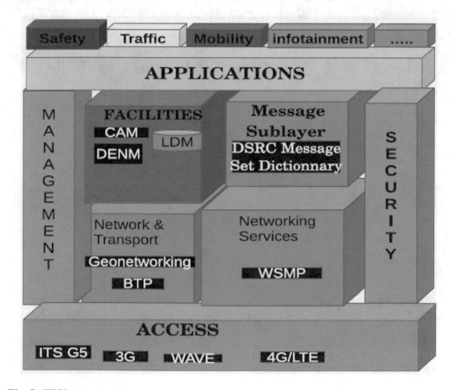

Fig. 5 ETSI system architecture

layer, two extra layers were added which are the administration and security layers, alongside the presence of ITS committed stack incorporating the Geo-organizing and tending to [17]. Never the less such engineering doesn't explain the cooperation cycle in the event of handover and different components that include outer gadgets communication. The organization of IoT in the VANET setting empowered promising arrangements and administrations which prompted the introduction of the present Web of Vehicles IoV which comprises 4 principal parts which are:

- End focuses Vehicles, cell phones or any associated gadgets inside the organization.
- Infrastructure: Street Side Joins RSUs, WI-FI Hotspots, Cell networks suck like 3G, LTE… and so forth bases stations.
- Operations: Strategy authorization, stream-based administration, and security
- Services: Public cloud, private cloud, and venture cloud.

Cisco has proposed a 4 layers IoV network engineering comprising of implanted frameworks and sensors, multi administration edge, center, server farm, and cloud as appeared in Fig. 6 [17].

IoV contains various correspondences application situations i.e., device to device (D2D), vehicle to vehicle (V2V), vehicle to the side of the road unit (V2R), vehicle to the framework (V2I), side of the road to the side of the road (R2R), vehicle to everything (V2X), notwithstanding numerous others introduced in Fig. 7 [17].

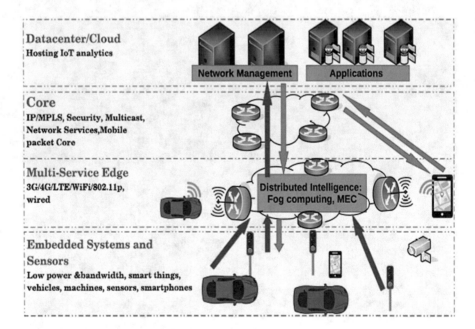

Fig. 6 Cisco four layers architecture

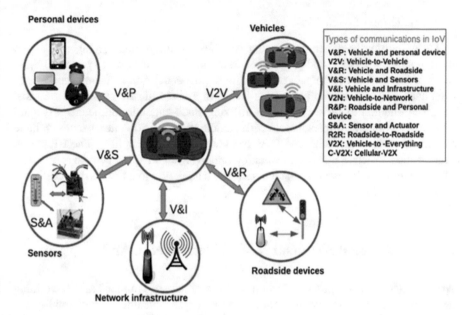

Fig. 7 IoV environment with D2D communications

Numerous tales investigates have been directed utilizing Deep learning strategies in IoV based applications; a portion of the explores are summed up as follows.

A. Security Systems with DL

Shielding IoV from assaults turned out to be recognizably significant due to the extensive development of IoV in the earlier years and the developing significance for the coming years. A proposed approach for making sure about IoV foundation for having dependable interchanges, utilizing Deep learning strategies made out of regulated and solo learning [18]. DL is applied to screen dangers; results show that the framework's observing exactness is obviously superior to customary frameworks [16, 18].

B. Smart Car Cameras-based DL Mobile Cloud System

Deep Learning has become an exceptionally mainstream innovation in the field of picture handling, data recovery, and different applications where savvy information preparing is required [12]. Portable distributed computing shows promising outcomes, putting the preparation cycle and model vault in cloud stage moving the computational cycles to the cloud making it quicker and substantially more made sure about, and acknowledgment and information gathering measures are moved to cell phones [19]. The proposed structure utilized git convention to guarantee the achievement of information transmission, utilizing savvy vehicle camera's information recorded recordings while driving, applying DL techniques show higher recognition rates and stable activity even in insecure association conditions.

C. Prediction of Collision

Deep learning has been performing very well at improving driving encounters by improving wellbeing and proficiency in the intelligent transport system (ITS). A proposed model named impact forecast model dependent on GA-enhanced neural organizations for dynamic [18]. DL's job is to assist with the choice of the most proficient method to serve the driver, utilizing Back-engendering neural organizations in assessing the impact hazard with vehicle to foundation interchanges, vehicle to vehicle correspondences, and GPS framework announcing [20]. The BP neural organization structure utilizes information created from VISSIM with an assortment of elements considered alongside a hereditary calculation to advance the coefficient cluster and limits.

5 Convolutional Artificial Neural Networks (CANN)

Artificial neural networks (ANN) resemble the structure of a human brain, containing an interconnected network of neurones or nodes responsible of undertaking the computational processes in a multi-layered environment [21]. Convolutional artificial neural networks (CANN) is a famous type of deep neural networks named after the mathematical linear operation called convolution, CANN can have multiple layers with or without parameters that has the best performance in dealing with Image processing problems [22].

5.1 The IoV Multilayer Perceptron's (MLPs)

Perceptron is a pattern recognition machine for optical character recognition invented in the 1950's. The perceptron gets multiple inputs are fully connected to an output layer consisting of multiple McColluch and Pitts PEs. McColluch-Pitts are a processing element sums the products followed by a nonlinearity threshold (see Eqs. 18 and 19).

$$y = f(net) = f\left(\sum_i w_i x + b\right) \tag{18}$$

McColluch-Pitts Equation: where w_i are weights and b is a bias term and f is the activation function is commonly referred to as signum function defined by:

$$f(net) = \begin{cases} 1 \text{ for net} \geq 0 \\ -1 \text{ for net} < 0 \end{cases} \tag{19}$$

In the perceptron concept each input x_j is multiplied by the weight w_{ij} is an adjustable constant before being inputted to the jth processing element in the output layer, as follows [23]:

$$y_j = f(net_j) = \left(\sum\nolimits_j w_{ij} x_i + b_i \right) \tag{20}$$

Multilayer Perceptron (MLP) showed promising performance results in the field of IoV for its pattern recognition and classification qualities. Having the need for a safe and reliable autonomous driving as an example has triggered the use of MLPs as a base to prediction algorithms [21]. MLP framework divided into two models. First model, L and model used to predict target lane for the surrounding vehicles. Second model, Trajectory model: gives trajectories for the predicted surrounding vehicles motion given measurement inputs. Then the MLP is used to provide probabilities based on real-world traffic data i.e., MLP train on real data to give predictions to improve the driving experience [24].

5.2 CANN IoV Architectures

Convolutional artificial neural networks (CANN) are comparable to the traditional artificial neural networks (ANN) for their self-optimizing neurons through learning; each layer of neurons receives an input and performs the processing [20]. CANNs are used for services that require pattern recognition with in images. CANN is a deep learning approach designed to perform processing tasks to multidimensional data such as Signals, images, and videos, it is a multilayer network where in each layer there is a set of neurons responsible of performing different types of processing [21].

CANN architecture consists of three types of layers as shown in Fig. 8. In the initial layer, the input layer does not perform any processing functionality. Convolutional layers, which determines the output of neurons through the calculation of the scalar products of their weights and regions connected to the input layer. An activation function referred to as "ReLu" is the rectified linear unit such as sigmoid is normally applied to the output of the activation resulted from the previous layer. The input is usually a raw data such as an image. Pooling layers will then reduce the number of parameters with in the activation. Fully connected layers perform the same functions of a regular ANN attempting to produce class scores from the activation for the purpose of classification to improve ReLu performance [21]. The next equation defines the convolutional operation.

$$\phi_{i,j} = (I \otimes \Theta)_{i,j} = \sum\nolimits_{x=-m/2}^{m/2} \sum\nolimits_{y=-n/2}^{n/2} I_{i-x,j-y} \Theta_{x,y} \tag{21}$$

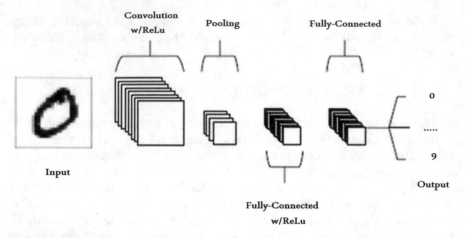

Fig. 8 General CANN architecture

Where: I is the bidimentional input data (usually an image). Θ is the convolutional mask with $m \times n$ size. $\phi_{i,j}$ is the resulted output produced from convolving the region from I centred at i, j (pixel) with the mask Θ.

The previous equation can be written in a dot product format in an m $m \times n$ dimensional space. Let $\vec{v}_{i,j} = [I_{i-x_1,j-y_1}, ..., I_{i,j}, ..., I_{j-x_m,j-y_n}]$ be a vector generated by concatenating columns corresponding to I. With the same context let $\vec{\Theta} = [\theta_{x_1,y_1}, ..., \theta_{x_m,y_n}]$ is the vector also generated by concatenating columns constituting the convolutional mask Θ. There for the convolutional equation corresponds to the dot product:

$$\phi_{i,j} = \vec{\Theta}^{T} \vec{v}_{i,j} \tag{22}$$

Convolutional equation can be written using the matrix–vector product, assuming a convolutional layer with i number of neurons for each neuron constitutes to $m \times n$ convolutional mask Θ_k, $k = 1$, can write the convolution step of each $\vec{\Theta}_k$ with $\vec{v}_{i,j}$ by fixing the local region $\vec{v}_{i,j}$ of I [21]

$$\begin{bmatrix} \vec{\Theta}_1^{T} \\ \cdot \\ \cdot \\ \cdot \\ \vec{\Theta}_i^{T} \end{bmatrix} \vec{v}_{i,j} = \begin{bmatrix} \phi_{i,j}^1 \\ \cdot \\ \cdot \\ \cdot \\ \phi_{i,j}^i \end{bmatrix} = \vec{\phi}_{i,j} \tag{23}$$

The convolution step constituting to linear transformation $T(\vec{v}_{i,j}) : \mathbb{R}^{m.n} \to \mathbb{R}^i$ taking vectors from $\mathbb{R}^{m \times n} \to \mathbb{R}^i$ when applied to all vectors $\vec{v}_{i,j}$ of I (where I = $q \times p$ dimension) the convolutional layer is considered a linear embedding of the input data into i dimensional feature space. The most applied supervised learning method is the stochastic gradient descent (SGD), employs a loss function on the weights of convolutional kernels stated in the following equation [25].

$$w_{t+1} = w_t - \eta_t \nabla_w (y_i - f(x_i, w))^2 p(x_i, y_i) \tag{24}$$

5.3   CANN IoV Algorithms

CANN algorithms are used in many IoV application areas such as detection and recognition, image classifications, and security aspects in terms of intrusion detection enable to ensure the reliability and security of the overall system. Using the images sent from the input layer, detection and recognition algorithms usually contain four stages (see Fig. 9). Stage 1, image pre-processing. Stage 2, detecting the RoI Region-of-interest is an area showing a precise part of the vehicle i.e., tail-light, detecting vehicle LED pattern. Stage 3 is object recognition. Stage 4 represents decision making. The main goal of this desired algorithm is object detection, using the data in the first stage for learning, where the last stage main objective is to recognize objects [26].

The entire algorithm aims to use the vehicle LED signals and cameras as transmitter and receiver, respectively for V2V and V2I communication. The vehicles in V2V are determined as forward vehicle and following vehicle, whereas the role of forward vehicle is to transmit the information and following vehicle is to receive the information. Vehicles can also receive traffic information such as traffic condition, safety information, and etc. from traffic lights [26]. Tail-lights LEDs of the forward vehicle is used to transmit to the following vehicles using S2-PSK modulation, then CANNs are used by the following vehicle to decode the information. V2C vehicle to cloud communication uses the cellular technology to maintain the internet connection, in both forward and following vehicles can share information with in the cloud

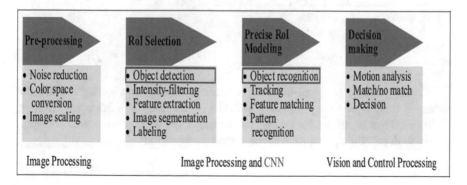

Fig. 9 General CANN recognition and detection algorithm

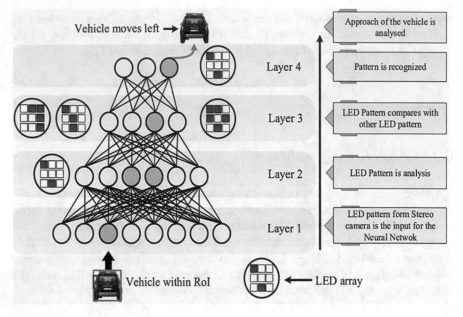

Fig. 10 CANN based pattern recognition and detection algorithm

server is responsible for information processing and broadcast this information back to IoV network [26, 28].

As the input data is coming from cameras installed along the road to detect vehicles, the calculation of distances between vehicles is important to consider the risk factor of the surrounding vehicle behaviour. It is determined by the following steps (see Fig. 10). Step 1, Image acquisition as input [27]. Step 2, Image rectification using linear transformation. Step 3, Segmentation for recognition, detection, and objects measurement in the image. Step 4, Stereo matching algorithms for depth calculation such as sum of absolute differences (SAD), correlation, normalized cross correlation (NCC), sum of squared differences (SSD) algorithms are defined through the following equation where it calculate the intensity differences for each centre pixel (i, j) in a $W(x, y)$ window [26, 29].

$$SAD(x, y, d) = \sum_{(i.j) \in W(x,y)}^{N} |I_L(i, j) - I_R(i - d, j)| \qquad (25)$$

Where: I_L and I_R are pixel-intensity functions, $W(x, y)$ is a square window that surrounds the position (x, y) of the pixel. The coordinates of a 3D point P(xp, yp, zp) from 2D image can be determined by the use of depth map estimation for stereo camera with parallel optical axes in the following equations

$$\frac{z_p}{f} = \frac{x_p}{x_i} = \frac{x_p - b}{x_r} = \frac{y_p}{y_i} = \frac{y_p}{y_r} \qquad (26)$$

$$x_p = \frac{x_i z}{f} = b + \frac{x_r z}{f} \tag{27}$$

$$y_p = \frac{y_i z}{f} = \frac{y_r z}{f} \tag{28}$$

Where: f is the focal line, b is the baseline. (x_i, y_i) And (x_r, y_r) are corresponding image points. The rectified image can produce depth calculation from the disparity map generated from stereo camera in the following equation:

$$d = x_i - x_r = f\left(\frac{x_p + \frac{b}{2}}{z_p} - \frac{x_p - \frac{b}{2}}{z_p}\right) = \frac{fb}{z_p} \; Or \; z_p = \frac{fb}{d} \tag{29}$$

Every algorithm takes into account the general algorithm stated in the previous paragraph depending on the proposed application with minor updates in order to cope with the desired result with the maximum efficiency and performance [26].

6 Recurrent Neural Networks (RNNs)

RNN is a type of neural network designed to analyse sets of data by configuring hidden units within the network, in some applications the output depends on the computation of previous layers [26]. RNNs are significant in terms of keeping records of previous inputs i.e. saves the history. Variations of RNNs have been produced to overcome the problem of vanishing gradient, are long short-term memory (LSTMs) and gated recurrent units (GRUs) they have more benefits over the traditional RNNs for their capability in maintaining long term interrelations and nonlinear dynamics for input datasets with time series. In LSTM specifically the same weight is kept across all layers in order to control the number of parameters required for learning process [27]. The typical RNN architecture is designed using the connection weight matrix $W = [w_{ij}]$ where w_{ij} represents the connection weight from node i to node j. When w_{ij} = zero that means there is no connection between nodes. Generally neural networks can be divided into feed forward network, recurrent networks, lattice network, layered feed forward networks, and cellular networks shown in Fig. 11 [28].

A. **Feed forward Network:** in this topology there is no connection between neurons within the same layer, connection between neurons is aimed in one direction. A feed forward network is usually arranged in the form of layers. The case is different in the fully connected layered feed forward network in all nodes is connected to each other i.e., any node is connected to its adjacent node in the forward layer examples include multilayer perceptron and radial basis function network [28].

B. **Recurrent Network:** examples include Boltzmann machine and Hopfield model in at least one feedback connection is available [28].

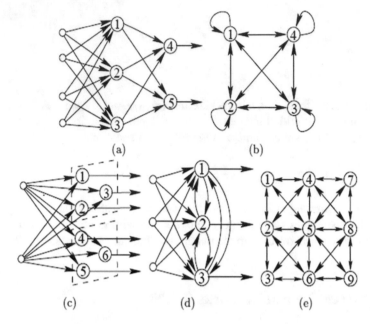

Fig. 11 General neural networks architecture, **a** Layered feed forward network, **b** Recurrent network, **c** Lattice network, **d** Layered feed forward network with lateral connections, **e** Cellular network

C. **Lattice Network:** contains one or more number of dimensional arrays of neurons where each array has a correlating set of input nodes. Layered feed forward network with lateral connections. Connection between units is lateraled between units in the same layer [28].

D. **Cellular Network:** contains neurons are equally spaced called cells, communications between nodes are restricted to the immediate neighbourhood, cells that are close together are connected by a mutual interconnection, each cell is enthusiastic by its own signals and signals coming from the close cell.

There are many neural networks applications operating in the field at the meantime, such like function approximation, classification, clustering and vector quantization, associative memory, optimization, and feature extraction and information compression [28, 29]. Researches have been studying specific details in the driving experience using RNNs in anticipating driver's behaviour. RNNs are able to disseminate data in a chain-like neural network, while handling consecutive information it takes into account the present input x along with the past hidden state h at each time step. At each time step an arrangement of gates Rd control the yield of the module as old hidden state element ht−1 d and its contribution in the present time step x_t. Utilizing d to represent the memory measurement in the LSTM, where all vectors in this architecture share the same measurement. The following equations represent the LSTM transition function:

$$i_t = \sigma(W_i.[h_{t-1}, x_t] + b_i) \tag{30}$$

$$f_t = \sigma(W_f.[h_{t-1}, x_t] + b_f) \tag{31}$$

$$q_t = \tanh(W_q.[h_{t-1}, x_t] + b_q) \tag{32}$$

$$o_t = \sigma(W_o.[h_{t-1}, x_t] + b_o) \tag{33}$$

$$c_t = f_t \odot c_{t-1} + i_t \odot q_t \tag{34}$$

$$h_t = o_t \odot \tanh(c_t) \tag{35}$$

Where: f_t is the forget gate, i is the input gate, o is the yield gate, c_t is the memory cell, h_t is the current hidden state, σ is the strategic sigmoid function having a yield in $[0, 1]$, tanh represents the hyperbolic tangent that has a yield in $[-1, 1]$, and $\odot$ is the element wise augmentation. LSTM is especially intended for time-arrangement information to adapt to long term conditions [29].

7 Deep Reinforcement Learning for IoV Network

Mischances are caused basically by the driver's misjudgement and botch. IoVs has given arrangements to promote the driving encounter, being an empowering calculates for savvy cities applications. IoV is ceaselessly creating innovations and structures in arrange to diminish activity congestions and mishaps changing vehicles from standard driving instruments to brilliant instruments [30]. Support learning gives design where an organization learns from a previous encounter in arranges to extend a few compensate flags. In learning to control, a specialist specifically by accepting high-dimensional input such as vision and discourse in a greatly repetitive errand known as revile of dimensionality. There are numerous proposed arrangements for this issue such as utilize of direct work guess, various levelled representation, and state conglomeration [31].

Multi-state choice-making handle in a probabilistic environment requires an exact controlling handle. Markov choice handle (MDPs) offers a measures formalism clear technique. MDP may be a discrete-time stochastic control process. Where, at each time step, the method is in a few state x and the decision creator chooses an attainable activity a based on that the method moves to an unused state x' and gives the choice producer a comparing remunerate $r(x, a, x')$. The plausibility that the method moves into modern state x' is decided by the chosen activity. It is characterized by the state move work $T(x'|x, a)$ satisfies the Markov property. MDPs fundamental issue is to discover an approach for the choice producer, given the current state x and activity a considering another state x' is completely free of all past states [31]. Characterized by work π for a particular activity x' is completely autonomous of all past states [31]. Characterized by work π for a particular activity $\pi(x)$ for the choice creator to select when state x is. MDP's fundamental reason is to discover an arrangement for π to expand the taking after total work condition:

$$R = \sum_{n=0}^{\infty} r(x_n, a_n)\gamma^n \tag{36}$$

Where n is the time step and γ could be a markdown calculate between and 1. Choosing the γ to calculate will choose how the method is attempted i.e., having $\gamma = 0$ will result in making the specialist short-sighted while giving γ esteem drawing nearer to 1 will make it rummage around for a future tall compensate. A classic MDP operation can be illuminated by two approaches, esteem cycle or arrangement cycle. In any case, these approaches assume that the decision-maker has precise esteem for the move work and remunerate for all states within the environment. Whereas within the real operation a choice creator may not know the move work. Q-learning was presented to overcome this issue; it may be a shape of reinforcement learning instructs the choice creator how to act in an MDP environment when the move work and/or compensate are obscure [31]. Each state is allotted a beginning esteem called Q-value, and the proper Q-value can be evaluated utilizing online incremental upgrade stochastic Q-learning calculation appeared within the taking after condition.

$$Q(x_n, a_n) = [r(x_n, a_n)]a(n) + Q(x_n, a_n) - Q(x_n, a_n) + \gamma_{a_{n+1}}\max\{Q(x_{n+1}, a_{n+1})\} \tag{37}$$

where: $r(x_n, a_n)$ is the single step reward and $a(n)$ step-size learning rate among 0 and 1. Note that when $a(n) = 0$ the Q-values are not upgraded and subsequently nothing is learnt.

The over procedure is called the value-iteration calculation and it meets the ideal action-value work $Q(x_n, a_n) \to Q^*(x_n, a_n)$ as $n \to \infty$. An issue called the "curse of dimensionality" ordinarily happens within the classical Q-learning calculation since it takes an expansive sum of time to meet. To overcome this restriction of MDPs, useful guess procedures are utilized. Neural systems are particularly able of finding great highlights for tall dimensional input information. Neural systems can be spoken to inside the action-value work take into thought the current framework state and activity as input and gives the appropriate Q-value as a yield; this is often commonly known as profound support learning. A Q-network with weights θ may be a shape of neural systems is prepared to memorize the parameter θ of the action-value work $Q(x_n, a_n; \theta)$ by minimizing the grouping of misfortune work where the ith misfortune work $L_i(\theta_i)$ is expressed within the taking after condition [31]

$$L_i(\theta_i) = \mathbb{E}\left[max_{a_{n+1}} Q(x_{n+1}, a_{n+1}; \theta_{i-1}) - Q(x_n, a_n; \theta) + r_n\right]^2 \tag{38}$$

where: θ_i is the NN parameters at the ith update. Along with θ_{i-1} is the parameter from past state are kept settled when optimizing the misfortune work $L_i(\theta_i)$. The part $\{r_n + max_{a_{n+1}} Q(x_{n+1}, a_{n+1}; \theta_{i-1})\}$ is the target for cycle i, hence the objective is to keep this taken a toll expression as little as conceivable, this can be done by the utilize of angle plunge calculation. Which it rehashes the computation of the angle $\nabla_{\theta_i} L_i(\theta_i)$ denoted as in the following notation

$$\nabla_{\theta_i} L_i(\theta_i) = \mathbb{E}\big[(r_n + max_{a_{n+1}} Q(x_{n+1}, a_{n+1}; \theta_{i-1}) - Q(x_n, a_n; \theta))\nabla_{\theta_i} Q(x_n, a_n; \theta)\big] \tag{39}$$

But the customary slope plummet calculation is moderate when preparing colossal sums of datasets, there for utilize of Stochastic Angle Plunge SGD is presented to overcome this issue [31].

7.1 Energy Efficiency and Operation of Scheduling

VANETs and IoV play as an application of brilliantly transportation framework (ITS) with its communication structures like vehicle to vehicle (V2V) and vehicle to the framework (V2I) hand-off on genuine-time data transmitted between vehicles and roadside units (RSUs) introduced along the street side. In able to form VANETs more dependable and productive to move forward the driving involvement and security, through the V2I communication structure the information is transmitted between the vehicle and RSUs making the require for a planning approach to organize the V2I communication that protects the batteries within the RSUs and drag out the lifetime of the organize pointing to satisfactory levels of Quality-of-Service (QoS). RSUs sent have batteries that energize in time interims utilizing vitality collecting by means of sun powered and wind or by physically reviving [32].

A Deep Fortification learning operator is connected within the RSUs to apply planning calculations in arranges to organize the communication and draw out the organized lifetime. Numerous calculations have been presented for planning arrangement is spoken to in MDP, such as action-value work, and Q-learning. The execution of the planning approach system is spoken to in Fig. 12, where an assignment planning framework is proposed to have vitality proficient in a general framework that increment the lifetime of the organization [33].

The profound Q-network (DQN) specialist learns to adjust the control utilization. It has way better execution over the other calculations. Other calculations expressed are arbitrary vehicle choice calculation, prioritizing withdrawing vehicles (PDV), eager control preservation calculation (GPC), and ravenously prioritize leaving vehicles (GPDV). Assignment de-correlation is arrange dependable for the planning approach in arrange to organize assignments and diminish the control utilization [33].

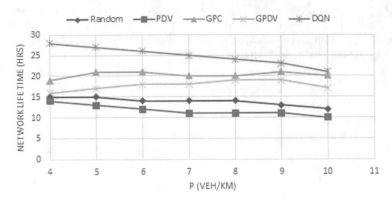

Fig. 12 Deep support operator planning calculations assessment in terms of organize lifetime based on assignment planning whereas it is characterizing the time until IoT-GW cuts off considering ρ vehicle speed

7.2 Resources Allocation and Management

For the criticalness of IoV benefit provisioning, the Quality-of-Service (QoS) or Quality-of-Experience (QoE) for vehicle clients must be kept up at worthy levels. Hence, different asset assignment approaches have been proposed for IoV communications to move forward with the arrange execution. Program character-ized networking-based asset assignment plans were presented to guarantee vehicle clients QoS. A highly effective and dependable communication structure called device-to-device empowered V2V (D2D-V2V) is presented with promising advance-ments since it is able to diminish transmission idleness and control utilization by giving neighbourhood message spread and making strides range effectiveness for its nearness and decreasing pick up properties [34].

An effective exchange AC (ETAC) learning approach is presented to handle the clever asset allotment issue in an IoV arrangement. A shrewd asset administration approach is vital to supply the V2X communication with clever choice-making to fulfil different QoS necessities such as inactivity, and unwavering quality necessities of V2V communication or least information rate necessity for V2I joins. The RL portion of the arrangement embraces the Markov choice handle and an actor-critic (AC) system is proposed for the clever asset allotment, proposing of a modern remu-nerate work Markov for cleverly asset administration in IoV organize is expressed within the taking after conditions.

$$r = \underbrace{c_1 \left(\sum_{m \in M_{nor}} R_m^{d,nor} + \sum_{k \in K} R_k^c \right)}_{\text{Part 1}} \tag{40}$$

$$-c_2\left(\sum_{m \in M_{uni}} \left(p_m^{outage} + p_m^{delay}\right)\right) \tag{41}$$

$$\underbrace{\phantom{-c_2\left(\sum_{m \in M_{uni}} \left(p_m^{outage} + p_m^{delay}\right)\right)}}_{\text{Part 2}}$$

$$-c_3\left(\sum_{m \in M_{nor}} \left(R_m^{d,nor,tar} - R_m^{d,nor}\right) + \sum_{k \in K} \left(R_k^{c,tar} - R_k^{c}\right)\right) \tag{42}$$

$$\underbrace{\phantom{-c_3\left(\sum_{m \in M_{nor}} \left(R_m^{d,nor,tar} - R_m^{d,nor}\right) + \sum_{k \in K} \left(R_k^{c,tar} - R_k^{c}\right)\right)}}_{\text{Part 3}}$$

where part 1 speaks to the whole information rate, parts 2 and 3 are quickly taken a toll on capacities. The coefficient c_i, $i \in \{1, 2, 3\}$ are the weights of the 3 parts, utilized to adjust the utility and fetched. M_{uni} Speaks to the D2D-V2V match sets. Within the IoV arrange each operator chooses an approach π to maximize the remunerate $Q^{\pi}(s, a)$ denoting the state-action work, could be a summed markdown compensate for starting the organize s with a given arrangement π expressed within the taking after the condition

$$Q^{\pi}(s, a) = E\left\{\sum_{t=1}^{\infty} \gamma^t r_t(s_t, a_t)|s_0 = s, \pi\right\} \tag{43}$$

The point of brilliantly asset administration is to find the approach π best depict the maximization of arranging objective compensate, expressed within the taking after the condition

$$J(\pi) = E\{Q^{\pi}(s, a)\} = \int_S d(s) \int_A \pi(s, a) Q^{\pi}(b, a) dads \tag{44}$$

where d(s) denotes as the state conveyance work, and π(s,a) is a stochastic approach with the state's over the current activity a, uncovers the conditional likelihood thickness of the activity a at the state s. Given the over circumstance, the approach can be optimized numerically by applying the esteem emphasis strategy with the support devices such as Q-learning and actor-critic [34].

The likelihood of fulfilled V2V and V2I expanded when applying the ETAC approach giving a supreme vehicle speed, with the thought of shifting a number of gadget vehicle clients could be a vehicle working inside the D2D-V2V Communication structure. It outflanking other support learning approaches such as Q-learning (see Fig. 13).

An AC operator can be included in arrange to apply the shrewd asset administration to fulfil the vehicle client's QoS. The faultfinder preparation: the point of the faultfinder preparation of the AC learning specialist is to assess the quality of the arrangement that the learning framework is supposed to rummage around for [34]. The on-screen character Prepare: the stochastic arrangement slope (PG) strategy more often than not utilizes the performing artist portion to overhaul parameterized policies in arrange to optimize the approach completely to improve the objective work within the condition:

$$J(\pi) = E\{Q^{\pi}(s, a)\} = \int_S d(s) \int_A \pi(s, a) Q^{\pi}(b, a) dads \tag{45}$$

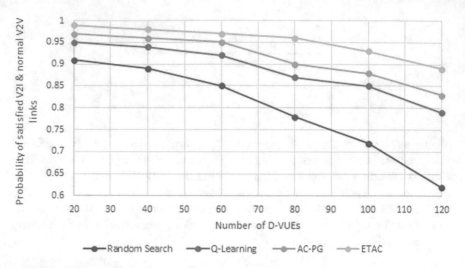

Fig. 13 D2D-V2V communication in ETAC agent performance

The policy can be initialized by the parameter vector $\theta = (\theta_1, \theta_2, \ldots .\theta_u)$ is represented by $\pi_\theta(s, a) = \Pr(a|s, \theta)$ thus the gradient of the policy with respect to θ is stated in the following equation. Gibbs distribution represents the parameterized stochastic policy $\pi_\theta(s, a)$

The approach can be initialized by the parameter vector $= (\theta_1, \theta_2, \ldots .\theta_u)$ is spoken to by $\pi_\theta(s, a) = \Pr(a|s, \theta)$ in this way the slope of the arrangement with regard to θ is expressed within the taking after condition. Gibbs dispersion speaks to the parameterized stochastic arrangement $\pi_\theta(s, a)$

$$\nabla_\theta J(\pi_\theta) = E\{Q^\pi(s, a)\} = \int_S d(s) \int_A \pi_\theta(s, a) Q^{\pi_\theta}(b, a) dads \qquad (46)$$

$$\pi_\theta(s, a) = \frac{\exp(\theta^T.\Phi(s, a))}{\sum_{a' \in A} \exp(\theta^T.\Phi(s, a'))} \qquad (47)$$

Where, $\Phi(s, a)$ is the long-run vector. Another policy parameter vector is over-hauled based on the slope of the objective remunerate given within the taking after the condition

$$\theta_{t+} = \nabla_\theta \beta_a J(\pi_\theta) + \theta_t \qquad (48)$$

where: β_a is the learning rate of the actor.

ETAC proposes the capacity of the vehicle clients to moment learned procedures from more encounter VUE, giving it an autonomous learning preparation. By utilizing the esteem work parameter vector v is upgraded within the pundit portion, and the approach parameter θ is overhauled by the performing artist. Both of these vectors can be overhauled amid the learning handle iteratively and at the same time [34].

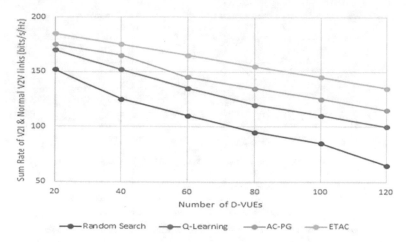

Fig. 14 ETAC performance evaluation taking the communication links sum rate within the V2V-D2D network

To overcome the standard AC approach issues, the qualification follow component is utilized to improve the productivity of the learning handle, and the advantage work coordinates with the pattern surmised is utilized to lessen the change within the gradient also to progress the work estimation exactness within the pundit portion, and the activity technique exchange learning is given to improve the in general learning quality by lifting the merging speed [34, 36]. An execution assessment of the ETAC has appeared in Fig. 14. By considering the shifting number of D-VUEs, ETAC has appeared superior comes about than other reinforcement learning approaches with an improved sum rate.

7.3 Optimization Performance

In IoV arrange structure, the gigantic sum of information is being traded between the distinctive IoV components vehicles, RSUs, Drivers, and framework. The information ought to be put away and trades in a secured environment to upgrade activity security and effectiveness. Two sorts of operations are considered in IoV to organize information capacity and information sharing. A piece chain framework is proposed to work in a supportive learning approach, pointing to optimize the execution of the IoV organize.

The blockchain system comprises of two primary parts. To begin with portion, Square makers: with the presumption of N number of hubs i.e., piece producers' candidates, K piece makers, ΦS set of hubs, γn (in token) and c_n (in GHz) for stake and computational assets. Moment portion, agreement models: applying common-sense byzantine blame tolerant (PBFT) calculation for agreement is considered as an exceptionally strong convention. Where a client issues a square with a number of

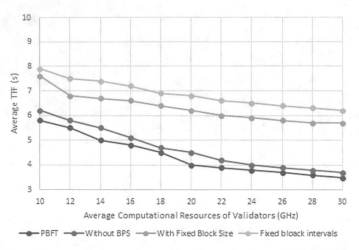

Fig. 15 Practical Byzantine fault tolerant (PBFT) performance assessment in terms of average and latency resources of computational

operations capacity or sharing and approves it through broadcasting it to other validators to reach an agreement. The method includes trading and confirming messages (see Fig. 15). The time changing transmission joins is displayed through finite-state Markov channels (FSMC). For the message confirmation as it were the taken a toll of cryptographic operation is computed, counting signature confirmation, message confirmation codes era MACs, and MACs confirmation [35].

Among diverse plans applying the piece chain framework, the normal latency/time to finality (TTF) diminishes with the increment of normal computational assets since that agreement can become more rapidly among more computationally able validators. It moreover takes the slightest time least normal TTF for the proposed plot to affirm exchanges when compared. The Proposed Profound Fortification learning DRL approach for execution optimization comprises of state space, activity space, and reward work.

A. **State space:** State space is denoted at decision epoch t ($t = 1, 2 \ldots$), mean traffic size is represented by χ, stake distribution Υ, computing competence of node $c = \{c_k\}$, transmission rate of the links between devices is $R = \{R_{i,j}\}$. $S^{(t)} = [\chi, \Upsilon, c, R]$

B. **Action space:** For maximizing the operational throughput, piece chain parts ought to be changed to adjust to the exceedingly energetic environment of IoV arrange. Modification incorporates block makers a, block size S^B and block periods T^I. The action space at decision epoch t is begin as follows $A^{(t)} = [a, S^B, T^I]$. where: the block maker pointer is $a = \{a_n\}$, $a_n \in \{0, 1\}$, $z_n \in \Phi_s$ with $a_n = 1$ meaning node z_n is for a block maker while $a_n = 0$ then; block size $S^B \in \{0.2, 0.4, \ldots, \dot{S}\}$, and lastly the block periods $T^I \in \{0.5, 1, \ldots, \dot{T}\}$.

C. **Reward work:** The remunerate function's primary part is to guarantee the maximization of organized throughput whereas keeping up the decentralization, conclusion, and security of the square chain framework [35].

8 Deep Learning Driven IoV Networks

Vehicular arrange of associated vehicles utilizing remote systems that give tall portability broadband get to have picked up much consideration from the industry and researchers. For the developing number of activity mishaps, the require for fasters and more astute explanatory frameworks are critical, hence numerous researchers proposed a colossal number of inquiries about in this field, utilizing profound learning procedures for its demonstrated comes about and tall execution in zones of collision location, investigation, and notice [36]. Comparable to the OSI show, the IoV design comprises of numerous layers such as the Network–layer, and Application-Layer depending on the proposed design of the shrewd transportation framework ITS application.

8.1 DL Driven Network-Level Internet of Vehicle Data Analysis

Creating multidimensional layered engineering that incorporates the brilliantly computing, real-time enormous information analytics, and IoV measurements. The web of vehicle measurement is separated into layers, to bolster the information investigation operations [36]. These layers are expressed as:

A. **Recognition layer**, incorporate sensors, actuators, vehicles, and keen phones.
B. **Framework organize layer,** contains RSUs, AP, BS, and switches, it is the spine of the IoV organize giving the communication framework to supply the gadgets with the vital network as well as information preparing and information storage.
C. **Counterfeit insights layer,** comprises of all the computational calculations and designs,
D. **Communication layer**, give the specified communication such as 5G, 4G/LTE [37]. Different Profound learning procedures are utilized for information investigation picture classification, proposing beginning v3 arrange design –based is received for binary image classification accomplished the most excellent picture classification comes about, and it could be a CNN demonstrate employments GPU designed computer [36]. Beginning v3 utilizes two profound learning arrange architectures:

- Densely associated convolutional systems (Dens-Net) increment the profundity of high-dimensional neural networks

- Squeezes-and-excitation systems (SE-Net) are capable for sifting the final yield includes to cancel or expel highlights that are not required for the current task.
- Feature shapes of inputs are altered to fit within the beginning v3 show since Dens-Net and SE-Net cannot be straightforwardly connected [36, 37].

8.2 DL Driven App-Level Internet of Vehicle Data Analysis

The application layer work is to supply information and control APIs to permit applications to utilize the apportioned information store and decide the arrangement setup. It is mindful for providing the IoV organize with security and non-safety applications through the common communication models in IoV, i.e., V2V, V2I, V2 and etc. [37]. Real-time information gathering and investigation is required in IoV systems to guarantee prompt and fitting activities, for illustration emergency vehicle requiring genuine-time information on activity data to maintain a strategic distance from activity jams and spare lives.

Keeping up the OoS prerequisites is vital within the IoV applications in arrange to preserve the efficient operation of IoV administrations. Applications generally work within the application layer accept the internet administrations from the benefit layer utilizing push/pull or publish/subscribe information spread procedures, the thrust technique includes pushing overhauls straightforwardly from the benefit layer to the application layer whereas drag methodology pulls data from the benefit layer [37, 38, 41]. The publish/subscribe methodology get to lets applications to enlist their intrigued to be informed approximately the benefit overhauls [38].

There are three primary categories for applications i.e., real-time, close real-time, and group information requirements-based applications, thrust and publish/subscribe are best fitted for these sorts of applications, whereas drag or publish/subscribe best fit close real-time applications for their endure of delay in the reaction such as course arranging applications, and versatile activity flag planning applications. Noteworthy information is required by the group information necessity; in this way, it has moo QoS prerequisites in terms of idleness [38]. Handle offloading and security applications as real-time QoS prerequisite is required for real-time applications, these applications offer assistance in complex information handling such as video handling on IoV arrange.

8.3 DL IoV Systems Clustering Procedures and Protocols

Clustering's primary objective is to bunch and categorize comparable information together into a cluster based on the similitude. Classical clustering strategies are ordinarily assembled into partition-based strategies, density-based strategies, and

progressive strategies, classification is most appropriate when centring on the organized design utilized for clustering instead of the clustering misfortune [40]. There
is numerous progress within the field of clustering and numerous calculations are
presented such as AE-Based profound clustering, K-Means, and DBSCAN.

A.   AE-Based DL Cluster Algorithm

Auto-encoder may be a sort of Neural Organize is used for unsupervised learning,
its objective is to play down remaking, is the mean-square mistake or the cross-
entropy between the yield and input, disciplines arrange for made a yield diverse
from input [41]. It is spoken to by the taking after the condition

$$\min_{\phi,\theta} L_{rec} = min \frac{1}{n} \sum_{i=1}^{n} \left\| x_i g_\theta \left(f_\phi(x_i) \right) \right\|^2 \tag{49}$$

where, ϕ represents the encoder parameter and θ represents the decoder parameter.
Encoder function $h = f_\phi(x)$ lookup the original data x into a latent notation h, and
a decoder that construct a restructure $r = g_\theta(h)$.

The recreated r must be as comparable as conceivable to x. Put in intellect that both
encoder and decoder can be developed by fully-connected neural organize or CNN.
The recreation misfortune of an auto-encoder comprises of as it were the contrasts
between the input layer and yield layer, but the misfortunes of all layers can be
optimized within the AE-based profound learning within the taking after equation:

$$L = (1 - \lambda)L_c + \lambda L_{rec} \tag{50}$$

The AE-based profound clustering calculation is outlined in Fig. 16. Numerous
strategies have been presented based on the AE-based profound clustering such
as profound clustering arrange (DCN), profound implanting organize (Sanctum),
profound subspace clustering systems (DSC), profound implanted regularized
clustering (Delineate), and profound ceaseless clustering (DCC) [39].

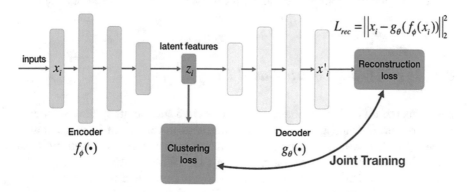

Fig. 16 Auto-encoder based deep cluster procedure architecture

B.   K-Means Cluster

K-Means clustering calculation builds clusters that are circular in shape, it is one of the speediest and least difficult calculations, and it is additionally a partition-based clustering calculation. It isolates objects in a dataset into a settled number of K disjoint subsets [40]. In each cluster, the calculation increases the homogeneity within the cluster utilizing by decreasing the square blunder. It is spoken to within the taking after condition:

$$E = \sum_{i=1}^{K} \sum_{j=1}^{n} |dist(x_j, c_i)|^2 \tag{51}$$

Where the square mistake breaks even with the remove squared between each protest x and the Cruel (or middle) of the cluster, and c signifies the individual middle of each cluster.

C.   Density-Based Spatial Cluster of Applications with Noise (DBSCAN)

Density-Based calculations consider clusters as thick zones of objects that are isolated by less thick zones [40]. These clustering calculations is predominant to partition-based calculation since it isn't confined to finding circular melded clusters, but can discover clusters of irregular shapes. DBSCAN is based on the concept of thickness reachability and density-connectivity, depend on two parameters: epsilon is the remove around a question q that characterizes its eps-neighbourhood, and the least number of focuses has work in connection with the other parameter epsilon.

The q question is said to be a central question when the number of objects inside the epsilon neighbourhood is at the slightest least number of focuses, at that point all objects are said to be density-reachable from q inside the epsilon neighbourhood [40]. Too, protest p is said to be density-reachable in case it is inside the epsilon neighbourhood of a protest that's straightforwardly density-reachable or thickness reachable from question q, meaning that the objects inside an epsilon -neighbourhood are reachable to each other. In this way, objects p and q are density-connected in the event that another question o exists being density-reachable from both objects p and q.

8.4   IoV Network Control Based DL

Next-generation communication counting both wired and remote systems is required for a brilliant instrument to control the enormous development in organizing an activity for their changing nature of organizing scenes such as versatility, cloud computing and huge information preparing [43]. Within prepare of making the systems more brilliant by making them learn how to course arrange activity between switches to optimize the arrange execution. Deep learning has appeared promising comes about in arranging control, by partitioning the system into stages. Stage 1 is an introductory stage, getting the related information for preparing the Deep framework. Stage 2 is the preparing stage, based on the collected information on the past stage, applying administered learning calculations to prepare the Deep learning-based engineering. Stage 3, speaks to the running stage where we put the learned calculation into an activity where activity designs are utilized as an input that yields a modern way of course after learning the organized conduct based on Deep learning algorithms [41].

Virtualization is proposed as a modern headway to the IoV organize, to have more control over the organize assets and to rise the arrange execution. The single root I/O virtualization (SR-IOV) is proposed to permit I/O gadgets to be shared by numerous virtual machines (VMs) without diminishing the runtime execution of the generally organize. SR-IOV is able of making Virtual Capacities permit visitors to have coordinate get to [42]. It is an proficient network device that gives the good thing about coordinate I/O throughput and diminishes CPU utilization whereas altogether expanding the versatility and sharing capacities of the gadget [41, 42].

8.5   Predictive Analytics "Regression" Issues in DL

ML employments directed learning in having an input variable x and yield variable y employing a learning algorithm in between to memorize the mapping work from the input to the output = f(x). Basically, having an input and anticipating a yield after wrapping up the learning prepare. Approaches to directed machine learning incorporate, straight and calculated relapse, classification, choice tree, and bolster vector machine [43]. Relapse alludes to the operation of displaying the relationship between one or more free variable and a subordinate variable, its primary objective is expectation. Issues more often than not happen when relapse is gathered to anticipate numerical values such as costs, number of vehicles, foreseeing request estimating, or length of holding up at a line and etc. [44].

A.   **Simple Linear Regression:** the demonstrate gives presumptions almost both the cruel work communicated by $E(y|u\dashv)$ and the fluctuation work, with the presumption that for a few change $u = u(x)$ can be composed as

$$E(y|u) = \beta_0 + \beta_1 u \qquad (52)$$

Expressing the connection between y and u, straightforward straight relapse is working with as it were one indicator.

B.   **Multiple Regression:** issues of relapse happened has activated the ought to consider working with more than one indicator, having the subordinate variable y depend at the same time on indicators $x_1, x_2, ..., xn$. We start with n indicators $x_1, ..., x_n$ building a set of k terms based on the indicators u0, u1, ..., uk−1 to have the numerous straight relapse cruel work [45].

$$E(y|u_1, \ldots, u_{k-1}) = \sum\nolimits_{k=1}^{K} \beta_{k-1} u_{k-1} \tag{53}$$

Deep learning models have been proposed to tackle the relapse issue, for its demonstrated exhibition in PC vision assignments, for example, picture grouping, object location, and picture investigation [43]. The overall design comprises of various convolutional layers, trailed by not many completely associated layers, and a characterization Delicate Max layer, this engineering is alluded to as convolutional neural organization, known for its capacity to tackle relapse issues, for this situation the Delicate Max layer is supplanted with a completely associated relapse layer with straight or sigmoid actuations. Deep relapse calculations have given outcomes is old style vision relapse issues, for example, human posture assessment, and facial milestone discovery. Two basic models are proposed alluded to as VGG-16 and ResNet-50, VGG-16 is notable for its demonstrated abilities in working in Picture Net, while ResNet-50 has far superior execution with more limited preparing time. The two designs are prepared on Picture Net datasets. VGG-16 comprises of 5 squares, holding 2 or 3 convolutional layers and a maximum pooling layer, VGG-16 can be defined by [46]:

$$-\left(-CB^1 + \sum\nolimits_{i=2}^{5} CB^i + Fl + \sum\nolimits_{i=1}^{2} FC^i + SM\right) \tag{54}$$

$$-\left(-CB^1 + \sum\nolimits_{i=2}^{5} CB^i + GAP + SM\right) \tag{55}$$

where: CB^i defines the ith legacy block, FC^i defines the ith fully connected layer, Fl defines to the level layer that converts the 2D component map into a vector, SM defines to the delicate max layer. GAP defines to the worldwide normal pooling layer [47]. The ResNet-50 enlarges the organization's profundity and lessens the number of boundaries, thereby having a more limited preparation time. All the convolutional squares of ResNet-50 barring the primary layer have personality association makes them remaining convolutional blocks [48–52].

9 ML Future Directions and Challenges in IoV Applications

Machine Learning is expected to elevate the IoV applications as many advances are developed rapidly to enhance the operation of IoV applications, all in aims to improve the user's experience [53]. With the introductions of smart vehicles are capable of computer, data storage and communications the trending machine learning architectures are vital in the process of learning, smart vehicles are equipped with different types of sensors, multi-interface cards, and on-board wireless devices, all working together to have an overall system that is reliable and secured.

Smart vehicle services are expected to reach full autonomy i.e. self-driving by the year 2030, with many applications ahead such as in safety applications including driving safety, and, context awareness. The second application is No safety applications including video streaming in IoV specifically on ITS, augmented reality and in infotainment services [54]. Machine learning has a great role in the future to improve the overall experience by applying its architectures and algorithms in areas like network security and privacy, scheduling and load balancing, offloading, and resource management.

Future vehicular networks are expected to combine communications, computing and resource cashing. Thus, deep machine learning is expected to lead the researches in the coming years in order to improve QoS requirements. There is still much work to be done in improving the learning process having it to be less complex [55]. Also in dimensionality reduction by the use of compressed sensing theory, and semi-supervised learning where the dimensionality of data is reduced by measuring the similarities between training data and samples. Although deep machine learning is considered to be on its beginning it is obvious that advances made in developing machine learning approaches will shape the future of machine learning and artificial intelligence [56–58]. In the process of applying DL to IoV researchers are often faced with a number of challenges which are posing opportunities for more improvements is order to come up with a robust system which satisfies QoS requirements. These challenges are represented as general such as:

- Data Volume, in which tons of network parameters need to be estimated properly, creating a high volume of data in need for careful processing [60].
- Computational complexity is high in an environment that is features with high mobility and continuous change [61].
- Incomplete perception, which is due to the limited sensing capabilities and information loss caused by limited transmission capability in the network layer [62].
- Delayed control, which is the delay between measuring a system's state and acting upon it. Control delay is always present in real systems due to transporting measurement data to the learning agent, computing the next action, and changing the state of the actuator [63].
- Multi-Agent Control, for distributed and semi-distributed architectures it is challenging to enable efficient collaboration and fair competition among multiple agents [64].
- Cost of applying DL to IoV, it has been proven that training a DL is expensive having many factors contribute to the cost which are: the size of dataset, model size, and training volume. The researchers at AI21 labs have quantitatively estimated the costs of training differently sized BERT models on the wikipedia and book corpora where they have obtained the cost of one training run, and a typical fully-loaded cost [65].

10 Conclusion

With the rapid development of deep learning approaches, architectures, and algorithms with consideration of the growing IoV applications and services. These triggered more and extensive research to improve the life style. Deep learning approaches for IoV were discussed in this chapter including convolutional neural networks (CNN), recurrent neural networks (RNN), deep reinforcement learning (DRL). These approaches have proven a high performance when applied to IoV applications and services, i.e., in autonomous driving, traffic monitoring, accident prevention, traffic guidance systems, safe navigation, electronic toll collection, and safe navigation. In future, the machine learning promises for optimization solutions to the QoE and QoS for smart future vehicular networks. The improvements will enhance the quality of data streaming for both entertainments and traffic services. However, the deployment of ML approaches on IoV will face different challenges due to large amount of exchanged data and different resources accessibility.

References

1. Shetty, D., Harshavardhan, C.A., Jayanth Varma, M., Navi, S., Ahmed, M.R.: Diving deep into deep learning: history, evolution, types and applications. Int. J. Innov. Technol. Explor. Eng. (IJITEE) 9(3), 2835–2846 (2020)
2. Yao, S., Zhao, Y., Zhang, A., Hu, S., Shao, H., Zhang, C., Su, L., Abdelzaher, T.: Deep learning for the internet of things. Computer 51(5), 32–41 (2018)
3. Pathan, A.-S.K., Saeed, R.A., Feki, M.A., Tran, N.H.: Integration of IoT with future internet. J. Internet Technol. (JIT) 15(2), 145–147 (2014)
4. Verbraeken, J., Wolting, M., Katzy, J., Kloppenburg, J., Verbelen, T., Rellermeyer. J.S.: A survey on distributed machine learning. arXiv:1912.09789, 20 December 2019
5. Ibrahim, S., Saeed, R.A., Mukherjee, A.: Resource management in vehicular cloud computing, chap 4. In: Grover, J., Vinod, P., (eds.) Vehicular Cloud Computing for Traffic Management and Systems, pp. 75–97. IGI Global, USA (June 2018)
6. Amodei, D., et al.: Deep Speech 2: end-to-end speech recognition in English and Mandarin. In: International Conference on Machine Learning, vol. 48. JMLR: W&CP (2016)
7. Bojarski, M., Del Testa, D., Dworakowski, D., Firner, B., Flepp, B., Goyal, P., Jackel, L.D., Monfort, M., Muller, U., Zhang, J., Zhang, X., Zhao, J., Zieba, K.: End to end learning for self-driving cars. arXiv:1604.07316v1, 25 April 2016
8. Khandani, A.E., Kim, A.J., Andrew, W.L.: Consumer credit-risk models via machine-learning algorithms. J. Bank. Finance 34, 2767–2787 (2010)
9. Peteiro-Barral, D., Guijarro-Berdiñas, B.: A survey of methods for distributed machine learning. Prog. Artif. Intell. 2(1), 1–11 (2013)
10. Qiu, J., Qihui, Wu., Ding, G., Yuhua, Xu., Feng, S.: A survey of machine learning for big data processing. EURASIP J. Adv. Sig. Process. 2016(1), 67 (2016)
11. Abdelgadir, M., Saeed, R.A., Babikir, A.A. Mobility routing model for vehicular ad-hoc networks (VANETs), smart city scenarios. Veh. Commun. 9, 154–161 (2017)
12. Liu, L., Özsu, M.T.: Encyclopedia of Database Systems. Springer, US (2018)
13. Li, P.: Optimization algorithms for deep learning. Department of Systems Engineering and Engineering Management, The Chinese University of Hong Kong
14. Ruder, S.: An overview of gradient descent optimization algorithms. arXiv:1609.04747v2 [cs.LG], 15 June 2017

15. Duchi, J., Hazan, E., Singer, Y.: Adaptive sub gradient methods for online learning and stochastic optimization. J. Mach. Learn. Res. **12**, 2121–2159 (2011)
16. Zeiler, M.D.: ADADELTA: an adaptive learning rate method. arXiv:1212.5701v1 [cs.LG], 22 December 2012
17. Kingma, D.P., Ba, J.L.: Adam: a method for stochastic optimization. arXiv:1412.6980v9 [cs.LG], 30 January 2017
18. Hsu, R.C.-H., Wang, S., (eds.): Internet of vehicles – technologies and services. In: 1st International Conference, IOV 2014, Beijing, China, 1–3 September 2014 (2014)
19. Ahmed, Z.E., Saeed, R.A., Mukherjee, A.: Challenges and opportunities in vehicular cloud computing. In: Jyoti Grover, P., Vinod, C.L. (eds.) Vehicular Cloud Computing for Traffic Management and Systems, pp. 57–74. IGI Global (2018). https://doi.org/10.4018/978-1-5225-3981-0.ch003
20. Sharma, S., Ghanshala, K.K., Mohan, S.: A security system using deep learning approach for internet of vehicles (IoV). In: 2018 9th IEEE Annual Ubiquitous Computing, Electronics & Mobile Communication Conference (UEMCON) (2018)
21. Abdelgadir, M., Saeed, R.A.: Evaluation of performance enhancement of OFDM based on cross layer design (CLD) IEEE 802.11p standard for vehicular ad-hoc networks (VANETs), city scenario. Int. J. Sig. Process. Syst. **8**(1), 1–7 (2020)
22. Chen, C.-H., Lee, C.-R., Walter Chen-Hua, L.: A mobile cloud framework for deep learning and its application to smart car camera. In: Hsu, C.-H., Wang, S., Zhou, A., Shawkat, A. (eds.) Internet of Vehicles – Technologies and Services, pp. 14–25. Springer International Publishing, Cham (2016)
23. Chen, C., Xiang, H., Qiu, T., Wang, C., Zhou, Y., Chang, V.: A rear-end collision prediction scheme based on deep learning in the Internet of Vehicles. J. Parallel Distrib. Comput. **117**, 192–204 (2017)
24. O'Shea, K., Nash, R.: An introduction to convolutional neural networks. arXiv:1511.08458v2 [cs.NE], 2 December 2015
25. Albawi, S., Mohammed, T.A.: Understanding of a convolutional neural network. In: ICET 2017, Antalya, Turkey (2017). https://doi.org/10.1109/ICEngTechnol.2017.8308186.
26. Principe, J.C., Euliano, N.R., Lefebvre, W.C.: Multilayer perceptrons, chap. 3. In: Neural and Adaptive Systems: Fundamentals Through Simulation (1997)
27. Yoon, S., Kum, D.: The multilayer perceptron approach to lateral motion prediction of surrounding vehicles for autonomous vehicles. In: 2016 IEEE Intelligent Vehicles Symposium (IV), Gothenburg, Sweden, 19–22 June 2016 (2016)
28. Ferreira, M.D., Corrêa, D.C., Nonato, L.G., de Mello, R.F.: Designing architectures of convolutional neural networks to solve practical problems. Exp. Syst. Appl. **94**, 205–217 (2018)
29. Amirul Islam, M., Hossan, T., Jang, Y.M.: Convolutional neural network scheme–based optical camera communication system for intelligent Internet of vehicles. Int. J. Distrib. Sens. Netw. **14**(4), 155014771877015 (2018)
30. DiPietro, R., Hager, G.D.: Deep learning: RNNs and LSTM. In: Handbook of Medical Image Computing and Computer Assisted Intervention (2020)
31. Du, K.-L., Swamy, M.N.: Neural Networks and Statistical Learning. Springer, London (2019)
32. Virmani, S., Gite, S.: Performance of convolutional neural network and recurrent neural network for anticipation of driver's conduct. In: 2017 8th International Conference on Computing, Communication and Networking Technologies (ICCCNT) (2017)
33. Eltahir, A.A., Saeed, R.A.: V2V communication protocols in cloud assisted vehicular networks, chap 06. In: Grover, J., Vinod, P. (eds.) Vehicular Cloud Computing for Traffic Management and Systems, pp. 125–150, June 2018. IGI Global, USA. https://doi.org/10.4018/978-5225-3981-0. ISBN13: 9781522539810, ISBN10: 1522539816
34. Ning, Z., Dong, P., Wang, X., Guo, L., Rodrigues, J.J.P.C., Kong, X., Huang, J., Kwok, R.Y.K.: Deep reinforcement learning for intelligent internet of vehicles: an energy-efficient computational offloading scheme. IEEE Trans. Cogn. Commun. Netw. **5**(4), 1060–1072 (2019)
35. Atallah, R.F., Assi, C.M., Khabbaz, M.J.: Scheduling the operation of a connected vehicular network using deep reinforcement learning. IEEE Trans. Intell. Transp. Syst. **20**(5), 1669–1682 (2019)

36. Atallah, R., Assi, C., Khabbaz, M.: Deep reinforcement learning-based scheduling for road-side communication networks. In: 2017 15th International Symposium on Modeling and Optimization in Mobile, Ad Hoc, and Wireless Networks (WiOpt) (2017)
37. Cheng, M., Li, J., Nazarian. S.: DRL-cloud: deep reinforcement learning-based resource provisioning and task scheduling for cloud service providers. In: 23rd Asia and South Pacific Design Automation Conference (ASP-DAC) (2018)
38. Yang, H., Xie, X., Kadoch, M.: Intelligent resource management based on reinforcement learning for ultra-reliable and low-latency IoV communication networks. IEEE Trans. Veh. Technol. **68**(5), 4157–4169 (2019)
39. Liu, M., Teng, Y., Yu, F.R., Leung, V.C.M., Song, M.: Deep reinforcement learning based performance optimization in blockchain-enabled internet of vehicle. In: 2019 IEEE International Conference on Communications (ICC) (2019)
40. Chang, W.-J., Chen, L.-B., Su, K.-Y.: DeepCrash: a deep learning-based internet of vehicles system for head-on and single-vehicle accident detection with emergency notification. IEEE ACCESS **7**, 148163–148175 (2019)
41. Al-Hmoudi, M.I., Saeed, R.A., Hasan, A.A., Khalifa, O.O., Mahmoud, O., Sellami, A.: Power control for interference avoidance in femtocell network. Aust. J. Basic Appl. Sci. (AJBAS) **5**(6), 416–422 (2011)
42. Darwish, T.S.J., Bakar, K.A.: Fog based intelligent transportation big data analytics in the internet of vehicles environment: motivations, architecture, challenges, and critical issues. IEEE Access **6**, 15679–15701 (2018)
43. Iqbal, R., Butt, T.A., Omair Shafiq, M., Talib, M.W.A., Umar, T.: Context-aware data-driven intelligent framework for fog infrastructures in internet of vehicles. IEEE Access **6**, 58182–58194 (2018)
44. Min, E., Guo, X., Liu, Q., Zhang, G., Cui, J., Long, J.: A survey of clustering with deep learning: from the perspective of network architecture. IEEE Acess **6**, 39501–39514 (2018)
45. Erman, J., Arlitt, M., Mahanti, A.: Traffic classification using clustering algorithms, Pisa, Italy, 11–15 September 2006 (2006)
46. Kato, N., Fadlullah, Z.M., Mao, B., Tang, F., Akashi, O., Inoue, T., Mizutani. K.: The deep learning vision for heterogeneous network traffic control: proposal, challenges, and future perspective. IEEE Wirel. Commun. **24**, 146–153 (2016)
47. Senan, S., Hashim, A.H.A., Saeed, R.A., Daoud, J.I.: Evaluation of nested network mobility approaches. J. Appl. Sci. **11**(12), 2244–3349 (2011)
48. Dong, Y., Yu, Z, Rose, G.: SR-IOV networking in Xen: architecture, design and implementation, Xen is a trademark of XenSource, Inc. https://www.usenix.org/legacy/events/wiov08/tech/full_papers/dong/dong.pdf
49. Brownlee, J.: Machine learning algorithms. Logistic regression for machine learning, 12 August 2019. https://machinelearningmastery.com/logistic-regression-for-machine-learning/
50. Zhang, A., Lipton, Z.C., Li, M., Smola, A.J.: Book: Dive into Deep Learning Release 0.7.1, 02 July 2020. https://d2l.ai/d2l-en.pdf
51. Lindley, D V., Novick, M.R., Pearl, J., Simpson, E.H.: Linear Hypothesis: Fallacies and Interpreti e Problems (Simpson's Paradox). In: International Encyclopedia of the Social & Behavioral Sciences. Elsevier Science Ltd. (2001). ISBN 0-08-043076-7
52. Lathuilière, S., Mesejo, P., Alameda-Pineda, X., Horaud, R.: A comprehensive analysis of deep regression. arXiv:1803.08450v2 [cs.CV], 13 February 2019
53. Hassan, M.B., Ali, E.S., Mokhtar, R.A., Saeed, R.A., Chaudhari, B.S.: NB-IoT: concepts, applications, and deployment challenges, chap. 6. In: Chaudhari, B.S., Zennaro, M., (eds.) LPWAN Technologies for IoT and M2M Applications. Elsevier, March 2020. ISBN 9780128188804
54. Ahmed, Z.E., Saeed, R.A., Ghopade, S.N., Mukherjee, A.: Energy optimization in LPWANs by using heuristic techniques, chap. 11. In: Chaudhari, B.S., Zennaro, M., (eds.) LPWAN Technologies for IoT and M2M Applications. Elsevier, March 2020. ISBN 9780128188804
55. Saeed, R.A., (ed.): WiMAX, LTE, and WiFi interworking. J. Comput. Syst. Netw. Commun. **2010**, 2 (2010). Article ID 754187

56. Raza, S., Wang, S., Ahmed, M., Anwar, M.R.: A survey on vehicular edge computing: architecture, applications, technical issues, and future directions. Wirel. Commun. Mob. Comput. **2019** (20119). Article ID 3159762
57. Hassan, M.B., Ali, E.S., Nurelmadina, N., Saeed, R.A.: Artificial intelligence in IoT and its applications. In: Intelligent Wireless Communications. IET Book Publisher (2020)
58. Miotto, R., Wang, F., Wang, S., Jiang, X., Dudley, J.T.: Deep learning for healthcare: review, opportunities and challenges. Brief. Bioinform. **19**(6), 1236–1246 (2017)
59. Eom, J., Kim, H., Lee, S.H., Kim, S.: DNN-assisted cooperative localization in vehicular networks. Energies **12**(14), 2758 (2019)
60. Lei, L., Tan, Y., Zheng, K., Liu, S., Zhang, K., Shen, X.: Deep reinforcement learning for autonomous internet of things: model, applications and challenges. IEEE Commun. Surv. Tut. **22**(3), 1722–1760 (2020)
61. Sagar, R.: Why deep learning is a costly affair. Anal. India Mag. (2020)
62. Robinds, H., Monro, S.: A stochastic approximation method. Ann. Math. Stat. **22**, 400–407 (1951)
63. Nesterov, Y.: A method for unconstrained convex minimization problem with the rate of convergence o($1/k^2$). Doklady ANSSSR **269**, 543–547 (1983). Translated as Soviet. Math. Docl.
64. Saeed, R.A., Khatun, S., Ali, B.M., Khazani, M.: A juoint PHY/MAC cross-layer design for UWB under power control. Comput. Electr. Eng. (CAEE) **36**(3), 455–468 (2010)
65. Ruder, S.: An overview of gradient descent optimization algorithms. arXiv:1609.04747, January 2016

Intelligently Reduce Transportation's Energy Consumption

Andreas Andreou, Constandinos X. Mavromoustakis, George Mastorakis, Evangelos Pallis, Naercio Magaia, and Evangelos K. Markakis

Abstract Reducing vehicle energy demand has been the subject of academic research for several years. Distributing energy equally and efficiently among vehicle wheels addresses challenges that may have an impact on energy consumption. Therefore, this chapter aims to contribute through the proposal of a novel system by integrating smart panels. The effects of road's change of gradient and anomalies motivate us to work on more torque at the wheel joints that demand more energy procurement. Though the supply affects all the wheels equally, which interprets for the engine as more consumption. We approach the emerging issue by developing a V2SP (Vehicle to Smart-Panel) system in fluid dynamics. Based on fundamental mathematics equations of fluid flow from Navier Stokes, we implement an algorithm that could run in real-time on the grid that will be integrated on panels and interact accordingly with

A. Andreou (✉) · C. X. Mavromoustakis
Mobile Systems Laboratory (MoSys Lab), Department of Computer Science,
University of Nicosia, Nicosia, Cyprus
e-mail: andreou.andreas@unic.ac.cy

C. X. Mavromoustakis
e-mail: mavromoustakis.c@unic.ac.cy

G. Mastorakis
Department of Management Science and Technology, Hellenic Mediterranean University,
Agios Nikolaos, 72100 Crete, Greece
e-mail: gmastorakis@hmu.gr

E. Pallis
Department of Electrical and Computer Engineering, Hellenic Mediterranean University,
71500 Estavromenos, Heraklion, Crete, Greece
e-mail: pallis@pasiphae.eu

N. Magaia
LASIGE, Department of Computer Science, Faculty of Sciences, University of Lisbon,
1749-016 Lisboa, Portugal
e-mail: ndmagaia@ciencias.ulisboa.pt

E. K. Markakis
Department of Electrical and Computer Engineering, Hellenic Mediterranean University,
71500 Estavromenos, Heraklion, Crete, Greece
e-mail: markakis@pasiphae.eu

© Springer Nature Switzerland AG 2021
N. Magaia et al. (eds.), *Intelligent Technologies for Internet of Vehicles*, Internet of Things,
https://doi.org/10.1007/978-3-030-76493-7_9

a vehicle's operational system. The objective is the distribution of the appropriate energy among wheels based on their needs.

Keywords Energy consumption · Internet of things IoT · Internet of vehicles IoV · Smart-panels

1 Introduction

The rapid increase in urban transport results in an increase in energy consumption. Consequently, air pollution levels affect multiple biological systems involved in functional decline and the increased risk of disease in the ageing population. The chapter proposed a novel IoV solution to reduce energy consumption with an innovation integrated into hybrid vehicles based on pressure-driven interpretation from the road to the smart panels. Although it may not be obvious, there is a direct link between energy consumption and air pollution in an ecosystem. When we consume less energy, we reduce the number of toxic fumes emitted by vehicles, preserve the earth's natural resources. Therefore, the chapter's research work enables innovation by pursuing economic energy consumption through the novel contribution of smart panels to allow the optimal distribution of power among the wheels. The proposed framework elaborates the in-vehicle IoT connection between the suspension system by integrating Smart-Panels with the vehicle data management system. We enable a novel module of IoV ecosystem called Vehicle to Smart-Panel (V2SP), as there is no reference-based to an in-vehicle IoT connection in literature.

1.1 Internet of Vehicles (IoV)

IoT integrations focus on large-scale IoV research, an emerging vehicle technology that integrates IoT and an intelligent transport framework [1]. The gradient of the road effects fuel consumption as shown by the Travesset-Baro et al. who also proved that the slope has a noticeable impact on fuel efficiency [2]. Most research works consider road gradient as negligible, which leads to inaccurate measurements and numerous errors for the coefficient of correlation [3, 4]. Wyatt et al. prove that a level street profile, as opposed to an inclining one, could prompt errors in assessing the genuine energy utilisation and CO_2 discharges of vehicles [5]. According to the vehicle's fuel usage, models have developed to estimate the energy required to deal with external contrast factors, including blockage motion, air friction, angle blockage, and blockage acceleration [6]. Therefore, this chapter's contribution is to investigate the impact of the road's gradient and anomalies on energy consumption and enable an innovative V2SP system to reduce battery energy consumption.

Figure 1 illustrates the communication patterns among IoV systems, for instance, Vehicle-to-Grid (V2G), Vehicle-to-Infrastructure (V2I), Vehicle-to-Person (V2P),

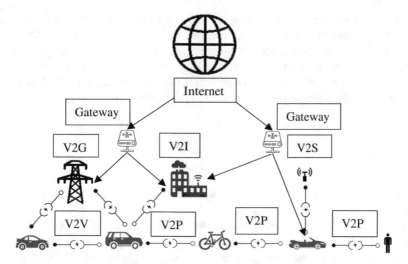

Fig. 1 Communication in IoV systems

Vehicle-to-Vehicle (V2V) and Vehicle-to-Sensor (V2S). Multi-services such as location services, resource sharing, public emergency, transportation and smart grid require different broadband speeds. Due to the enormous volume of data generated, high-level coordination, and flexible network management needed. Real-time communication in an IoV ecosystem enables the motivation to search for solutions for delay-sensitive vehicular applications.

1.2 Literature Review

Khayat et al. introduced the Vehicular Ad Hoc Network, abbreviated as VANET, discussing the five categories of routing protocols [7]. Yu et al. investigated the IoV-based vehicle-mounted media scenarios and constructed in-vehicle infotainment solutions for mobility services [8]. Xie et al. proposed an innovative and collaborative vehicular state-of-the-art network architecture to extend computing and storage resources for the edge of the network [9]. Schouveiler et al. experiment on circular and spiral waves by the observation in the flow between a rotating and a stationary disk. The Spatiotemporal characteristics of the waves studied with Fourier transforms of these velocity signals [10]. Computation, communication and storage are out of paramount importance in IoV context. These requirements motivated Zhaolong et al. to develop an energy-efficient framework for Mobile Edge Computing-Enabled IoV to reduce energy consumption for roadside units that provide extensive network coverage along highways and main streets [11]. Furthermore, Liu et al. proposed and implemented an energy consumption model emphasising how the ambient temperature affects electricity consumption [12]. The blockchain is a protected, dependable, and inventive instrument for dealing with various vehicles searching for network. Nevertheless, following blockchain standards, the number of exchanges that need to be updated raises essential issues for vehicles such as those that can procure the most accessible energy. To address this, Sharma et al. present a productive model, suitable for meeting the energy demands of IoV that has been enhanced with blockchain, ideally controlling the number of exchanges through group traffic [13].

2   Fluid Dynamics

Fluid flow flows if an obstacle creates vortices that therefore produce forces in the engagement. Flat viscous flows managed by constant Navier - Stokes conditions with inhomogeneous Dirichlet limit information in a (virtual) square containing a deterrent. The diversity of the solution is relevant to the appearance of the forces within a symmetrical context. The fluid formula development could explain the road pressure on wheels based on realistic fluid results derived from fundamental mathematical formulas in physics. Navier–Stokes equation enables our algorithms to geared a visual representation of reality [14]. The algorithm runs in a real-time environment and emphasises stability and velocity that could be interpreted by random time steps.

The Navier–Stokes equations were developed between 1750 and 1850 and represent in fundamental mathematics how fluid flows. The equation exploited around 1950 by applying solutions embedded with algorithms in a computing environment by researchers. Therefore, solvers of the equations simulate realistic fluid-flow through the pressure on a grid, which sparks our motivation to implement applied mathematics algorithms to create visual effects in computation applied mathematics. The objective is to interpret these effects as pressure through the vehicle's computer system, which will adjust the information to the required power. The following equations are the Navier–Stokes equations for velocity in a reliable vector notation and the density equation moving through the velocity field.

$$\frac{\partial u}{\partial t} = -(u.\nabla)u + v\nabla^2 u + f$$

$$\frac{\partial p}{\partial t} = -(u.\nabla)p + k\nabla^2 p + S$$

At a given point in time, the liquid state proves to be a velocity carrier an ability that launches a velocity carrier to any end in space. The Navier–Stokes equations are an accurate representation of the progress of a velocity field over long distances. Given the current state of speed and the current regulation of forces, our conditions reveal how the rate will change in a small amount of time. The requirement states that the speed adjustment is due to the three essentials to the first equation's right-hand side. The density field's growth through the liquid's velocity field can also be represented

Fig. 2 Density exchange

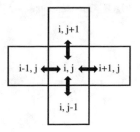

continuously by an exact arithmetic state. The reader isn't relied upon to completely comprehend these conditions.

Notwithstanding, it ought to be apparent to anybody that the two conditions above look a lot of the same, and this likeness was instrumental in improving our algorithm. Navier–Stokes equations defined as the change of density that occurs for three reasons over a single time step dt. On the right, we find the three terms that represent these three reasons as follows: in the beginning, density should follow velocity, in the second it should be interpreted as density should be said at a specific rate, and last it states that density increases according to sources.

2.1 Physical Approach

Velocity field refers to the distribution of what we called in natural language speed in a given region. The presentation of that field is visually remarkable as soon as it moves objects, i.e. liquid. The conversion of the velocity into the component force enables the motion's interpretation into Euclidean space and bi-dimensional space. Through the Cartesian coordinate system, it defines points uniquely in a plane. Objects with limited mass/volume due to the few particles they contain are low-density objects to adapt to velocity vectors. Therefore, fluid particles are replaced by fluid density in our work as it is prohibitively difficult to model any low-density liquid particle. Density values range within the interval [0, 1] and designate the quantity of particles in the area studied. We called substance movement diffusion due to the random movement of fluid particles from a field with high concentration to a low particle concentration field. As shown in Fig. 2, each cell interacts and exchange of density with each neighbouring cell [20]. Hence the diffusion solver calculates these density conversions between cells.

2.2 Fluid Flow in the Grid

Fluid flow formed in Fig. 3 in the two-dimensional square grid containing an additional column and a row of cells to increase the boundaries. The centre cell among

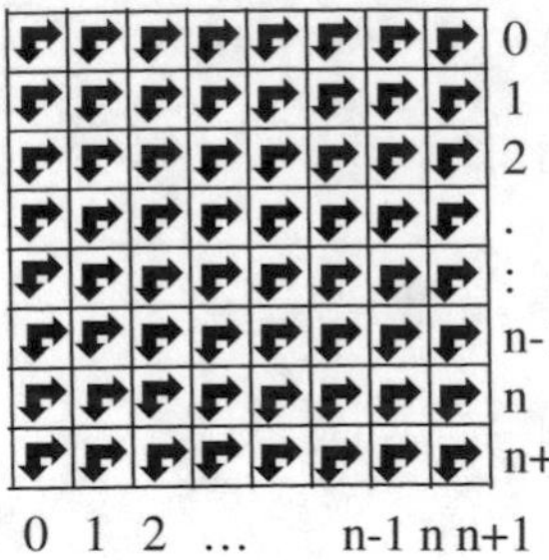

Fig. 3 Computational grid

grid cells presents the constant vectors of velocity and density. We start with some underlying condition for velocity and density and update its values, as indicated by cases that occur through pressure. The forces will move the liquid, while sources will inject densities when the smart panel exerts pressure. Therefore, the simulation represented by a set of speed and density images, and the time interval between them will be the variable dt.

3   Methodology

In general, vehicles powered by a single-engine or motor with a drive-line transferring that power to the wheels, generating torque at the wheel hubs. Although it will be beneficial for energy conservation, the wheels can spin at different speeds to one another to adapt cornering and variations based on the pavement's surface. Therefore, we propose a novel in-wheel power evaluation system to reduce energy consumption to motors without differential that delivers torque directly and independently to the wheels—the innovation based on fluid mechanics, a framework of fluid behaviour at rest and in motion. The bet was to transfer and interpret the pavement's pressure level from the wheels to the system to be more specific. Thus, we develop an app that translates the five most valuable relationships in fluid mechanics problems: kinematic, stress, conservation, regulating, and constitutive into pixels.

We propose an open differential engine to enable wheel speed and pressure to be different across the axles. Based on the friction force across tire and pavement that is dissimilar on one wheel to the other, each spin will transmit the diverse pressure to a panel able to interpret the amount of force into liquid pixels through the developed application. After that, the panels will transfer the information to the vehicle's central system through a virtual private network. Each unit among wheels comprises the panel integrated into a single package housed entirely within the wheel rim). Still, it would also be possible to incorporate the panels within the suspension system of the vehicle. The motor's main objective is to provide accelerating and braking torque among the wheels according to the requirement that derives from the road surface irregularities.

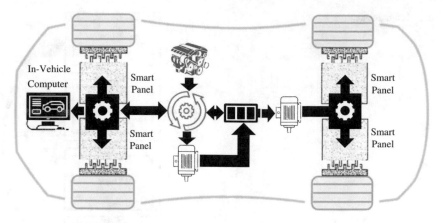

Fig. 4 Hybrid vehicular system architecture

Figure 4 presents the proposed operating system by intelligently integrating panels throughout the power transmission and distribution system into a hybrid vehicle. As shown, two electric motors generate force in the form of torque applied on the motor's shaft and an internal combustion engine. The battery is charged through regenerative braking and by the internal combustion engine. The panels' data will be transferred to the vehicle computer for interpretation and process to enable optimal battery power distribution through the machines.

3.1   Implementation

The development of the pseudocode for the solver begins with the initial state for density and the velocity. Based on the changes that occur due to the road's pressure, they inform their values (Fig. 5). For obvious reasons, the prototype allows us to transfer forces and density through the connected black surface tablet that functions through a dedicate pen. We used One by WACOM, which interprets the compressive forces that allow the liquid to move while the densities assimilate the injection into the grid. The following pseudocode shows the implementation of intake, which is the transport of fluid through mass motion. Instead of moving the cell forward in the velocity field, we look for the elements that end up right after the cells' reverse time. The Gauss–Seidel method enhanced our algorithmic technique as it simplified the process of inverting the matrix [15, 21]. The Poisson equation improves our algorithm for calculating the height field [16].

Fig. 5 Fluid density

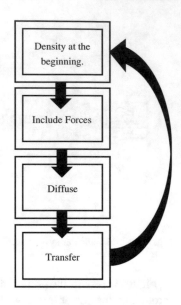

Pseudocode:

```
Smart Panel;

String[] fluid_names = {"hydraulics", "Pneumatics"};
float[][] fluid_properties = {{0.0016f, 0.0091f},
                              {0.0001f,0.0000001f}};
panel void settings(){
   size(N*SCALE,N*SCALE);
}
panel void setup(){
   //Initialization with hydraulics
   //diffusion, viscocity, dtime

   fluid = new Fluid(fluid_properties[0][0], fluid_prop-
erties[0][1], dt);
}
panel int Constrainchange (int maxV, int val){
   if(val >= maxV){
       return maxV - 1;
   }
   if(val<=0){
      return 1;
   }
   return val;
}
```

```
panel void TabletPen(){
    float amountX = tabletX - ptabletX;
    float amountY = tabletY - ptabletY;
    int maxX = PApplet.parseInt(width/SCALE) - 1;
    int maxY = PApplet.parseInt(height/SCALE) - 1;
    if (penPressed && (penButton == UP)){
      fluid.AddDensity(ConstrainTablet(maxX,tab-
letX/SCALE) , ConstrainTablet(maxY, tabletY/SCALE) ,
1000);
      fluid.AddVelocity(ConstrainTablet(maxX,tab-
letX/SCALE) , ConstrainTablet(maxY, tabletY/SCALE) ,
amountX*10, amountY*10);
    }
    if(tabletPressed && (penButton == Down)){
      fluid.AddDensity(ConstrainTablet(maxX,tab-
letX/SCALE) , ConstrainTablet(maxY, tabletY/SCALE), -
1000);
    }
}
panel void draw(){
  background(0);
  TabletPen();
  fluid.step();
  fluid.RenderD(true,true);
}
final int N=128;
final int GRIDSIZE = (N+2)*(N+2);
final int SCALE = 8;
final float dt = 0.05f;
panel int IX(int i, int j){
    return i + (N+2)*j;
}
class Fluid{
    float dt;
    float diff;
    float visc;
    int size;
    float[] u;
    float[] v;
    float[] u_prev;
    float[] v_prev;
    float[] dens;
    float[] dens_prev;
  Fluid(float diffusion, float viscosity, float dtime){
    this.diff = diffusion;
    this.visc = viscosity;
    this.dt = dtime;
```

```
    this.dens_prev = new float[GRIDSIZE];
    this.dens = new float[GRIDSIZE];
    this.u = new float[GRIDSIZE];
    this.u_prev = new float[GRIDSIZE];
    this.v = new float[GRIDSIZE];
    this.v_prev = new float[GRIDSIZE];
}
panel void step(){
    VelStep(u,v,u_prev,v_prev,visc,dt);
    DensStep(dens,dens_prev,u_prev,v_prev,diff);
}
panel void SetBnd(int b, float[] x){
    for (int i=1 ; i<=N ; i++ ) {
        x[IX(0 ,i)] = (b==1) ? -x[IX(1,i)] :
x[IX(1,i)];
        x[IX(N+1,i)] = (b==1) ? -x[IX(N,i)] :
x[IX(N,i)];

        x[IX(i,0 )] = (b==2) ? -x[IX(i,1)] :
x[IX(i,1)];
        x[IX(i,N+1)] = (b==2) ? -x[IX(i,N)] :
x[IX(i,N)];
    }
    x[IX(0 ,0 )] = 0.5f*(x[IX(1,0 )]+x[IX(0 ,1)]);
    x[IX(0 ,N+1)] = 0.5f*(x[IX(1,N+1)]+x[IX(0 ,N )]);
    x[IX(N+1,0 )] = 0.5f*(x[IX(N,0 )]+x[IX(N+1,1)]);
    x[IX(N+1,N+1)] = 0.5f*(x[IX(N,N+1)]+x[IX(N+1,N )]);
}
panel void Advect(int b, float[] d, float[] d0,
float[] u, float[] v){
    int i, j, i0, j0, i1, j1;
    float x, y, s0, t0, s1, t1, dt0;
    dt0 = dt*N;
    for(i=1;i<=N;i++){
        for(j=1;j<=N;j++){
            x = i-dt0*u[IX(i,j)]; y = j-dt0*v[IX(i,j)];
            if (x<0.5f) x=0.5f; if (x>N+0.5f) x=N+
0.5f; i0=PApplet.parseInt(x); i1=i0+1;
            if (y<0.5f) y=0.5f; if (y>N+0.5f) y=N+
0.5f; j0=PApplet.parseInt(y); j1=j0+1;
            s1 = x-i0; s0 = 1-s1; t1 = y-j0; t0 = 1-t1;
            d[IX(i,j)] =
s0*(t0*d0[IX(i0,j0)]+t1*d0[IX(i0,j1)])+
s1*(t0*d0[IX(i1,j0)]+t1*d0[IX(i1,j1)]);
```

```
            }
          }
        SetBnd(b,d);
      }
   panel void Diffuse(int b, float[] x, float[] x0,
float diff){
      float a = dt*diff*N*N;
      for(int k=0;k<20;k++){
          for(int i=1;i<=N;i++){
              for(int j=1;j<=N;j++){
                  x[IX(i,j)] = (x0[IX(i,j)] + a*(x[IX(i-
1,j)] + x[IX(i+1,j)] + x[IX(i,j-1)] +
x[IX(i,j+1)]))/(1+4*a);
              }
          }
          SetBnd(b,x);
      }
   }
   panel void AddDensity(int x, int y, float amount){
      int index = IX(x,y);
      this.dens[index] += amount;
   }
   panel void AddVelocity(int x, int y, float amountX,
float amountY){
      int index = IX(x,y);
      this.v[index] += amountX;
      this.u[index] += amountY;
   }
   panel void Routine(float[] u, float[] v, float[] p,
float[] div){
      int i,j,k;
      float h;
      h = 1.0f/N;
      for(i=1;i<=N;i++){
          for(j=1;j<=N;j++){
              div[IX(i,j)] = -0.5f*h*(u[IX(i+1,j)]-
u[IX(i-1,j)]+ v[IX(i,j+1)]-v[IX(i,j-1)]);
              p[IX(i,j)] = 0;
          }
      }
      SetBnd(0,div); SetBnd(0,p);
      for ( k=0 ; k<20 ; k++ ) {
          for ( i=1 ; i<=N ; i++ ) {
              for ( j=1 ; j<=N ; j++ ) {
```

```
                    p[IX(i,j)] = (div[IX(i,j)]+p[IX(i-
1,j)]+p[IX(i+1,j)]+ p[IX(i,j-1)]+p[IX(i,j+1)])/4;
            }
        }
        SetBnd(0, p);
    }
  for ( i=1 ; i<=N ; i++ ) {

        for ( j=1 ; j<=N ; j++ ) {
            u[IX(i,j)] -= 0.5f*(p[IX(i+1,j)]-p[IX(i-
1,j)])/h;
            v[IX(i,j)] -= 0.5f*(p[IX(i,j+1)]-p[IX(i,j-
1)])/h;
        }
    }
    SetBnd(1, u); SetBnd(2, v);
  }
  panel void DensStep (float[] x, float[] x0, float[]
u, float[] v, float diff ){
    //AddSource(x, x0, dt );
    Diffuse ( 0, x0, x, diff);
    Advect ( 0, x, x0, u, v);
  }
  panel void VelStep(float[] u, float[] v, float[] u0,
float[] v0, float visc, float dt){
    float[] tmp;
    //AddSource(u,u0,dt); AddSource(v,v0,dt);
    tmp = u0; u0 = u; u=tmp;
    Diffuse(1,u,u0,visc);
    tmp = v0; v0 = v; v=tmp;
    Diffuse(2,v,v0,visc);
    Routine(u,v,u0,v0);
    tmp = u0; u0 = u; u=tmp;
    tmp = v0; v0 = v; v=tmp;
    Advect(1,u,u0,u0,v0); Advect(2,v,v0,u0,v0);
    Routine(u,v,u0,v0);
  }
  panel void RenderD(boolean dark, boolean code){
    noStroke();
    colorMode(HSB,100);
    for(int i=0;i<N;i++){
      for(int j=0;j<N;j++){
        int x = i * SCALE;
        int y = j * SCALE;
```

```
                float dr = this.dens[IX(i,j)];
                if(dark == true && code ==
      true){fill(dr,255,dr);}
                else if (dark == false && code ==
      false){fill(255,dr,255);}
                else if (dark == true && code ==
      false){fill(0,dr,dr);}
                else if (dark == false && code ==
      true){fill(dr,dr,255);}
                square(x,y,SCALE);
              }
          }
      }
  }
    static panel void main(String[] passedArgs) {
       String[] appletArgs = new String[] {
  "sketch_201122a" };
       if (passedArgs != null) {
         PApplet.main(concat(appletArgs, passedArgs));
       } else {
         PApplet.main(appletArgs);
       }
    }
  }
```

3.2   Visual Presentation

A set of snapshots of density and velocity transformation within the grid is the presentation of the simulation. The variable dt gives the instantaneous differential of time. Figure 4 shows the different interactions due to various pressure concerning the interval of time and the snapshots on the smart panels. We develop an application through java that conveys the tension through time to pixels presented as different liquid colours extending in the black background. Fundamental quantities of colours roughly red colours turn blue and various green shades as the observed sample thickness increases. The liquid's green hue is an intrinsic property caused by selective absorption and scattering of white light (Fig. 6).

Out of paramount importance is to interpret the circular waves due to pressure using fundamental mathematic equations. Based on the sine wave as we present in Fig. 7, we can convert pure periodic circular movement into a sine wave.

Fig. 6 Interpretation of
pressure on smart panels

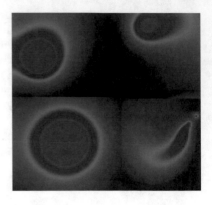

Fig. 7 Sine wave

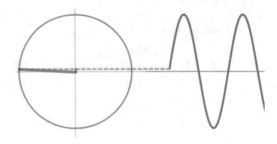

3.3　Mathematical Approach

Our conversion approach based on the wave equation's solution for a membrane
integrated on the panels in a circular disc shape. As shown below, the Wave equation
is a fundamental second-order linear partial differential equation that translates liquid
waves in circular furtherance.

$$\frac{\partial^2 u}{\partial t^2} = c^2 \left(\frac{\partial^2 u}{\partial r^2} + \frac{1}{r}\frac{\partial u}{\partial r} \right)$$

Let $u(r, t) = f(r)g(t)$ based on the chain rule we obtain from the equation of the
wave that:

$$\frac{g''(t)}{c^2 g(t)} = \frac{f''(r) + \frac{1}{r} f'(r)}{f(r)} = A$$

If we substitute $A = -a^2$ and $b = ca$ we derive the following linear differential
equations:

$$g''(t) + b^2 g(t) = 0$$

$$f''(r) + \frac{1}{r}f'(r) + a^2 f(r) = 0$$

The second linear differential equation becomes a Bessel equation [17] of order zero by the substitution of $k = ar$ as follow:

$$k^2 f''(k) + k f'(k) + (k^2 - 0^2) f(k) = 0$$

With general solution $f(k) = c_1 J_0(k) + c_2 Y_0(k)$. After that, we develop an orthogonal eigenfunctions system by setting $a = \frac{\lambda_n}{R}$, λ_n is the nth zero terms of Bessel's function $J_0(k)$ we obtain $f_n(r) = J_0\left(\frac{\lambda_n}{R}r\right)$. If we then substitute $b_n = c\frac{\lambda_n}{R}$ we obtain $g_n(t) = \lambda_n cosb_n t + \mu_n sinb_n t$ thus,

$$u_n(r, t) = (\lambda_n cosb_n t + \mu_n sinb_n t) J_0\left(\frac{\lambda_n}{R}r\right)$$

In proportion to the initial conditions λ_n and μ_n specified as follow:

$$\lambda_n = \frac{2}{R^2 J_1^2(\lambda_n)} \int_0^R f(r) J_0\left(\frac{\lambda_n}{R}r\right) r dr$$

$$\mu_n = \frac{2}{b_n R^2 J_1^2(\lambda_n)} \int_0^R g(r) J_0\left(\frac{\lambda_n}{R}r\right) r dr$$

We evaluate the sum of the following infinite Fourier–Bessel [18] series:

$$u(r, t) = \sum_{n=1}^{\infty} (\lambda_n cosb_n t + \mu_n sinb_n t) J_0\left(\frac{\lambda_n}{R}r\right)$$

Based on the Sturm–Liouville [19] problem, $r \in [0, R]$ is the weight function that corresponds to the $f_n(r) = J_0\left(\frac{\lambda_n}{R}r\right)$ functions.

4 Summary

The present chapter presents an innovative solution to the significant reduction of energy consumption based on interpreting the pressure derived from the road's slope and anomalies—the proposed method is based on fluid dynamics represented on grids named smart panels. The integration of the smart-panels via in-vehicle IoT allows the calculation of the distribution for the required energy between the wheels, to create torque and rotation separately. Navier–Stokes equations have deployed to enhance our approach to implementing the appropriate algorithm to be integrated with the grids installed above the wheels.

5   Conclusion and Future Work

We elaborated four smart-panels on a vehicle suspension system with a sensor grid for pressure measurement by the fluid float modelling to optimise energy consumption. The results require a simulator that we intend to study in future research work. The implementation could be developed through the Simulink MATLAB graphical programming environment to model, simulate, and analyse dynamic multi-sector systems. The pseudocode presented in this research work could extend by solving the Navier–Stokes equations in three dimensions, which could allow the further implementation of the current novel idea.

Acknowledgement This research work was funded by the Smart and Health Ageing through People Engaging in supporting Systems SHAPES project, which has received funding from the European Union's Horizon 2020 research and innovation programme under grant agreement No 857159. Parts of this work were also supported by the Ambient Assisted Living (AAL) project vINCI: "Clinically-validated INtegrated Support for Assistive Care and Lifestyle Improvement: The Human Link" funded by Cyprus Research and Innovation Foundation in Cyprus under the AAL framework with Grant Nr. vINCI /P2P/AAL/0217/0016.

References

1. Liu, K., Yamamoto, T., Morikawa, T.: Impact of road gradient on energy consumption of electric vehicles. Transp. Res. Part D: Transp. Environ. **54**, 74–81 (2017)
2. Travesset-Baro, O., Rosas-Casals, M., Jover, E.: Transport energy consumption in mountainous roads. Transp. Res. Part D: Transp. Environ. **34**, 16–26 (2015)
3. Levin, M., Duell, M., Waller, S.: The effect of road elevation on network wide vehicle energy consumption and eco-routing. Transp. Res. **2427**, 26–33 (2014)
4. Hyodo, T., Watanabe, D., Wu, M.: Estimation of energy consumption equation for electric vehicle and its implementation of driving behavior. In: Proceedings of the 13th World Conference on Transport Research, Rio de Janeiro (2013)
5. Wyatt, D.W., Li, H., Tate, J.E.: The impact of road grade on carbon dioxide (CO2) emission of a passenger vehicle in real-world driving. Transp. Res. Part D: Transp. Environ. **32**, 160–170 (2014)
6. Boroujeni, B.Y., Frey, H.C.: Road grade quantification based on global positioning system data obtained from real-world vehicle fuel use and emissions measurements. Atmos. Environ. **85**, 179–186 (2014)
7. Khaya, G., Mavromoustakis, C.X., Mastorakis, G., Maalouf, H., Batalla, J.M., Pallis, E., Markakis, E.K.: Intelligent vehicular networking protocols. In: Convergence of Artificial Intelligence and the Internet of Things, pp. 55–59. Springer, Cham (2020). https://doi.org/10.1007/978-3-030-44907-0
8. Yu, Z., Jin, D., Song, X., Zhai, C., Wang, D.: Internet of vehicle empowered mobile media scenarios: in-vehicle infotainment solutions for the mobility as a service (MaaS). In: MDPI and ACS Style, vol. 12, no. 18, (2020)
9. Xie, R., Tang, Q., Wang, Q., Liu, X., Yu, F., Huang, T.: Collaborative vehicular edge computing networks: architecture design and research challenges. IEEE Access **7**, 178942–178952 (2019)
10. Schouveiler, L., Gal, P.L., Chauve, M.P., Takeda, Y.: Spiral and circular waves in the flow between a rotating and a stationary disk. In: Experiments in Fluids, pp. 179–187. Springer-Verlag (1999). https://doi.org/10.1007/s003480050278

11. Ning, Z., Huang, J., Wang, X., Rodrigues, J.J., Guo, L.: Mobile edge computing-enabled internet of vehicles: toward energy-efficient. IEEE Netw. **35**(5), 98–205 (2019)
12. Liu, K., Wang, J., Yamamoto, T., Morikawa, T.: Exploring the interactive effects of ambient temperature and vehicle auxiliary loads on electric vehicle energy consumption. Appl. Energy **227**, 324–331 (2018)
13. Sharma, V.: An energy-efficient transaction model for the blockchain-enabled internet of vehicles (IoV). IEEE Commun. Lett. **23**(2), 246–249 (2019)
14. Gazzola, F.S.G.: Steady Navier-Stokes equations in planar domains with obstacle and explicit bounds for unique solvability. Arch. Ration. Mech. Anal. **238**, 1283–1347 (2020)
15. Albreem, M.A.M., Vasudevan, K.: Efficient hybrid linear massive MIMO detector using gauss-seidel and successive over-relaxation. Int. J. Wireless Inf. Netw. **27**, 551–557 (2020)
16. Abide, S.: Finite difference preconditioning for compact scheme discretisations of the Poisson equation with variable coefficients. J. Comput. Appl. Math. **379**, 01 (2020)
17. Everitt, W.N., Kalf, H.: The Bessel differential equation and the Hankel transform. J. Comput. Appl. Math. **208**(1), 3–19 (2007)
18. Bhattacharyya, A., Singh, L., Pachori, R.B.: Fourier–Bessel series expansion based empirical wavelet transform for analysis of non-stationary signals. Digital Sig. Process. **78**, 185–196 (2018)
19. Al-Gwaiz, M.A.: Sturm-Liouville Theory and its Applications. Springer, Berlin (2008). https://doi.org/10.1007/978-1-84628-972-9.pdf
20. Zhang, X., Song, Q.: Predicting the number of nearest neighbors for the k-NN classification algorithm. Intell. Data Anal. **18**(3), 449–464 (2014)
21. Ndanusa, A.: Convergence of preconditioned Gauss-Seidel iterative method for matrices. Commun. Phys. Sci. **6**(1) (2020)

Security and Privacy

Blockchain-Based Internet-of-Vehicle

Alkhansaa A. Abuhashim and Chiu C. Tan

Abstract Internet-of-vehicle (IoV) is a heterogeneous environment involving information exchange between system components such as road infrastructure devices, vehicular embedded sensors and other vehicular elements. In IoV network, devices communications and data exchange need to be secure, efficient and transparent to achieve the platform's goals. Blockchain technology is proved to provide decentralization, immutability, security and transparency properties due to the features of its distributed ledger. In this chapter, we introduce the integration of blockchain technology into IoV networks to support the essential data exchange and storage requirements such as decentralizing, security and transparency. We describe blockchain terminologies and how blockchain can be adopted in IoV environments. Then, we explain blockchain support for IoV data sharing. Besides, blockchain support for IoV trust and verification is also clarified. We also explore the most popular blockchain-based IoV applications the researchers have developed.

1 Introduction

Blockchain technology has become a subject of significant interest and a field of considerable innovations for millions of connected Internet of Things (IoT) devices. Cryptocurrencies such as Bitcoin [1] and Ethereum [2] are adopted in various platforms, including the economy [3], healthcare [4], transportation [5], education [6] and many other domains [7]. The technology has a secure-by-design ledger in which a chain of blocks storing a network transaction is permanently stored in the network nodes. This decentralized design gives the blockchain the immutability that supports the security property.

Blockchain technology is an effective technology to support security concern for Internet of Vehicle (IoV) devices. It adds a layer of transparency, efficiency, security

A. A. Abuhashim (✉) · C. C. Tan (✉)
Department of Computer and Information Sciences, Temple University, Philadelphia, PA, USA
e-mail: tug48402@temple.edu

C. C. Tan
e-mail: cctan@temple.edu

© Springer Nature Switzerland AG 2021
N. Magaia et al. (eds.), *Intelligent Technologies for Internet of Vehicles*, Internet of Things,
https://doi.org/10.1007/978-3-030-76493-7_10

and immutability to the existing IoV systems. However, integrating blockchain with IoV has numerous challenges that are combined with the decentralized property of the blockchain. The large number of connected devices in IoV and the massive data generated by these embedded devices increase the payload of blockchain ledger. The following are some challenges that are combined with the integration of blockchain with IoV:

- *Scalability*, the high demanded of hardware and devices capabilities that are required to maintain the blockchain ledger and operations, the blockchain is not a suitable alternative for limited resources enterprises [30].
- *Resource constraints*, blockchain requires high capabilities of software and hardware since blockchain algorithms drain a lot of energy. Therefore, blockchain-based IoV applications should consider the limitations of IoV embedded devices in applications' designs [12].
- *Privacy concerns*, blockchain ledger stores vehicles transactions and user data in a public database which exposes sensitive data to the public access [8].

This chapter is organized as follows. Section 2 introduces preliminary background about the blockchain technology, the integration architecture of blockchain-based IoV systems and the integration of AI and blockchain into IoV. Then, Sect. 3 describes blockchain support for IoV data sharing as one of the essential applications for the Internet-of-Vehicular environment. Section 4 depicts blockchain support for IoV trust management to authenticate and efficiently verify users and traffic information. After that, Sect. 5, presents popular blockchain-based IoV applications that have been the focus of researchers' studies. Section 6 discusses a number of open issues and future research opportunities to improve the internet-of-vehicle environments. Section 7 reviews the lesson learned from this chapter and the conclusion.

2   Background

This section presents a preliminary background information related to the blockchain and IoV. First, we provide an overview of the general purpose blockchains. Then, we describe the blockchains for IoV frameworks. Lastly, we explore applying Artificial Intelligence (AI) to the blockchain-based IoV network.

2.1   Overview on Blockchains

Blockchain, as an immutable distributed ledger, automates and secures passing digital information from node A to node B. Party A initiates a transaction and broadcasts it to blockchain peer-to-peer network to be verified by the network. Verified transactions are added as a block to the chain, forming a blockchain ledger that stores a historical chain of blocks. To send a transaction to the blockchain, a peer node, e.g., IoV device, broadcasts data using its blockchain address as a public key and a password as a secret key.

The following is a description of the main components of recent blockchain platforms. Section 2.1.1 describes blockchain security properties that distinguish the technology by its well-known features. Section 2.1.2 illustrates the structure of the chain of blocks of the distributed ledger, including blockchain blocks and transactions.

2.1.1 Security Properties

Blockchain has the substantial features of immutability and security through decentralized computing of distributed, shared and immutable databases that store a fault-tolerant chain of blocks. All participating peers in the blockchain network maintain a copy of the blockchain ledger to preserve transparency between the system's nodes. Further, blockchain has no central authority that manages the ledger; instead, network nodes represented by distributed virtual machines participate in accessing and updating the current state of the blockchain network.

One of the core parts of the blockchain is consensus protocol by which all nodes of the blockchain network agree on the current state of the distributed ledger. Prof-of-work (POW) is one of the most well known consensus algorithm applied by Bitcoin and Ethereum which are the most popular blockchain platforms. It demands a high computational power from some network peer called miners to solve a complex computational puzzle to validate and append new blocks to which are blockchain ledger [1]. The higher the computational resource a miner has, the higher probability it mines the current block by winning the puzzle. The winner miner broadcasts the correct answer to the puzzle to other miners, and all transactions are accepted as a block after they are validated.

To tamper any hashed block, i.e., reversing transactions that have already taken place, it needs any malicious miner or a group of miners to do a tremendous computational power, called 51% attack. In this attack, the malicious node(s) should have 51% of the total network mining power to manipulate any block hash, which is extremely challenging and would not be cost worthy.

Indeed, there are lightweight consensus mechanisms for IoV environment to pick a miner peer every time a block is mined. These specialized mechanisms for IoV have been proofed to be suitable in such environments, as described in Sect. 4.

2.1.2 Blockchain Structure

In a general blockchain platform, each block contains a pack of transactions verified and signed based on the verification of previous blocks and the correctness of the current block of transactions. This structure prevents the data stored as transactions from being modified by any malicious party. Any block has two parts: block header,

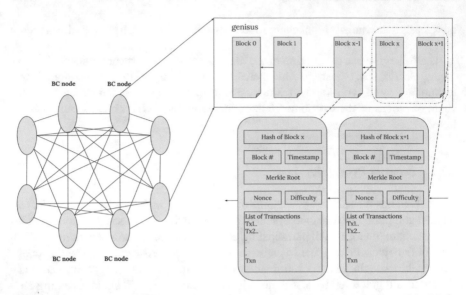

Fig. 1 Blockchain structure. The block has two parts: (1) block header: the metadata of the block, (2) block body: stores ordered transactions.

that has metadata of the block, and block body that stores ordered transactions that are mined into that block see Fig. 1.

When the current block is verified and mined, it is cryptographically chained to the previous blocks in the blockchain ledger. Block header consists of:

- *Parent block hash:* represents the hash of the previous block's header.
- *Block number:* is an ascending number of the current block, The first (genesis) block has a number of zero.
- *Timestamp:* refers to the time the block is generated.
- *Nonce:* is a hash counter used for Proof of Work algorithm.
- *Merkle root:* is the hash value of all transactions of the block
- *Difficulty:* indicates the difficulty level of the current block. It can be calculated from the current timestamp and the previous block's difficulty.

The nonce and difficulty are attributes used in Proof of Work (PoW), which is an essential process that ensures the security property in the blockchain. IoV block structure is described next in blockchain-based IoV subsection.

2.2   Blockchain-Based IoV

Various blockchain-based IoV scenarios have been developed since the blockchain introduces several advantages to IoV environments. We describe the blockchain-based IoV model that provides IoV applications with the previously mentioned features. Section 2.2.1 introduces the data structure of IoV block that stores a list of IoV messages. Section 2.2.2 explores different messaging techniques in blockchain-based IoV platforms. It also presents smart contracts that can be used for interacting with the blockchain, formulating platforms' rules and triggering transactions. Section 2.2.3 explains the most general and representative blockchain-based IoV architecture.

2.2.1 IoV Block Creation

Vehicular block data structure keeps the necessary fields of block structure in the standard blockchain. If the blockchain is used to secure vehicle-to-vehicle (V2V) and vehicle-to-infrastructure (V2I) communications in a decentralized and trusted environment, *a list of state changes* should be stored as a part of the block [9]. The state changes contain all updated information of IoV entities. For example, a vehicle's updates are stored in *a list of state changes* which has the following fields:

- *Position:* vehicle's current location.
- *Direction:* moving direction of the vehicle.
- *Timestamp:* the time of updated state.
- *Type of state changes:* one of predefined types of vehicle's state change such as Turn, Accelerate and Brake.
- *Value:* if the state change type is measured by a number, the value is the quantity to represent the accurate measurement.

On the other hand, if the blockchain is used to exchange traffic information or share road conditions, the block keeps event messages instead of transactions [10–12]. The event message includes: event type, event ID, event timestamp, event location, driving direction and other related information, see Fig. 2. Transmitted data between IoV entities takes different forms, which are presented in the following subsection for blockchain IoV Data Messaging and interaction.

2.2.2 Data Messaging and Interaction

Blockchain messaging in Intranet-of-Vehicles takes various forms, for example, when a vehicle detects an event in the road such as a traffic jam or an accident, the form of sending the traffic information is formulated based on the communication techniques adopted by IoV platform. Blockchain nodes and vehicles can communicate through *request and response messages* [13]. The vehicle sends a registration request to RSU, which contains the request message, current location and timestamp of the vehicle. RSU forwards the request, after encrypting it using the vehicle's private key, to authority nodes for registering and verifying the vehicle in the blockchain. The response to RSU from authorized node contains the response message, all locations and timestamp of the vehicle. After that, RSU responds to the vehicle to confirm the registration process by a response message.

Fig. 2 IoV block structure.

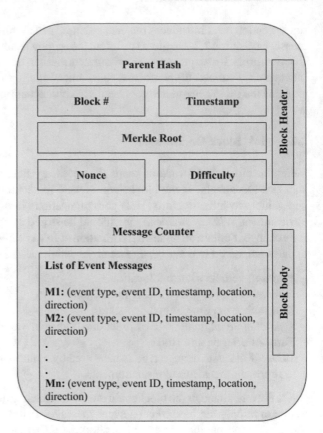

In [24], there are two types of messages: *beacon messages and safety event messages*. Beacon messages are vehicular entity updates such as broadcasting periodical messages for vehicle status and position to update its information. The safety event messages are sent when emergency events are occurred, such as traffic accidents. Event messages can be prioritized based on their urgent level to be able to deal with each event in a timely manner.

Another blockchain communication is accomplished via transmitting *data packages* between application nodes as in [14]. Request Packet (RQP) is sent from a user to others in the network to announce for traffic information. When a member of the network intends to join the announcement, she sends a Reply Packet (RPP) to increase the voting for that announcement. After the announcer reaches a certain number of voters, who confirm her announcement, Announcement Packet (ANP) is sent to the network along with a signature to approve the announcement.

For interacting with the blockchain, many recent blockchain designs support the idea of *smart contracts* to widely brace more application and platform functionalities. A smart contract is a program run and stored in the blockchain to define a set of digital commitments as agreements compiled and deployed to the network. Contract

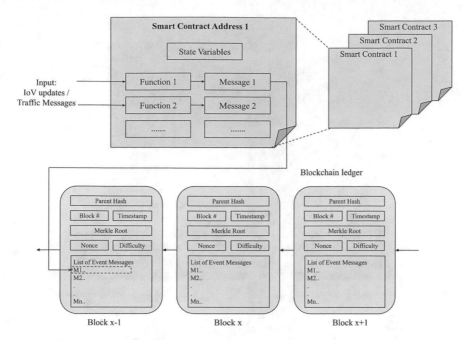

Fig. 3 IoV blockchain and smart contracts interaction.

functions are executed by contract participants to trigger transactions that are mined permanently into the blockchain. When the contract compiled and deployed by its owner, virtual machine nodes can access the contract by calling its function(s) with the matching parameters. With that said, smart contracts in IoV environment are considered as general procedures to control IoV devices and automate the system's operations. Figure 3 shows how the smart contract's state variables and functions relate blocks and stored in the blockchain ledger [11].

2.2.3 Generic Blockchain-Based IoV Architecture

In the conventional centralized Internet-of-Vehicle network, RSU plays a role of networking services. Because of its high communication abilities, such as its capacity, compared with vehicles, they can be used in exchanging the information and providing services to vehicles and other users in IoV. In contrast to the conventional centralized IoV, the RSU role in IoV is not only used for communication purposes, but it is also utilized in the provision of the blockchain network for IoV [2]. Each network entity is registered in the blockchain ledger as a network participant by having a blockchain address that represents its public key. A participant pays fees to the blockchain miners for mining blocks and/or funding the running and maintaining the infrastructure. Mining might be done by RSUs, vehicles, other system components or external providers.

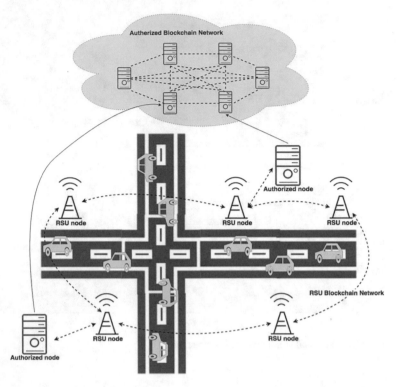

Fig. 4 Blockchain-based IoV architecture.

The government's authorized parties or identified authorized vehicles are distributed as authorized blockchain nodes in different city areas to create a blockchain network, as in [13]. These nodes represent a higher blockchain layer to verify RSU nodes in the vehicular network, Fig. 4. RSU nodes is another blockchain layer for vehicle registration and information verification. RSU registration starts by sending a registration request to an authorized node in the higher layer. RSU is registered in the authorized blockchain network after it is verified and validated by the authorized distributed nodes using a consensus protocol. After authorized nodes register the RSU in the authorized blockchain, it sends an acknowledgment to RSU for joining the network.

When a new vehicle moves on a road, it sends a registration request to a nearby RSU node. The RSU nodes verify vehicle identification in a distributed manner to record the vehicle in the RSU blockchain network. Road information such as traffic jams, road constructions and other road-related details is sent by vehicles to RSUs to be validated and stored in the RSU blockchain layer. This is a generic blockchain-based IoV architecture; other works such as [15, 17] and [18] propose similar blockchain-based IoV models with special-purpose functionalities that will be clarified in the following sections.

2.3 Blockchain-Based Supporting AI in IoV

The innovation of smart vehicles requires intelligent decision making. AI models have been developing by vehicle industries to invest in smart IoV environment. AI and blockchain bring the advantages of the two technologies into the IoV such as securing the IoV, automating smart vehicles and enabling decentralizing. AI requires a sufficient computation capability for AI algorithms. Further, by integrating blockchain adds a higher computational load into IoV infrastructure which affects the performance of the whole system.

[12] proposes a system architecture that integrates the two technologies to support IoV infrastructure. AI models with a blockchain network in IoV environment has three layers, as the following:

- **Data generation.** The framework starts by generating data with IoV infrastructure components, executing AI models and performing blockchain related tasks. Vehicles connect to the RSUs (vehicular layer) to send data generated by IoV devices.
- **Vehicular edge.** After the data is being generated, it is transferred to computational servers to be processed. This layer provides a secure transmission to the cloud servers, next layer, through RSU clusters that maintain blockchain ledger. RSU clusters only store the metadata of vehicles' data in the blockchain to increase the scalability of the framework. RSU has three roles in its cluster represented by smart contract. A smart contract is an *authentication* smart contract for authenticating vehicles and creating a record for the participant to be recognized in the cluster. Another smart contract is *data management* smart contract for classifying the data to be forwarded to the right data center. The third smart contract is *AI models management* which is triggered in a frequent time to update embedded vehicles AI models from the cloud layer.
- **Fog and Cloud servers.** RSUs' blockchains are connected to fog servers which are supported by cloud servers. The layer performs framework's computations and updates AI models for IoV.

Integrating AI and blockchain into IoV is in its early stages since there are a small number of researches addressing the integration of these technologies. The integration will be a great opportunity for the researches to improve the blockchain-based AI in IoV network.

3 Blockchain Support for Data Sharing

Building an effective vehicular data sharing is one of the significant issues in the Internet-of-Vehicle environment. There is a lack of user's enthusiasm to send or respond to network messages. IoV users are usually not interested in sharing data or providing traffic information since participating in the network costs them driving

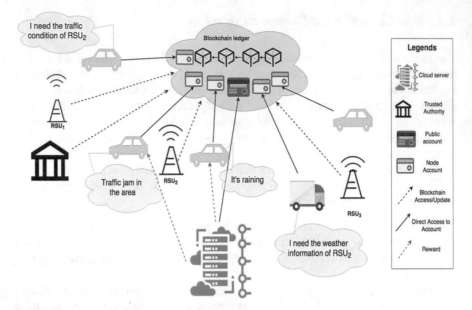

Fig. 5 A general design for blockchain-based data sharing model for IoV.

time and effort, and most importantly, it is vulnerable to privacy leakage. Therefore, motivating IoV users to share road information, securing the system's transactions, and keeping users' privacy are from the most challenging factors in IoV sharing data.

Various blockchain-based data-sharing platform designs support all (or often most) of the mentioned data-sharing challenges. In [15], users are divided by a cloud server into groups based on their nearest RSUs, where they connect, see Fig. 5. Each entity in the network has a blockchain account with an address to manage the user's data and balance. The cloud server supervises the requester's announcement and participants' replies by assigning the announcement to the related RSUs then verifying the collected data to be sent to the requester. Personal information, such as the user's name and vehicle plate number, is registered in the cloud server, which is the trusted node. However, to protect the user's privacy, only the account address is sent with the transaction to the blockchain. When the user is new, the related account needs to be charged with credits. After sharing traffic information and engaging others' announcements, the user gets sufficient credits to request information on her own. Accordingly, communications between vehicles, RSUs and the cloud server are recorded in transactions to be audited and added as blocks to the blockchain by RSUs, after reaching the agreement state, as described in Sect. 4.2.

There are two main challenges in blockchain-based data sharing for IoV platforms. The first challenge is how to encourage IoV users to participate in traffic information with the network. The second challenge is how to enforce rules and manage information exchange between IoV network nodes. The following is a blockchain-based incentive protocol that operates to incentivize sharing traffic information between users in the IoV environment. The other part of this section shows how smart contracts help in data sharing by managing the communications between different network entities.

3.1 Incentive Announcement Protocol

The objective of the incentive mechanism is to motivate the user to participate in road and traffic information by honestly forward true announcements. Forwarded transactions of vehicle-to-vehicle and vehicle-to-infrastructure are validated by trusted entities; then, if the transactions are validated and confirmed, they are added as blocks on the system's blockchain ledger. Every user is registered with a blockchain account charged with credits or coins, which represents her reputation points or credited money. Users can make credits or coins by announcing traffic information or replying to an announcement of others. The primary goal of requesting information costs to be spent as points or coins is to reward the participants for sharing traffic information. Besides, it is essential to protect the network from any malicious node from launching too many requests, i.e., Denial-of-service (DOS) attacks and meaningless requests. To further prevent the system from misbehaving nodes with height repetition/coin levels from compromising the system with repetitive requests, the system authority may set a daily (or hourly) request limits for vehicles that should not be exceeded by the vehicles in each time window. DOS can not be occurred by participants responding to requested information since each participant has one chance to respond to each request. Malicious users can not modify their (or other's) credits since the transactions are recorded and verified in the blockchain database.

Incentive protocol can be supported by the IoT cloud server to manage the vehicle's announcement and emergency notifications. When certain information is required, such as inquiring road traffic by an ambulance for an emergency, there are five phases for the incentive protocol of sharing traffic information: *Platform Initialization phase*, *Registration phase*, *Event Announcement phase*, *Event Responding phase* and *Rewarding Participants Phase*:

1. *Initialization phase,* this phase is for initializing a set of system's parameters and platform's entities such as system's cloud server, Service Manager (SMs) and RSUs. Besides, if the system requires specific values or thresholds to be set for its entities, they should be placed in this phase. Additionally, blockchain ledger should be launched by creating the genesis block which is the first block (Block 0) of system's blockchain.
2. *Registration phase*, this phase is for registering vehicles' On-Board-Units (OBUs), RSUs and SMs. The registration process generates an account for the registered entity with all related information such as an address, public key (PK), private key (SK) and other system-specific data. The account address is a unique address for the entity in the blockchain network; PK is the entity identifier (ID) in the net-

work, and SK is the entity's password. In adition, the registration phase includes the authentication process, see Sect. 4, for any new registered untrusted entity.

3. *Event Announcement phase,* the cloud server publishes a description for an event announcement (e.g. emergency event) through a smart contract to be shared in the blockchain network. If system's announcements demanding specific settings for each event, such as a threshold number of voting, the cloud server indicates these variables when the smart contract is published. Further, in this phase, the announcer specifies how much she pays to receive the requested information. Given that, the mapping between the number of participants and how much each one is receiving will be calculated by a smart contract.

4. *Event Responding phase,* Using cloud server's smart contract, vehicles who intend to participate in the announcement offer information that will be stored in the blockchain. In this phase, users' responses are verified and validated as described in Sect. 4.

5. *Rewarding Participant phase*, After a certain number of participants share traffic information as responding to an announcement, the cloud server distributes the announcement incentive to the participants. Incentives could be increasing in participant's reputation points or gaining real money as coins. Based on the vehicle's reputations (and/or other criteria set by system administration), the vehicle is either selected or eliminated from the announcement winning list.

The initialization phase is done once in the system's lifetime, while the registration phase is performed when a new entity joins the platform. *Event Announcement phase* could be *an announcement request* when a vehicle broadcasts a request for traffic information such as enquiring a road status for an ambulance. Another scenario is *an announcement posting* when a vehicle intends to share traffic information, such as reporting for an accident. In the first scenario for requesting an announcement, the participants in the *Event Responding phase* will respond to the requester's announcement by individual responses. In this case, if the framework incentive relies on coin-based, the requester will pay coins for the participants to get the traffic information. On the other hand, if the framework is a reputation based incentive, the requester can ask for traffic information when she has enough points to ask for traffic information. For this reason, if the user's car is new, it has to contribute in a sufficient number of announcements to collect enough credits to request announcements itself. Suppose it requests high priority information and still not meeting the request requirements. In that case, the response depends on the system authority's rules if the requester can receive the response and get a minus credit to be paid at a particular time or before the next request(s), if not admitted, the vehicle will be blocked from using any services. Otherwise, the request is set to be rejected until the participant acquires a certain level of coins/reputation. These rules are set in the initialization phase to be applied to all new vehicles joining the system.

For the announcement posting scenario, the participants in the Event Responding phase will vote for the announcement if the event has indeed happened. The reward for the announcement is distributed between the voters, and the announcer receives the major portion of the incentives if the following two conditions are satisfied: (1)

the announcement is verified and committed by RSUs and the cloud server, (2) the number of voters exceeds the system's threshold, which is specified by system's authority in the initialization phase.

[14] proposes a blockchain-based incentive protocol to motivate users to participate in traffic information via network announcements. In addition to the basic component of blockchain-based IoV: OBUs, RSUs and cloud server, the system consists of *Trusted Authority* for maintaining users' keys and identities, and *Trace Managers* that are distributed in different areas to trace malicious vehicles. For sharing traffic information between vehicles, transactions are validated by RSUs. If the transactions are validated, a cloud application server confirms the transactions and adds them to blocks in the blockchain.

The framework is a Credit network in which responding to other announcements requires spending coins. The coins of the framework are representing repetition points which prevent long-period DOS attacks. A participant will receive several coins in replying to an announcer request, and each user has a daily limit of replies to protect the system from abuse of replies. When a requester puts more coins in her announcement request, the announcement will receive replies sooner than if it is with fewer coins. An expiration day is set to each user's account for the credits to keep the network active and prevent coins accumulation. If the coins are reserved and not spent until that day, some of the coins will be transferred to the public network. Indeed, the system prevents corrupting account balance by illegally increasing the coins since the blockchain ledger maintains the user's account. In addition, the process of altering data in the ledger is not an easy process in the blockchain.

A time frame for some announcements should be predefined, especially for emergency events [16]. The following is an incentive scenario that adopts a cloud server as the information manager of the framework. If the participants intend to participate in the announcement, Fig. 6 demonstrates the steps of the framework's interactions as the listing:

1. The cloud server publishes an announcement description via a smart contract to be accessed by the network.
2. The participants access the smart contract to view the announcement description and offer the inquired information.
3. After the predefined announcement time, the cloud server executes the smart contract to choose the elected vehicle(s) that provide(s) the traffic information.
4. The vehicle(s) provide(s) the complete information that is stored in the blockchain.
5. When the requested information is published using the smart contract, the cloud server rewards the participated vehicle(s), with the help of the RSUs in the mining process, a certain amount of credits defined in the published smart contract.
6. The cloud server considers the vehicle(s) to participate in the announcement based on the vehicle's credit level and the integrity of the shared data.

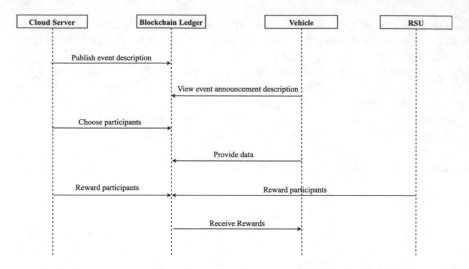

Fig. 6 Cloud server incentive protocol, that utilizes the cloud server as the information manager of the framework.

3.2 Smart Contracts in Incentives and Data Sharing

Smart contracts can help IoV environment to manage information exchange transactions and enforce rules in a decentralized manner without any central authority. Participants can execute and verify their transactions by communicating self-enforcing contracts that define the environment's rules and borderlines. A smart contract can implement the incentive announcement system's phases that should be adopted to motivate participants to privately announce and respond to traffic information messages in a non-trusted environment [17]. Each phase is applied by invoking one or more smart contract function(s) that enforce(s) system's requirements in each stage.

When requester R requests for traffic information as a request packet (RQP), she asks RSUs to broadcast her request to vehicles in the inquired area. This process is a *request phase* which can be initiated by calling a smart contract function and sending RQP with other related details, such as RQP timestamp and R's ID, to be stored in the blockchain. After the request phase, the *announcement phase* is executed by RSUs who verify participants' information and responses by *verify* function in the smart contract.

For the system's incentive, a *payment phase* would be implemented by two smart contract functions, a function to manage to send message fees to all honest participants for sharing traffic information, and a function to pay service fees to RSUs for managing announcement transactions. A deposit fee should be paid by participants when they join the announcement to prevent false messages. For an honest participant, the deposit fee would be refunded with the message fee in the payment phase. On the other side, non-honest participants would go through a *claim phase* in which they lose their deposit fee.

```
contract incentiveAnnouncementSC {

  uint public station_id;
  struct userRecord{
    bytes32 message;
    uint balance;
    ...
  }
  mapping (Address => userRecord) users;
  uint fee;
  address RSUAddr;
  bytes32 majority;

  function verify(message) public return (bool) {
    // verify and authenticate the shared msg and
    // return true if it is verified,
    // Otherwise, return false
    if (ecrecover(message,sig) == msg.sender) {
      users[msg.sender].message = message;
      return true;
    }
    return false;
  }

  function computeMsg() {
    // compute the majority msg based on participants
    // threshold set by authority as one of the
    // system's parameter
  }

  function RSUFee() public payable {
    //transfer fee to RSUs addresses
    RSUAddr.trasfer(fee);
  }

  function MsgFee() public payable {
    // transfer fees to all participants in the shared msg
    for (address userAdd : users) {
      if (users[userAdd].message == majority)
        users[userAdd].transfer(fee);
    }
  }
  ...
}
```

Fig. 7 Incentive announcement smart contract for motivating the participants to shared traffic information by transferring fees to the winning participants [17].

Figure 7 shows an overview of the incentive announcement smart contract written in Solidity, which is a programming language for implementing smart contracts in Ethereum. It defines the user data as a *struct* data structure and then maps user struct to the user's address in the blockchain by using a *mapping* data structure. Then, all IoV incentive announcement functionalities should be implemented by the smart contract functions. For example, some of the functions of the incentive announcement smart contract:

– *verify()* function to validate the shared message by calling ecrecover() Solidity function that allows the smart contract to validate the incoming message is signed by the user,

- *computeMgs()* function to find the message shared by the majority of users after the number of participants reaches the authority's threshold set in the initialization phase,
- *RSUFee()* function to transfer the service fees to all participated RSUs, and
- *MsgFee()* function to transfer message fees to all participated users.

[15] enables a smart contract for automating data sharing and providing a self-driven secure incentive mechanism. Major information exchange processes between vehicles are performed by utilizing smart contract functions. The smart contract function can be either a function for setting the system's parameters and components or a function for managing users' requests and responses. To illustrate, *Init* function initializes all the required variables of the user's request, such as a list of the request participants, a threshold for the number of participants, if applied, etc. Another method is *Add* function, which adds participants into the previously defined list. After deploying the smart contract, initializing request components, and adding and verifying participants into participants list, the smart contract will be ready to be accessed by the platform's entities to deploy a verified response to the user's request. Users' requests are implemented by *Order* and *Submit* functions to allow the requester to Order information from participants and the participants to Submit their responses to the blockchain. To confirm users' requests, *Fetch* function is performed after reaching sufficient shared information from the participants. Finally, *Commit* function would do responses verification and participants' payment.

4   Blockchain Support for IoV Trust Management

Sharing transportation information via vehicle-to-vehicle (V2V) communication or OBU and RSUs collaboration via vehicle-to-infrastructure (V2I) communication should be transmitted correctly, securely and efficiently. Connected vehicles share crucial transportation information such as vehicle accidents, traffic jams and road conditions. While the Internet-of-vehicle environment is untrusted, there is an urgent need to protect the communication between connected vehicles from any malicious party. Moreover, traffic event validation, such as traffic warning, can be achieved in a decentralized manner of blockchain to obtain integrity. The following four security and privacy requirements should be met for trust management in the IoV network to protect IoV users from false event announcements and malicious activities from any untrusted parties [18].

1. *Integrity and correctness* of sender's (i.e. vehicle) data which should be accurate and not modified or discarded.
2. *Privacy* of user's identity which should not be disclosed when transmitted with the communicated data. (Discussed in Sect. 3 and 6)
3. *Authentication* of each identity in the network.
4. *Efficiency* of transferring and computing the transportation information in real time under security, privacy and system requirements.

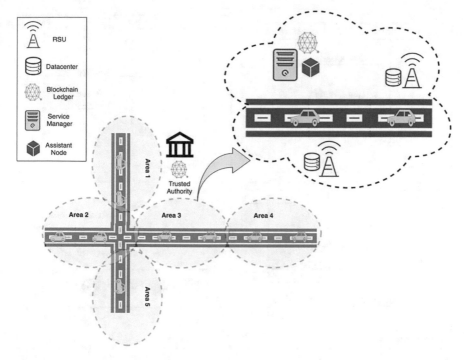

Fig. 8 Lightweight authentication mechanism architecture, which consists of multiple areas, each area has a service manager (SM) to manage many datacenters

The following subsections target most of those requirements. The privacy of IoV users was presented in Sects. 3 and 6. Section 4.1 is a lightweight protocol to authenticate IoV users and secure the communication between vehicles in the IoV platform in a lightweight manner. It satisfies the authentication and efficiency requirements. Integrity and correctness requirements are addressed by Sect. 4.2, Traffic Events Verifications, which is verification protocols to obtain the integrity of distributed traffic events.

4.1 Lightweight Authentication Mechanism

Due to the high demand for authenticating untrusted entities and communicating efficiently in the high mobility of vehicles, a blockchain lightweight authentication protocol is essential in the IoV environment. In such environment, a vehicle sends a message to request a vehicular service and authenticate the user's communication. The platform, as proposed by [19], consists of multiple areas in which every area has a service manager (SM) to manage many datacenters, see Fig. 8.

The system operates in five components:

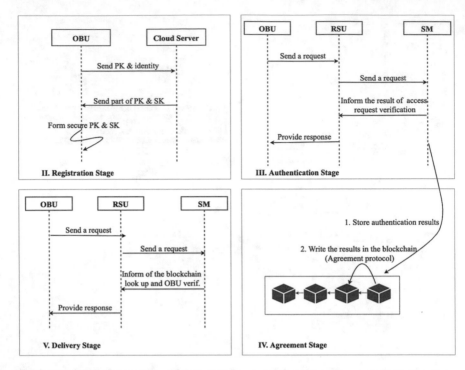

Fig. 9 Blockchain lightweight authentication framework has five stages: an initialization stage, which is performed once in the system's lifespan, and four repetitive stages: registration stage *(II)*, authentication stage *(III)*, agreement stage *(IV)* and delivery stage *(V)*

1. *Trusted authority (TA)*, the trusted party maintains the system's integrity by recognizing unauthorized vehicles. It also has the authority to register SMs and OBUs.
2. *SM*, the public blockchain manager for each area. It controls several datacenters and validates each OBU passing the area.
3. *Assistant node (AN)*, the entity that collaborates with SM to store the results of authentication processes in the blockchain.
4. *RSU*, the supervisor of a datacenter and the manager of vehicles' requests.
5. *OBU*, the device of user's vehicle that should be registered in the TA.

To reduce the platform's blockchain computation and communication overhead, the blockchain network is formed by only TA, SMs and ANs that maintain a copy of blockchain ledger in each.

The framework has an initialization stage, which is performed once in the system's lifespan, and four repetitive stages: registration, authentication, agreement and delivery phase, see Fig. 9 for the repetitive stages.

- *Initialization stage:* The first stage is performed once by the TA when the system is launched by defining the system's parameters, security settings and the PK and SK for TA.

- **Registration stage:** This stage is for registering SMs and OBUs by TA, which sets secured PKs and SKs for these entities.
- **Authentication stage:** Each SM authenticates OBUs in the area after vehicles' OBU communicate with their nearest RSU. The RSU forwards the vehicle's message to the area's SM, which performs authentication operations on OBU's message, PK and SK. Then, SM broadcasts the authentication results to all ANs for the agreement protocol to store the results in the blockchain. When the vehicle is authenticated, SM reports the RSU to respond to OBU's request.
- **Agreement stage:** While this chapter focuses on blockchain and the related topics for IoV, the *Agreement stage* is described in more details in Sect. 4.2.2.
- **Delivery stage:** The last stage for the framework is the delivery stage. When a vehicle moves from one area of RSU to another, it needs to submit an access request to the new RSU. The new RSU sends the request to SM for the area. SM authenticates the vehicle by searching its public key in SM's local database. If it is not found, it first checks the blockchain for the list of illegitimate OBUs. If the vehicle is on the list, the connected SM rejects its request. In the case of the public key of OBU is not in the list and SM can not authenticate OBU's information, SM will report TA to inform all other SMs about the illegal vehicle to be added in the illegitimate list. Otherwise, if SM verifies the vehicle's signature and the request timestamp and the authentication succeeds, SM asks RSU to respond to OBU's request.

4.2 Traffic Events Verifications

Traffic event validations, such as traffic warning, can be achieved in a decentralized manner of blockchain to acquire the integrity of events. When the collected data related to the announced event reaches a predefined number of votes as a threshold, RSUs verify participants' traffic information. If the event announcement is confirmed to be not a false warning, RSU broadcasts the announcement to vehicles around the area of the event. Announcement's transactions are verified and added by RSUs, or any trusted party such as the cloud server, to the blockchain to be stored permanently. The following are verification protocols for IoV shared traffic information. The first subsection is Prof of Event (PoE) protocol that protects the IoV platform from malicious users who attempt to publish wrong events and warning announcements. The second subsection is some agreement protocols suitable for the IoV environment to reach out to a consensus state between the platform's nodes for creating and adding a block to the blockchain.

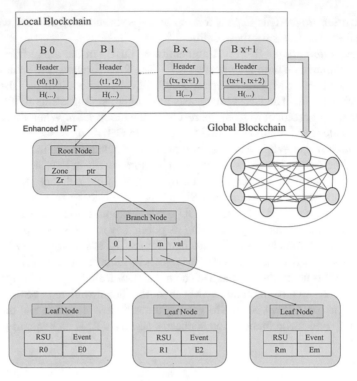

Fig. 10 Enhanced MPT tree blockchain. The block of the local blockchain contains a header, life timestamp and a hash of the enhanced MPT tree root.

4.2.1   Prof of Event (PoE)

Prof of event (PoE) is a mechanism to identify the network's malicious behaviors and the distribution of false traffic information through a two-pass validation process for any distributed event [18].

a) Two-phase Transaction: Traffic warning messages need to be delivered in a specific region and time. To timely broadcast transactions to all possible nodes in the IoV platform, the transactions are transmitted into two sequential stages. First, a traffic transaction is stored in a local blockchain-based on its geographical areas, and then it is propagated into the framework's global blockchain.

To obtain the two phases, an enhanced Merkle Patrica Trie (MPT) tree data structure is designed as in Fig. 10. The block of the local blockchain contains a header, life timestamp and a hash of the enhanced MPT tree root. Tree's root node represents the zone area, which covers many RSUs, and points to a branch node which is an index to all RSUs in the zone. The leaf node, which represents an RSU in the root node zone, has two components: (1) RSU id and (2) event description. For the same zone, the first event announcer creates the block that contains all events transactions for all RSUs in the zone. The block generator sets block lifetime and includes the root hash, which is the hash of events signatures and public keys of events participants. This information is sent to the zone's RSUs so that the block generator is verified, and then the new block is added to the local blockchain. After a specified period e.g., x hour(s), each local blockchain should be sent to the framework's global blockchain by the last block generator of each local blockchain. When all local blockchains are transmitted and chained in the global blockchain, all event descriptions are permanently stored for public access. This process for storing the local blockchains in the system's global blockchain is managed by a trusted party in the framework, such as the trusted authority or the cloud server. The following is PoE's trust verification to protect the network from spreading wrong traffic information and storing false events in the blockchain.

b) Trust Verification: Each newly created local block should be committed by all RSUs of the same zone. On the other hand, to transmit a local block to the global blockchain, the last block generator is responsible for verifying the root hashes of all enhanced Merkle trees with their corresponding event descriptions. If the verification fails for any reason, e.g., a malicious last local block generator sends a faulty block to the global blockchain, the global block generator drops the event that causes the failure and recalculates a new hash value of the enhance Merkle Tree of the new block excluding the dropped event. To speed up an event announcement, especially for emergency events, e.g., car accident, the event is announced to the close vehicles until the event is validated to be announced globally. Consequently, this two-level of validation for traffic event guarantees that false and fake events are hardly shared by malicious vehicles or block generators.

4.2.2 Agreement Protocols

Instead of utilizing the complex Proof of Work or other blockchain popular consensus protocols to pick a winner node every time a block is added to the blockchain ledger, more IoV specialized agreement protocols have been proven to be suitable in such environments.

a) Leader Rotation Mechanism: A leader rotation mechanism is another suitable consensus mechanism for the IoV environment [15]. The leader rotation process operates in a cloud server environment, which gets RSU's request to become the next consensus rotation leader, see Fig. 11. The candidate RSU broadcasts the next block information and transactions to all other RSUs in the network to verify the block. To commit the block in the blockchain, RSUs share the block verification results. The agreement is reached when all RSUs send acknowledgments of the block integrity to the server. Finally, the server confirms the block and adds it to the blockchain.

b) mod Consensus Process: Some IoV platform designs involve *assistant nodes (ANs)*, which are entities that collaborate with SM to store systems results in the

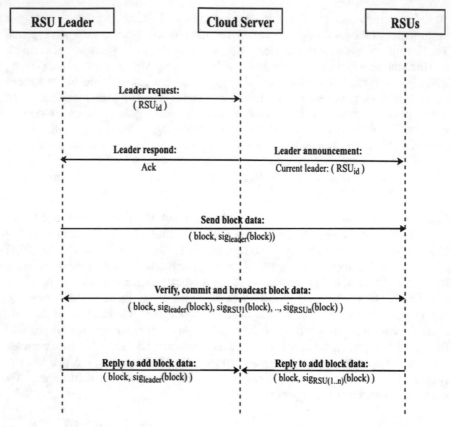

Fig. 11 Leader rotation mechanism, to process RSU's request to become the next consensus rotation leader

blockchain [19], see Sect. 4.1. ANs store the results of OBUs authentication in blockchain ledger after the consensus process passes. Block generator (or miner) x_i is chosen from the list of the ANs by $i = (height \bmod k) + 1$, where x is an AN serial, k is the total number of ANs and height is the height of the current block. ANs store authenticated information received from SM, as described in Sect. 4.1, in their memory to be monitored. Generator x_i broadcasts the block to other ANs, called voters, the following details: VRq, height, block, $sig_{x_i}(block)$ where VRq is vote request from x_i to other ANs. When the block is distributed, each voter v_r replays with: VRs, height, block, $sig_{v_r}(block)$ where VRs is voting response for v_r. The block is confirmed when more than $(k - f_n)$ ANs vote and sign the block. f_n represents the maximum number of failing ANs allowed that is equal to $[(k - 1)/3]$.

5 Blockchain Support for Applications

With the rapid development of IoV technologies, many beneficial road applications have been developed. The various implementations of road applications help creating a smart environment for vehicles, authorities, insurance companies and other vehicular related parties. Introducing the emergent technologies in the road infrastructure has been inquired globally to provide safe and efficient road services. The following are some popular IoV applications that have been the focus of researchers' studies.

5.1 Accident Detection

Blockchain-based accident detection is significant in providing accident warnings and calls on the roads to minimize the related consequences such as other accident occurrences, injured people and road conjunctions. The detection model can be divided into two main scenarios: online detection and offline detection models [20]. The online detection model assumes accident witnesses and road entities, such as RSUs, are connected to the model's network. To inform on an accident and verify the integrity of the accident warning, one or more of the previously described protocols, such as the incentive announcement protocol, can be combined and implemented.

In the offline detection model, neither cars nor the road infrastructure is connected to the network where an accident happens in a disconnected road where GPS does not work. The accident participant notifies the accident after she connects with any road entity. To verify the accident, blockchain is implemented to store the participant and other witnesses' signed information. Many benefits are gained by confirming that the accident happened: roads and cars damage are estimated, injured people are minimized and traffic jams are reduced.

5.2 Forensics Application

Adopting blockchain in IoV forensics applications enhances the application's transparency and traceability by providing accurate auditing for vehicle related records. Collecting data around vehicles in an accident location helps to investigate the reasons behind the accident. Smart contracts play an essential role in forming rules as decentralized agreements between vehicles, authorities, insurance companies and other related parties. Block4Forensic [21] framework is a comprehensive vehicular forensics solution that joins all associated stakeholders and collects all necessary data for vehicles. The collected data is used in post-accident investigations to locate the faulty party.

Figure 12 shows the framework's stakeholders and the role each party plays in the system. There are four different roles for the nodes: leader, validator, monitor units

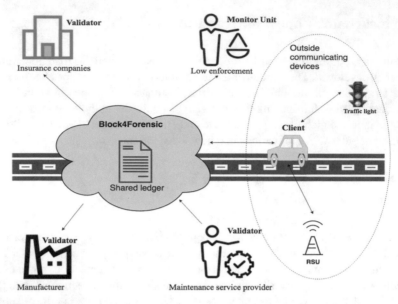

Fig. 12 Block4Forensic system model with its stakeholders and the role each party plays in the system.

and clients. The leader is randomly selected from the validator nodes to validate clients' signed transactions to be added to the shared ledger as a new block. On the other side, monitor units are law enforcement authorities who access the shared ledger as part of post-accident investigations but they are not involved in block creation processes. HACIT project [22, 23] offers forensics inquiries by allowing vehicular users to be able to access the history of transactions.

5.3 Charging Application for Electronic Vehicles

With the increase in the number of electronic vehicles (EV), charging pile management for EVs is necessary. Charging management includes obtaining adequate charging settings or minimizing the required charging resources, i.e., time and money, by assessing scheduling strategies. LNSC (lightning network and smart contract) [25] is a blockchain decentralized security model that securely manages the trading between EVs and charging piles. The model protects the transactions between EVs and charging stations in four phases: registration, scheduling, authentication and charging phases.

ChargeItUp [26] is a smart contract Ethereum system to send machine-to-machine transactions for refueling autonomous vehicles in vehicle-to-charging stations. More trading frameworks for EVs are proposed by researchers to ensure the security and efficiency of data trading [27] and charging trading [28] in IoV networks.

5.4 *Carpooling and Platooning*

Carpooling is a ride-sharing solution to reduce carbon emissions, minimize traffic congestion and improve commercial benefits. Carpooling mostly developed by utilizing cloud servers to allow passengers to be able to locate drivers, which increases response latency and useless communication overhead. In [29], a blockchain-based carpooling scheme is developed to improve the efficiency and protects the security and privacy of carpooling user processes. It uses blockchain vehicular fog nodes to ensure data audibility and to match a driver with a passenger where the blockchain formed by RSUs. The system model anonymously authenticates users' communication by establishing a secret communication key between the passenger and the driver.

Similarly, vehicle platooning is a technique for merging a group of vehicles to increase roads' capacities by reducing the distance between each vehicle in the platoon. The platoon is led by a head vehicle while the remaining vehicles are considered platoon members [30]. A blockchain-enabled platooning model [31] provides a secure and decentralized platooning management in IoV environment. It creates a platoon by using path information matching to match vehicles with others correctly. Platoon head is chosen based on the repetition level, while platoon members pay service fees for platoon's head for the provided services. The blockchain model basically utilizes a smart contract to record transactions and offer secure payment among platoon head and members.

6 Open Issues and Future Directions

Utilizing blockchain as a tamper proof, decentralized and secure-by-design ledger provide the vehicular environment layers of security and transparency for the data and communications. However, integrating blockchain with IoV leads to several opeen challenging issues that need to be targeted by researches to improve the best of the two technologies. In this section, we present a number of future research opportunities to improve the internet-of-vehicle environments.

Privacy Preserving: As the blockchain ledger is publicly stored in network nodes, user's data is exposed to the public access. For that reason, IoV users are usually not interested in sharing data or providing traffic information since participating in the network is vulnerable to privacy leakage. Various blockchain-based IoV research propose several solutions and system designs to target the problem as in [14, 22, 23, 30]. However, several privacy issues are still required to be addressed to protect user's private data such as real identification, real location and location history from any malicious user. In some scenarios, privacy can be preserved bu adding some noise to IoV data to hide, for example, user's location. For that reason, preserving the privacy is a trade-off between accuracy and privacy level which may introduce difficulties in data accuracy for some IoV frameworks. Thus, privacy and accuracy balancing is an important privacy issue to be considered by the researches. Lastly, we believe smart contracts will play an important role in protecting IoV users privacy.

Resource Constraints: Although IoV smart devices have higher computational and storage capabilities than lightweight IoT smart devices. However, managing IoV infrastructure with the high volume of data being sent and received in the vehicular environment still needs to be improved because vehicular devices drained fast in network high demanded computations and communications which would affect their performance [30]. On the other side, since blockchain algorithms consume a lot energy for computations, it requires higher capabilities of software and hardware. RSUs and vehicles could be engaged in the mining process to reach network consensus and create blocks to be stored in blockchain ledger. These processes are not trivial, demanding processing power and adequate storage capabilities. Therefore, managing how blockchain networks utilize and support IoV infrastructure should be considered in the future research directions of blockchain-based IoV application designs. Furthermore, optimizing computation, communication and storage cost is also crucial to be consider in order to make the technology more feasible for the stakeholders [12].

Scalability: With the rapid growth of the IoV network, more IoV nodes join the system, leading to a vast amount of data that has to be recorded. The more the data is captured, the more improvement in IoV safety and efficiency. However, this brings a scalability issue in IoV nodes' communication and storage. Subsequently, integrating the blockchain with IoV could add an additional layer of scalability problem since the technology raises the resource constraints challenge mentioned before. [32] suggests a performance optimization framework for blockchain-based IoV to improve blockchain throughput for the transactions. The framework maximizes the throughput by modulating block size and interval and selecting block procedures. It guarantees other blockchain properties, which are decentralization, latency and security. Indeed, few works address the blockchain-based IoV's scalability challenge demanding high attention from the research communities.

7   Lessons Learned and Conclusion

It is a significant benefit to using blockchain technology to enhance the security, automation and decentralization of IoV platforms. In this chapter, the *integration of blockchain technology and IoV platforms* is introduced as a decentralized and secure solution. With blockchain and IoV integration, the *functionalities of RSUs* should be revisited and supported for the mining and verification processes. In this regard, RSUs may require more processing power and sufficient storage to support the extensive operations for mining the blocks into the blockchain ledger.

One of the essential applications for the Internet-of-Vehicular environment is *blockchain support for IoV data sharing*. The core of the data sharing of IoV is the blockchain-based *incentive protocols*, which are described as part of participating information between IoV users. As part of the incentive announcement protocols, utilizing *smart contracts* in the incentive and data sharing mechanisms is significant for transaction exchange and data management.

Blockchain support for IoV trust management is presented in the chapter to authenticate and efficiently verify users and traffic information. Several popular blockchain-based IoV applications such as accident detection, forensics application, carpooling and platooning have been the focus of researchers' studies.

As a conclusion, blockchain technology will significantly play a significant role in the transportation systems, despite the challenges mentioned in the previous section. Hopefully, these obstacles will attract the research communities' attention to better improve the area of blockchain and IoV.

References

1. Nakamoto, S.: Bitcoin: a peer-to-peer electronic cash system (2009). http://www.bitcoin.org/bitcoin.pdf
2. Wood, G.: Ethereum: a secure decentralised generalised transaction ledger. Ethereum Proj. Yellow Paper **151**(2014), 1–32 (2014)
3. Huckle, S., Bhattacharya, R., White, M., Beloff, N.: Internet of things blockchain and shared economy applications. Procedia Comput. Sci. **98**, 461–466 (2016)
4. Azaria, A., Ekblaw, A., Vieira, T., Lippman, A.: MedRec: using blockchain for medical data access and permission management. In: Proceedings of 2nd International Conference Open Big Data (OBD), pp. 25–30 (2016)
5. Cruickshank, A., Cao, Y., Asuquo, P., Ogah, C., Sun, Z.: Blockchain-based dynamic key management for heterogeneous intelligent transportation systems. IEEE Internet Things J. **4**(6), 1832–1843 (2017)
6. Turkanović, M., Hölbl, M., Košič, K.: Heričko, M., Kamišalić, A.: EduCTX: a blockchain-based higher education credit platform. IEEE Access **6**, 5112–5127 (2018)
7. Casino, F., Dasaklis, T., Patsakis, C.: A systematic literature review of blockchain-based applications: current status classification and open issues. Telematics Inform. **36**, 55–81 (2019)
8. Tripathi, G., Ahad, M., Sathiyanarayanan, M.: The role of blockchain in internet of vehicles (IoV): issues, challenges and opportunities. In: International Conference on contemporary Computing and Informatics (IC3I) (2019)
9. Singh, M., Kim, S.: Branch based blockchain technology in intelligent vehicle. Comput. Netw. **145**, 219–231 (2018)
10. Kchaou, A., Abassi, R., Guemara, S.: Toward a distributed trust management scheme for VANET. In: Proceedings of the 13th International Conference on Availability (2018)
11. Zhou, Q., Yang, Z., Zhang, K., Zheng, K., Liu, J.: A decentralized car-sharing control scheme based on smart contract in internet-of-vehicles. In: IEEE 91st Vehicular Technology Conference (2020)
12. Hammoud, A., Sami, H., Mourad, A., Otrok, H., Mizouni, R., Bentahar, J.: AI, blockchain and vehicular edge computing for smart and secure IoV: challenges and directions. In: IEEE Internet of Things Magazine (2020)
13. Saini, A., Sharma, S., Jain, P., Sharma, V., Khandelwal, A.: A secure priority vehicle movement based on blockchain technology in connected vehicles. In: Proceedings of the 12th International Conference on Security of Information and Networks (2019)
14. Li, L., Liu, J., Cheng, L., Qiu, S., Wang, W., Zhang, X., Zhang, Z.: Creditcoin: a privacy-preserving blockchain-based incentive announcement network for communications of smart vehicles. IEEE Trans. Intell. Transp. Syst. **19**(7), 2204–2220 (2018)

15. Chen, W., Chen, Y., Chen, X., Zheng, Z.: Toward secure data sharing for the IoV: a quality-driven incentive mechanism with on-chain and off-chain guarantees. IEEE Internet Things J. **7**(3), 1625–1640 (2019). https://doi.org/10.1109/JIOT.2019.2946611

16. Yin, B., Wu, Y., Hu, T., Dong, J., Jiang, Z.: An efficient collaboration and incentive mechanism for internet of vehicles (IoV) with secured information exchange based on blockchains. IEEE Internet Things J. **7**(3), 1582–1593 (2020)

17. Yang, Y., Chen, J., Zheng, X., Liu, X., Guo, W., Lv, H.: Blockchain-based incentive announcement system for internet of vehicles. In: IEEE International Conference on Parallel & Distributed Processing with Applications, Big Data & Cloud Computing, Sustainable Computing & Communications, Social Computing & Networking (ISPA/BDCloud/SocialCom/SustainCom), pp. 817–824 (2019)

18. Yang, T., Chou, D., Tseng, W., Tseng, H., Liu, C.: Blockchain-based traffic event validation and trust verification for VANETs. IEEE Access **7**, 30868–30877 (2019)

19. Yao, Y., Chang, X., Mišić, J., Mišić.: BLA: blockchain-assisted lightweight anonymous authentication for distributed vehicular fog services. IEEE Internet Things J. **6**(2), 3775–3784 (2019)

20. Davydov, V., Bezzateev, S.: Accident detection in internet of vehicles using blockchain technology. In: 2020 International Conference on Information Networking (ICOIN) (2020)

21. Cebe, M., Erdin, E., Akkaya, K., Aksu, H., Uluagac, S.: Block4forensic: an integrated lightweight blockchain framework for forensics applications of connected vehicles. IEEE Commun. Mag. **56**(10), 50–57 (2018)

22. Decoster, K., Billard, D.: HACIT: a privacy preserving and low cost solution for dynamic navigation and forensics in VANET. In: 2018 4th International Conference on Vehicle Technology and Intelligent Transport Systems (VEHITS) (2018)

23. Kevin, D., David, B.: HACIT2: a privacy preserving, region based and blockchain application for dynamic navigation and forensics in VANET. In: International Conference on Ad Hoc Networks in International Conference on Ad Hoc Networks. Springer, LNICST, vol. 258, pp. 225–236 (2019). https://doi.org/10.1007/978-3-030-05888-3_21

24. Shrestha, R., Bajracharya, R., Shrestha, A., Nam, A.: A new-type of blockchain for secure message exchange in vanet. Digital Commun. Netw. **6**(2), 177–186 (2019)

25. Huang, X., Xu, C., Wang, P., Liu, H.: LNSC: a security model for electric vehicle and charging pile management based on blockchain ecosystem. IEEE Access **6**, 13565–13574 (2018) https://doi.org/10.1109/ACCESS.2018.2812176

26. Pedrosa, A., Pau, G.: ChargeltUp: on blockchain-based technologies for autonomous vehicles. In: The 1st Workshop on Cryptocurrencies and Blockchains for Distributed Systems, pp. 87–92 (2018)

27. Chen, C., Wu, J., Lin, H., Chen, W., Zheng, Z.: A secure and efficient blockchain-based data trading approach for internet of vehicles. IEEE Trans. Veh. Technol. **68**(9), 9110–9121 (2019)

28. Xia, S., Lin, F., Chen, Z., Tang, C., Ma, Y., Yu, X.: A Bayesian game based vehicle-to-vehicle electricity trading scheme for blockchain-enabled internet of vehicles. IEEE Trans. Veh. Technol. **69**(7), 6856–6868 (2020)

29. Li, M., Zhu, L., Lin, X.: Efficient and privacy-preserving carpooling using blockchain-assisted vehicular fog computing. IEEE Internet Things J. **6**(3), 4573–4584 (2018)

30. Mollah, M., Zhao, J., Niyato, D., Guan, Y., Yuen, C., Sun, S., Lam, K., Koh, L.: Blockchain for the internet of vehicles towards intelligent transportation systems: a survey. IEEE Internet Things J. **8**, 4157–4185 (2020)

31. Chen, C., Xiao, T., Qiu, T., Lv, N., Pei, Q.: Smart-contract-based economical platooning in blockchain-enabled urban internet of vehicles. IEEE Trans. Ind. Inf. **16**(6), 4122–4133 (2020)

32. Liu, M., Teng, Y., Yu, F., Leung, V., Song, M.: Deep reinforcement learning based performance optimization in blockchain-enabled internet of vehicle. In: ICC IEEE International Conference on Communications (ICC), pp. 1–6 (2019)

Vehicle Guidance System Based on Secure Mobile Communication

Christoph Maget

Abstract Vehicle guidance systems are considered key to improve capacity and safety of transport systems. Information and communication technologies enable both vehicle speeds and distances to be optimized without being limited to human reaction times. As the major two competing topological approaches for communication networks, viz. ad-hoc and cellular networks, exhibit specific advantages and disadvantages in different applications, there is still no standardized solution in the offing. Established encryption methods have either proven to be insecure or lack real-time capabilities when used in distributed automation systems, where the only proven secure concept for encryption – perfect security – has not been employed so far due to practical shortcomings. Meeting existing standards, a communication architecture for vehicle guidance systems allowing for perfectly secure encryption and observing real-time requirements for wireless communication is presented. Its core components are a central instance authenticating all participants, generating and distributing the required keys as well as a transmission infrastructure based on relay stations. Different sensitivity analyses show that one-time pad cryptography can keep up with or even outperform the AES in the presented use case. The keys required for a sufficiently long operating time can be stored on common storage media.

1 Introduction

Due to the increasing need for mobility, the capacity limit of existing transport systems has been reached or exceeded. Extension or new construction is elaborate and expensive. If equipped with suitable sensors and actuators, automation systems can significantly reduce the response times in technical systems. The driving behavior of vehicles equipped with suitable automation systems is thus no longer limited by human responsiveness. Consequently, higher speeds and shorter vehicle distances can be realized, which increase the capacity of existing roads.

C. Maget (✉)
FernUniversität in Hagen, Fakultät für Mathematik und Informatik, Universitätsstraße 47, 58097 Hagen, Germany
e-mail: christoph.maget@studium.fernuni-hagen.de

© Springer Nature Switzerland AG 2021
N. Magaia et al. (eds.), *Intelligent Technologies for Internet of Vehicles*, Internet of Things, https://doi.org/10.1007/978-3-030-76493-7_11

Such a vehicle guidance system must be embedded in a technical infrastructure that provides all the necessary resources and equipment. This infrastructure is formed by "intelligent transport systems" (ITS). According to the relevant directive of the European Union, an ITS is intended to increase security and efficiency of the transport infrastructure through better coordination [1]. Uniform interfaces and communication protocols are essential for the correct interaction of all entities. Consequently, various standards have emerged in order to standardize communication and to provide the various functionalities [2]. At the same time, an intelligent transport system must meet high security requirements and avoid abusive behavior. Since an ITS both communicates digitally and interacts within the physical world, this is a typical example of a cyber-physical system (CPS) that can be described with the Reference Architectural Model Industry 4.0 (RAMI 4.0) [3].

Communication between vehicles ("Vehicle-to-vehicle", V2V) enables the fully automatic exchange of status and control messages. The data exchange with the infrastructure ("Vehicle-to-infrastructure", V2I) provides additional information related to the environment that cannot be perceived by vehicle sensors alone [4]. "Vehicle-to-everything" (V2X) is often used as a generic term to describe communications in the internet of vehicles (IoV).

A vehicle guidance system is intended to achieve fully autonomous steering of motor vehicles. The degree of automation is divided into six levels, which differ in a different degree of automation [5]. Communication with real-time conditions between vehicles is an indispensable part of vehicle guidance systems on the way to full automation. Only cooperation between vehicles leads to increased traffic safety and more efficient road use in ITS through intelligent route finding [6].

2 Problem and Contribution

Networking is required for the interaction of automated vehicles. Wireless technologies must be used to connect moving objects and all communications must be handled via the air interface. The advantage is a significantly reduced cabling effort. This facilitates, on the one hand, flexible integration of additional components in existing networks and thus dynamic scaling of the entire automation system.

On the other hand, there are numerous challenges to be solved if a suitable communication architecture is to be developed. First, a suitable topology has to be selected that meets all the requirements of a vehicle guidance system. There are essentially two competing approaches in transport systems: WiFi based solutions (IEEE 802.11p) and cellular based solutions (Cellular V2X, C-V2X) [7]. Second, the implementation of adequate cryptographic protection must guarantee data security and integrity, since the physical access to the air interface cannot be restricted. All communications can therefore also be received and changed by unauthorized third parties. Third, message formats must be analyzed and specified that enable a uniform and unambiguous exchange of information.

As a possible solution, a uniform communication architecture for automation systems is developed and evaluated, which enables encryption with perfect secrecy and meets real-time requirements for wireless communicating components in ITS. The requirements for a vehicle guidance system are derived from existing standards and the usability in a uniform communication architecture is assessed.

2.1 Requirements for a Vehicle Guidance System

There is not yet a comprehensive standard available for a vehicle guidance system. However, its individual parts are well specified concerning topology, security, and message format. The essentials of relevant standards are briefly summarized in the following.

ETSI TR 102 962 *Intelligent Transport Systems (ITS); Framework for Public Mobile Networks in Cooperative ITS (C-ITS)* requires the integration of V2X in existing cellular networks. Example scenarios are given for LTE, GSM, and EDGE.

EN 303 613 *Intelligent Transport Systems (ITS) - LTE-V2X Access layer specification for Intelligent Transport Systems operating in the 5 GHz frequency band* specifies the frequency bands that are approved for use by ITS. In particular, it defines the operation of an ITS in the dedicated frequency band at 5 GHz.

ISO 24534-3:2016 *Intelligent transport systems - Automatic vehicle and equipment identification - Electronic registration identification (ERI) for vehicles - Part 3: Vehicle data (ISO 24534-3:2016)* is a standard for defining uniform digital identification features for vehicles, including electronic registration. The use case of electronic toll collection mentioned therein requires central administration and billing. Central administration counteracts collisions when assigning unique identifications.

ISO/IEC 27000:2018 *Information technology - Security techniques - Information security management systems - Overview and vocabulary* is the central series of standards for information and communication security. The series specifies the management, control, maintenance and improvement of an information security management system (ISMS). An ISMS has to ensure and improve the IT security of a product in all phases of its life cycle.

SAE J 3061:2016-01-14 *Cybersecurity Guidebook for Cyber-Physical Vehicle Systems* describes a framework for safeguarding the information security of a connected vehicle over its entire life cycle and provides an overview of cybersecurity methods through numerous references to other standards such as the ISO 26262.

ISO/IEC 15408-1:2020-06 *Information technology - Security techniques - Evaluation criteria for IT security - Part 1: Introduction and general model* is a current draft standard that provides a concrete evaluation model for the implemented concepts and principles concerning IT security. For this purpose, so-called "safety profiles" are assigned to the individual functional and safety components. A vehicle guidance system must therefore provide a subdivision into separate components and equip these components with a rights and role model as well as safety profiles.

ETSI EN 302 637-2 *Intelligent Transport Systems (ITS); Vehicular Communications; Basic Set of Applications; Part 2: Specification of Cooperative Awareness Basic Service* specifies information that each vehicle sends as a parameter regarding its current status. The size of a message varies with the message content, which can be defined modularly and in the form of containers.

ENV 12313-4:2000 *Traffic and Traveller Information (TTI) - TTI Messages via Traffic Message Coding - Part 4: Coding Protocol for Radio Data System - Traffic Message Channel (RDS-TMC) - RDS-TMC using ALERT Plus with ALERT C* specifies traffic and travel information to be carried in the inaudible area of the FM signal in the form of specific codes (Radio Data System - Traffic Message Channel, RDS-TMC). These codes are linked to traffic messages that have been precoded in terms of content and geography and are stored as full text in a database on board the vehicle. A transmitted and appropriately processed code word triggers the recipient to display a detailed message whose components are fed from the database. In this way, the driver can be informed about the location and the type of any traffic disruption in her national language.

To sum up existing standards, the most important design decisions for a vehicle guidance system regarding topology and communication are to employ

1. An encryption system with sufficient security to ensure confidentiality, integrity, and availability,
2. A centralized management of digital identities (dID),
3. Measures to guarantee real-time constraints, and
4. An optimized message format for information transmission about the status of vehicles and roads.

In the following section, the current state of the art is analyzed with respect to these requirements.

2.2 State of the Art

The development of a vehicle guidance system and its integration into existing infrastructures requires numerous design decisions concerning network topology, information security, and message format. This topic is therefore subject to intensive research.

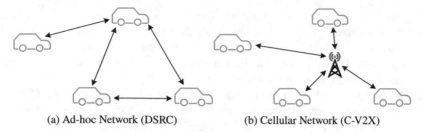

(a) Ad-hoc Network (DSRC) (b) Cellular Network (C-V2X)

Fig. 1 Topology in vehicular networks

2.2.1 Network Topology

Concerning network topology, two concepts compete: Ad-hoc networks ("dedicated short range communication", DSRC) and cellular networks ("cellular vehicle to everything", C-V2X). The two concepts are visualized in Fig. 1. Both concepts imply technical advantages and disadvantages and are subject to commercial and legal constraints [8].

Cellular Networks carry out all communications via a central access point. The central access point is also referred to as a "relay". In infrastructure mode, important parameters such as the distances between the participants and the relays, the load of the individual relays, and the signal propagation times can be determined. These are essential prerequisites for communications with real-time constraints. The infrastructure mode is already being used successfully in networks for voice and data communication between mobile users (e.g. Global System for Mobile Communications, GSM), which are, however, only soft real-time systems. The necessary identifiers (ID) are assigned and managed centrally using a Subscriber Identity Module (SIM), an Universally Unique Identifier (UUID) or a Globally Unique Identifier (GUID). Great potential is seen in an expansion of the GSM standard towards 5G, especially for real-time applications. For this purpose, existing networks are to be expanded by so-called "cloudlets" which provide the required functions [9]. Thanks to its use as an emergency call system (eCall) [10], C-V2X has already proven itself in a safety-relevant traffic engineering application with real-time requirements.

Ad-hoc Networks connect the communication participants directly to each other. No additional infrastructure is required and all messages can be transmitted either directly between the participants or indirectly with participants as communication nodes. One advantage of ad-hoc mode is the independence from additional infrastructure. This also increases the reliability of individual routes through possible redundancies [11]. However, protecting the transmitted data is challenging and requires complex measures to protect data content and privacy: The number and length of connections between sender and receiver can vary or even tear off. This generally non-deterministic network structure does not meet the real-time conditions that are required by vehicle guidance systems. In ad-hoc networks, however, a central authority is required to assign and monitor the network addresses used by the participants. If considered as distributed peer-to-peer (P2P) networks that are based on underlying physical networks, the hardware address can perform this functionality. However, this does not offer any protection against deliberately changed or allegedly claimed addresses.

An ad-hoc network for vehicles is also called "vehicular ad-hoc network" (VANet). A range in the frequency band from 5.85 GHz to 5.925 GHz was reserved for VANets based on IEEE 802.11 and standardized as IEEE 802.11p [12].

It is worth noticing that most established communication networks are managed by central instances so far:

- Internet: The Internet Corporation for Assigned Names and Numbers (ICANN) and its sub-division Internet Assigned Numbers Authority (IANA) are hierarchically organized and manage the parameters required for the operation of the internet at the top level. They provide authenticated participants with IP address ranges and top-level domains, thus avoiding address conflicts through appropriate organization.
- Computer networks: The IEEE defines and manages unique manufacturer IDs and makes them available in a central database to assign a Media Access Control (MAC) address to network interfaces.
- Global System for Mobile Communications (GSM): Subscriber Identity Module (SIM), International Mobile Equipment Identity (IMEI) and International Mobile Subscriber Identity (IMSI) are managed centrally by a telecommunications company and assigned to the mobile subscribers.
- Electronic banking: The Society for Worldwide Interbank Financial Telecommunication (SWIFT) operates a dedicated, centrally managed telecommunications network (SWIFTNet) for the standardized exchange of messages between financial institutions. Account information is managed centrally at the connected banks. Access authorizations for the accounts are distributed to customers using chip cards. Customers use them to authenticate themselves and generate a transaction authentication number (TAN) for authorizing financial transactions. A TAN often comprises a data volume of six digits.

The central administration prevents IDs from being assigned twice. Complex procedures for decentralized consensus are not necessary with a central administration in place. However, special precautions are required to ensure security and reliability, since every central administration is a potential single point of failure (SPOF).

2.2.2 Information Security

Concerning secure communication, also two concepts compete: Symmetric and asymmetric cryptography. Asymmetric cryptography can be broken, because private keys can be calculated from public keys, given sufficient computing power.

Although the security of symmetric cryptography also depends on key length [13], it comprises unbreakable methods with perfect secrecy.

Symmetric Cryptography uses the same key for encryption and decryption. Known methods are block ciphers such as the Advanced Encryption Standard (AES) and the Data Encryption Standard (DES) as well as stream ciphers such as A5. The processes are characterized by their speed and the relatively small key length compared to the message. Although the three presented methods have been evaluated many times during development and have been widely used, the DES and A5 are already considered broken [14, 15]. Modern algorithms such as AES are currently considered secure, but are also characterized by a limited key length. It can therefore be assumed that these algorithms will also prove to be unsafe, at the latest with the advent of quantum computers.

Asymmetric Cryptography uses distinct keys for encryption and for decryption. One key forms the private key to be kept secret, the other forms the public key to be passed on. Known methods are the RSA method [16] and Elgamal [17]. The practical application of asymmetric cryptography requires comparatively large computing power and is therefore preferred for the exchange of symmetric keys. If an attacker knows one of the two keys (which can be assumed for the public key), she can use it to calculate the other key, since the mathematical relationship is known. The security is based solely on the fact that this calculation would require a (currently) impracticably long period of time or computing power. Asymmetric cryptography must always involve a trustworthy third party to prevent a man-in-the-middle (MITM) attack in communication between two participants. This instance is involved at different points in the transmission protocol in order to authenticate the participants and their keys [18].

Perfect Secrecy itea subtype of symmetric cryptography which is proven to be unbreakable [19]. If a cipher text is received, an attacker can only conclude that there is an exchange of messages. However, it is not possible for her to draw conclusions about the content of the message that go beyond mere guessing. For such an encryption method, however, keys are needed that are at least as long as the message itself. On the one hand, the latter is seen as an obstacle. On the other hand, it is worth mentioning that perfect secrecy also means quantum-safe security when properly deployed.

Encryption is not only used to ensure the confidentiality, but also necessary to proof the integrity of a message.

2.2.3 Message Format

A vehicle guidance system needs to cover two kind of information: Route-related and vehicle-related data.

Route-related Information such as increased travel times due to traffic jams or road closures is transmitted as route-related data. A common procedure is to add traffic and travel information to the broadcast radio in the inaudible area of the radio signal (Radio Data System - Traffic Message Channel, RDS-TMC). Specific codes are linked to traffic reports that have been coded beforehand in terms of content and geography and are stored as full text in a database on board the vehicle. A transmitted and appropriately processed codeword triggers the recipient to display a detailed message. The components of the message are read from the contents of the pre-stored database. This is a unidirectional broadcast, a return channel from the vehicle to the transmitter is not provided.

Vehicle-related Information contains parameters describing the current status of vehicles as vehicle-related data. Data formats are e.g. specified with the "Cooperative Awareness Message" (CAM) [20] and the "Decentralized Environment Notification Message" (DENM) [21]. The message size varies with the content, which can be defined modularly using containers. Messages can either be sent with a fixed frequency (CAM) or event-driven (DENM). Messages in CAM or DENM format can be both received and sent.

In contrast to RDS-TMC, there is still no implementation of a service based on a CAM or a DENM yet. The standards of the ETSI are also still incomplete and refer to further specifications in the future.

2.2.4   System Integration

Different architectures try to connect vehicles with other entities of the internet of things (IoT) [22], which seems an expedient approach. In order to maintain the organization and the efficiency of the emerging IoV, however, these architectures must be supplemented by a suitable administrative structure. LASAN [23] provides a promising organizational approach for authentication and authorization in inter- and intra-vehicle communication. However, the asymmetric cryptography used still has to be replaced by verifiably secure procedures in order to meet the high security requirements in the automotive sector. Inadequate organization has proven to be a vulnerability for side-channel attacks such as time spoofing or sybil attacks [24]. Approaches for integrating the required functions into a vehicle guidance system can be divided into three classes.

Fully autonomous vehicles replace the steering by human drivers with computers. All necessary information about the environment is gathered by sensors on board vehicles. Digital communication that replaces human interaction is not intended [25]. Fully autonomous vehicles are thus not able to anticipate the driving behavior of other vehicles. They can only react based on the recognized maneuvers of others.

Connected Vehicles are capable to digitally exchange information. Those messages can contain data about the current or the intended status of a vehicle or the infrastructure. The latter enables the mutual anticipation of driving behavior and thus a further optimization of vehicle interactions. Communication is mostly implemented with ad-hoc networks, also referred to as VANet [26].

Connected Vehicles Depending on Roadside Equipment require additional infrastructure to form reliable networks. This additional communication equipment is often referred to as "roadside unit". Different schemes exist that try to optimize the placement of those roadside units [27]. As this additional roadside equipment is expensive in construction and maintenance, financial issues arise. Savings in road construction are therefore purchased with additional costs for roadside units.

Consequently, efforts to develop a vehicle guidance system always need compromises that may result in the violation of any standards. Mediating neutrally between these positions, we follow a strictly standard-driven approach for a vehicle guidance system.

2.3 Methodological and Statistical Challenges of Vehicular Networks

The literature review shows that vehicle guidance systems are an established field of research. However, existing approaches for networking almost without exception favor decentralized concepts following the ad-hoc structure where IT security is based on asymmetric cryptography. The security level of the encryption depends on the key length in this case. Confronted with increasing power of attackers, the key length is then only increased *reactively*, instead of *proactively* using appropriate cryptographic procedures. Moreover, nodes in such decentrally administered VANets can deliberately or inadvertently delay packet forwarding. This increases the likelihood of message delays or losses and circumvents real-time message transmission. The availability of network meshes in VANets is thus subject to statistical influences due to complex vehicle movements and interactions.

On a cursory level, there are reasons for such a combination of techniques like the aforementioned scalability. A consistent comparison with the requirements for vehicle control systems developed in this chapter shows, however, that essential requirements are not or only insufficiently met with existing technologies. The goals of the further development of a vehicle control system are thus derived from the following methodological challenges:

- Asymmetric cryptosystems are conceptually insecure and prone to widespread attacks.
- Symmetric cryptosystems that are currently considered secure cannot permanently meet the high IT security requirements in ITS.
- Retrofitting security features increases the complexity and makes existing systems difficult to operate and maintain.
- Vehicle networks based on the ad-hoc topology are complex and not real-time capable, despite isolated approaches.
- Decentralized user administration is complex and makes scaling difficult.

Consequently, the development target is a uniform communication architecture that enables perfect secrecy and fulfills real-time conditions throughout all processes for message exchange. The cryptographic methods must resist known and future attacks in order to guarantee permanent independence from subsequent improvements to those methods. At the same time, the communication architecture must be scalable to any fleet size without compromising on real-time capability, security and availability.

2.4 Research Question

The literature review reveals that existing solutions cannot yet meet the requirements of a vehicle guidance system that were derived from existing standards. Consequently, the question arises whether a communication architecture is possible that can fulfill the following aspects:

- There is consensus concerning network topology.
- Encryption with perfect security is deployed.
- The message format is optimized towards vehicle-related as well as route-related data.

With our contribution we will elaborate whether existing concepts can be integrated to a vehicle guidance system that fulfills those aspects and meets the relevant existing standards. The resulting communication architecture is referred to as "SIKAF".

3 Introducing SIKAF: The Communications Architecture for Vehicle Guidance Systems

Following the approach to perform a structured analysis of existing standards, we can derive specifications for all relevant parts of a vehicle guidance system. The following requirements for a vehicle guidance system will be incorporated in SIKAF:

- The data exchange between vehicles must be handled via a central access point.
- The network addresses of the individual vehicles must be generated, distributed and managed by a central authority.
- The central authority must act as a moderator for the message exchange.
- The encryption must resist all current and future attacks.
- Symmetric encryption must be used to secure message exchange between vehicles and to guarantee real-time constraints.
- The keys must be generated by a trustworthy, sovereign authority.
- A secure communication channel must be used for key distribution.

It is thus challenging to combine all those requirements into a single system.

Fig. 2 Organizational structure of SIKAF

3.1 Organization

Organizationally, SIKAF is broken down to specific components. The components are assigned to dedicated levels and can thus be integrated into existing structures. Figure 2 shows the classification of the SIKAF levels and their reference to the automation pyramid and the RAMI 4.0. This classification avoids a break of SIKAF with proven structures and transfers the traditional automation pyramid to the internet of things within the transport sector.

3.1.1 Central Authority

The vehicles connected by SIKAF generally have no information about a communication partner at the first approach. Identifiers as well as keys for symmetric cryptography must therefore be managed and provided by a third party. This institution has to

- be trustworthy because it knows the keys,
- be efficient because keys must be generated in large amounts and in short periods of time, and
- establish a secure channel to all participants to distribute keys and identifiers.

In SIKAF, this institution is affiliated with a government agency that already identifies and manages vehicles in the physical world. These existing functions are expanded to include the administration of digital identities (dID) and the issuing of cryptographic keys. The government agency has subordinate bodies that have to be contacted by vehicles regularly (e.g. for technical inspection). A secure channel to the vehicles for key distribution and key replenishment can be established via these subordinate bodies. As a public authority, this authority is also authorized to extend or limit the rights of vehicles in the physical traffic system for global optimization. Trustworthiness to the authority is given by democratic legitimation.

3.1.2   Relay for Mobile Communications

Data transmission is carried out via a dedicated mobile network in SIKAF. The entirety of the communication infrastructure required for the transmission of messages is referred to as a "relay". The relay consists of the components

- access points for the communication participants,
- connections between the access points, which forward the messages reliably and in real-time, and
- a central main frame computer for encryption and routing tasks.

The relay is equipped with a dedicated, tamper-proof communication network, so that each message transmission in this network (but not to the vehicles) can be processed in real-time. There is also a permanent, secure channel to the central authority, through which all required cryptographic information can be exchanged and provided.

3.1.3   Vehicles

Vehicles are the sources and sinks of all communications in SIKAF. As actuators of the traffic system with numerous mechanical components, they are exposed to constant wear. It is therefore necessary and mandatory to carry out regular inspections, for which the vehicle must get into physical contact with the central authority or one of its subordinate bodies. In addition to checking the hardware of the vehicle, these inspections can be used to carry out software updates. The latter also includes updating data such as the cryptographic keys.

3.2   Implementation

With the identified prerequisites, we are able to design the communication architecture for a vehicle guidance system. Figure 3 shows the overall architecture.

3.2.1   Topology

Only infrastructure mode is able to meet all requirements with an acceptable level of complexity: Digital identities are generated and managed in a central registry to avoid ambiguities. All vehicles communicate with each other through a relay. The transmission distances and the number of hops are therefore predictable and make SIKAF real-time capable.

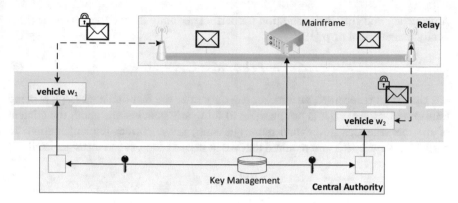

Fig. 3 Communication architecture SIKAF

3.2.2 Information Security

The identified requirements only allow to use encryption with perfect secrecy, which is a type of symmetric encryption. The need for symmetric encryption also results from the time requirements, as in real-time applications like a vehicle guidance system only symmetric encryption methods may be used to guarantee predictable time behavior. Every message is encrypted by the sender and sent to the relay. The relay decrypts the message, forwards it internally and, finally, sends it newly encrypted to the receiver. The infrastructure mode can guarantee a reliable communication channel and to meet real-time requirements thanks to a predictable hop count.

3.3 Cryptography

Confidentiality, integrity and availability (also referred to as "CIA triad") must be achieved with cryptographic measures.

3.3.1 Perfect Secrecy

The Vernam Cipher [28] is implemented in SIKAF as unbreakable encryption method. To theoretically prove this "perfect secrecy", consider the set of all plaintexts P and the set of all possible cipher texts C. With the probability $Pr(p)$ one obtains any plaintext $p \in P$, with the probability $Pr(c)$ any cipher text $c \in C$. The conditional probability to get the plaintext p when the ciphertext c was chosen is $Pr(p|c)$. An encryption method is perfectly secure if the following applies to all plain texts $p \in P$ and to all cipher texts $c \in C$:

$$Pr(p|c) = Pr(p) \tag{1}$$

A necessary and sufficient condition for perfect certainty is that $Pr(c|p)$ is stochastically independent of p [19]:

$$Pr(c|p) = Pr(c) \ \forall c \in C \wedge \forall p \in P \tag{2}$$

If a ciphertext is received, an attacker can only infer the fact that a message is being transmitted. However, it is not possible to draw any conclusions about the content of the message that go beyond mere guessing, given correct implementation. In particular, keys $k \in K$ are required that are at least as long as the message itself:

$$|k| \geq |p| \tag{3}$$

All keys have to be generated with a true random generator.

3.3.2   Key Generation

A real random process must be used when generating keys. The keys have to be created at a central location, because sufficient resources are permanently required for this. At SIKAF, the central authority carries out the key production. Figure 4 shows the overall key generation and distribution process. Technically, the following processes can be used to generate real random numbers.

Key Generation Using a Physical Process
Real random bits are generated by a combination of sampled processor noise and the color value of randomly recorded images. The resulting bit streams are XOR combined to form a key stream.

Key Generation Using a Chaos Process
Another possibility for generating random numbers is the use of recurrence in a chaos process in which only the initialization vector x_0 has to be provided as seed value.

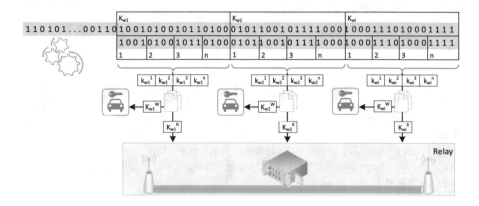

Fig. 4 Key generation and distribution

Further numbers can then be generated using the equation

$$x_{n+1} = rx_n(1 - x_n) \tag{4}$$

where $r = const$. The difference to the physical process is that only the seed value is the secret to be exchanged and the actual key can then be generated from it for all participants.

The random bits are taken from a constant bit stream in both cases. They are first divided into blocks K and supplemented by the participant's respective digital identification (dID). A key block K_{dID} is therefore initially available for each vehicle. The key blocks are then divided into packets and each packet is given an index n. The resulting packets K_{dID_n} are duplicated to $K_{dID_n^R}$ and $K_{dID_n^W}$, since a key must be available on the vehicle and a matching key on the relay. The following applies:

$$K_{dID_n^R} = K_{dID_n^W} \ \forall n \tag{5}$$

The packages $K_{dID_n^W}$ are prepared for distribution to the vehicles, the packages $K_{dID_n^R}$ for use by the relay.

3.3.3   Key Distribution

Each sender must be able to encrypt any message for each recipient. However, not every vehicle can hold symmetric keys for all other vehicles. In this case, the resulting complexity class $\mathcal{O}(n^2)$ would make the scaling, i.e. the possibility of flexibly adding more vehicles to the communication architecture, disproportionately difficult. Since the relay serves as an intermediate station for communications and the messages are encrypted only on the radio link between the vehicle and the relay, it is sufficient if each vehicle receives an individual key set $K_{dID_n^W}$. The relay has the corresponding counterpart $K_{dID_n^R}$ and is therefore able to decrypt the messages of all vehicles. $K_{dID_n^R}$ must be distributed to the relay and $K_{dID_n^W}$ to the vehicles. There is a constant, secure channel from the central authority to the relay R, for example via a dedicated fiber optic cable. The central authority can process key distribution and key replenishment to R via this channel. The keys are distributed to the vehicles during their physical contact with a subordinate element of the sovereign authority. The following options can be considered:

Transfer during production in the factory of an initial key supply to the read-only memory in the vehicle. There is sufficient key replenishment from the central authority to the vehicle plant assumed.

Maintenance or fuel consumption is necessary for motor vehicles at regular intervals to ensure the operational and traffic safety of the vehicle. Part of the inspection is the examination of recorded sensor values via standardized diagnostic protocols (on-board diagnosis, OBD), which requires a wired connection to the vehicle electronics.

A physical connection to the vehicle is also established when refueling. In the case of electric vehicles, the use of digital computers is necessary anyway due to the charge regulation. The supply infrastructure required in both cases can be equipped with a secure channel to the key-generating central authority. Both connections can then be used to transfer the keys to the read-only memory on board the vehicle.

License plates are commonly issued and renewed in many countries by a government agency. The physical plates are either completely or partially replaced in order to extend the validity. During this process, a unidirectional, secure channel from the key-generating authority to the vehicle is available, which can be used to transfer new keys to the read-only memory.

Car sharing or rental car needs transfer and management of the vehicle key. In the case of central administration, a secure channel to the central authority can be established from this point. The key (or another physical access token) can be used as a carrier to transfer the data required to use SIKAF to the read-only memory on board the vehicle. This process also authenticates the actual driver. The corresponding supply facilities are connected to the public authority via a secure channel, for example by a dedicated fiber optic cable.

On site, a secure transmission channel to the vehicle can also be established by using physical data storages. In particular, no wireless transmission channel is required. Necessary updates of the libraries or other software components are carried out through the same process. The current trends in favor of electromobility and the shared use of vehicles enhance the presented options to ensure key replenishment. With this type of key transfer, a robust implementation of encryption with perfect secrecy is achievable. Further security-related measures have to aim at minimizing the risk of side-channel attacks. The reception of all keys is acknowledged in SIKAF to enable non-repudiation.

3.3.4 Key Revocation

Communication participants who violate the requirements of the communication architecture or pose a threat to the overall system are wholly or partially withdrawn from SIKAF. This prevents the affected communication participants from spreading damaging messages. SIKAF can take the necessary measures to block affected nodes on the relay side only. If the rights are withdrawn from a vehicle $id \in dID$, the $K_{id_n^R} \forall n$ keys are labeled as blocked on the relay. The vehicle id can therefore no longer use $K_{id_n^W}$ for encrypted data transmission or for authentication because the use of a blocked key is recognized by the relay. The relay replaces the original message with a corresponding blocking message to inform the message recipient about the blocking of the transmitter. This procedure means that no notification of the blockage is required to the other vehicles. In addition, there is no time-consuming recall of keys or certificates necessary. When supplying the vehicles with new versions of the keys, a list of blocked participants is distributed at the same time, so that a current version of this blocked list is stored for all vehicles.

To digitally block a vehicle, violations or dangers must be evident. The possibility of blocking individual participants must not lead to arbitrary censorship. The blocking of a participant *id* can be canceled by SIKAF by labeling $K_{id_n^R}$ as valid again. All of the operations required for this can also be carried out on the relay side alone.

3.3.5 Data Transmission

In SIKAF, all subsections of the data transmission s_i are designed for real-time data transmission ($t(s_i) \leq t_{krit}$). The following also applies to the entire message transmission:

$$\sum_i t(s_i) \leq t_{krit} \tag{6}$$

where t_{krit} represents an upper time interval to be defined for the transmission route and the respective application. SIKAF always transmits messages indirectly via the relay R, since the distance between two vehicles w_n and w_{n+1} is generally too large for a direct radio transmission. After the transmission from w_n to R, the transfer from R to w_{n+1} takes place.

Transmission Between Vehicle and Relay
The positions of all access points r_m are known to SIKAF. The positions of the vehicles w_n are continuously determined by them and transmitted to the mainframe computer via SIKAF. The Euclidean distances $|s_i|$ can be obtained for all i from $s_i = w_n - r_m$. With the distance, the propagation speed of electromagnetic waves, and the protocol-specific processing time the transmission time of a message packet on the radio links can be determined. This applies to both the transfer from the vehicle to the relay and to the transfer from the relay to the vehicle.

Processing Time: Vehicle
The baseband processors of the vehicles used at SIKAF are able to modulate the message to the carrier signal in real-time. A real-time operating system is run on the implemented electronic control units (ECU) and microcontrollers. The total duration of the message processing on board the vehicles is therefore always predictable.

Processing Time: Relay
The access points are connected to each other and to the mainframe via a dedicated channel, which can carry out all relay-side communications in real-time and without bandwidth restrictions. The total duration of the relay's internal transmissions is therefore always predictable.

The analysis of all processes and routes shows that the duration of the data transmission in SIKAF is always predictable and SIKAF can therefore be used for real-time communication.

3.3.6   Multicast and Broadcast

Messages can be sent as uni-, multi- and broadcast in SIKAF. Whereas unicast has already been described in detail and broadcast is the trivial solution through unencrypted transmission,[1] a special process is used to multicast a message by the sending vehicle. An area in the message header is reserved for listing the recipients. The packet then contains the key index for the relay in plain text and the message itself together with the intended recipient group encrypted in the reserved area of the packet. The sender transmits the message to the relay encrypted like an unicast. The relay then handles the multicast by encrypting the message individually for all recipients and routing the message to them. In this way the sender transmits the message only once, while the relay, which has greater computing capacity and transmission power, carries out the individual encryption and delivery of the message. The process is summarized in Fig. 5. Vehicle 8 transmits a message therein to vehicles 3, 5 and 7. In principle, all vehicles, including vehicle 6 in the example, can receive the message since it is transmitted via the air interface. By encryption with the individual keys of the authorized recipients, however, only these are capable to decrypt the message. If the recipient group is larger than the header area reserved in the address block, the sender must iterate the broadcast process by sending several packets of the same type to the relay, which in turn executes the multicast iteratively. If no multicast is carried out, the area reserved for the multicast addresses in the message is filled with zeros or is available for additional message content.

3.3.7   Filtering

Another function of the relay is to filter unwanted messages. The relay can be used as a moderator for all communications between vehicles and, if necessary, classify messages as "spam", since the message content is available in plain text at the relay. Affected messages are then either discarded, not forwarded, or justify a blocking of the participant in charge. SIKAF decides on the classification of a message as spam via

- whitelisting through a comparison with a library or
- blacklisting by checking for improper content.

The long-term or high-frequency transmission of messages can also be an indicator for spam and a reason for further checking. Messages with security-relevant content are never filtered, because given correct encoding, they will always pass the filter due to the whitelisting. The filtering prevents the limited transmission capacities of the air interface from being overloaded. The resources are exclusively reserved for the necessary transfers. SIKAF prevents arbitrary filtering ("censorship") by generating the required blacklists and whitelists from the central authority. In this way there is mutual control between the evaluating authority and the executing relay.

---

[1] Integrity check required.

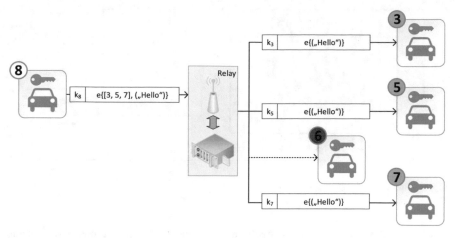

Fig. 5 Multicast

4 Results

The concept outlined is implemented in both hardware and software. This is to show that SIKAF is feasible and can meet all expectations.

4.1 Hardware for Prototype

A prototype implementation demonstrates that SIKAF can be practically implemented with the identified requirements. At this stage of a feasibility study, the hardware and software components can be chosen deliberately, which is why modules that are available easily and inexpensively are used. If mass production is possible, SIKAF could also be adapted to existing vehicle systems or integrated into them.

For each hierarchical level of SIKAF, individual prototypes are made in order to fulfill the modular structure of SIKAF. For all prototypes, a pre-assembled development board in the form of a single-board computer is used. The wiring is depicted in Fig. 6. The required peripheral components are already partially integrated on the board, which makes development considerably easier. The developer board used here provides significantly more resources in terms of computing power and interfaces than would be required for the development of a prototype. As a result, price and power consumption are higher than with an optimally integrated solution. A microprocessor from the ARM Cortex-A family is used for all prototypes. To implement wireless communication, existing radio interfaces based on the Bluetooth and WiFi standards are used. For Bluetooth, a chipset from the BCM family (manufacturer: "Broadcom") is used; for the WiFi module, a chipset from the RTL family (manufacturer: "Realtek") is used. A flash drive with SD form factor is used as mass storage.

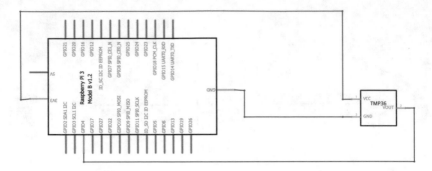

Fig. 6 Prototype wiring

The use of the developer boards enables a comparatively simple implementation and testing of various functions without the need to develop new components for each function. After a successful test, the successful implementation can be rescaled to the target platform in the vehicle and ported there. It does not necessarily have to be a dedicated hardware platform if the required functions can be integrated into existing systems, for example into on-board units (OBUs) of vehicles. The drawings of the prototypes are shown in simplified form, for example, short-circuit jumpers and pre-resistors have not been depicted as they have no significance for the communication connections.

4.2 Formal Language for Messages

The messages exchanged with SIKAF consist of a formal language, L_{sikaf}. The required alphabet Σ^{sikaf} combines three parts:

- An alphabet of the message content $\Sigma^{sikaf'}$ to describe events and their quantification,
- An alphabet for generating timestamps with an accuracy of milliseconds $\Sigma^{timestamp_{ms}}$ and
- An alphabet for using a standardized data format Σ^{json}

SIKAF uses the JavaScript Object Notation (JSON) because this notation meets the requirements for both computer processability and human readability. The messages can thus be formed from the alphabet

$$\Sigma^{sikaf} = \Sigma^{sikaf'} \cup \Sigma^{timestamp_{ms}} \cup \Sigma^{json} \tag{7}$$

Every alphabet and therefore Σ^{sikaf} is a set of characters. It should be noted that these characters are not letters in the sense of a natural language, but rather keywords to describe the message content and defined characters to form the data format. A word in the sense of this formal language is then the message that is exchanged with SIKAF.

4.3 Message Size

Information exchanged with SIKAF can relate either to vehicles or to routes. Vehicle-related messages contain information about the current and the intended state of a vehicle, in particular regarding the vehicle trajectory. Route-related messages contain information about the current status of individual route sections. The latter information is reported either by vehicles traveling on this section of the route or by radio beacons that are located temporarily or permanently on a section of the route.

For each message content, the exact amount of data in bits (bit) is specified for coding the respective information. The exact message format can also be fixed in a JSON schema. Wherever possible, a specific data type (e.g. "double") is not used. As will be shown, all content can always be represented using integers and an appropriate convention.

In the following we introduce examples for route-related and vehicle-related messages with specific content and respective data size.

4.3.1 Route-Related Message

The route-related message has three main components:

- Information about the sending vehicle in order to be able to trace the origin of each message,
- The actual information on the condition of the route in order to provide this information to other vehicles, and
- The cryptographic parameters of the message in order to achieve IT security.

Table 1 shows the content and respective size of a route-related message. Each route-related message begins with general information in the *Message Specifications* section.

Exemplarily, the composition of these 119 bit of the message specifications (specs) is shown in Table 2: Transmitting the version assures correct decoding of all contents. There are only two types of messages, which can therefore be specified with a Boolean variable. The IRIG time code enables the time stamp to be displayed with 60 bits and ensures direct compatibility with the time signal from the Global Positioning System (GPS). The vehicle's digital identification (dID) is linked to its assigned rights. All further information about the vehicle can then be obtained from a database using this dID. By using 32 bit on the dID (represented in the hexadecimal system) the specifications of $2^{32} = 4.294\,967\,296 \times 10^9$ vehicles can be stored, which significantly exceeds the current and expected number of existing vehicles.

Table 1 Size of route-related messages

Content	Total size	Coded[a]	There of encrypted[b]
Message specs	119 bit	94 bit	32 bit
Position	172 bit	33 bit	0 bit
Road surface	5 bit	5 bit	0 bit
Sight	9 bit	1 bit	0 bit
Speed limit	11 bit	3 bit	0 bit
Obstacle	6 bit	6 bit	0 bit
Detour	34 bit	34 bit	32 bit
Parking spaces	18 bit	2 bit	16 bit
Free text	142 bit	2 bit	140 bit
Message verified	10 bit	2 bit	0 bit
Checksum	20 bit	0 bit	20 bit
Key index	80 bit	0 bit	0 bit
Sum	626 bit	182 bit	240 bit

[a] Precoding possible
[b] Encryption possible

A recipient does not have to be specified in this case because the recipient of route-related messages is always the relay. The validity period specifies time constraints on when the message must be transmitted. The transmission frequency has, along with the message size, a considerable influence on the amount of data to be encrypted and thus the required amount of keys. All values in curly brackets can be precoded. This results in the pre-encodable data volume of 94 bit. Only the sender's dID is suitable for encryption, hence 32 bit. The other entries of the route-related message such as position, road surface, et cetera are composed accordingly: The message events are located using both geographic and map-related coordinates. This is followed by modular event containers. The container contains information on whether there is an event on the route, the value with which the event is quantified and whether the information is to be encrypted. Any further information can be transmitted in a free text area, whereas the use of codes is not possible, at least on initial use. Any information that is longer than the intended free text can be transmitted in concatenated form by sending several messages. 10 bit are used as an indicator on whether the route-related information was already verified by other vehicles.

The last part of the message is for the cryptographic information. Using a known cyclic redundancy check (CRC), a checksum of the message is generated to check authenticity and integrity of the message. For this purpose, the checksum is generated by the sender, encrypted and transmitted together with the key index used (not the key itself). The receiver decrypts the checksum and compares it with the result of the known cyclic redundancy check, with which it reproduced the checksum from the received message. If the values of the checksums match, the recipient can assume the authenticity and integrity of the message. With the length of the checksum there is a conflict of objectives between cryptographic security and data economy. Based on

Table 2 Route-related message: message specifications (119 bit)

Code	Value	Size
0...1023	SIKAF Version	10 bit
{0, 1}	{route-, vehicle-}related	1 bit
{0x000...00, ..., 0xfff...ff}	IRIG Timestamp	60 bit
{0, 1}	Sender anonymized {no, yes}	1 bit
{0x00000000, ..., 0xffffffff}	dID Sender	32 bit
0	Validity Unlimited	11 bit
1...2047	Validity Period (s)	
0...15	Messages per Second (s^{-1})	4 bit

the example of the TAN in financial transactions, a length of six digits is assumed to be sufficient. If further message content is encrypted, keys with a continuous index are used. So only the first and the last key index have to be transmitted together with the message and the recipient can assemble the key from his supply accordingly.

4.3.2 Vehicle-Related Message

The vehicle-related message contains information about the sending vehicle, in particular about its current and predicted trajectory. The mutual behavior of vehicles can be anticipated with this information. Table 3 shows the content and the respective size of a vehicle-related message. As with the route-related message, a further distinction is made between the pre-coded and the encrypted portion of the content.

Each message starts with the message specifications (specs) which contain again the dID, an exact time stamp in IRIG timecode format and the validity in milliseconds. The physical properties of the vehicle are stored in a database and linked to the dID. In contrast to the route-related message, it is mandatory to specify the recipient's dID

Table 3 Size of vehicle-related messages

Content	Total size	Coded[a]	There of encrypted[b]
Message specs	152 bit	127 bit	64 bit
Attributes	7 bit	7 bit	6 bit
Trajectory	224 bit	33 bit	223 bit
Fuel	12 bit	5 bit	11 bit
Free text	142 bit	2 bit	140 bit
Checksum	20 bit	0 bit	20 bit
Key index	80 bit	0 bit	0 bit
Sum	637 bit	174 bit	464 bit

[a] Precoding possible
[b] Encryption possible

within the message specifications in order to be able to route the message correctly. Each further message content starts with a Boolean variable to indicate whether the respective content is encrypted.

Attributes indicate whether it is an emergency vehicle with sovereign rights, whether dangerous goods are loaded, and which mode of transport the vehicle is assigned to (e.g. motorized private transport, mPrT). The largest message block is the one for the trajectory, with which information on the current and intended driving behavior is transmitted. A precise spatial location of the vehicle at the time of message generation is represented both in a geographic coordinate system and a linear referencing system. This double encoding of the location is also used for redundancy, since the position is a security-relevant information. Precise data on the current speed vector, consisting of magnitude and direction, allow the coordination and optimization of vehicle movements. Each vehicle also sends information about the intended driving behavior up to a time horizon of 1.5 s in order to enable the surrounding vehicles to anticipate the driving situation. There is no separate specification of the acceleration, since the speed is the decisive parameter for coordination. The acceleration value can be derived from the time series of the speeds if necessary. As with the route-related message, a longer than the intended free text can be transmitted by concatenating and sending several vehicle-related messages.

4.4 Required Amount of Keys

As the used encryption with perfect secrecy needs keys that are at least as long as the message to be encrypted (see Eq. 3), the massive key requirement was identified as a possible practical disadvantage and thus as an obstacle when used in a vehicle guidance system. It should therefore be checked how large the amount of keys actually is in order to be able to dimension the key supply adequately.

It should be noted, that keys are not only required for encryption, but especially for decryption. If a vehicle receives the messages from all other vehicles, this leads to a greatly increased key requirement, since in this case there is an $n : 1$ transmission (see Fig. 7 left). This problem also occurs when the relay R only passes through the messages of all vehicles, as is provided by current concepts that work according to the infrastructure mode. The advantage of SIKAF, however, is that the relay preprocesses the messages and only forwards the consolidated information to the receiver.

The key replenishment for the relay can be regarded as uncritical due to the fixed connection to the central authority. If the information can be bundled by the relay and sent with a combined message (see Fig. 7 right), then only one key per message is required by the receiving vehicle. In this case there is a 1: 1 transmission.

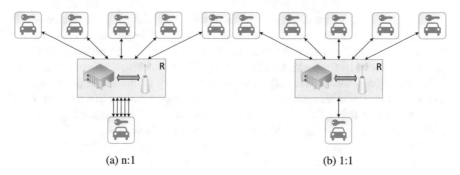

(a) n:1   (b) 1:1

Fig. 7 Transmission types

The following parameters have an influence on the required amount of keys $|K|$ in a vehicle:

- Data amount D of the message part to be encrypted ($[D]$ = bit)
- Transmission frequency f_t of encrypted messages ($[f_t]$ = Hz)
- Receive frequency of encrypted messages ($[f_r]$ = Hz)
- SIKAF activity time t_A ($[t_A]$ = s)

The activity time t_A can be composed of partial activity periods t'_A: $t_A = \sum t'_A$. The amount of data D, the transmission frequency f_t and the reception frequency f_r are not constant, but depend on the current driving and environmental situation. The driving and environmental situation in turn depends on the location and time of the vehicle affected. Since the location can be assumed as a function of time, the following applies:

- $D = D(t)$
- $f_t = f_t(t)$
- $f_r = f_r(t)$

An additional distinction in the parameters data quantity $D(t)$, transmission frequency $f_t(t)$ and reception frequency $f_r(t)$ must be made into vehicle-related and route-related messages because the frequencies and the amount of data can vary depending on the situation. The distinction is identified with the additional indices b for route-related and w for vehicle-related. The required amount of keys $|K|$ can then be calculated using these input variables using the following equation:

$$|K| \geq \sum_{t'_A} \int_{t'_A} \left[\left(f_{t_b}(t) + f_{r_b}(t) \right) \cdot D_b(t) + \left(f_{t_w}(t) + f_{r_w}(t) \right) \cdot D_w(t) \right] dt \quad (8)$$

The (partial) activity time must be determined appropriately in order to determine the total amount of the keys used and thus the dimension of the read-only memory. If the read-only memory is not sufficient, shorter (partial) activity periods or additional maintenance intervals can be provided.

There are generally no empirical values available for the parameters mentioned, since no reference implementation of the presented vehicle guidance system exists yet. Therefore, three scenarios are considered below, one of which is characterized by the minimum, one by the maximum and one by the average required amount of keys. The assumptions made for the individual parameters are explained respectively. The parameter that affects all scenarios equally is the activity time of SIKAF t_A. First of all, the operating time of a vehicle would be required for this, but for which hardly reliable sources can be found. The mileage of motor vehicles (VK, $[VK] =$ km) is well documented. Together with the average travel speed v_R in the different scenarios, the activity time t_A' can be determined according to

$$t_A' = \frac{VK}{v_R} \tag{9}$$

With these prerequisites different scenarios can be considered.

It should be pointed out once again that, according to Sect. 3.3.1, the keys must be at least as long as the message to be encrypted (see Eq. 3).

Scenario 1: Minimum Amount of Keys Required (for Automatic Vehicle Guidance) If the presented vehicle guidance system is used, theoretically any vehicle speeds for which the vehicle mechanics are designed can be realized. For the first scenario, a constant cruising speed of 130 km/h is chosen, since this meets additional constraints such as noise protection and passenger comfort. It also corresponds to the current speed limit in many countries. According to Eq. 9 and assuming a mileage of 13700 km per year [29], the SIKAF activity time during one year is $t_A' = \frac{13700\,\text{km}}{130\,\text{km/h}} \approx 105\,\text{h}$. With automatic vehicle guidance, the vehicle is controlled solely by vehicle-related messages. The transmission of route-related messages would therefore in principle not be necessary, since all the required variables are determined by the vehicle guidance system. However, to ensure functional safety, the vehicle must also be able to operate autonomously. For this reason, the required information of road conditions must be available at all times, which is transmitted with route-related messages. In this scenario, the smallest intended message frequency of $f_{t_b} = f_{r_b} = f_{t_w} = f_{r_w} = f_{min} = 1\,\text{Hz}$ is set and only the checksum with $D_b = D_w = D_{min} = 20\,\text{bit}$ is assumed as the message part to be encrypted. A further reduction is not possible, since otherwise the authenticity of the messages cannot be ensured. According to Eq. 8 in conjunction with Eq. 3, the required amount of keys is

$$|K_{min}| \geq t_A' \cdot 4 \cdot f_{min} \cdot D_{min} = 105\,\text{h} \cdot 3600\,\text{s}\,\text{h}^{-1} \cdot 4 \cdot 1\,\text{Hz} \cdot 20\,\text{bit} = 30\,240\,000\,\text{bit} \tag{10}$$

with a typical mileage of one vehicle in one year. Consequently, an amount of keys of about 3.60 MiB is required in each vehicle[2] .

---

[2]With 1 MiB = 1024 KiB = 1024 × 1024 B, 1 B=8 bit.

Scenario 2: Average Amount of Keys Required (for Manual Vehicle Guidance)

In the case of manual vehicle control by a human driver, the kinetic parameters are subject to further restrictions, in particular vehicle distances and speeds are limited to specific intervals. At present the speed of travel in the central Europe region for motor vehicles over land is given as 60 km/h, but in urban areas it can be lower than the speed of bicycles (15 km/h) [29]. In this scenario, an average value of 36 km/h is assumed, which is, however, easily adaptable. There is no reception of vehicle-related messages required with a high frequency since the vehicle is controlled manually: $f_{t_w} = f_{r_w} = 0$ and $D_w = 0$. For this, the information about the road network must be kept up to date by the route-related messages so that the driver can optimally plan her route: $f_{t_b} = f_{r_b} = f_{avg} = 5$ Hz. Only the checksum of the message is assumed to be encrypted, since the route-related information should be accessible to all road users. The authenticity of the messages is therefore ensured with a data size of again $D_b = D_{avg} = 20$ bit. According to Eq. 9, the activity time of SIKAF results in the mileage from one year to $t'_A = \frac{13700\,\text{km}}{36\,\text{km/h}} \approx 381$ h. According to Eq. 8 in conjunction with Eq. 3, the required amount of keys is

$$|K_{avg}| \geq t'_A \cdot 2 \cdot f_{avg} \cdot D_{avg} = 381\,\text{h} \cdot 3600\,\text{s}\,\text{h}^{-1} \cdot 2 \cdot 5\,\text{Hz} \cdot 20\,\text{bit} = 274\,320\,000\,\text{bit} \tag{11}$$

with a typical mileage of one vehicle in one year. Consequently, an amount of keys of about 32.70 MiB is required in each vehicle.[3]

Scenario 3: Maximum Amount of Keys Required (for Automatic Vehicle Guidance)

In order to determine an upper limit of the required amount of keys, the parameters are set to the respective maximum values. The vehicle-related and route-related messages are updated with a frequency of $f_{t_b} = f_{r_b} = f_{t_w} = f_{r_w} = f_{max} = 10$ Hz, what is currently accepted as an upper limit according to existing standards [20, 21]. It is further assumed that the complete messages are encrypted with a size of $D_b = 626$ bit (route-related) and $D_w = 637$ bit (vehicle-related). Again, a constant cruising speed of 130 km/h is assumed, since the vehicle is guided completely automatically. According to Eq. 9 there is again $t'_A = \frac{13700\,\text{km}}{130\,\text{km/h}} \approx 105$ h and according to Eq. 8 in conjunction with Eq. 3, the required amount of keys is

$$\begin{aligned}
|K_{max}| &\geq t'_A \cdot (2 \cdot f_{max} \cdot D_b + 2 \cdot f_{max} \cdot D_w) \\
&= 105\,\text{h} \cdot 3600\,\text{s}\,\text{h}^{-1} \cdot (2 \cdot 10\,\text{Hz} \cdot 626\,\text{bit} + 2 \cdot 10\,\text{Hz} \cdot 637\,\text{bit}) \\
&= 9\,548\,280\,000\,\text{bit}
\end{aligned} \tag{12}$$

with a typical mileage of one vehicle in one year. An amount of keys of around 1.11 GiB is required in each vehicle.[4]

The presented data model largely predefines the content of the messages exchanged with SIKAF, whereas the possibility of free-form transmission of content in the free text area of a message remains. The specified data model makes it easier to check

---

[3] With 1 MiB = 1024 KiB = 1024 × 1024 B, 1 B=8 bit.

[4] With 1 GiB = 1024 MiB = 1024 × 1024 KiB = 1024 × 1024 × 1024 B, 1 B=8 bit.

Table 4 Necessary amount of keys for one-year operation in different scenarios

Scenario	1: Minimum	2: Average	3: Maximum	4: CAV
Transmit/Receive	1 Hz	5 Hz	10 Hz	10 Hz
Packet size (Vehicle)	20 bit	0 bit	637 bit	637 bit
Packet size (Route)	20 bit	20 bit	626 bit	626 bit
Activity time	105 h	381 h	105 h	7300 h
Key amount	3.60 MiB	32.70 MiB	1.11 GiB	77.28 GiB

messages for unwanted content, and these can be marked during transmission or sorted out as spam. The unique digital transmitter identification (dID) enables the vehicle causing a problem to be identified and blocked in SIKAF if spam is sent continuously. Additional information can be divided into several datagrams, but according to the protocol, a datagram must always contain entire information. The distribution of content to several datagrams in a form that requires sequence control is not permitted.

The calculations showed that it is possible to store keys necessary for encryption with perfect secrecy for several years on disks of common dimensions. The results are summarized again in Table 4. The table also includes a fictitious scenario with connected and autonomous vehicles (CAV). A daily operating time of 20 h is assumed for these CAVs, thus 7300 h per year. The required keys are again calculated using Eq. 8 in conjunction with Eq. 3. Finally, it should be emphasized that these are sample calculations. The variation of each parameter can lead to considerable deviations in the results. Last but not least, this includes the length of the checksum, which also determines the security level of message integrity.

4.5 Processing Time

To validate time requirements and scalability, the encryption process was implemented in a test environment.[5] Messages of different lengths were encrypted several times and either this number of encryptions or the data size was kept constant. The results from this simulation are depicted in Fig. 8.

The most important results are

1. the confirmation of a linear relationship between the time required for encryption, the message length and the number of encryptions and
2. no indication of negative scaling effects.

---

[5]Intel(R) Core(TM) i7-4510U CPU @ 2.00 GHz, 8 GB RAM, 64-bit operating system, Python 3.8.3, pycryptodome 3.9.7, pandas 1.0.3.

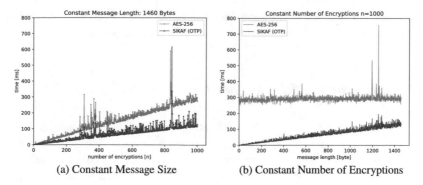

(a) Constant Message Size   (b) Constant Number of Encryptions

Fig. 8 Processing time for encryption

With a constant message size of 1460 B, the time required for encryption increases linearly in both SIKAF and AES with the number of encryptions (Fig. 8 (a)). SIKAF has a speed advantage here.

The time required for AES encryption is independent of the message length in this specific scenario, i.e. the test environment with the programming language used, the libraries imported, and the selected parameter set, while SIKAF increases linearly. This can be due to the implementation of the AES algorithm or to the system used if, in the case of the AES, the encryption of the generated blocks can be assigned to different CPU cores. Such a parallelization is not possible with the encryption in SIKAF and is also not provided for in order to be able to use SIKAF also on systems with limited computing power in a foreseeable time. With a constant number of encryption processes, the time advantage thus seems to shift in favor of the AES-256 for longer messages (Fig. 8 (b)). However, SIKAF does not transmit datagrams larger than 1460 B. Moreover, looping through 1000 encryptions in this test is an order of magnitude that is not to be expected when exchanging a message in SIKAF. In the relevant range up to 1460 B, the speed of encrypting is higher with SIKAF than with AES encryption. With the exception of individual artifacts (evident as peaks), which occur in an equivalent manner with AES-256, no scaling problems are observed at SIKAF. The creation of the artifacts is attributed to specific plain text-key combinations or background tasks that require the processor.

5 Conclusion

We presented a vehicle guidance system whose communications are managed by a specialized communications architecture named SIKAF. The architecture enables message routing with an optimized hop count and optimal security by encryption with perfect secrecy. The amount of keys required for a sufficient operating time can be stored in vehicles on common data storage devices. It prevents any stochastic processes and, thus, enables predictable timing for all processing steps.

5.1 Classification of the Architecture

Like any communication architecture, SIKAF can be systematized according to the respective aspects of topology, cryptology and administration. Topologically, it is an infrastructure network with a central access point, referred to here as a relay. The cryptographic maxim is symmetric encryption, which enables the implementation of perfect secrecy. For the necessary administration, a trustworthy authority is set up. The presented architecture thus contrasts direct connections based on the ad-hoc principle, which is common in numerous other approaches to vehicle networking and is preferably secured with asymmetric cryptography. As has been shown, however, this principle does not meet the high safety requirements in vehicle guidance systems.

5.2 Possibilities and Limitations of the Architecture

It was shown that the presented architecture can meet the requirements for safety, processing time and complexity regarding communication in a vehicle guidance system. In addition, the almost arbitrary scalability, the independence from the manufacturer and the consistent focus on the internet of things are decisive advantages compared to existing concepts. SIKAF is therefore suitable for direct use in existing traffic systems. However, communication between two vehicles cannot take place without a dedicated infrastructure. If this infrastructure is not available, communication from vehicle to vehicle is not possible with the presented concept. Given the importance and increasing spread of communication technologies, however, this aspect is not expected to be a decisive obstacle to the possible introduction of SIKAF.

Some may also criticize the lack of end-to-end encryption in SIKAF. However, as has been shown, it is not conceptually possible to guarantee encryption with perfect secrecy and, at the same time, guarantee hard real-time for continuous data exchange without a trustworthy central authority.

The prototypes were developed with consumer electronics that are not approved for industrial use. The feasibility of the architecture could therefore only be demonstrated in the form of a proof of concept. The debate as to whether these components used are completely inadequate in terms of availability and reliability or whether they can even be used in an industrial control system is controversial. Experience from the present work shows that there is a large number of users of these components, which means that extensive documentation and the experience of a large user community is available. Due to this widespread use, long-term product maintenance is worthwhile for the manufacturer, especially with regard to spare parts supply and software updates. Overall, the widespread use adds up to a significant number of operating hours, which makes it easier to find systematic malfunctions.

6 Discussion

The architecture presented has transferred the concept of perfectly secure encryption to intelligent transport systems. It thus offers provably secure encryption even with the advent of quantum computers. Further concepts and applications of vehicle control systems can be set up on this basis. To increase the reliability of radio transmission, SIKAF should use a dedicated frequency band. It would therefore be appropriate to reconsider the decision to reserve the 5 GHz frequency band for insecure ad-hoc networks with a PKI. This band could be used more effectively for demonstrably safe systems like SIKAF.

With the transport sector an application was chosen that affects all "fallacies of distributed computing" [30]. With the prototypical implementation of the presented architecture it could be shown that a correct implementation is possible in a simple manner and thus a resolution of these "fallacies" is achievable.

6.1 Possible Extensions

Hybrid vehicles, which act both as participants and as relays, could be used as a possible substitution for the additional infrastructure required. Instead of the dedicated access points, these vehicles would form the access points to the relay and simultaneously initiate message processing and forwarding. Such concepts are primarily developed under the heading of "georouting" [31] and continue to form hierarchically structured networks in infrastructure mode. However, central aspects are still unclear. This includes the required share of hybrid vehicles in the total collective, the necessary measures to ensure network availability and last but not least communication security.

The application of vehicle guidance systems is not limited to road vehicles like in the present use case. The use of autonomous drone systems for the transport of passengers and goods is a much-discussed topic [32]. The exact design of such transport systems is still the subject of research. However, there is largely consensus on the need for automatic guidance of these aviation systems, since the amount of objects can no longer be monitored by human pilots. SIKAF is predestined for this application for two reasons: On the one hand, SIKAF can demonstrably meet the high safety requirements in aviation. On the other hand, there is far more regular contact with (ground) infrastructure in aviation than is the case with motor vehicles. The key replenishment can be ensured in a particularly simple manner by this contact.

SIKAF can also add perfectly secure encryption and interoperability to existing IoT architectures [33], given that only control messages with small amounts of data are exchanged.

6.2 New Concepts for Decentralized Autonomous Systems

In the present work, a vehicle guidance system was viewed as a scalable automation system that was nevertheless closed off from other actors. With the further expansion of the industrial internet of things, a direct interaction of mutually unknown, digital entities is conceivable. In this case, there is no longer permanent access to information from a central authority available. Solutions have to be found for authentication, authentication and authorization even in fully autonomous environments [34]. In a figurative sense, the principle of *trust* must be transferred to an open cyber-physical system.

6.2.1 Generating and Managing Trust

The exchange of information, especially if it is security-relevant for the individual or a collective, requires trust in the respective communication partner. In a social context, trust is acquired through positive, subjective experiences with previous behavior. Even in a cyber-physical system, evidence must be provided that the information transmitted by the recipient is not improperly used. For a technical application, the size of trust must be quantified by definable, measurable input variables and made available to potential communication participants [35]. A suitable cost function can be found for quantification, which defines the trustworthiness of each participant. In addition, a decentralized and consistent database management system is required, which enables every communication participant at any time to determine the trustworthiness of the respective communication partner without a central database.

6.2.2 Blockchain

A database system that can meet these special requirements has only recently established through the concept of the blockchain [36]. The blockchain is a decentralized database with the properties required to store and provide a trust value in autonomous cyber-physical systems. The concept of the blockchain can be transferred to a vehicle guidance system if three essential hurdles are solved. First, the blockchain in its present form cannot meet real-time conditions: The creation of new blocks takes an unpredictable period of time, so a complete calculation with given time limits cannot be guaranteed. Second, the blockchain can comprise a very large data volume if it is used for a long time. This results in great demands on the required bandwidth and transmission time. The necessary redundant storage at all participants also requires large storage capacities. Third, the problem of flooding the network with competing blocks can arise, particularly as the number of connected vehicles increases. In this case, the database would be in an inconsistent state for a certain time and thus violate the integrity. With current implementations, scaling the blockchain is therefore only possible to a limited extent.

If these obstacles can be resolved, the principle of the blockchain is a promising possibility for decentralized information management in cyber-physical systems, since fully autonomous operation can be achieved without the need for a central instance.

References

1. Directive 2010/40/EU of the European Parliament and of the Council (2010)
2. Williams, B.: Intelligent Transport Systems Standards. Artech House, Boston (2008)
3. DIN SPEC 91345:2016-04, Reference Architecture Model Industry 4.0 (RAMI4.0). Beuth Verlag, Berlin (2016)
4. Sommer, C., Dressler, F.: Vehicular Networking. Cambridge University Press, Cambridge (2015)
5. SAE J 3016:2018-06-15, Taxonomy and Definitions for Terms Related to Driving Automation Systems for On-Road Motor Vehicles. Beuth Verlag, Berlin (2018)
6. Bagloee, S., Tavana, M., Asadi, M.: et al.: Autonomous vehicles: challenges, opportunities, and future implications for transportation policies. J. Mod. Transp. **24**, 284–303 (2016)
7. Tahir, M.N., Mäenpää, K., Sukuvaara, T.: Evolving wireless vehicular communication system level comparison and analysis of 802.11p, 4G, 5G. In: International Conference on Communication, Computing and Digital systems (C-CODE), pp. 48–52 (2019)
8. Medhi, D., Ramasamy, K.: Network Routing: Algorithms, Protocols, and Architectures. Morgan Kaufmann, Cambridge (2017)
9. Soyata, T: Enabling Real-Time Mobile Cloud Computing Through Emerging Technologies. IGI Global, Hershey (2015)
10. Oorni, R., Goulart, A.: In-vehicle emergency call services: eCall and beyond. IEEE Commun. Mag. **55**(1), 159–165 (2017)
11. Kim, S., Noh, W., An, S.: Multi-path ad hoc routing considering path redundancy. In: Proceedings of the Eighth IEEE Symposium on Computers and Communications. ISCC 2003, vol. 1, pp. 45–50 (2003)
12. Jiang, D., Delgrossi, L.: IEEE 802.11p: towards an international standard for wireless access in vehicular environments. In: IEEE Vehicular Technology Conference, pp. 2036–2040. IEEE (2008)
13. Paar, C., Pelzl, J.: Understanding Cryptography. Springer-Verlag, Berlin (2010)
14. Electronic Frontier Foundation: Cracking DES: Secrets of Encryption Research, Wiretap Politics and Chip Design. O'Reilly (1998)
15. Kalenderi, M., Pnevmatikatos, D., Papaefstathiou, I., Manifavas, C.: Breaking the GSM A5/1 cryptography algorithm with rainbow tables and high-end FPGAs. In: International Conference on Field Programmable Logic and Applications, pp. 747–753 (2012)
16. Rivest, R., Shamir, A., Adleman, L.: A Method for Obtaining Digital Signatures and Public-Key Cryptosystems, Technical report, Massachusetts Institute of Technology (1977) Accessed 07 Jul 2019
17. Elgamal, T.: A public key cryptosystem and a signature scheme based on discrete logarithms. IEEE Trans. Inf. Theor. **31**(4), 469–472 (1985)
18. Hasrouny, H., Samhat, A., Bassil, C., Laouiti, A.: VANet security challenges and solutions: a survey. Veh. Commun. **7**, 7–20 (2017)

19. Shannon, C.: Communication Theory of Secrecy Systems. Bell Syst. Tech. J. **28**(4), 656–715 (1949)
20. ETSI EN 302 637-2, Intelligent Transport Systems (ITS); Vehicular Communications; Basic Set of Applications; Part 2: Specification of Cooperative Awareness Basic Service Online (2019). Accessed 02 Jan 2020
21. ETSI EN 302 637-3, Intelligent Transport Systems (ITS); Vehicular Communications; Basic Set of Applications; Part 3: Specifications of Decentralized Environmental Notification Basic Service (2019). Accessed 02 Jan 2020
22. Khattak, H., Farman, H., Jan, B., Din, I.: Toward integrating vehicular clouds with IoT for smart city services. IEEE Netw. **33**(2), 65–71 (2019)
23. Mundhenk, P., et al.: Security in automotive networks: Lightweight authentication and authorization. ACM Trans. Des. Autom. Electron. Syst. **22**(2), 1–27 (2017)
24. Bittl, S., et al.: Emerging attacks on VANET security based on GPS time spoofing. In: 2015 IEEE Conference on Communications and Network Security (CNS), pp. 344–352 (2015)
25. Levinson, J., et al.: Towards fully autonomous driving: Systems and algorithms. In: 2011 IEEE Intelligent Vehicles Symposium (IV), pp. 163–168 (2011)
26. Hartenstein, H., Laberteaux, K.: VANET: Vehicular Applications and Inter-Networking Technologies. Wiley, Chichester (2010)
27. Lee, J., Kim, C.: A roadside unit placement scheme for vehicular telematics networks. In: Kim, T., Adeli, H. (eds.) Advances in Computer Science and Information Technology, Springer-Verlag, Berlin, pp. 196–202 (2010)
28. Vernam, G.: Secret Signaling System. Patent US 1310719 A (1919)
29. Statista, de.statista.com (2020). Accessed 12 Sep 2020
30. Rotem-Gal-Oz, A.: Fallacies of Distributed Computing Explained, Technical Report, Sun Microsystems (1997). Accessed 20 Jul 2020
31. Kumar, S., Verma, A.: Position based routing protocols in VANET: a survey. Wireless Pers. Commun. **83**, 2747–2772 (2015)
32. Lee, P.U., Idris, H., Helton, D., Davis, T., Lohr, G., Oseguera-Lohr, R.: Integrated trajectory-based operations for traffic flow management in an increasingly diverse future air traffic operations. In: IEEE/AIAA Digital Avionics Systems Conference, pp. 1–9 (2019)
33. Guth, J., et al.: A Detailed Analysis of IoT Platform Architectures: Concepts, Similarities, and Differences. In: Di Martino, B., Li, K., Yang, L., Esposito, A. (eds.) Internet of Everything, pp. 81–101, Springer, Singapore (2018)
34. Hoeper, K., Gong, G.: Pre-authentication and authentication models in Ad Hoc networks. In: Xiao, Y., Shen, X., Du, D. (eds.) Wireless Network Security, Signals and Communication Technology, Springer, Berlin, pp. 65–82 (2002)
35. Trung, S.D.: On Trustworthiness Recommendation. PhD thesis, FernUniversität in Hagen (2017)
36. Nakamoto, S.: Bitcoin: a peer-to-peer electronic cash system (2008). Accessed 14 May 2017

Attack Models and Countermeasures for Autonomous Vehicles

Man Chun Chow, Maode Ma, and Zhijin Pan

Abstract With the rapid development of smart transportation, autonomous vehicles (AVs) are becoming one of the most anticipating means of transport. However, as the complexity of autonomous vehicles is increasing, it is intuitive that it would bring along with more possible attacks and higher potential risks. For example, by tampering the in-car sensors or hacking into any of the electronic control units (ECUs) in the vehicle, it could severely affect the driving performance or even cause life-threatening situations to users. Moreover, since AVs will also be the Internet of Vehicles (IoVs) that connect to the vehicular network in the future, the network security of the intra-vehicular and inter-vehicular links should also be carefully studied. To identify and mitigate the security risks involved in AV holistically, in this chapter, we provide a comprehensive taxonomy for attack surfaces and countermeasures for defense. Specifically, four different attack surfaces are defined, namely ECUs, sensors, intra-vehicular links, and inter-vehicular links. For each of the attack surfaces, various common attack vectors are discussed in detail. Subsequently, we also provide a survey of the latest major existing work for defending the attacks on each surface. We hope this chapter can be a guide for the general public to understand the security aspect of AVs, as well as to encourage future researchers to improve the security in AVs.

1 Introduction

Since the concept of self-driving cars evolved in the 1920s, and the first fully autonomous car prototype invented in the 1980s, autonomous vehicles (AV) became a promising research area for both automobile industry and academia. In recent years, the advancement in the wireless network, sensor technology and artificial intelligence have facilitated the rapid development of AVs. Nowadays, AVs are no longer just a self-driving car, but a smart, network-connected Internet of Vehicles (IoV). These

M. C. Chow · M. Ma (✉) · Z. Pan
School of Electrical and Electronic Engineering, Nanyang Technological University, Singapore, Singapore
e-mail: emdma@ntu.edu.sg

© Springer Nature Switzerland AG 2021
N. Magaia et al. (eds.), *Intelligent Technologies for Internet of Vehicles*, Internet of Things, https://doi.org/10.1007/978-3-030-76493-7_12

connected vehicles not only drive autonomously, but they will also connect with the surrounding vehicles and infrastructures, and then make the best driving decision based on the information collected from sensors and the vehicular network. In fact, according to the estimation in [1], even if there is only 10% of the people in the United States use AVs, it will potentially save more than 1000 lives and 200,000 traffic accidents per year due to the smarter driving decisions in AVs. Also, since AVs can plan the route efficiently, it is expected to save more than 700 h of total traffic time. Therefore, AV is the believing technology that increases road capacity, improves road safety, and provides faster and more convenient transportation in the future.

While AVs are bringing us convenience by making complicated driving decisions without human intervention, it also comes with various new threats. Since driving decisions have to be made intelligently and spontaneously, AVs are usually integrated with numerous complicated sensors, sub-systems and algorithms. As the complexity of the system grows, it is more likely to introduce new security issues and vulnerabilities. For instance, in 2015, security researchers found that there are some firmware vulnerabilities in Tesla Model S that allow hackers to halt the vehicle randomly [2]. Besides, AVs nowadays use sensor fusion and image recognition to detect road conditions. Security researchers recently found that if any part of the sensors or road signs is deliberately blocked, the car could be misled to make dangerous driving decisions [3]. Also, in 2020, another security researcher found that some refurbished media control units from old Tesla cars were stored with numerous sensitive data such as call logs and website cookies, and these data can be easily recovered by malicious hackers [4]. Therefore, to mitigate all the risks aforementioned and reduce the chance of causing dire consequences, it is essential to identify all potential attacks in AVs systematically and discuss their countermeasures carefully. However, most of the existing research work only focuses on providing security solutions for some specific attacks, while only a little work has been done to provide the taxonomy of attack surfaces and provide a survey on their countermeasures. In this chapter, we would like to provide readers with a holistic understanding of the security aspect of AV technology. This chapter can also act as guidance for researchers to start the investigation on the security of AVs and develop a more robust and practical defensive technology in the future. The contribution of this chapter is summarized as follows:

(1) We classify the potential risks in AV technology with four major attack surfaces: the electronic control units (ECU), intra-vehicular links, sensors, and inter-vehicular links.
(2) We explain the definitions of each attack surface and point out the possible attacks in detail.
(3) We discuss the major and the latest defense techniques in AV technology.
(4) We list out the open issues and suggest future research directions.

The rest of the chapter is organized as follows: Sect. 2 discusses the attack surfaces and examples of potential threats. Section 3 focuses on the latest and major existing solutions for defense. Section 4 suggests the open issues and future research directions. Finally, a conclusion is drawn in Sect. 5 to provide a summary of the security aspects of AV technology.

2 Attack Surfaces and Potential Threats

This section describes the major attack surfaces of a modern AV. Although AV comprises numerous components and subsystems, these components can be roughly categorized into the following four attack surfaces shown in Fig. 1: Electronic Control Units (ECUs), sensors, intra-vehicular links and inter-vehicular links.

2.1 Electronic Control Units (ECUs)

ECUs are the embedded subsystems that control the major components in the car, such as engine, cabin environment, and media control unit. In a modern AV, there are more than 100 ECUs embedded, and the number of ECUs expects to increase in the future. By collecting external environment information from sensors input and the vehicular network, these ECUs can also provide computation and make decisions based on telematics. However, as modern ECUs are usually programmed based on state-of-the-art operating systems, it could be vulnerable to the following attacks:

2.1.1 Invasive Attacks

Invasive attacks involve modification of the software, or physically damaging or inserting extra components to the existing hardware. These are some common invasive attacks:

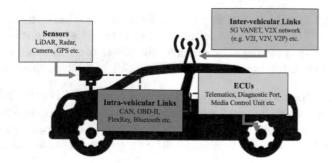

Fig. 1 Attack surfaces of autonomous vehicles

Physical Damage: One of the most apparent attacks is destroying the vehicle physically. Examples are dismantling or destroying some crucial ECUs such as telematics modules, actuators, or the immobilizer.

Code Modification: If the integrity of the firmware in ECUs is not protected with digital signature or cryptographic hash functions, it could be easily modified by third parties using the physical connections such as Joint Test Action Group (JTAG) bus, Onboard Diagnostic Port (OBD-II) or Control Area Network (CAN). Also, malicious firmware could be installed using the flawed upgrade procedures in the existing code [5]. In this way, ECUs could be malfunctioned or even create physical damages to other components in the vehicle.

Code Injection: Some firmware in ECUs is programmed based on modern operating systems like Linux, Android, or other real-time operating systems (RTOSs). As these operating systems support multitasking, by injecting extra malicious code, the ECUs can behave normally while sending the sensitive car information to hackers at the same time. Even worse, hackers could compromise the car by controlling the ECUs remotely or ask for ransom for unlocking some components. Consequently, the privacy and safety of car users could be severely affected.

Diagnostic Port Attack: Since 1996, all vehicles in the United States require to equip with an Onboard Diagnostic Port (OBD-II). However, this diagnostic port could be dangerous as it allows technicians to access to all ECUs in a vehicle with an ODB scanner readily available in the market. As some hackers or organization can abuse this port as an entry point of retrieving sensitive data or even injecting malicious code into the vehicle, this attack is still a prominent security threat to AVs.

2.1.2 Non-invasive Attacks

Some attacks can be performed without physically tampering the firmware or the hardware of the vehicle:

Side-Channel Attack: Rather than finding the weakness in the software or components in the ECUs, side-channel attacks exploit some extra information generated while the vehicle is running. For example, the power consumption, the generated sound, and the electromagnetic leaks from the ECUs or other components.

Brute-Force Attack: if the sensitive information stored in ECU or the onboard unit (OBU) is encrypted with a weak password, adversaries can guess the password and retrieve the content easily. By systematically attempting all possible passwords or common phrases, the correct password can be found in a controllable time.

2.2 Sensors

Sensors in AV are responsible for sensing the environment surrounding the vehicle. There are four predominant sensors in the vehicle: Radar, LiDAR, camera, and GPS. Light Detection and Ranging (LiDAR) is a 360-degree sensor that uses light beams to measure the distance between the vehicle and other obstacles. Similarly, radar is a radio-based sensor that determines the distance using the millimeter-wave. The camera captures live 360-degree images and videos of the surroundings to recognize road signs and other visible objects. Global Positioning System (GPS) determines the position of the vehicle using the latitude, longitude and altitude information calculated by the triangulation of satellite signals. Combining all these sensors, AV should be able to precisely identify objects like pedestrians, road signs and surrounding vehicles to make driving decisions autonomously. Nevertheless, as these sensors can be easily seen in the vehicle, they are also prone to the following attacks:

Spoofing Attack: It refers to the behavior that adversaries falsify the sensors to receive incorrect information. For example, LiDAR spoofing [6] happens when attackers replay laser pulses at a specific position to pretend to have an obstacle farther than the location of the spoofer. By using digital radio frequency memory repeater which stores the previously received signal and replays it, the radar may also see a "ghost vehicle". For GPS, it can also be spoofed by fabricating satellite signals that are stronger than the genuine signals, resembling the original GPS signals; or replaying the satellite signals which was captured previously. For the camera, it can also be spoofed by vandalizing the road signs and markers in the road.

Jamming Attack: By physically blocking the sensors or creating jamming signals to override the original signals, jamming attacks can paralyze the sensors from collecting information from the surroundings. For example, cameras without infrared (IR) filters can easily be blinded using high-brightness IR LEDs or IR lasers. Radar can also be jammed by transmitting high power jamming radio signals, such that it is difficult for the receiver to distinguish the reflected signals from the obstacles. For GPS, there are also some easily accessible GPS jamming devices in the market, which can disrupt the GPS reception by creating jamming signals at L1, L2 and L5 frequency bands.

Cloaking Attack: Unlike the jamming attack that creates additional jamming signals to override the reflected signals, cloaking is a technique of absorbing the reflection wave using specially crafted materials such as plastic foams or special coating. For example, military drones may use cloaking to hide from being detected by the radars.

Adversarial Attack: There is an increasing trend of using deep neural network (DNN) and sensor fusion to perform object detection. While DNN provides more accurate detection with higher error tolerance, it is vulnerable to some specially crafted images or situations that aim to attack the weaknesses of the neural network. For example, researchers found that adversarial attacks are more effective than simply spoofing the LiDAR in the vehicle, and it can improve the attack success rate to 75% during the experiment [6].

2.3 Intra-vehicular Links

Intra-vehicular links are the wired and wireless connections between sensors, ECUs, telematics modules, and end-user equipment within a vehicle. For example, the Control Area Network (CAN) bus is the wired network designed for connecting all ECUs. Similarly, FlexRay is another wired automotive network designed to replace the CAN network with faster speed and a more robust signaling mechanism [7]. In recent years, Apple has also developed CarPlay to allow smartphones to connect with the in-car entertainment system through wired or Bluetooth connection. Since these intra-vehicular links are comparable to the modern computer network, they would also be vulnerable to the following preexisted communication vulnerabilities:

Sniffing/Eavesdropping: By listening to and collecting information from the wired or wireless channel passively, eavesdropping can collect sensitive data (e.g. location, personal data) easily if the content is not encrypted and the identity of sender and recipient are visible. For example, in CAN bus, data will be broadcasted to all ECUs in the vehicle at the same time without any security mechanism such as message encryption and authentication by design. Therefore, if there is any malicious node eavesdropping the CAN bus, it should be able to retrieve information related to all ECUs easily.

Denial of Service (DoS) Attack: It aims to paralyze or flood the network with multiple devices. For example, DoS in CAN bus can target either the CAN network or a single ECU. For the CAN network, adversaries can control a node to flood the network and occupy the transmission line for a long time. It makes other nodes fail to perform their regular tasks as their real-time signals are blocked from transmission. For the case of single ECU, the DoS attack is also known as the "bus-off attack". It makes the CAN network launch its error handling procedures indefinitely and put the target ECU into the error active state, which disable the data receiving and transmitting infinitely.

Impersonation/Masquerade Attack: If the network is not protected using a proper authentication mechanism, adversaries may modify the sender information in the packets to impersonate the sender. For example, malicious nodes connected to the CAN bus could fabricate any message and pretend to be from a legitimate node.

Replay Attack: After recording the message previously sent from a legitimate node, a malicious node can send the same message again without modifying it to pretend to be a legitimate node. If the protocol does not have a proper mechanism such as nonce, sequence number, or timestamp (e.g. the CAN protocol or some flawed keyless entry system), the same message can be replayed as many times as the malicious node desires.

Man-in-the-Middle (MITM) Attack: It means an adversary can be the middleman in the connection but neither the sender nor receiver knows about the existence of this middleman. It usually happens when the security protocol does not exist, or it has a flawed design. For example, in the in-car Wi-Fi network, adversaries can set up an evil twin access point (AP) that broadcasts the same service set identifier (SSID) with the stronger signal strength to tempt other devices to choose it. If the evil-twin relays the messages honestly, both data senders and data receivers will never know about the existence of this evil-twin, while it is intercepting all the transmitted data.

2.4 Inter-vehicular Links

Inter-vehicular communication links refer to the vehicle-to-everything (V2X) communication, which is also known as the vehicular ad-hoc network (VANET). It includes all outgoing network traffic such as vehicle-to-vehicle (V2V), vehicle-to-infrastructure (V2I), vehicle-to-pedestrian (V2P) and the vehicle-to-internet of things (V2IoT) communication links. Since AVs must reach other entities and exchange information using the inter-vehicular links, the security vulnerabilities can be classified into the following two main categories, including the vulnerabilities in communication, and the vulnerabilities in the target devices.

2.4.1 Vulnerabilities in Communication

Like the communication vulnerabilities in the intra-vehicular links, VENET also exhibits similar characteristics to the modern wireless network. As a result, it is also vulnerable to the preexisted communication vulnerabilities as mentioned in the previous section, such as the eavesdropping, denial-of-service, impersonation, replay and man-in-the-middle attacks.

2.4.2 Target Device and Service Vulnerabilities

For every inter-vehicular communication, vehicles would connect with some target devices (such as the surrounding vehicles or traffic lights) or services (such as cloud service) frequently to exchange information. If any of these targets are compromised, the vehicles could be misled by the fraudulent information received from others, even if the communication between the two parties is legitimate. If the target is an AV, its security vulnerabilities would be the same as aforementioned. However, if the target is a cloud service or a distributed service, it could also be vulnerable to the following attacks:

Physical Damage: By deliberately tampering the RSUs and dismantling communication equipment in the roadside or the vehicles, the V2I and V2V communication would be interrupted. Also, if the cloud servers providing traffic condition data are destroyed due to natural disasters such as earthquakes or tsunamis, the driving decisions of AVs would also be affected.

Sybil or Bogus Information Attack: If there are a lot of malicious users (e.g. fake users or botnet) submitting false information to the cloud or distributed service, it could mislead the service provider to make false judgments. For example, researchers found that by creating ghost vehicles and fake accidents in a mapping service [8], the navigation system in AVs could be misled to avoid these fake traffic events, causing more traffic congestion or even life-endangering situations.

Compromised Key Attack: Most AVs should have a list of default trusted certificate authority (CA) for connecting to cloud services. However, if the private keys of these services are stolen and the key revocation mechanisms in the AVs are not triggered timely, these AVs could be dangerous because of its trust in the compromised cloud services. To avoid this single point of failure, distributed services (e.g. blockchain) and better key revocation mechanisms should be considered.

2.5   Attack Summary

To summarize, in this section, we have studied numerous common attacks in each of the attack surfaces of AVs. We have discussed the definition of the attacks and listed our taxonomy graphically in Fig. 2. However, as the security requirement and the architecture of AVs are always changing, it is possible to have more new attack methods in the future. Therefore, our listed attacks in this section may not be exhaustive.

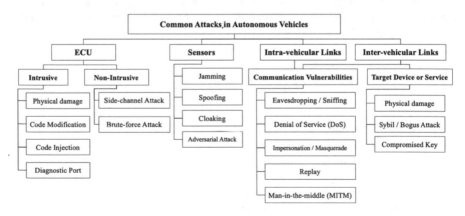

Fig. 2   Common attacks for each attack surface of AV

3 Existing Countermeasures

We have covered various common attacks in each of the attack surfaces of AVs. To build a safe and robust AV/IoV in the future, numerous researchers have proposed different ways to improve various security aspects in AVs. In this section, we discuss the latest and major existing work for defending different attack surfaces of AVs and point out their security improvements briefly.

3.1 Solutions for ECU Attacks

3.1.1 Prevention of Code Modification and Injection

Baza et al. [9] have proposed a blockchain-based distributed firmware update scheme tailored for AVs. In the proposal, multiple vehicle manufacturers cooperate to build a distributed blockchain network that stores the message authentication code (MAC) of the latest firmware. Then, the manufacturers choose some reliable AVs to distribute the latest firmware to them to upgrade. After finishing the upgrade, these vehicles also send the latest firmware to other surrounding vehicles using the V2V network. For the receiver AV, after receiving the firmware, it first checks its integrity by comparing the calculated MAC code with the blockchain. Then, after finished upgrading, it sends back a zero-knowledge proof token to the sender AV. This token allows the firmware distributor AV to receive some rewards in the blockchain for crediting their hard work on distributing the firmware. To avoid malicious users from getting the raw binary of the firmware for reverse engineering, the proposal has also employed attribute-based encryption (ABE) to encrypt the firmware according to the model of the vehicles. In this way, only legitimate vehicles can decrypt their firmware. This proposal can prevent the denial-of-service (DoS) attack and single point of failure of the firmware server, which could be essential if some crucial updates have to be pushed to a large number of cars in a short time. With MAC and ABE employed for protecting the binary, it also protects the ECUs and onboard computer by preventing malicious code modification and injection.

Yu et al. [10] have proposed a security scheme protecting the integrity and confidentiality of the firmware in the code memory. In the proposal, it has suggested the code should be obfuscated during the transmission from OBD-II port to the ECU. Then, to prevent the content from being modified or eavesdropped, it has also suggested that all information transmitted through the OBD-II port must be encrypted and verified using a digest generated with a cryptographic hash function. Moreover, since the processor in the ECU cannot execute the encrypted code directly, it has further suggested decrypting the code on-the-fly using a decryption program stored in cleartext and protected in the hardware. Finally, to distinguish potential threats in the future, the ECU should also create a bait environment to attract potential attackers.

After the attacker has finished intrusion and installed malware into the bait environment, further observation and analysis can thus be carried out based on its behavior. This proposal prevents code injection and code modification by ensuring the integrity and confidentiality of the received data. Assuming OBD-II is the default protocol of inter-vehicular communication, this scheme should also prevent eavesdropping effectively by encrypting and obfuscating the message during transmission.

To facilitate safety firmware upgrade of ECUs using a standard protocol, Mansor [11] has proposed an improved E-safety Vehicle Intrusion proTected Applications Protocol (EVITA +). Similar to the proposal in [12], this protocol uses various cryptographic techniques such as public–private key pairs, digital signatures, random tokens, timestamps and acknowledgments to build a secure channel between the diagnostic tool, the central communication unit (e.g. the OBU) and the ECU. Besides normal firmware upgrade, the proposal has also suggested a rollback and backup mechanism to prevent bricking of ECU (i.e. the device cannot start normal operations due to the corrupted firmware) when the communication is interrupted or inconsistent. Using formal verification tools such as Scyther and CasperFDR to perform security analysis, the proposed protocol can provide mutual authentication of multi parties, as well as to protect data integrity and confidentiality of the firmware during data transmission. Consequently, it can prevent malicious code modification during the installation and transmission of the ECU's firmware.

3.1.2   Mitigation of Counterfeit ECU

El-Said [13] has proposed a mechanism for detecting the counterfeit ECUs using the Physical Unclonable Function (PUF). In the proposal, during the manufacturing of ECUs, a challenge value is presented to the PUF microchip to collect a response value. After that, the response value will be hashed using SHA-1, and its result will be locked into another microchip inside the ECU. When the ECU is shipped to manufacturers or service providers, the same challenge will then be used to check if the result is consistent with the locked value. If the result is different, some microchips inside the ECU are likely to be counterfeits. This mechanism prevents service providers from sourcing malicious ECUs into their component supply chains. However, since both chip manufacturers and many service operators share the same challenge value, this value could be leaked easily. Besides, the PUF value of ECU can no longer be checked easily after finished assembling. Consequently, while it ensures the hardware security of ECUs, this mechanism should always be used together with other software security services to further guarantee the integrity of ECUs.

Kurachi et al. [14] have proposed an asymmetric key-based ECU rekeying method for ensuring the replaced ECUs are genuine units, even if the dealers or service providers are attackers. In the proposal, it assumes that all vehicles are equipped with a secure gateway in the CAN network to authenticate the connections between ECUs and the onboard unit. Then, to prevent malicious shop dealers replace counterfeit ECUs into the vehicle, the shop dealer needs to enter the signed public key of the replacement unit to the gateway during the installation of the new ECU. Then, the gateway will verify if the key is issued from legitimate manufacturers. After verifying the public key, whenever the vehicle starts to work, the gateway will hand-shake with all ECUs and exchange session keys using the public keys previously provided by the shop dealer. In this way, the shop dealer cannot replace the counterfeit ECU easily, and it does not require internet connections for verifying the ECUs. The performance evaluation in an embedded system processor shows that it only takes 229 ms to complete a signature verification, which is practical for daily usage. However, if the private key of the manufacturers or the private keys of some ECUs were compromised, the proposed system could be vulnerable without any key revocation mechanism.

3.1.3   Summary

The existing schemes have improved the security vulnerabilities in upgrading the code in ECUs. For instance, in the software aspect, [9] detects malicious firmware using blockchain as a reliable table for reference, [10] discovers new attacks using a bait environment, and [11] ensures message authenticity between the diagnostic port and the ECU during firmware upgrade. For the hardware, [13] ensures the integrity of the ECU hardware with PUF, and [14] uses gateway ECU and keys to detect counterfeit ECUs. In fact, to improve the security of ECU holistically, all these schemes should be considered since they have covered all common attacks. However, as [9] and [10] assume ECUs and the OBU are computationally powerful, and [13] and [14] require extra components to work, these schemes could inevitably increase the unit price of a vehicle. Hence, the cost of improved security in ECUs should be carefully studied and balanced.

3.2   Solutions for Sensor Attacks

3.2.1   Radar

The delay injection attack is a variant of spoofing attacks that replays the signal with delay to create an illusion that the object is much far away than its actual distance. To detect jamming attacks and delay injection attacks targeted to active sensors such as radar, Dutta et al. [15] have proposed a mechanism to detect these attacks with a challenge-response authentication method, and an algorithm to estimate the correct sensor measurement using the recursive least square method. For the challenge-response authentication method, instead of sending the active signal continuously like the conventional radar, it adjusts the signal frequently using a binary modulation signal. In this way, the radar signal will not be continuous as it will stop emitting at some selected time points. Then, by measuring the reflected results, if it is found to be a continuous signal, we can identify that the radar is under the jamming attacks. Similarly, if the received signal is shifted or missed due to the time incurred by adversaries' hardware, we can then identify that the radar is under the delay injection attacks. After that, to estimate the correct distance under these attacks, a recursive least square estimation algorithm with the time complexity of O(n2) will be executed. The simulation with two vehicles shows that the proposed mechanism can accurately detect delay injection attacks without any false positives or false negatives, and it can estimate the relative velocity well using the novel algorithm within 13 ms. The proposed mechanism is suitable for other active sensors such as LiDAR, however, if the adversaries can replay the signal without shifting or missing events (i.e. it can take the samples of the incoming signals much faster than the sending rate of radar), this mechanism would fail to detect the attacks properly.

Guan et al. [16] have proposed a jamming attack detection method based on cryptographic hash functions for the frequency-modulated continuous-wave (FMCW) radar. Conventionally, to prevent jamming attacks, radar employs pseudorandom functions to emit different random signals to suppress the interference in a single channel or time frame. In the proposal, the radar uses SHA256 hash function to replace the pseudorandom function, so that it provides randomness to the adversaries but remains predictable to the radar itself. Assuming the adversaries can perform active jamming by recording and replaying the signals using a digital radio frequency memory (DRFM), the experiments show that the proposed method can improve the overall jamming detection performance by more than 30%. However, as the proposed scheme does not have detailed security analysis for other attacks, it can only guarantee to provide jamming attacks prevention.

3.2.2 LiDAR

To detect spoofing attacks that attackers inject counterfeit light pulses into the LiDAR sensor to create fake obstacles, Matsumura et al. [17] have proposed an AES based side-channel fingerprinting method for LiDAR to identify the counterfeit signals. In the proposal, an authentication fingerprint signal is modulated with the side-channel information produced from the AES encryption circuit. Then, this amplitude modulated (AM) signal will emit to the surroundings, and the photodiode on LiDAR will receive the reflected signal. Finally, the reflected signal will be demodulated and inspected with an oscilloscope. As it is assumed that the side-channel information modulated to the light pulses is hard to produce without the secret key, if the received reflected signal is incorrect, it is likely to be a counterfeit spoofing signal. The experimental results show that the proposed method can detect most of the spoofing attacks with a fake distance longer than 30 cm. To further improve the performance, the sampling rate of the oscilloscope can increase, but it could also produce higher costs.

To prevent the received data from malicious modifications, Changalvala et al. [18] have proposed a semi-fragile data hiding-based technique which inserts a binary watermark into the LiDAR sensor data. Specifically, a data watermark is added to the received data using the 3-dimensional quantization index modulation (3D-QIM) technology. If the attacker intercepts the communication between the sensor and the decision-making unit and modifies some part of the transmitted data, the decoding algorithm in the decision-making unit would find out some parts of the received data cannot be verified. In this way, the decision-making unit (e.g. a central server or the OBU) knows that the received signal has tampered, and it can then locate which part of the data is the culprit using the algorithm. The performance evaluation based on a benchmark KITTI dataset shows that the proposed method can achieve a very high detection success rate when the additive noise in the received LiDAR signal is under control. Also, the security evaluation shows that it can detect data spoofing attacks and insider attacks in which the connection between the sensors and the decision-making unit (e.g. CAN or VANET) is assumed to be secure but the adversaries exist within these connections.

3.2.3 Camera

The camera captures continuous images in the visible light spectrum to identify different objects and road signs. To prevent it from attacks such as camera blinding attacks (a variant of jamming attacks), Petit et al. [19] have proposed some countermeasures to control the environmental light and improve the auto exposure control. In the proposal, it has suggested that blinding occurs when the environment is too bright so that the received image is overexposed, or a malicious bright light source is emitted intermittently so that it takes extra time for the auto exposure control to adapt the brightness to the environment. To prevent these two scenarios from worsening the image quality and causing temporary blinding, it has suggested that multiple cameras taking the same image at different angles should be used. Also, removable near-infrared-cut filters and photochromic lenses should be installed in front of the cameras, such that they can filter out the excessive near-infrared lights and other unwanted lights. These countermeasures should effectively improve the quality of the received pictures so that to reduce the chance of suffering from blinding attacks.

To predict and identify camera blinding attacks correctly, Yadav et al. [20] have proposed a sequence modeling and predictive analytics system using a convolutional encoder-decoder neural network. The proposal has suggested that all camera frames should be passed into a neural network called the Motion-Content network (MCnet). Subsequently, the neural network can predict the next scene based on the inputted frames. To identify the blinding attacks, the structural similarity algorithm is employed to calculate the similarity score by comparing the latest frame with the previous frames and predicted frame. If the similarity score is less than a predefined threshold value, the camera is likely under blinding attacks. The experimental results show that the proposed system can predict the next frames accurately and detect blinding attacks with 96.5% true positive. However, since the proposed system may not differentiate blinding attacks and natural blinding effects, potential false positives could happen. Also, although MCnet is computationally less expensive than other neural networks, it could still consume more power and computational time than conventional approaches.

3.2.4   GPS

To detect GPS spoofing in real-time using the encrypted military GPS signals, O'Hanlon et al. [21] have proposed a detection scheme that uses two GPS receivers to compare and detect spoofing signals. In the proposal, the receivers A is put in a secured location without GPS spoofing, and receiver B is put at arbitrary places that could be attacked by adversaries. Then, assuming the P(Y) code GPS signal encrypted and implemented by the US military cannot be easily spoofed, the scheme compares the GPS and GPS P(Y) signals from the two receivers to calculate the spoofing detection statistics. If the statistic value is high, it means both receivers can sense the identical P(Y) code signals. However, if the statistic value is small, it indicates that either one of the receivers does not have the P(Y) code signals so that it could be under the spoofing attack. The field test carried out in with two sensors put in different places shows that the probability of detecting all kinds of GPS signals is close to 100%. However, since it assumes that there is a statistically non-spoofable relationship between the GPS signals and the P(Y) signal, it could be dangerous if the governor who controls the P(Y) signal chooses to change the rules or even exploits this scheme by spoofing the P(Y) messages. Also, since all the experiments are carried out in a static environment, its feasibility should be further investigated in the mobility environment like AVs.

Psiaki et al. [22] have proposed a GNSS/GPS spoofing detection system with two antennas for detecting the direction of signal sources. In the proposal, it assumes that the spoofing signal is emitted from one fake GPS emitter so that all these signals are from a single direction. Conversely, for the non-spoofed case, since there are many satellites in the sky, the GPS signals should be emitted from many different directions. Then, having at least two antennas installed at different positions, it can calculate the carrier phase difference, and then determine the direction of all satellite signals. The proposal has also suggested a real-time version of the system that can detect the spoofing attack using real-time software radio receivers. The field test carried out on a yacht cruising around Italy shows that it can detect GPS spoofing within 0.4 s with a relatively low probability of false alarm. The proposed system has been tested to be suitable for the high-speed moving environment such as cruise and AVs. Also, its cost can be adjusted by having two GPS receivers at the same time, or only one receiver but connecting to two antennas with a switch.

Shafiee et al. [23] have proposed a multi-layer neural network-based GPS spoofing detection system. The system takes three main features of the received GPS signals, namely the early-late phases, deltas, and signal levels. For the neural network, the system uses a Multi-Layer Perceptron (MLP) neural network with three neurons in the input layer, two neurons in the hidden layer, and one neuron in the output layer. In the learning phase, to optimize accuracy, the network is trained with Levenberg Marquardt (LM) learning algorithm. Testing results of the NN show that it can detect more than 98% of the GPS spoofing signals, and the detection time can be shorter than 0.5 s. Also, it outperforms other existing proposals that employ conventional K-nearest neighborhood and Bayesian classifiers. However, since the experiments are carried out in a static environment, the system may not be suitable for the mobility environment in AVs. Also, since the neural network could be vulnerable to adversarial attacks, the potential threats of MLP and this system should be further investigated.

3.2.5 Summary

As one of the crucial parts of AVs, sensors are particularly vulnerable to attacks such as jamming and spoofing. To detect these attacks effectively in the active sensors like radar and LiDAR, [15, 16] and [17] modify the sensor output and check the reflected signals to distinguish normal signals, spoofing signals, and jamming signals. Although these proposals are computationally lightweight to be applied, multiple strategies should be combined to make it more difficult for adversaries to create new spoofer devices easily. On the other hand, to prevent the received signals from malicious modification and fabrication, [18] adds watermarks to the received LiDAR data. There are also some proposals such as [19] and [22] that use additional hardware to detect spoofing attacks. For GPS, it can also use the secret military signal to detect spoofing [21]. Finally, [20] and [23] use neural networks to detect attacks. However, since neural networks could consume more time and energy during the training and running phases, their total cost should be carefully studied. While most of the schemes can only passively detect attacks, [15] is the only scheme that tries to actively reduce the effect of attacks by recovering the original signals.

3.3 Solution for Intra-vehicular Link Attacks

3.3.1 Intrusion Detection

To identify the malicious packets and therefore detect potential intruders in the CAN network, Kang et al. [24] have proposed a deep neural network (DNN) based intrusion detection system (IDS). This DNN involves two major phases, the training phase, and the detection phase. In the training phase, a list of normal packets and attack packets are inputted to the network to do supervised learning. It helps the DNN to learn the characteristics of attack packets internally. To shorten the training time and improve the overall training results, the pre-trained deep brief network (DBN) is chosen as the initial weights of the neural network. After that, in the detection phase, the neural network can process all packets in the CAN network to identify potential attack packets in real-time. The experimental results showed that the proposed neural network outperforms the standard support vector machine (SVM) and the conventional artificial neural network (ANN). Also, the simulation shows that the proposed neural network can provide a real-time response to the attack with an accurate detection ratio of 98% on average while maintaining the computational complexity to be small with a limited number of layers. However, there is a crucial shortcoming of the proposed network: since the replayed packets are identical to the legitimate packets, the system may not detect these packets correctly. Consequently, the proposed neural network could not prevent replay attacks.

Song et al. [25] have proposed a lightweight IDS algorithm based on the analysis of time intervals of CAN messages. Specifically, since every ECU has its regular frequency and time interval, by measuring its message rate carefully, it can be the signature of all messages sent from that ECU. In the proposal, when a new message appears on the CAN bus, the system records the time interval between the message and the last message. If it has a shorter time interval than normally expected, the system will determine that the ECU has message injection attacks. Also, if the message interval is smaller than 0.2 ms, the system would increase the score representing the likeliness of DoS attack by one. If the score is higher than a threshold value, it would then be treated as a DoS attack. The experiments show that the proposed system can detect message injection attacks in a millisecond with more than 93% accuracy if the threshold value sets to higher than 3. However, since this system could only detect message injection attacks and DoS attacks, other communication attacks such as replay attacks cannot be prevented or detected with this system.

Waszecki et al. [26] have proposed an in-vehicle network traffic monitoring algorithm which detects the denial of service and jamming attacks (i.e. message flooding) under the CAN network. This proposal has suggested using the computationally lightweight leaky bucket algorithm to detect the traffic on the CAN network in real-time. Then, to prevent this monitoring framework from suffering the risk of a single point of failure, it also suggested an automatic task distribution mechanism based on the "graph-based system architecture" to all ECUs. In this architecture, ECUs and the CAN bus are classified as resources, and they are connected through some architectural links. Then, all these pairs will also be chosen fairly, according to the formulation of the integer linear program (ILP). To prove the feasibility of the proposed scheme, a case study on an automobile with more than 100 ECUs was carried out. It shows that all monitoring tasks can be evenly distributed with proper tuning of parameters, and the detection algorithm works as expected with low computational overheads. Although this mechanism detects jamming attacks in CAN networks efficiently without significantly burdening a single component, a lot of research work is yet to be done to support the detection and prevention of other attacks such as spoofing attacks or the jamming attack added with random delays.

Hafeez et al. [27] have proposed an artificial neural network (ANN) that identifies the fingerprints of the unique parametric signals emitted from different ECUs and therefore authenticates them without extra communication overhead. Since there are some unavoidable hardware imperfections in every ECU during the manufacturing process, ECUs cannot produce a perfect rectangular waveform during signal transmissions. Hence, the proposal has exploited these channel imperfections and collected seven different parameters, including the overshoot, peak time, settling time, peak value, steady-state value, damping ratio, and natural frequency, to fingerprint and therefore differentiate the signal from different ECUs. After extracting the signals and training the ANN with eight different CAN channels and more than 900 records, the evaluation shows that the overall channel detection accuracy of the

ANN is higher than 97%. This proposal can effectively detect spoofing attacks in the CAN network, such that it allows only the authenticated ECUs can control some specific ECUs. However, to implement the proposal to AVs completely, all ECUs in the vehicle should equip with additional hardware for ANN training. That could induce long training time after replacing any components and high costs on ANN training hardware.

3.3.2 Message Authentication

Protecting confidentiality and authenticity of the data in the inter-vehicular channels, Agrawal et al. [28] have proposed a secure CAN with Flexible Data (CAN-FD-Sec) protocol that uses gateway ECU (GECU) to provide secure group communication in the CAN bus. Instead of having an individual key for each ECU to communicate securely, the proposal has suggested that there should be a computationally powerful ECU that generates group session keys and re-encrypt the messages across different groups. The proposal has a total of 5 phases. The first phase is the initialization that preloads two keys into each ECU. The second phase is the key loading phase that the GECU should be loaded with 2n keys (assume n is the total number of ECUs). The third phase is the session key generation phase that each ECU creates its session key with GECU, and then the GECU distributes a group session key to all ECUs in the same group using their session keys. The fourth phase is the authentication encryption phase that all ECUs in the same group can communicate using the group session key. If an ECU in group A wants to talk to another ECU in group B, it can be achieved by sending the encrypted message to GECU, and then let the GECU to re-encrypt the message using the group B's session key. The final phase is the key update phase that updates the group session key periodically and after a new user joining or leaving the CAN network. The simulation shows that this proposal can maintain below 2 ms of transmission delays in different connection cases. Also, it can effectively prevent eavesdropping and replay attacks since it uses session keys for all communication.

3.3.3 Summary

Intra-vehicular links are vulnerable to various communication attacks such as eavesdropping, DoS attack, and message fabrication attacks. To detect the DoS attack and ensure the unforgeability of the message in the intra-vehicular links (e.g. CAN network), [25] and [26] have proposed real-time algorithms for checking the characteristics of all data traffic. Although they are computationally lightweight, they are mostly built based on many assumptions that may not be valid all the time. To improve the situation, [24] and [27] use neural networks to identify the message issuers and therefore detect potential attacks. However, these neural network-based schemes could consume more computational resources in OBUs. Alternatively, cryptographic based authentication schemes (e.g. [28]) could also be considered. These schemes aim to encrypt the message and ensure mutual authentication between message issuers and receivers in rationally low costs. Although [28] requires extra gateway ECUs that could increase the unit cost, it provides a much higher level of security than other IDS-based schemes in terms of message confidentiality, unforgeability, and mutual authentication.

3.4 Solutions for Inter-vehicular Link Attacks

3.4.1 Improving Reliability of Target Devices and Services

To prevent bogus information attack in VANET, Celes et al. [29] have proposed a solution using position verification techniques to find out attackers in the network. This proposal has suggested that when a vehicle sends a sensor data message to the road-side unit (RSU), the RSU will select some nearby vehicles as verifiers and ask them to verify the truthfulness of the message. In a normal situation, all verifiers should agree that the message is valid. If some verifiers disapprove of the message, the RSU takes a simple majority from all selected verifiers to decide if the message is trustworthy. However, if most of the verifiers disapprove of this message, the RSU will conclude that it is very likely to be a bogus information attack. The simulation using the NS2 network simulator shows that since bogus attacks can be detected effectively at the early stage within a small area, it prevents the false information from spreading across the network. Although the proposed scheme cannot fully stop bogus attacks because there could be more than one malicious vehicle in the area (e.g. the vehicles are controlled by botnets), it makes the attack much harder to achieve.

Combining blockchains and software-defined network (SDN) into fog computing VANET, Gao et al. [30] have proposed a paradigm sharing the managerial responsibilities of the network to blockchain and SDN instead of a single trust authority. In the proposal, it assumes that all vehicles are the end-users. They can connect with surrounding vehicles and RSUs to form a fog computing zone. Then, these fog computing zones will be managed by RSU hubs, and the SDN will allocate channel resources to all these computing zones. The proposal also assumes that four different entities are managing the consortium blockchain: the authentication server which handles device registration, the data management server which assists the SDN controller in data management, the access controller, and the policy management server which contains many smart contracts. For any vehicle that wants to access the VANET, it firstly broadcasts its information to the authentication server to get a certificate. Upon successful inspection, the information or the vehicle will be recorded in the blockchain. In case that the vehicle has received some data from others, it can decide the trustworthiness of the message by running the trust system that evaluates the reputation scores based on the verdicts from other vehicles. The proposed paradigm prevents data spoofing in the fog computing VANET while maintaining a high packet delivery ratio for moving vehicles.

Luo et al. [31] have proposed a trust management method based on Dirichlet distribution and blockchain to provide vehicle location anonymity and determine the trustworthiness of vehicles in the VANET. Specifically, while using location-based services, service providers could abuse the location information from the data requesters. Hence, to hide the identity of the data requester vehicles, the proposed scheme creates an anonymous cloaking region that covers at least k participants, so that adversaries can only trace back the requester with the confidence level of 1/k. Then, to ensure that the cloaking region can provide anonymity effectively, the scheme also identifies the malicious vehicles while constructing the cloaking region using counterparties' current behavior and their historical trust information. Since the historical information of all vehicles is stored with a blockchain, it is immutable to all parties. Thus, it is trustworthy for all vehicles to reference it. The performance evaluation shows that the probability of location privacy leakage is lower and the percentage of malicious vehicles in a cloaking region is also reduced compared with existing schemes. In this way, Sybil attacks that could potentially harm the location privacy of AVs while using location-based services can be reduced effectively.

Protecting the integrity of sensory data during transmission and immutability during data sharing, Kong et al. [32] have proposed an efficient, privacy-preserving and verifiable sensory data collection and sharing scheme. The scheme comprises of permissioned blockchain, 2-DNF (disjunctive normal form) cryptosystem and identity-based signcryption. In the proposal, it assumes that all vehicles are installed with air-quality sensors and they will transmit the collected encrypted data to RSUs. For each RSU, it aggregates the received encrypted data and uploads the results to the server. For the server, it receives the aggregated results from RSUs, recovers the original data, and sends back to RSUs. Finally, RSUs can verify the recovered data and then share it with other RSUs and vehicles by recording to a permissioned blockchain. Unlike open access blockchains that use Proof of Work (PoW) as verification and consensus, the proposed permission blockchain is only accessible by RSUs, and all RSUs will assist verifying the sensory data blocks sent from other nodes. The consensus can only be made when the number of successful verifications of a sensory data block passes a threshold value so that it can be uploaded to the blockchain. The security analysis shows that the proposed scheme can protect the anonymity, location privacy of vehicles. Also, the uploaded sensory data is free of data spoofing as it is verifiable by the signcryption scheme and immutable because of the properties of the blockchain.

3.4.2 Improving Communication Security

To authenticate AVs and secure the V2X communication using biometric information, Raiyn [33] has proposed an authentication scheme based on an iris recognition system. In the proposal, users' iris information is collected during the training phase, and then they can use the system to authenticate with the central server in the testing phase. With various machine learning and data mining algorithms embedded in the

Waikato Environment for Knowledge Analysis (WEKA), the proposal has demonstrated a satisfactory result of identifying and classifying different iris correctly. Then, to secure the V2X communication, this proposal has further suggested an algorithm that extracts the features of the iris and encodes it with a binary code. After that, by combining this binary code with an encryption key distributed by the central server, a new identity encryption key is generated. Using it to encrypt all data transmission, the central server ensures that the data sender in the AV is an authenticated person. Consequently, this proposal guarantees that only the owner of the vehicle can start secure communication. Also, by checking the iris frequently, it can provide extra information such as the sleepiness and tiredness of the driver. However, since the central server or the data pool must store the iris information of all drivers, the security of the central server must be carefully protected to prevent any privacy disclosure.

Wang et al. [34] have proposed a two-factor lightweight privacy-preserving authentication scheme (2FLIP) for securing VANET connections using the decentralized certificate authority (CA) and biometric password. In the proposal, each vehicle equips with a tamper-proof device that stores the system keys and use it to sign and verify messages. During initialization, all vehicles must register themselves to CA. After that, when the driver starts the vehicle, he needs to prove his identity to the telematics module using biometric passwords such as face or fingerprints. Then, the vehicle will generate an instant access token to logon to the telematics module. If the login is successful, the telematics module will download keys from the CA for message signing in the future. When the vehicle wants to broadcast messages to others, it will first redo the login phase to update the pseudo-identity, sign the message using MAC hash functions, and broadcast it to the nearby vehicles. All vehicles nearby can verify the message with CA. Also, CA can decide to revoke the privileges of the vehicle, or even trace the message sender using the signature. The performance evaluation shows that the proposed scheme is computationally lightweight because lightweight cryptographic hash functions and fast MAC are chosen for message signing and verification between vehicles. Also, the security evaluation shows that the proposed scheme provides various security features such as message unforgeability, message nonrepudiation, device anonymity, traceability for CA, and secure key updates. However, since the work still relies on a trustworthy CA to authenticate and verify all vehicles, it could be vulnerable if the CA is compromised.

To provide efficient device revocation and authenticity validation of anonymous message, Shao et al. [35] proposed a threshold authentication anonymous authentication protocol for VANET using group signatures. Constructed with bilinear pairings, the new group signature scheme assumes that there are four major parties, namely the central authority (CA), the tracing manager (TM), many RSUs and many OBUs, each of them has different responsibilities. For CA, it authenticates the public keys and issues public-key certificates to RSUs. For TM, it authenticates the public keys and issues public-key certificates to OBUs. For RSUs, they manage the covered OBUs by issuing a group certificate to them, such that the OBUs can communicate with others by signing the message with the group certificate and its private key. For other OBUs to accept the message, it needs to verify if the number of signatures of the same message has exceeded its threshold value. In case that there are some malicious messages, TM can also reveal the identity of the signer with the signature, and then revoke the privilege of that user by sending the updated revocation list to all RSUs. The security analysis shows that the scheme can provide efficient revocation, message unforgeability, device anonymity and traceability. However, since the performance evaluation also shows that the computational time for message verification is long (about 1.23 s for each message), if many messages are queuing to be verified, it could take a considerable amount of time. Consequently, some improvement work such as fast or batch verification should be considered in the future.

Vehicular Cloud (VC) is the concept of using underutilized vehicles to form a cloud computing and storage platform in the VANET. It has low costs and it increases the computation and storage capabilities of all vehicles. To form VC dynamically and securely in the VANET, Zhang et al. [36] have proposed a secure and privacy-preserving communication scheme with pseudo-identities and dynamic identity-based authenticated asymmetric group key agreement (DIBAAGKA). In the proposal, it has a total of five stages. The system setup stage initializes parameters for identity-based cryptosystems in the trusted authority (TA). Then, in the enrollment stage, all vehicles and RSUs updates the pseudo-identities and keys with TA. After that, all vehicles can form VCs and exchange messages securely with the surrounding users during the "VC initialization and maintenance stage" and the "secure message delivery stage". If there are some malicious vehicles in the VC, their real identity can be found by TA in the trace phase. The security analysis shows that it can guarantees data confidentiality, integrity, authentication and privacy protection of all vehicles, and the performance analysis shows that it can form VCs with a small group of users efficiently and securely.

To prevent man-in-the-middle (MITM) attacks in VANET, Ahmad et al. [37] have proposed a MITM attack-resistant trust model in connected vehicles (MARINE). In the proposal, instead of using the conventional cryptography and public-key infrastructure (PKI), it uses a novel trust model that identifies the trustworthiness of users by considering both entity trust and content trust. Specifically, the scheme evaluates the trustworthiness of the sender node (i.e. entity trust) by considering its previous interactions and the recommendation from the neighboring vehicles. Then, to trust the message sent from the node (i.e. content trust), three aspects including the information quality, node's message forwarding capability and opinions from neighbors are jointly considered. These calculations can also be assisted with infrastructure (e.g. RSU). Finally, the received message would be accepted only if both two trusts are computed successfully and higher than the threshold. The simulation shows that the proposed scheme can achieve more than 15% improvements regarding the detection of MITM attacks than the existing proposals. However, since the scheme only prevents MITM attacks effectively, some other techniques should also be considered during actual deployment.

He et al. [38] have proposed an identity-based conditional privacy-preserving authentication scheme for VANET. Conditional privacy in the proposal refers to protecting the combination of identity privacy and the traceability of the message. Specifically, the proposed scheme has suggested that all vehicles should be registered to a trusted authority to get a real-identity, password and a private key and save them to the tamper-proof device. Then, if a vehicle wants to send a message, its tamper-proof device will generate a pseudo-identity and sign it using that identity and its private key. For vehicles and RSUs who received messages from others, they can validate the message by running a batch verification function to make sure the data sender is valid. The security analysis shows that the proposed scheme prevents various attacks such as impersonation, message modification, replay attacks, MITM attacks and stolen verifier table attack (an attack reveals the identity of other vehicles by stealing the lookup table). The performance evaluation also shows that it outperforms most of the existing schemes for more than 70% because it does not need the computationally expensive bilinear pairing operations.

3.4.3 Summary

Security in the inter-vehicular links involves two parts, namely the protection of target devices and services, and the protection in communication vulnerabilities. To improve the reliability of target devices and services, the existing schemes use surrounding vehicles to vote for the validity of messages [29, 35, 37]. Also, some proposals use blockchain to evaluate the trustworthiness of the surrounding vehicles [30, 31] or the broadcasted data [32]. Although these schemes prevent bogus information attacks and Sybil attacks effectively in the target services (e.g. traffic information or sensory data services), due to the use of some advanced cryptosystems and blockchains, their computational and storage overheads are considerably high. On the other hand, to improve communication security, biometric-based solutions [33, 34] are proposed to encrypt the messages and ensure the drivers are legitimate users. However, as storing the biometric information of users at cloud servers could raise public concerns about personal privacy, the practicability of these schemes is questionable. Finally, some proposals [36, 38] aim to protect the anonymity of vehicles using pseudo-identity and other novel techniques. These schemes are more practical because they can prevent most of the communication attacks such as location tracking, message fabrication, and replay attacks with low computational overheads. However, since these schemes do not guarantee message confidentiality, it could still be a security concern. Consequently, more research work is still needed to improve security in the inter-vehicular links holistically.

4 Future Directions

Through our taxonomy and analysis of the existing solutions, we have provided a broader view of the current development of the attack surfaces and the corresponding countermeasures in AVs. In this section, we conclude some open issues and then suggest some directions for improving the security of AVs in the future.

4.1 Open Issues

The current work has improved various security vulnerabilities in AVs. However, our chapter also revealed that there are still many challenges yet to be solved:

- Section 3.1 showed that improving the security in the ECUs could inevitably increase the unit cost of a vehicle, because extra components and higher computational power for running the cryptographic functions are required.
- Section 3.2 found that most of the sensor attack countermeasures can only detect intrusions but not recovering the original messages. Hence, applying these solutions still cannot comprehensively improve the reliability of sensors in AVs.
- Section 3.3 concluded that most of the intra-vehicular link attack countermeasures could only detect but not prevent DoS attacks from happening. This issue has to be addressed in the future because a tiny error in the ECU messages could also be lethal to the passengers in the vehicle.
- Section 3.4 revealed that the inter-vehicular links are vulnerable to many network attacks. To holistically protect the whole ecosystem from target services to communication messages, it requires higher computational costs. If this problem cannot be eradicated, it could reduce the incentive for AVs to adopt inter-vehicular communications.

4.2 Future Directions

As the existing solutions suffer from a number of shortcomings aforementioned, and many new attacks are emerging, AVs are still facing many security challenges. To unleash the full potential of AVs and make it safer for everyday use, in this chapter, we suggest four future research directions for improving the security in AVs: the lightweight protocols, sensor attack recovery, intrusion detection with artificial intelligence, and adversarial attack defenses.

Lightweight Protocols: Since vehicles have limited power and resources, the internally used security protocols for ECUs, sensors and intra-vehicular links must be computationally lightweight to reduce the energy consumption. Besides, for the inter-vehicular links, since there will be a tremendous number of fast-moving vehicles and infrastructures in the roadside, the authentication protocols used in V2X communication should also be lightweight to ensure two devices can exchange information as fast as possible. While most of the existing security schemes in VANET are still computationally expensive, novel lightweight protocols should be proposed to provide a better quality of services, higher spectral efficiency, and lower latency.

Sensor Attack Recovery: Most of the existing works for sensors can only provide intrusion detection. Whenever there are jamming or spoofing attacks, the in-car sensors can only report intrusions and stop functioning. Since AVs in the future are using sensors to make driving decisions primarily, it could be dangerous if the sensors stop working completely while they are under sensor attacks. Therefore, novel signal recovery, estimation, and prediction algorithms should be proposed to allow sensors to be more resilient to work under the attacks.

Intrusion Detection with Artificial Intelligence: Resembling human intelligence with deep learning networks and other innovative technologies, Artificial Intelligence (AI) is a promising way to solve many challenging problems including intrusion detection. Since deep learning networks can efficiently learn some hidden relationships between the input data and the output results, researchers have been using it to improve the accuracy of image recognition in the in-car sensors. In the future, as there will be more robust and efficient AI algorithms, it also opens the opportunity for ECUs and in-vehicle sensors to use them to predict and identify malicious attacks. For example, AI can effectively discover some distinctive patterns of suspicious signals which may not be easily detected by conventional algorithms. Also, by employing new machine-learning techniques such as transfer learning, the training cost of the network could be further reduced, making AI more favorable to the intrusion detection applications in the resource-constrained AV security.

Adversarial Attack Defenses: There will be more AVs using deep learning to provide obstacle detections and attack identification. However, since deep neural network could be vulnerable to adversarial attacks that exploit the weaknesses of neural network with specially crafted images or signals, it has to be carefully studied to avoid life-endangering situations. In the future, more in-depth research topics, for instance, the systematic ways for finding weaknesses in the deep neural network, some live examples of adversarial attacks in the road, and some countermeasures of the attacks, should be carried out.

5 Conclusion

In this chapter, we have suggested a systematic classification of four attack surfaces in AV, namely ECUs, sensors, intra-vehicular links, and inter-vehicular links. For each attack surface, we have listed various security threats and reviewed some latest and major research work for the defense. Subsequently, after understanding the shortcomings and difficulties of the current work, we have also suggested four open research topics in AV. We hope this taxonomy can be a guide for the general public to acknowledge the potential attacks and countermeasures, as well as to encourage security researchers to improve the attack defense technologies in AV in the future.

Acknowledgements This research is supported by A*STAR under its RIE2020 Advanced Manufacturing and Engineering (AME) Industry Alignment Fund – Pre-Positioning (IAF-PP), Singapore (Grant No. A19D6a0053).

References

1. Fagnant, D.J., Kockelman, K.: Preparing a nation for autonomous vehicles: Opportunities, barriers and policy recommendations. Transp. Res. Part A Policy Pract. **77**, 167–181 (2015). https://doi.org/10.1016/j.tra.2015.04.003
2. Nie, S., Liu, L., Du, Y.: Free-fall: hacking tesla from wireless to can bus. Black hat USA 2017, pp. 1–16 (2017) https://www.blackhat.com/docs/us-17/thursday/us-17-Nie-Free-Fall-Hacking-Tesla-From-Wireless-To-CAN-Bus-wp.pdf
3. Povolny, S., Trivedi, S.: Model Hacking ADAS to Pave Safer Roads for Autonomous Vehicles | McAfee Blogs, McAfee Labs (2020). https://www.mcafee.com/blogs/other-blogs/mcafee-labs/model-hacking-adas-to-pave-safer-roads-for-autonomous-vehicles/. Accessed 19 Jun 2020
4. Ruffo, G.H.: Tesla Data Leak: Old Components With Personal Info Find Their Way On eBay, Insideevs (2020). https://insideevs.com/news/419525/tesla-data-leak-personal-info-ebay/. Accessed 22 Jun 2020
5. Jo, H.J., Choi, W., Na, S.Y., Woo, S., Lee, D.H.: Vulnerabilities of android OS-based telematics system. Wireless Pers. Commun. **92**(4), 1511–1530 (2017). https://doi.org/10.1007/s11277-016-3618-9
6. Cao, Y., Zhou, Y., Chen, Q.A., Xiao, C., Park, W., Fu, K., Cyr, B., Rampazzi, S., Morley Mao, Z.: Adversarial sensor attack on LiDAR-based perception in autonomous driving. In: Proceedings of the ACM Conference on Computer and Communications Security, pp. 2267–2281 (2019). https://doi.org/10.1145/3319535.3339815
7. Minuth, J.: FlexRay^{TM} electrical physical layer: Theory, components, and examples. In: Communication in Transportation Systems, pp. 117–175 (2013). https://doi.org/10.4018/978-1-4666-2976-9.ch005
8. Wang, G., Wang, B., Wang, T., Nika, A., Zheng, H., Zhao, B.Y.: Defending against sybil devices in crowdsourced mapping services. In: Proceedings of the 14th Annual International Conference on Mobile Systems, Applications, and Services, MobiSys 2016, pp. 179–191 (2016). https://doi.org/10.1145/2906388.2906420
9. Baza, M., Nabil, M., Lasla, N., Fidan, K., Mahmoud, M., Abdallah, M.: Blockchain-based firmware update scheme tailored for autonomous vehicles. In: IEEE Wireless Communications and Networking Conference, WCNC, vol. 2019, pp. 1–7 (2019). https://doi.org/10.1109/WCNC.2019.8885769
10. Yu, L., Deng, J., Brooks, R.R., Yun, S.B.: Automobile ECU design to avoid data tampering. In: ACM International Conference Proceeding Series, vol. 06–08 (2015). https://doi.org/10.1145/2746266.2746276
11. Mansor, H., Markantonakis, K., Akram, R.N., Mayes, K.: Don't brick your car: Firmware confidentiality and rollback for vehicles. In: Proceedings - 10th International Conference on Availability, Reliability and Security, ARES 2015, pp. 139–148 (2015). https://doi.org/10.1109/ARES.2015.58
12. Idrees, M.S., Schweppe, H., Roudier, Y., Wolf, M., Scheuermann, D., Henniger, O.: Secure automotive on-board protocols: A case of over-the-air firmware updates. Lecture Notes in Computer Science (including subseries Lecture Notes in Artificial Intelligence and Lecture Notes in Bioinformatics), vol. 6596 LNCS, pp. 224–238 (2011). https://doi.org/10.1007/978-3-642-19786-4_20
13. El-Said, M.: ECU counterfeit mitigation using holistic approach in modern automotive echo system. In: 2020 IEEE 17th Annual Consumer Communications and Networking Conference, CCNC 2020, pp. 16–17 (2020). https://doi.org/10.1109/CCNC46108.2020.9045510
14. Kurachi, R., Takada, H., Adachi, N., Ueda, H., Miyashita, Y.: Asymmetric key-based secure ECU replacement without PKI. In: Proceedings of IEEE International Symposium on High Assurance Systems Engineering, vol. 2019, pp. 234–240 (2019). https://doi.org/10.1109/HASE.2019.00043

15. Dutta, R.G., Guo, X., Zhang, T., Kwiat, K., Kamhoua, C., Njilla, L., Jin, Y.: Estimation of safe sensor measurements of autonomous system under attack. In: Proceedings - Design Automation Conference, vol. Part 12828 (2017). https://doi.org/10.1145/3061639.3062241

16. Guan, Z., Chen, Y., Lei, P., Li, D., Zhao, Y.: Application of hash function on FMCW based millimeter-wave radar against DRFM jamming. IEEE Access 7, 92285–92295 (2019). https://doi.org/10.1109/ACCESS.2019.2928000

17. Matsumura, R., Sugawara, T., Sakiyama, K.: A secure LiDAR with AES-based side-channel fingerprinting. In: Proceedings - 2018 6th International Symposium on Computing and Networking Workshops, CANDARW 2018, pp. 479–482 (2018) https://doi.org/10.1109/CANDARW.2018.00092

18. Changalvala, R., Malik, H.: LiDAR data integrity verification for autonomous vehicle using 3D data hiding. In: 2019 IEEE Symposium Series on Computational Intelligence, SSCI 2019, vol. 7, pp. 1219–1225 (2019) https://doi.org/10.1109/SSCI44817.2019.9002737

19. Petit, J., Stottelaar, B., Feiri, M., Kargl, F.: Remote Attacks on Automated Vehicles Sensors: Experiments on Camera and LiDAR. Blackhat.com, pp. 1–13 (2015). https://www.blackhat.com/docs/eu-15/materials/eu-15-Petit-Self-Driving-And-Connected-Cars-Fooling-Sensors-And-Tracking-Drivers-wp1.pdf

20. Yadav, S., Ansari, A.: Autonomous Vehicles Camera Blinding Attack Detection Using Sequence Modelling and Predictive Analytics. *SAE Technical Papers*, vol. 2020-April, no. April, pp. 1–6, 2020, doi: https://doi.org/10.4271/2020-01-0719.

21. O'Hanlon, B.W., Psiaki, M.L., Bhatti, J.A., Shepard, D.P., Humphreys, T.E.: Real-time GPS spoofing detection via correlation of encrypted signals. Navig J. Ins. Navig. 60(4), 267–278 (2013). https://doi.org/10.1002/navi.44

22. Psiaki, M.L., O'Hanlon, B.W., Powell, S.P., Bhatti, J.A., Wesson, K.D., Humphreys, T.E., Schofield, A.: GNSS spoofing detection using two-antenna differential carrier phase. In: 27th International Technical Meeting of the Satellite Division of the Institute of Navigation, ION GNSS 2014, vol. 4, pp. 2776–2800 (2014)

23. Shafiee, E., Mosavi, M.R., Moazedi, M.: Detection of spoofing attack using machine learning based on multi-layer neural network in single-frequency GPS receivers. J. Navig. 71(1), 169–188 (2018). https://doi.org/10.1017/S0373463317000558

24. Kang, M.J., Kang, J.W.: Intrusion detection system using deep neural network for in-vehicle network security. PLoS ONE 11(6), 1–17 (2016). https://doi.org/10.1371/journal.pone.0155781

25. Song, H.M., Kim, H.R., Kim, H.K.: Intrusion detection system based on the analysis of time intervals of CAN messages for in-vehicle network. In: 2016 International Conference on Information Networking (ICOIN), pp. 63–68 (2016). https://doi.org/10.1109/ICOIN.2016.7427089

26. Waszecki, P., Mundhenk, P., Steinhorst, S., Lukasiewycz, M., Karri, R., Chakraborty, S.: Automotive electrical and electronic architecture security via distributed in-vehicle traffic monitoring. IEEE Trans. Comput. Aided Des. Integr. Circuits Syst. 36(11), 1790–1803 (2017). https://doi.org/10.1109/TCAD.2017.2666605

27. Hafeez, A., Topolovec, K., Awad, S.: ECU fingerprinting through parametric signal modeling and artificial neural networks for in-vehicle security against spoofing attacks. In: ICENCO 2019 - 2019 15th International Computer Engineering Conference: Utilizing Machine Intelligence for a Better World, pp. 29–38 (2019). https://doi.org/10.1109/ICENCO48310.2019.9027298

28. Agrawal, M., Huang, T., Zhou, J., Chang, D.: CAN-FD-sec: improving security of CAN-FD protocol. In: Hamid, B., Gallina, B., Shabtai, A., Elovici, Y., Garcia-Alfaro, J. (eds.) Security and Safety Interplay of Intelligent Software Systems: ESORICS 2018 International Workshops, ISSA 2018 and CSITS 2018, Barcelona, Spain, September 6–7, 2018, Revised Selected Papers, pp. 77–93. Springer, Cham (2019)

29. Celes, A.A., Elizabeth, N.E.: Verification based authentication scheme for bogus attacks in VANETs for secure communication. In: Proceedings of the 2018 IEEE International Conference on Communication and Signal Processing, ICCSP 2018, pp. 388–392 (2018) https://doi.org/10.1109/ICCSP.2018.8524540

30. Gao, J., Agyekum, K.O.B.O., Sifah, E.B., Acheampong, K.N., Xia, Q., Du, X., Guizani, M., Xia, H.: A blockchain-SDN-enabled internet of vehicles environment for fog computing and 5G networks. IEEE Internet Things J. **7**(5), 4278–4291 (2020). https://doi.org/10.1109/JIOT. 2019.2956241

31. Luo, B., Li, X., Weng, J., Guo, J., Ma, J.: Blockchain enabled trust-based location privacy protection scheme in VANET. IEEE Trans. Veh. Technol. **69**(2), 2034–2048 (2020). https:// doi.org/10.1109/TVT.2019.2957744

32. Kong, Q., Su, L., Ma, M.: Achieving privacy-preserving and verifiable data sharing in vehicular fog with block chain. IEEE Trans. Intell. Transp. Syst. 1–10 (2020). https://doi.org/10.1109/ tits.2020.2983466

33. Raiyn, J.: Data and cyber security in autonomous vehicle networks. Transp. Telecommun. **19**(4), 325–334 (2018). https://doi.org/10.2478/ttj-2018-0027

34. Wang, F., Xu, Y., Zhang, H., Zhang, Y., Zhu, L.: 2FLIP: A two-factor lightweight privacy-preserving authentication scheme for VANET. IEEE Trans. Veh. Technol. **65**(2), 896–911 (2016). https://doi.org/10.1109/TVT.2015.2402166

35. Shao, J., Lin, X., Lu, R., Zuo, C.: A threshold anonymous authentication protocol for VANETs. IEEE Trans. Veh. Technol. **65**(3), 1711–1720 (2016). https://doi.org/10.1109/TVT.2015.240 5853

36. Zhang, L., Meng, X., Choo, K.K.R., Zhang, Y., Dai, F.: Privacy-preserving cloud establishment and data dissemination scheme for vehicular cloud. IEEE Trans. Dependable Secure Comput. **17**(3), 634–647 (2020). https://doi.org/10.1109/TDSC.2018.2797190

37. Ahmad, F., Kurugollu, F., Adnane, A., Hussain, R., Hussain, F.: MARINE: man-in-the-middle attack resistant trust model in connected vehicles. IEEE Internet of Things Journal **7**(4), 3310–3322 (2020). https://doi.org/10.1109/JIOT.2020.2967568

38. He, D., Zeadally, S., Xu, B., Huang, X.: An efficient identity-based conditional privacy-preserving authentication scheme for vehicular Ad Hoc networks. IEEE Trans. Inf. Forensics Secur. **10**(12), 2681–2691 (2015). https://doi.org/10.1109/TIFS.2015.2473820

Routing Protocols

VASNET Routing Protocol in Crisis Scenario Based on Carrier Vehicle

Grace Khayat, Constandinos X. Mavromoustakis, George Mastorakis,
Hoda Maalouf, Jordi Mongay Batalla, Evangelos Pallis, Naercio Magaia,
and Evangelos K. Markakis

Abstract This chapter will tackle one of the most important disadvantages that
will take place in the Vehicular Ad Hoc Network during a crisis scenario. Vehicular
Ad Hoc Network which is abbreviated as VANET is a variation of Mobile Ad Hoc
Network (MANET). VANET routing protocols are a wide research area due to their
different classifications with the pros and cons of each. This chapter will summarize
the five VANET routing protocols categories. The common criteria among those
five-routing protocols are the requirement of infrastructure which is also known as

G. Khayat (✉) · C. X. Mavromoustakis
Department of Computer Science, University of Nicosia, Nicosia, Cyprus
e-mail: khayat.g@live.unic.ac.cy

C. X. Mavromoustakis
e-mail: mavromoustakis.c@unic.ac.cy

G. Mastorakis
Department of Management Science and Technology, Hellenic Mediterranean University,
Agios Nikolaos, Crete, Greece
e-mail: gmastorakis@hmu.gr

H. Maalouf
Department of Computer Science, Notre Dame University – Louaize, Zouk Mosbeh, Lebanon
e-mail: hmaalouf@ndu.edu.lb

J. M. Batalla
National Institute of Telecommunications and Warsaw University of Technology,
Nowowiejska Street 15/19, Warsaw, Poland
e-mail: jordim@tele.pw.edu.pl

E. Pallis · E. K. Markakis
Department of Electrical and Computer Engineering, Hellenic Mediterranean University,
Heraklion, Crete, Greece
e-mail: pallis@hmu.gr

E. K. Markakis
e-mail: markakis@pasiphae.eu

N. Magaia
LASIGE, Department of Computer Science, University of Lisbon, Campo Grande,
1749-016 Lisbon, Portugal
e-mail: ndmagaia@ciencias.ulisboa.pt

© Springer Nature Switzerland AG 2021
N. Magaia et al. (eds.), *Intelligent Technologies for Internet of Vehicles*, Internet of Things,
https://doi.org/10.1007/978-3-030-76493-7_13

a road-side unit (RSU) or base station (BS). The role of the RSU is to provide an internet connection. Packets will be uploaded to the RSU which will, in turn, be uploaded to the internet thus making them available for future download. In a crisis scenario, one or more RSU might be disconnected from the internet which will result in a disconnected network. The disconnected network will lead to the failure of packet upload and time out leading to loss. This chapter will propose a protocol to ensure a successful packet delivery with a disconnected RSU from the internet. The network is considered as a set of sensors that periodically upload data to the RSU which in turn uploads them to the internet. The sensors might be those of Wireless Sensor Network (WSN) or onboard sensors. The proposed protocol will take into consideration that vehicles are assumed to be capable of short-range wireless communication and hence can collect data from a nearby RSU. The main target of the proposed protocol is to ensure a successful data transfer during the time a certain vehicle known as the carrier vehicle is within the communication range of a road-side unit. A simulation carried in MATLAB studied the different effects of network parameters on the successful data transfer from the disconnected RSU to the carrier vehicle.

1 VANET Introduction

Vehicular Ad Hoc Network (VANET) is a special type of Ad hoc networks characterized by high mobility, self-organization, distributed communication, road pattern restrictions, and unbounded network. In VANET, the contributors in packet routing are vehicles or access points. The access points are also known as infrastructure since they are fixed and connected to the internet. The packets usually contain data about safety such as traffic, fire, or flooding. The data might also be about weather conditions or environmental conditions in addition to many other fields of interest.

VANET communication is split into two types [1]:

- Vehicle to Vehicle communication (V2V) where vehicles communicate among themselves directly. The vehicles are the source, destination, and intermediate hops.
- Vehicle to Infrastructure communication (V2I) where vehicles communicate with fixed equipment at the side of the road known as a road-side unit.

With technology evolvement, a new type of communication was added to VANET's communication family, Vehicle to Everything (V2X). In V2X the vehicle can communicate with any device having a transceiver and wireless communication capabilities such as mobile phones, tablets, smart TVs, smartwatches, sensors, satellites, etc. This new type of communication is the result of the evolvement of the Internet of Things (IoT) where anything that is connected to the internet can send and receive data [2] (Fig. 1).

VANET's vehicles have wireless transceiver and data sharing functionalities which will lead to a dynamic network. The vehicles might be the sender, the receiver, or even a router that will route the packets to their destinations. VANET applications

Fig. 1 VANET
communication

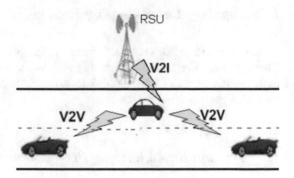

effectiveness mainly depend on packet routing with at least one intermediate node. Routing involves two activities which are determining the optimal path and transferring the information as packets. Packet transfer may take place in a single hop or multiple hops. Some applications in VANET present data that should be transferred without any delay such as road traffic conditions, road condition warnings, and weather information. Hence a major requirement in VANET networks is connectivity and ensure no packet loss with as minimal delay as possible [3].

The VANET basic architecture consists of vehicles connected wirelessly. The vehicles rely on infrastructure to ensure an end to end communication. The infrastructure is known as Roadside Units abbreviated as RSU. In a crisis case scenario, one or more of the RSUs might be nonfunctional thus resulting in a disconnected network. In this chapter, we will target ensuring an end to end communication by using one of the vehicles. The vehicle selection will be based on the protocol to be discussed below. This vehicle will take a copy of the data from the disconnected RSU and carry it to be uploaded to the first functional RSU it will pass across.

Sensors had been mentioned as a source of data to be transmitted to the base station or what is also known as RSU in the case of VANET. Therefore, in the next section, we will discuss wireless sensor network which is abbreviated as WSN.

Routing protocols target to find the best path for packet routing with the least communication time and network resources. Many challenges exist in VANET routing design such as security, privacy, connectivity, and quality of services. VANET routing protocols are classified into five groups based on their routing characteristics and techniques:

- Topology based routing protocols
- Position based routing protocols
- Broadcast based routing protocols
- Geocast based routing protocols
- Cluster based routing protocols

1.1   Topology Based Routing Protocols

Topology-based routing protocols transfer data packets using link information [4]. These protocols require the knowledge of nodes' topology information. A routing table is maintained to store the link information; source to destination path; based on which packets are forwarded.

1.2   Position Based Routing Protocols or Geographic Routing

This group of protocols relies on the vehicle's geographical position. The packet's routing path is based on the destination's and neighbor's positions. Therefore, those protocols require resources like GPS (Global Positioning System) or periodic beacon messages for location detection. The source node appends a header to the packet carrying destination location information. The nodes will forward packets to the geographically closest neighbors to the destination. This process is repeated until the packet is delivered to the destination [5].

1.3   Broadcast Routing

Broadcast Routing protocols are simple and used for applications such as weather, traffic, road situation, etc. Those protocols flood the packets to all available nodes within the broadcast domain. A major drawback of this flooding is the high consumption of network bandwidth due to replicated packets [6].

1.4   Geocast Routing

Geocast routing, also known as a location-based protocol, multicasts packets from source to destination. The packets are broadcasted within a specified geographic region [7]. Like position-based routing protocols, geocast routing also relies on position determining services such as GPS.

1.5   Cluster-Based Routing

These protocols split the network into groups known as clusters. The grouping is based on vehicle characteristics such as direction or velocity. Every cluster has one cluster head which is responsible to manage communication inside and outside the cluster [8]. Each cluster header will be establishing a link between its cluster and other clusters.

2 WSN Introduction

Millions of sensor nodes can be deployed and integrated using radio frequencies systems. Also, they may be equipped with effective power scavenging methods, such as solar cells, because the sensors may be left unattended for months and even years at a time [9]. This type of network is known as Wireless Sensor Network (WSN). WSN consists of a base station and several sensor nodes. The sensor nodes are usually scattered in a sensor field. Each sensor is expected to detect events of interest that are to be uploaded to the base station. WSN has different types of sensors such as seismic, low sampling rate magnetic, thermal, visual, infrared, acoustic, and radar [10]. Sensor nodes can be used for continuous sensing, event detection, event identification, location sensing, and local control. WSN applications can be categorized into military, environment, health, home, and other commercial areas. Environmental applications include tracking movements such as traffic or monitoring environmental conditions such as forest fire or flood [11] (Fig. 2).

There exist two types of routing protocols in WSNs:

- Flat Wireless Sensor Network
- Hierarchical Wireless Sensor Network

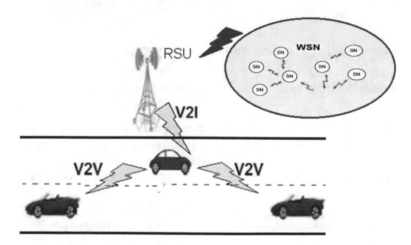

Fig. 2 VANET and WSN networks

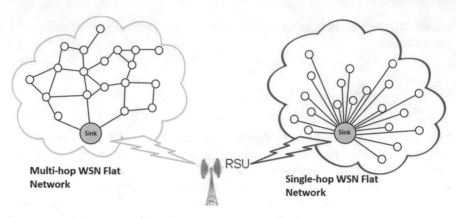

Fig. 3 WSN flat network

2.1 Flat Wireless Sensor Network

A flat wireless sensor network is a homogeneous network with identical nodes in terms of battery energy and hardware complexity [12]. One node is an exception to the mentioned characteristics. This node is known as the sink. The sink acts as a gateway responsible for forwarding the gathered information to the final user. There exist two types of WSN flat networks either multi-hop or single-hop as shown in Fig. 3 [13, 14].

2.2 Hierarchical Wireless Sensor Network

A hierarchical wireless sensor network is based on three layers of wireless devices [15]:

- A Sensor Nodes layer (SN) consists of low-power sensor nodes with limited functionalities. This layer does not offer multi-hop routing however, it routes packets via higher tier nodes. This layer is the lowest tier in the hierarchical wireless sensor networks as it relies on the upper layers to route the packets to their destination.
- A Forwarding Nodes layer (FN) consists of higher power radio forwarding nodes that route packets between radio links. This layer offers multi-hop routing capability to the nearby SN or other FN. The forwarding layer has two wireless interfaces; one communicates with lower-tier nodes (SN) and the other connects to higher tier nodes (FN and AP).
- An Access Point layer (AP) routes packets between radio links and the wired infrastructure. It is the highest tier in the network and has both wireless and wired interfaces. The wireless protocol used resembles that of 802.11b wireless LAN

Fig. 4 Hierarchical WSN

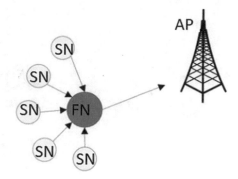

protocol. This tier provides multi-hop routing for packets from SN and FN within radio range, in addition to routing data to and from the Internet.

The hierarchal routing protocols can further be classified into two types (Fig. 4):

- 2-tier routing protocols
- 3-tier routing protocols

2.2.1   Nomadic Sensor Network (2-Tier Architecture)

The nomadic wireless networking paradigm tries to fit its network architecture into the general trend of extremely simple devices as opposed to particularly complex ones [16]. It is therefore assumed to be a 2-tier architecture, where the separation in the functionality of the nodes follows the technological constraints imposed by the size of the respective devices.

A Nomadic Sensor Network consists of [17]:

- *Sensor Nodes-tier*, composed of sparse, simple, and cheap nodes deployed in the environment, with functionalities of sensing and communicating. Those nodes are expected to be extremely low-powered, and extremely simple. Their only role will be the single-hop broadcasting of the sensed information to the nearby user nodes. They do not run any kind of complex protocols, except for short-range communication with user nodes.
- *User Nodes-tier*, corresponds to user devices such as cell phones, are assumed to be nodes of the network able to gather information from the sensor nodes deployed in the environment and to diffuse information to other user nodes in the communication range. These nodes are assumed to be capable of processing intensive operations and running complex information exchange protocols. These nodes will move in the environment because of the physical movement of the users and will collect information from sensor nodes, when in their communication range, and store this information in their device's memory.

The idea is to move the communication and computational burden on the user nodes that do not suffer from severe energy constraints as much as the sensor nodes. The

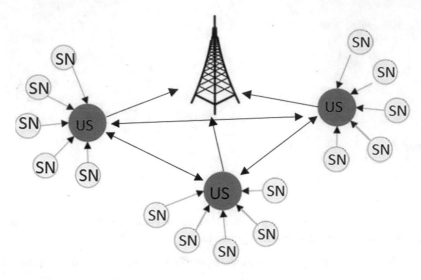

Fig. 5 Nomadic sensor network

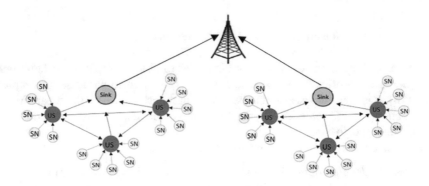

Fig. 6 WSN 3-tier network

WSN will enable a self-contained service, where the sensed information is provided to a user passing by in the proximity of the sensors. The user picks up the information then evaluates and exchanges it with other users it encounters during its movement (Fig. 5).

2.2.2 3-Tier Architecture

Given the broad and flat structure of WSNs, collecting information from the entire network at a single sink can result in immense network traffic near the sink. Therefore, a 3-tier architecture has been proposed targeting latency-tolerant applications for

WSNs [18]. Thus, the 3-tier architecture is a combination of the flat topology and the 2-tier topology (Fig. 6).

3 Vehicular Ad-Hoc and Wireless Sensor Network (VASNET)

A combination of Vehicular Ad Hoc Networks (VANET) and Wireless Sensor Networks (WSN) is termed as Vehicular Ad Hoc and Sensor Networks (VASNET). In VSANET, each vehicle acts as an intelligent node by sensing the real phenomena and collecting data. VASNET inherits its characteristics from WSN and VANET [19]. VASNET is divided into three layers [20]. The upper layer consists of monitor stations (MS) that will analyze the data received from the lower layers. Based on the analysis output a warning signal might be reported back to be broadcasted in a certain region. For example, if several high-temperature data entries had been recorded within the same region; the monitor station will automatically conclude the possibility of fire existence in that region. In such a case, the fire department within that region should be notified in addition to all inhabitants or vehicles that might have their trajectory towards the fire's region. These monitor stations are mainly connected by fiber optic cables to form the backbone of VASNET. The middle layer is also known as the region layer, consisting of the Roadside Unit or the Base Station (BS) that will be responsible for uploading the data to the monitor station. Finally, the lower layer is the field layer which consists of the sensors. These sensors might be deployed on the side of the highway known as roadside sensors (RSS) and/or onboard thus carried by the vehicles known as vehicular sensor nodes (VSN). These nodes are connected by short-range or medium-range wireless communication [21].

Therefore, VASNET consists of [20, 21]:

- Vehicular Sensor Nodes (VSN) which are carried by the vehicles example accelerometer. The sensor readings are to be sent to the base stations via short-range communication.
- Roadside Sensors (RSS) are placed at a fixed distance beside the road example temperature sensors to detect fires or humidity to report rainstorms or floods.
- Base Station (BS) is a fixed point through the roads which will aggregate all the data received from the VSN or the RSS.
- Monitor Station (MS) which will receive all the data from the BS, analyze it, and come up with conclusions. Monitor Station might require sending back alert messages to the BS in some cases such as flood detection or fire detection. Besides, the monitor station will be sending up periodic update messages required by some applications such as traffic or weather.

4   VASNET in Crisis Scenario

As Fig. 7 shows the coverage area of the RSU must overlap to ensure full coverage
for the network. In case of a crisis scenario such as fire or flooding one or more of
the RSU might be affected and thus being disconnected from the network [22]. The
disconnected RSU will result in packets undelivered to the monitor station knowing
that in such cases every single data is of high value due to the decisions to be made.

As Fig. 8 shows the RSU or the BS coverage area overlaps as per the network
design. In the crisis scenario, we are considering the middle RSU is disconnected
from the internet as an example resulting in a disconnected zone. The packets sent
from the WSN or the vehicles within the disconnected zone will fail in being uploaded
to the internet and thus not being delivered to the monitor station. The goal of the
below-proposed protocol is ensuring those packets are delivered to the monitor station
which will be crucial in analyzing the situation and taking next steps decisions. It is
highly important to clarify that even LTE/5G will rely on infrastructure to accomplish
their data transmission. This chapter is considering that the infrastructure is down
and nonfunctional in case of crisis. In this case, LTE/5G will not be a solution as
well since the eNB in the crisis scenario will not be nonfunctional.

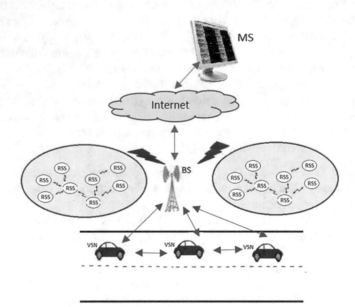

Fig. 7 VASNET

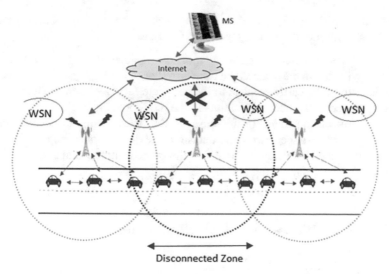

Fig. 8 VASNET in crisis scenario

5 VASNET with Dynamic RSU

The proposed protocol will be based on selecting a vehicle to carry the data from the disconnected zone to a connected one. Whenever the carrier vehicle is within the connected zone it will upload the carried packets to the RSU which will lead to successful delivery to the internet and thus to the monitoring station.

- SOS Phase: The first step in the proposed protocol is the "SOS Phase". Whenever an RSU is disconnected from the internet it should periodically broadcast SOS messages. The goal of this SOS message is to inform any vehicle passing within this RSU coverage zone that there is a disconnection state. Those SOS messages will keep on periodically being sent until the connection with the internet is reestablished.
- Handshaking Phase: Whenever a vehicle receives the "SOS Message" an "Ack SOS" reply will be sent back to the RSU which signifies the acknowledgment of SOS message reception and the will to help. The RSU upon receiving the "Ack SOS" reply will note the vehicle's id and stamp it with the current time.
- Data Transfer: The RSU will start transferring the buffered packets to the carrier vehicle. The carrier vehicle will in turn acknowledge by sending "Ack" message for the packets received. The RSU will keep track of the packets "Ack"ed to make sure that all the packets had been fully received by the carrier vehicle before leaving its coverage zone.
- Data Upload: The carrier vehicle will carry the packets received earlier until being in a connected zone again. Once in the connected zone, the carrier vehicle will upload all the carried data to the RSU, and thus clearing its buffer. As in the

previous step, the carrier vehicle will have to ensure the successful packet upload by the reception of "Ack" messages sent from the RSU.

Note that in case of an unreceived "Ack" message in either of the previous two cases the packet will stay in the buffer to be transferred or uploaded again. The successful data upload from the carrier vehicle to the connected RSU will ensure that the data will be uploaded to the internet and thus to the monitor station.

A very crucial constraint, in this case, is the delay. Minimal delay is highly accepted in crisis case scenario where decisions are dependent upon. Receiving this data with an acceptable delay; let us say in the range of minutes maximum; will be highly better than not receiving it at all.

6   VASNET with Dynamic RSU Simulation

6.1   Assumptions

The simulation for the proposed protocol was carried by MATLAB considering a network consisting of:

- Three RSU located on the side of the straight highway.
- The middle RSU is disconnected from the network.
- The three RSU are receiving data from the sensors located within the WSN.
- Vehicles with constant velocity are crossing the highway periodically.
- Several sensors are located within a predefined region known as WSN region

We will assume that the RSU in the network can be in direct contact with the passing vehicle which means they are one hop away.

6.2   Variables

Based on Sect. 6.1, Table 1 presents the different parameters used in the proposed algorithm. A subsequent explanation and usage of each of the defined parameters will be explained in Sect. 6.3.

6.3   Parameters Calculation

The distance spent by a potential carrier vehicle within the disconnected zone due to the RSU failure is noted by d. For simplicity, we will consider a straight path for the vehicles. Now, if we assume that the potential carrier vehicle moves at constant velocity (V_C) for a known distance d then time (T_{min}) which is the minimum time for

Table 1 Protocol's parameters

V_C	The velocity of the potential carrier vehicle
T_{min}	The minimum time a vehicle needs to stay within the communication range of the disconnected RSU
d	Distance within the uncovered region
R_s	Communication range of an RSU
T_{Total}	The total transfer time required by an RSU to deliver its data packets to the vehicle
d_{RC}	The distance between the RSU and the carrier vehicle

the potential vehicle to be within the coverage area of the disconnected RSU. Based on the distance, velocity, and time relationship $T = \frac{D}{V}$.

$$\text{Therefore: } T_{min} = \frac{d}{V_C} \tag{1}$$

T_{min} can also be calculated based on the packet size and transfer rate. For this calculation, we will consider the total time required for the packet to be transferred successfully from the RSU to the carrier vehicle as T_{Total}. As the packet size and the transfer rate are predefined variables in the network then we can conclude that T_{min} should satisfy the condition shown in Eq. 2:

$$T_{min} \geq T_{Total} \tag{2}$$

As the RSU coverage are is represented by a circular zone then the distance that separates the RSU from the carrier vehicle is of high importance. This parameter affects the successful packet transfer since the farther the carrier vehicle from the RSU the shortest is the time spent within the coverage area. Therefore, the next step is to calculate the minimum distance that should separate the RSU from the carrier vehicle to ensure full packet transfer.

Based on Fig. 9 and using Pythagoras famous law [22] the d_{RC} is calculated as:

$$R_s^2 = (d/2)^2 + d_{RC}^2 \tag{3}$$

Fig. 9 T_{max} calculation

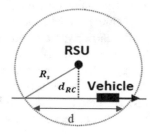

$$\Rightarrow R_s^2 = \left(\frac{V_C T_{min}}{2} \right)^2 + d_{RC}^2$$

$$\Rightarrow d_{RC}^2 = R_s^2 - \left(\frac{V_C T_{min}}{2} \right)^2$$

Thus, to be an eligible carrier node, an RSU must be at a distance d_{RC} from the carrier vehicle, where

$$d_{RC} \le \sqrt{R_s^2 - \left(\frac{V_C T_{min}}{2} \right)^2}. \tag{4}$$

And the velocity of the carrier vehicle will be:

$$V_C = \frac{2\sqrt{R_s^2 - d_{RC}^2}}{T_{min}} \tag{5}$$

In general, the distance between the RSU and the possible route the cars can take is predefined by the highway topology. Therefore, we can deduce the T_{min} as follows:

$$T_{min} = \frac{2\sqrt{R_s^2 - d_{RC}^2}}{V_C} \tag{6}$$

6.4   Results and Analysis

Based on the assumptions in Sect. 6.1 and on the theoretical formulas deduced earlier, the simulation will focus on studying the effect of each parameter on the packet delivery and network performance.

The first analysis to be done is based on Eq. 6. Figure 10 shows the minimum time required by the carrier vehicle to be spent within the coverage area of the disconnected network versus the distance d_{RC}. In the first analysis, we will consider a constant coverage radius for the RSU and study the impact on the velocity.

Figure 10 shows that as the distance d_{RC} increases the less time will be spent by the carrier vehicle within the coverage area of the disconnected RSU. This will lead to the ability of only transferring small packets respecting the time constraint $T_{Transfer} \le T_{min}$. The result also shows the impact of the velocity; as the velocity of the carrier vehicle increases less time will be spent by it in the coverage area of the disconnected RSU considering the same d_{RC}. As per the prior case, less time spent within the coverage area will lead to only successfully transferring small packets respecting the transfer time constraint.

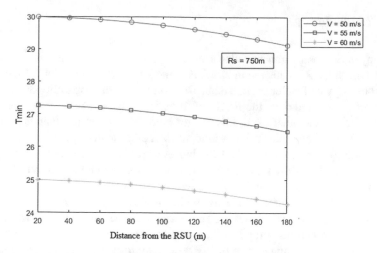

Fig. 10 T_{min} versus d_{RC} with variable vehicle velocity

The second analysis is also based on the same formula. Figure 11 studies the impact of the radius R_s having a constant velocity.

Figure 11 shows that as the distance d_{RC} increases the less time will be spent by the carrier vehicle within the coverage area of the disconnected RSU. As the disconnected coverage radius increases from 700 to 800 m with the same speed less time will be spent by the potential carrier vehicle in the coverage area. This will confine with the same prior conclusion; the transfer time is inversely proportional to both the velocity and the RSU coverage radius.

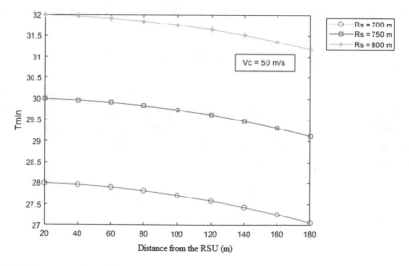

Fig. 11 T_{min} versus d_{RC} with variable disconnected RSU coverage radius

6.5   Redundancy

One copy of data being transferred to a carrier vehicle will not ensure successful data upload to both the RSU and the internet. For many reasons, the carrier vehicle might not successfully deliver the carried data. To ensure a higher probability of successful transfer of the carried data, the RSU will transfer its data to more than one carrier vehicle [23]. On the other hand, this cannot be done with no limitations otherwise the network will be flooded with redundant data and the disconnected RSU buffer will be filled by outdated data. The RSU buffer should be cleared out to receive new data as if the RSU is fully functional. This behavior should be respected in the case of disconnection or else no updated data will be received.

A new variable is added to the list which is T_{gap}. T_{gap} represents the interval time to resend the next copy. Having an interval time between copies increases the probability of successful delivery of the packet due to the dynamic changes of the VANET network. Sending copies of the same data resulted in two new constraints that should be respected by the algorithm.

Based on equation, 3 we can further conclude that

$$d = 2\sqrt{R_S^2 - d_{RC}^2} \tag{7}$$

where:

- d is the distance covered by the carrier vehicle within the coverage area of the disconnected RSU
- R_s is the radius of the coverage zone of the disconnected RSU
- d_{RC} is the distance separating the RSU from the carrier vehicle

The time spent within the coverage area of the disconnected RSU can then be deduced to be

$$T_T = \frac{2\sqrt{R_s^2 - d_{RC}^2}}{V_C} \tag{8}$$

The worst-case scenario is that the carrier vehicle enters the coverage area of the disconnected RSU with the start of the T_{gap}. Therefore, to ensure the successful packet transfer before the carrier vehicle leaves the coverage area of the disconnected RSU, the following constraint should be respected:

$$T_{gap} + T_T \leq T_{Total}. \tag{9}$$

$$\Rightarrow T_{gap} \leq T_{Total} - \frac{2\sqrt{R_s^2 - d_{RC}^2}}{V_C} \tag{10}$$

Table 2 Protocol's parameters with redundancy

V_C	The velocity of vehicles
d	Distance within the uncovered region
R_s	Communication range of an RSU
T_{Total}	The total transfer time required by an RSU to deliver its data packets to the vehicle
d_{RC}	The distance between the RSU and the carrier vehicle
T_{gap}	The time interval between the start transfer of two back to back copies of the same data
T_c	The total time spent by the carrier vehicle within the coverage area of the disconnected RSU
T_{Life}	The lifetime of the data before considering it to be outdated and thus dropped out of the disconnected RSU's buffer

where: T_{Total} is the total time required for successful packet transfer which can be calculated based on the transfer rate and packet size.

Due to the requirement of data reliability, the packet delay cannot be infinite otherwise the data is meaningless. Therefore:

$$N\left(T_{gap} - T_T\right) - T_T \leq T_{Life} \qquad (11)$$

where:

- N is the number of copies sent
- T_{Life} is the lifetime of the data perform considering it to be outdated. This highly depends on the type of data. This parameter is carried within the packet's header for tracking purposes.

As a conclusion, the updated parameters list is (Table 2):

Figure 12 shows the optimal number of copies to be transferred from the disconnected RSU to the carrier vehicle according to the variation of all other network parameters.

The target of Fig. 12 is to study the effect of the number of copies. Considering the transfer time is 3 s, for example, the figure shows a T_{life} of 10, 25, and 40 s for 2, 4 and 8 copies to be sent. As more copies are needed to be sent by the RSU the more the data must stay in the buffer thus increasing the T_{Life}. The optimal number of copies to be sent by the system highly depends on the type of data, for example, traffic packets might have a shorter lifespan if compared to packets carrying temperature data. The traffic data is time-sensitive information whereas the temperature is not at the same level of time sensitivity.

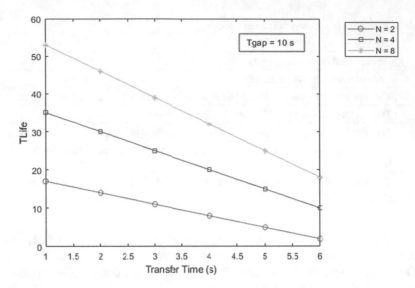

Fig. 12 T_{Life} versus the transfer time T_T

6.6 Throughput

Throughput is calculated by the ratio between the amount of data transferred and the total transfer time. Thus, the worst-case scenario for the throughput is due to the maximum transfer time for the same packet size [24].

Based on Eq. 11 the maximum transfer time is given by Eq. 12:

$$T_T = \frac{NT_{gap} - T_{Life}}{N + 1} \tag{12}$$

Therefore, the minimum throughput R for the proposed algorithm will be:

$$R_{min} = \frac{P(N + 1)}{NT_{gap} - T_{Life}} \tag{13}$$

where:

- R_{min} is the minimum throughput for the maximum transfer time due to the invertibility relationship between the two
- P is the packet size

The other parameters are already discussed in the previous sections.

Throughput is one of the most important metrics to study the system's effectiveness. Figure 13 represents the Throughput percentage versus the lifetime of the packets. The curve shows that as the lifetime of the packet increases the throughput increases since packets will not be dropped off the buffer due to timed out.

Fig. 13 Throughput % versus T_{Life}

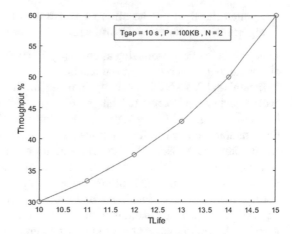

All of the above-mentioned parameters are inter-related. The decision of the optimal number of copies to be sent, the lifetime of packets, and other variable parameters highly depend on the nature of the network and the packet's information. In the same network and protocol, the number of copies to be sent varies depending on the type of data as mentioned earlier.

7 Conclusion and Future Work

This chapter introduced the Vehicular Ad Hoc Network (VANET) with its different classification types based on routing protocols. Also, this chapter introduced Wireless Sensor Networks (WSN) with its different classification types based on the network topology. VASNET is a combination of both VANET and WSN. In VASNET, the WSN sensors need to cooperate with the VANET's infrastructure to get connected to the internet thus uploading the data. As the car's technology evolved smart cars have a different variety of sensors that will act in the same way as the standalone sensors located in the WSN field. The data-driven by either the standalone sensors or vehicular sensors is of high value whenever in crisis case scenarios such as fire, flooding, windstorm, or any other life-threatening environmental situation. In such cases, one or more of the road infrastructure RSU might be disconnected from the network thus resulting in data timeout and loss. This chapter:

- Proposed a protocol ensuring that data buffered in a disconnected RSU is transmitted to a certain number N of vehicles denoted as carrier vehicles. These carrier vehicles will carry the data to the next connected RSU.
- Presented results of simulation done using MATLAB. The results obtained were analyzed to conclude the effect of several variable parameters in the algorithm. The analysis of those variables might be very helpful whenever setting up a new VASNET network.

- Spotted the light on the different constraints that should be respected by this algorithm to ensure a successful packet transfer to the carrier vehicle.

This protocol is very promising as currently, the buffered data has a high probability of upload to the internet therefore reaching the monitor station.

Traditional VANET is evolving into the Internet of Vehicles (IoV) due to the evolvement of the Internet of Things (IoT). Intelligent Transportation System (ITS) is a developing field for the automotive industries which will result in smart cities. For example, IoV can offer solutions for the flow control of traffic [25]. In 2018, Lucy Sumi and Virender Ranga proposed an algorithm to control the traffic thus helping emergency vehicles to arrive at their destination with minimal delay [26]. Similarly, the authors in [27] offered an active traffic management plan that helps in optimal route selection. Intelligent and collaborative routing can provide drivers the best routes to their destinations with minimal time and accident reduction. A self-learning algorithm such as Q-learning can be a great add-on to our proposed algorithm to find the shortest path from the source node which is the carrier vehicle to the destination node that is connected RSU. Therefore, VASNET network with artificial intelligence and machine learning technology added to the proposed algorithm is one of the optimal solutions to ensure data delivery in disconnected networks due to a crisis scenario.

Reference:

1. Khayat, G., Mavromoustakis, C.X., Mastorakis, G., Maalouf, H., Batalla, J.M.: Intelligent vehicular networking protocols. In: Convergence of Artificial Intelligence and the Internet of Things. Springer (2020)
2. Ghori, M.R., Zamli, K.Z., Quosthoni, N., Hisyam, M., Montaser, M.: Vehicular ad-hoc network (VANET): review. In: IEEE International Conference on Innovative Research and Development (ICIRD) (2018)
3. Qureshi, K.N., Abdullah, A.H.: Topology based routing protocols for VANET and their comparison with MANET. J. Theor. Appl. Inf. Technol. 58(3), 707–714 (2013)
4. Kaur, S., Gupta, A.K.: Position based routing in mobile ad-hoc networks: an overview. Int. J. Comput. Sci. Technol. 3(4), 792–796 (2014)
5. Misra, S.C., Woungang, I., Zhang, I., Misra, S.: Guide to Wireless Ad Hoc Networks. Springer, Heidelberg (2009)
6. Hamza, T., Nsiri, B.: A hybrid routing protocol for VANET using ontology. Procedia Comput. Sci. 73, 94–101 (2015)
7. Hosmani, S., Mathpati, B.: International Conference on Electrical, Electronics, Communication, Computer, and Optimization Techniques (ICEECCOT) (2017)
8. Culler, D., Estrin, D., Srivastava, M.: Overview of sensor networks. IEE Computer Society (2004)
9. Pinar, Y., Zuhair, A., Hamad, A., Resit, A., Shiva, K., Omar, A.: Wireless sensor networks (WSNs). In: IEEE Long Island Systems, Applications and Technology Conference (LISAT) (2016)
10. Garg, P.: Classification of sensors used in WSNs. Int. J. Comput. Sci. Netw. (IJCSN) 6(3), 379–383 (2017)
11. Albakri, A., Harn, L., Song, S.: Hierarchical key management scheme with probabilistic security in a wireless sensor network (WSN). Secur. Commun. Netw. 2019, 1–11 (2019)

12. Cao, N., Zhao, Y., Liang, J., Wang, T., Huang, T., Xu, D., Xu, Y.: The comparison of single-hop and LEACH protocols in wireless sensor networks. In: IEEE International Conference on Computational Science and Engineering (CSE) and IEEE International Conference on Embedded and Ubiquitous Computing (EUC) (2017)
13. Sajwan, M., Gosain, D., Sharma, A.K.: Comput. Electr. Eng. **67**, 96–113 (2018)
14. Ouafaa, I., Salah-Ddine, K., Jalal, L., Said, E.H.: Recent advances of hierarchical routing protocols for ad-hoc and wireless sensor networks: a literature survery. Bilisim Teknolojileri Dergisi (2016)
15. Guo, J., Koyuncu, E., Jafarkhani, H.: Energy efficiency in two-tiered wireless sensor networks. In: IEEE International Conference on Communications (ICC) (2017)
16. Kazemeyni, F., Johnsen, E.B., Owe, O., Balasingham, I.: MULE-based wireless sensor networks: probabilistic modeling and quantitative analysis. In: International Conference on Integrated Formal Methods (2012)
17. Carreras, I., Chlamatac, I., Woesner, H., Zhang, H.: Nomadic sensor networks. In: Proceedings of the Second European Workshop on Wireless Sensors Network (2005)
18. Jambli, M.N., Khan, A.S., Shoon, S.C.: A survey of VASNET framework to provide infrastructure-less green IoTs communications for data dissemination in search and rescue operations. J. Electron. Sci. Technol. **14**, 220–228 (2016)
19. Oiran, M., Murthy, G.R., Babu, G.P., Ahvar, E.: Total GPS-free localization protocol for vehicular ad hoc and sensor networks (VASNET). In: Thrid International Conference on Computational Intelligence, Modelling & Simulation (2011)
20. Piran, M.J., Rama Murthy, G., Praveen Babu, G.: Vehicular ad hoc and sensor networks; principles and challenges. Int. J. Ad Hoc Sens. Ubiquitous Comput. (UASUC) **2**(2), 38–49 (2011)
21. Khayat, G., Mavromoustakis, C.X., Mastorakis, G., Maalouf, H., Batalla, J.M., Mukherjee, M.: Tuning the uplink success probability in damaged critical infrastructure for VANETs. In: IEEE International Workshop on Computer Aided Modeling and Design of Communication Links and Networks (2019)
22. Stando, J., Lukawska, G.G., Guncaga, J.: From the Pythagorean theorem to the definition of the derivative function. In: International Conference on E-Learning and E-Technologies in Education (ICEEE) (2012)
23. Khayat, G., Mavromoustakis, C.X., Mastorakis, G., Maalouf, H., Batalla, J.M., Mukherjee, M., Pallis, E.: Retransmission-based successful delivery tuning in damaged critical infrastructures for VANETs. In: IEEE International Conference on Communications (ICC) (2020)
24. Minihi, R.N., Al, H.M., Sabbagh, H.-R., Al-Omary, A.: End-to-end throughput for VANET with and without cloud effect. Transp. Telecommun. **20**(1), 52–61 (2019)
25. Tolba, A.: Content accessibility preference approach for improving service optimality in internet of vehicles. Comput. Netw. **152**, 78–86 (2019)
26. Sumi, L., Ranga, V.: An IoT-VANET-based traffic management system for emergency vehicles in a smart city. In: Recent Findings in Intelligent Computing Techniques, pp. 23–31. Springer (2018)
27. Kumar, S., Singh, J.: Internet of vehicles over Vanets: smart and secure communication using IoT. Scalable Comput. Pract. Exp. **21**(3), 425–440 (2020)

Hybrid Swarm-Based Geographic VDTN Routing

Youcef Azzoug, Abdelmadjid Boukra, and Vasco N. G. J. Soares

Abstract Vehicular Delay Tolerant Network (VDTN) routing is referred to the hybridization of Delay Tolerant Networks (DTNs) with VANETs which mobilizes both knowledge-based and geography-based forwarding techniques. Numerous shortage are stated in existing VDTN routing protocols in both modes exposes such as the inaccurate location information and uncontrolled congestion due to bundles flooding. In this paper, we introduce a hybrid VDTN routing strategy combining a swarm-inspired algorithm, namely the Firefly Algorithm (FA) to enhance the decision-making of finding better next Store-Carry-and-Forward (SCF) relay vehicle in accordance with the use of geographical forwarding for better localization of closer nodes to the destination. Thus, the flooding of bundles is controlled by the movement of fireflies in early routing stages, then a reliable geographic routing is followed to better track closer SCF vehicles toward bundle's destination. We implement our approach using the Opportunistic Network Environment (ONE) simulator and compare it with few common DTN routers, namely Spray-and-Wait (SnW), ProPHET and Epidemic (ER) routers; the simulation results shows superior balance between average delivery delays and delivery probability with a reasonable overheads ratio and flooding levels.

Y. Azzoug (✉) · A. Boukra
USTHB Informatics Department, BP 32 El Alia, 16111 Bab Ezzouar, Algiers, Algeria
e-mail: aboukra@usthb.dz

V. N. G. J. Soares
Instituto de Telecomunicações, 6201-001 Covilhã, Portugal
e-mail: vasco.g.soares@ipcb.pt

Polytechnic Institute of Castelo Branco, 6000-767 Castelo Branco, Portugal

© Springer Nature Switzerland AG 2021
N. Magaia et al. (eds.), *Intelligent Technologies for Internet of Vehicles*, Internet of Things,
https://doi.org/10.1007/978-3-030-76493-7_14

Abbreviations

ACO Ant Colony Optimization
ASCF Adaptive Store-Carry-and-Forward
BRAVE Beacon-less Routing Algorithm for Vehicular Environments
DSDV Destination Sequenced Distance Vector
DTN Delay/Disruption Tolerant Network
ER Epidemic Routing
ETA Estimated Time of Arrival
FFRDV Fastest-Ferry Routing in DTN-enabled VANETs
GPS Geographic Positioning System
METD Minimum Estimated Time of Delivery
NP Nearest Point
NS Navigation System
ONE Opportunistic Network Environment simulator
ORWAR Opportunistic DTN Routing with Window-aware Adaptive Replication
PDR Packet Delivery Ratio
RSUs Road-Side Units
SCF Store-Carry-and-Forward
SnW Spray-and-Wait
TTL Time-To-Live
VDTN Vehicular Delay/Disruption Tolerant Network

1 Introduction

DTNs are partitioned wireless networks; such networks are exposed consistently to unstable connectivity due to the mobility of nodes' and changing conditions of ad-hoc communication. DTNs serve liaising sparse zones where it is nearby impossible to ensure reliable communication. DTN nodes are adapted specifically to extend the communication reach within challenged networks to cover the intermittent availability of network routers, where in order to gather a minimum message delivery, DTN mobile nodes employ a SCF routing mode by which each node stores every handled bundle from another node in its buffer cache for an undefined period until a better forwarding opportunity toward the destination arises to forward it the buffered bundle.

DTN routing has been classified following numerous criteria. For [1] DTN routing belongs to proactive and reactive styles. According to other taxonomies like in [2] DTN routing protocols are categorized in three main types which are: replication-based, knowledge-based and coding-based routing. The first distributes numerous copies of each message to ensure better delivery probability but exposes network resources monopolization and high overheads. For the second mode, bundle forwarding depends on available knowledge of network topology information to ensure better relay nodes selection; location-based services and source routing are better examples of knowledge-based category. The third mode uses different kinds of coding techniques to encrypt forwarded or flooded bundles. For [3] DTN routing is approached to single-copy and multiple-copy forwarding modes.

DTNs have been introduced for vehicular applications that extend classic VANETs applications to critical situations the case of safety applications, driving assistance, and minimum Internet access within mountainous and rural regions. So, a novel routing field, named VDTNs appeared since the need to assist communication in sparse networks when vehicles need to perform in DTN mode [4]. VDTN routing is the adaptation of DTN routing to meet the requirements of sparse VANET communication constraints, where vehicle-based DTNs are characterized by few differences with DTNs as the elimination of node's energy limitations, regular contacts between vehicles become tighter, VDTNs suffer VANETs' high mobility constraints and radio obstacles in urban scenarios [5]. The default SCF principle is applied similarly for VDTN routing by seeking a better relay vehicle that move closer towards the bundle's destination comparing to the current bundle's holder, comparing to the amongst available neighboring vehicles. Otherwise, the bundle host vehicle continues to carry the bundle [6].

Both geographic routing [7] and bio-inspired routing [8] have been introduced to DTN routing seen their utility on improving the SCF selection process which has the major impact of routing quality of any DTN router. In this book chapter, we introduce a hybrid notion of VDTN routing by combining the geographic DTN routing with the swarm-inspired SCF routing to enhance the quality of routing in VDTNs. The advantage of using geographic location of vehicles which resides in figuring out SCF vehicles moving closer to each bundle's destination is reinforced by seeking better reliable vehicles considering their expected trajectories and relative speed toward each bundle's destination. For swarm computation [9], it has been positively used for conventional DTN routing through the exposed literature in following section and the challenge of spreading this notion in VDTNs is still asking for more contribution with the support of bio-metaheuristic techniques.

The following of this book chapter is organized as coming: Sect. 2 spreads notable DTN protocols implemented for VANETs including adapted DTN routers, DTN-modified schemes for VDTNs moreover to the state-of-art of bio-inspired DTN routing techniques. Section 3 opens out on the discussed literature drawbacks and challenges. Section 4 introduces our proposition's principles and details the suggested hybrid swarm-based geographic VDTN solution. Section 5 demonstrates the simulations results of our approach comparing to few classic DTN protocols. Finally, Sect. 6 finalizes with the conclusion and the perspectives of this contribution.

2 Literature Review

DTN protocols take mobile nodes and network constraints such as buffer management, energy limitation and bandwidth consumption. Numerous taxonomies have discussed DTN routing: According to [2] three DTN routing modes exist, in this case: the flooding-based knowledge-based and coding-based routing. The first is based on the replication of uncountable number of copies for each message which usually entails network congestion. The second is a topology-based routing that seeks shortened path to destinations. The third uses coding techniques to encrypt messages. [10] discusses geography-based routing in DTNs that came to improve mobile information freshness noticed degradable in topology-based paradigm due to intermittent connectivity. Geography-based DTN forwarding is split to three categories namely: destination awareness, destination unawareness and hybrid form. The first require destination's location for forwarding operation. The second, proceeds forwarding regardless any information on destination's position. The third adopts the second strategy in early forwarding stages when destination's position is unknown while recovers for the first mode when such information in proximal from historical geographic information.

Amongst the notable early DTN protocols we mention:

Epidemic Router (ER) [11] the first conceived DTN protocol, a flooding-based algorithm that distributes an uncountable number of message copies to its contactable relay neighbors in the network. ER nodes exchange messages list (summary vector) when coming in contact with each other, the initiative is taken by the node having the smaller ID. The latter compare its vector with the received node to differentiate the bundles required for itself from the ones needed by its neighbor. First-time received bundles received from new neighbors are extracted and saved with their IDs, whereas the messages that do not exist in the neighbor's vector are sent to it.

Spray-and-Wait (SnW) [12] another flooding-based protocol that improves ER router. SnW liaises two forwarding phases: the first named spray phase in which the number of propagated messages by the source is limited to L copies sent to L different node considering that they do not have a genuine copy of the message. The wait phase is proceeded if the destination is not included between those L nodes. Then, each node keeps its copy until getting in contact with the destination as perform in Direct Delivery [36].

Probabilistic Routing Protocol using History of Encounters and Transitivity (PRoPHET) [13] the most well-known protocol DTN routing, a greedy prediction-based routing protocol based on delivery predictability (DP) concept which consists of implementing the encounter predictability which quantifies the encountering nodes' capacity to disseminate bundles (messages) to their destinations. A reply message is sent back in unicast mode to the encountering node in case it has a higher predictability value than the vehicle holding the bundle. The DP's evolution depends on three factors which are: direct probability [Eq. 1.1], transitivity [Eq. 1.2] and data aging [Eq. 1.3]. The first capitalizes the DP between two direct connections given A and B while the second is the liaison between two indirectly-connected nodes given A and C passing by a direct in-between node given B. the third factor tracks the DP's temporal evolution. Every PRoPHET node holds a meeting probability list regrouping contacts with every known vehicle. This policy seeks retrieving closer relay nodes toward for each bundle's destination which serves to shorten delivery delays.

$$P(a, b) = P(a, b)_{old} + (1 - P(a, b)_{old}) \times P_{ini} \qquad (1.1)$$

$$P(a, b) = P(a, b)_{old} \times \gamma k \qquad (1.2)$$

$$P(a, c) = P(a, c)_{old} + (1 - P(a, c)_{old}) \times P(a, b) \times P(b, c) \times \beta \qquad (1.3)$$

MaxProp [14] another reputed prediction-based protocol which relies on adopting a buffer priority classification to manage better the limited buffer storage capacity so that less important bundles are dropped when the buffer is full and the bundles with better opportunities are relayed. Thus, the bundles to drop and the others ones to forward are separated by referring a cost value affected for each destination. The priority is a probability-calculated formula that counts: contact with the destination, a hop list that store passed nodes by the bundle to locate its aging, which increases with hop counts. Thus, MaxProp prioritizes newer messages for dissemination. MaxProp includes the meeting probabilities when bundles are exchanged basing on the number of encounters between node pairs.

For notable knowledge-based DTN protocols we found:

Opportunistic DTN Routing with Window-aware Adaptive Replication (ORWAR) [15] a knowledge-based DTN protocol, energy-efficient model that seeks reducing overheads and bandwidth consumption by limiting the number of distributed bundle copies plus a controlled replication. This protocol exploits the speed, direction, location and radio range of nodes to approach the size of contact window which refers to the ratio size/bit of bundles. The latter metric influences the probability of partially transmitted messages thus low-sized messages are prioritized for transmission to exploit better the bandwidth.

Few enhanced versions of aforementioned versions came to fulfil their lacks, managing buffer size, reducing overheads, minimizing transmission duration and others. As examples we cite:

PRoPHET-based SnW [16] applies the probability-based concept on SnW with a buffer management strategy to control better buffer overflow. The spray phase performs ProPHET forwarding i.e. sending bundles to relay nodes having higher delivery probability (DP) for the related destination considering L copies for each bundle. L is reduced to 1 which marks the end of spray phase. For the wait phase, each node contacting a potential relay node exchange summary vector, and gives the only copy of each message in spray phase to it in case this neighbor has higher DP for the message's destination. The operation is repeated until meeting the destination. Moreover, received bundles' size is checked for coping with available space in buffer: bundle acknowledgment is sent to nodes asking destination's buffer size in order to forward each bundle, so the destination accept receiving the bundle if it has enough buffer space. This mechanism helps to raise up delivery ratio.

Grouping-Epidemic (G-ER) [17] a hybrid protocol designed for group mobility that combines table-driven routing for intra-group forwarding with the DTN forwarding between groups (inter-grouping). In G-ER, nodes showing strong connectivity as having similar motion and velocity act as one vehicle. Bundles forwarding inside each group is performed in a proactive paradigm using Destination Sequenced Distance Vector (DSDV) based bundles exchanging. This helps to reduces received bundles' copies to this group. Each group maintains one copy for each bundle and all summary vectors (SVs) are shared between nodes. For inter-group forwarding, each node has its group summary vector (GSV) which records messages contained in all group nodes' SVs. Inter-group $GSVs$ are exchanged periodically between their nodes. For outer group nodes, two-way $GSVs$ exchange is performed to reach more passed bundles between groups which speeds up relaying messages. The group ID is recorded in each bundle to avoid sending it numerous times to the same external groups. Two strategies are set for buffer management: the first for inter-group forwarding privileges messages passing by a smaller number of groups considering that bundles having longer passed grouped chain are sufficiently distributed in network thus, more likely to reach their destinations. The second for intra-group forwarding, messages are less buffered since are easier for reaching their destinations. An intra-group buffer sharing is set as well to exploit better available space in group buffers to reduce dropped bundles due to some fully buffers and their size limits.

Improved PRoPHET [18] is a hybrid version combining Epidemic and default ProPHET. The first used in early stages of bundles delivery cycle to speed up dissemination. In advanced phases, a switching mechanism to ProPHET is adopted based on two predefined thresholds for any message m named Hm (hop counter limit) and Nm (forwarded copies' count limit) where ProPHET is set when either Hm or Nm are surpassed in route. ProPHET is proceeded when it is considered that there are enough message copies distributed after a minimum distance from the source.

Vehicular DTN routing has emerged as an extension of DTN applications on VANETs exploiting the characteristics of Vehicular nodes on applying SCF principle on data routing on VANET environments. [19] discusses the main differences between DTN and VDTN routing reduced in five points: vehicular applications, Geographic Positioning System (GPS) awareness, frequent disconnection (high mobility), mobility forecasting and unlimited computational and storage abilities. VDTN routing have been introduced for topology-based, geography-based and cluster-based routing: [20] VDTN routing protocols are ranged into five classes namely: flooding-based, probability-based, information-based, infrastructure-based and incentive-based. For [4] DTN forwarding policies is differentiated based on six metrics which are: location information (GPS-assisted), infrastructure-assisted (RSUs), testing implementation (simulated environments), topology assumptions, target and forwarding metrics. VDTN routing are unattached from some DTN routing constraints notably the buffer storage and node energy limitations. VDTN protocols are usually geography-based taking advantage of GPS-assisted up-to-date location information of moving vehicles and awareness of readable mobility patterns in highways and cities moreover to radio obstacles.

For conventional VDTN routing, the related literature brought through numerous surveys different classifications, exposing challenges and the related realized forwarding policies, the case with [19] which discussed the exposed challenges to refine VDTN routing as mobility prediction and computation capabilities. [4] furnishes also a comparative analysis of VDTNs protocols following several performance indicators and follows with the major VDTN routing challenges such as the buffer management and the acquisition of vehicle location information. [21] cited the notable VDTN routers integrating the greedy forwarding, traffic awareness, real map-based routing and recovery routing. [22] associates the VANET routing challenges to the VDTNs evolution figuring the buffer management policies, messages lifetime and SCF contacts' unpredictability, and ranges VDTN routers into three classes: native replication, utility forwarding and hybrid replication- forwarding approach.

VDTN routing has taken over a decade since most classic DTN protocols as Epidemic and SnW have been applied to VANETs seen their independency from Vehicles' energy constraint which provide better performances. A generality of notable VDTN routing protocols is spread in almost all related literature surveys the case of [4] giving numerous examples DTN routers' versions for VANETs, namely the Probabilistic Bundle Relaying Scheme (PBRS) and Adaptive CSF (ASCF). Other instances of conventional VDTN routers are detailed in [21]. We open out a window of the remarkable works within the conventional VDTN routing as is spread ahead:

The evolution of DTN routing in VANETs is quite considerable since we can see numerous protocols realized to assist VDTNs such as:

Beacon-less Routing Algorithm for Vehicular Environments (BRAVE) [23] uses a beacon-less mechanism to improve the next-hop selection which is a dynamic process since each junction node can change the carried bundle's direction where two junctions are selected to cover the forwarding decision on each intermediary hop, performed using Dijkstra algorithm. With BRAVE, there is less counting on beacons particularly for nodes whose displayed radio reach is too low which does not permit enough reach for bundles. Instead, BRAVE implies four investigation bundle types to track best next-hop carrier/forwarder: $DATA$, $RESPONSE$, $SELECT$ and ACK. DATA is sent to available closer nodes to the next junction. Each node receiving this DATA reply with $RESPONSE$ indicating its speed and position toward next junction. Current bundle carrier chooses among received $RESPONSE$ bundles the vehicle providing the best report considering hop count position and speed, by sending $SELECT$ to all $RESPONSE$ senders. ACK which is sent back to bundle's carrier is used to confirm the reception of $SELECT$ indicating the good operation, otherwise the carrier vehicle retries $DATA$ broadcast twice before restarting the process in case of fail. BRAVE adjust waiting time to reduce overheads by restricting number of $RESPONSE$ to the node replying faster thus increasing reliability. Thus, a progress value is affected to each neighbor [Eq. 1.4] to judge its responding delay. On the basis of the latter variable, the waiting time value of each potential next carrier is generated in a way that prevents collision between neighbors belonging to the same zone within current hop's transmission range and prioritizes selecting higher progress values vehicle over short waiting time relay nodes as next forwarder.

$$P(n, d, c) = dist(c, d) - dist(n, d) \qquad (1.4)$$

The use of geography-based information has been profitable for VDTN routing overcome the lacks of topology-based DTN routing in either flooding-based
or knowledge-based routing to provide accurate information which improves a lot
the next-SCF node selection and message queue management. Few notable VDTN
protocols based on geographic routing are described below:

Geographical Opportunistic Routing (GeOpps) [24] the first geography-based
DTN protocol conceived basing on Navigation System (NS) assistance that provides
suggested routes for each solicited destination. This information is used to calculate
the Minimum Estimated Time of Delivery (METD) which sums up the bundle's
carrier's Estimated Time of Arrival (ETA) to the Nearest Point (NP) toward the
destination and the ETA from NP to the destination [Eq. 1.5]. GeOpps is a single-
copy algorithm mean that each vehicle carrying a bundle compares its METD with the
ones of the neighbors it encounters: if there exists a neighboring vehicle providing
a lower METD, then that neighbor holds and carries the bundle and the process
is repeated until reaching the destination or the expiration of the bundle's Time-
To-Live (TTL). METD is calculated also in case when neighbors are sharing the
same next NP. METD is calculated alternatively using average speed values to
predict vehicles with uncharacteristic routes and velocity changes. The accuracy of
METD is discussed in numerous exceptions as the driver's deviation from the initially
suggested route, vehicle's state (stopped, switched off, etc.), disrespect of suggested
route's rules, etc.

$$MET D = ET A(currentposition, NP) + ET A(NP, Destination) \quad (1.5)$$

Geographic Spray (GeoSpray) [25] a hybrid single-copy and multiple-copy
VDTN protocol that mobilizes geographical location information to make routing
decision, likewise GeOpps. Instead, it adopts a multiple-copy message retransmission
in early forwarding stages to detect alternative routes, then follows with a single-
copy replication mode to exploit more contact opportunities. GeoSpray modifies
the spray phase founded in Spray and Wait, where each source restricts replicating its bundles' copies to nodes that moves closer towards the destination's way
and/or arrive sooner. Then, every bundle carrier unicast its copy following GeOpps
policy. When each vehicles pair (X, Y) meet, they perform five (05) mutual operations: the explicit deletion of known delivered bundles of both X and Y, delivery of
stored bundles whose destination is X/Y, process of carrier choice amongst X and
Y, replication/forwarding decision making and scheduling of bundles replication.
In the third step, X calculates the METD values for Y's stored bundles and vice
versa. Then, the next step uses these values to collect lower-METDs bundles to X
(or Y) for either replication or forwarding depending on the remaining number of
each bundle's copies. The last step is set to deal with short contact opportunities and
limited bandwidth; thus, a scheduling order is set for the collected bundles, based
on increasing METDs and two tied-METD tiebreakers involving ascending number
of replications then descending remaining TTL. In GeoSpray, delivered bundles are
cleared from buffer using active receipts sent to meet vehicles which helps to save

storage area and avoids the transmission of already delivered bundles. More storage portion is allocated for bundles whose have been replicated less and are closer to their destinations.

GeoVDM [26] an enhanced geography-based VDTN protocol that adopts two parameters to define the next-hop selection policy: the first is the METD defined in GeOpps and GeoSpray but to detect vehicles moving quicker to the bundle's destination, while the second is a new metric, namely: the distance between the NP and bundle's destination $(Dist)$. METD is restricted to select the quickest next carrier, whereas Dist looks for the closest next relay vehicle to the destination. GeoVDM sets up an initially unlimited multi-copy scheme with a custody transfer mechanism, which allows a bundle carrier to relay its own bundle copy along with the responsibility of deletion or replication to the next relay vehicle in case it moves quicker to the destination. That serves limiting extra bundles replication. Custody mechanism is restricted only to the vehicle that will pass by to the final destination. Moreover, explicit ACK messages are mobilized to clear successfully delivered bundles. GeoVDM sets a beaconing process to gather fresher neighborhood information tracking particularly METDs and Dists values of all detected neighbors which helps to accelerate the next-hop selection operation.

VDTN-ToD [27] introduces a new metric for routing decision called the Trend of Delivery (ToD) based on fuzzy logic that serves to evaluate to what levels the nodes' mobility contributes to bundles delivery to a fixed destination. ToD is calculated on the basis of three parameters, namely: sense (ϖ), distance (ψ) and velocity (τ). The first (ϖ) measures the angles formed between the vehicle's way and the destination's direction; it takes four (04) possible ranges (great, good, bad and awful). The second (ψ) is bounded into four (04) intervals (very close, close, far and very far) seen the distance between the neighbor and the bundle's destination which is compared to the neighbor's transmission range. The third (τ) is calculated seen the (cos) of the previous angle θ [Eq. 1.6]. The best of collected ToD neighbors' values is compared to the current hop's ToD when three possible decisions about each bundle are made: copy of the bundle to the next vehicle, transfer of the bundle's custody to the next vehicle or keep the bundle's in the current hop's buffer. Based on ToD calculation, the future positions of vehicles can be estimated and following the difference between the real (Pr) and estimated (Pp) neighbor's position. The latter vehicle notifies its entourage with Positioning messages (mp) indicating its speed, position and mp's sending instant, when each vehicle's routing table is updated after receiving mp. ToD is used also to arrange buffer forwarding priorities where bundles having greater ToD remains longer in buffer.

$$v_{i;x} = v_i \times \cos\theta \qquad (1.6)$$

Fastest-Ferry Routing in DTN-enabled VANETs (FFRDV) [28] a geography-based VDTN algorithm that defines two phases for accelerating bundles transporting namely the ferry selection and message forwarding phases. In the first, all roads are divided to blocks, every current carrier node (IF) select the next bundle's transporter (Ferry) on the understanding that it provides faster speed, otherwise the IF carries the bundle to the next block and restarts the selection until reaching the next intersection. There, the transmission direction of bundles is either kept the same if the next intersection along the same road is nearer the destination or changed toward shorter path. The second phase seeks optimized number of Ferry hops which converges to a shorter dissemination delay which is calculated seen processing time (Tp) and transmission time (Tt): the best next Ferry provides a minimum Tp and Tt values hence the number of forwarded bundle's copies are limited.

We can figure out, seen the properties of DTNs, that VDTN routing furnishes the necessary conditions for sparse VANETs seen the complication of applying classic reactive, proactive routing or greedy forwarding in disconnected vehicular motion areas, hence the SCF principle can be very effective to fill unreachable network gaps where next SCF forwarder selection process can be optimized via nature-inspired approaches imitating the intelligent behaviors of existing living being particles in nature. That's what will be proven through this contribution and with the existing bio-inspired DTN routing protocols to show the possibility of globalizing this experience to VDTNs. Actually, there are some realized bio-inspired DTN-based routing protocols that were adapted to VANET to solve trivial DTN routing limitations such as connectivity loss and uncontrolled long delivery delays. Meanwhile, numerous propositions have been proposed for classic DTN ad-hoc routing, we describe next few examples:

CGrAnt [29] a hybrid swarm-inspired DTN routing protocol which hybridizes the Cultural Algorithm (CA) and the greedy Ant Colony Optimization (ACO) to find better positioned relaying grouped SCF message forwarders. DTN routing can be brought to multimodal optimization, where each potential solution seeks several next-SCF nodes. The CA is an EA that is formed from a social population and a belief space. CGrAnt exploits the adaptation and self-organization found in ants swarm plus the CA's ability to reutilize population experiences and control global knowledge mechanisms to discover common behavior patterns of DTN nodes. CGrAnt incorporates either basic metrics got from population space or composite metrics which are the result of other metrics manipulation.

AntProPHET [30] a hybrid ACO-inspired geographic DTN protocol whose probabilistic metric is founded from PRoPHET scheme while applies the foraging behavior of ants when seeking food nest to select better SCF nodes, seeking the improvement of delivery probability with less generated overheads. AntProPHET proceeds the PRoPHET predictability calculation method between nodes by approaching its quantification using up-to-date pheromone concentration in order to estimate better each available candidate relay node's visibility toward the routed bundle's destination. Pheromone updates between a given relaying node and each of its stored bundle destination follow the model of city-based Travel Salesman problem.

BeeAntDTN [31] another bio-inspired DTN routing scheme inspired, similarly to AntProPHET, from bees' colony to collect information and ants' colonies to seek better forwarding trajectories. Initially, each node sets up and records connectivity degree table, collects connectivity information by flooding two types of artificial bees, namely Forward bees ($Fbees$) and Return bees ($Rbees$) as request and reply exploration bundles respectively. Each host keeps updated three other data tables:

the first $TableHops$ stores three nodes offering the three highest direct connectivity degree with each relay node. The second $TableEnergy$ stores remaining energy of nodes on reception of $Fbees$ bundles. The third, $TableBees$ saves path duration between source-destination pairs and their global connectivity degrees. After that, the ant swarm model is implemented to discover shorten trajectories with better relay nodes by calculating the visibility between a given neighboring SCF node and the destination with aid of regrouped $TableBees$ information. BeeAntDTN is positively compared to PRoPHET, ER and SnW routers.

Social Grouping Based Routing (SGBR) [32] a social-based DTN protocol inspiring from grouping behavior stated in taxi cars to help reducing the amount of flooded bundle copies and reduce delivery delays. SGBR is based on the meeting frequency between nodes to build progressively the connectivity predictability between a node and a group of nodes. Following this approach, nodes restrict sending bundle copies to nodes of other groups using binary SnW while keep selective single-copy forwarding inside the group since the exchanges between intra-group nodes increases overheads without advancing bundles to their destinations hence the delivery delays are not shortened. SGBR builds the historic of meeting between two nodes based on last meeting time to update their connectivity degree on which nodes are regrouped. A predefined connectivity threshold is set to compare connectivity degrees in order to take forwarding decision while a dropping threshold is set to avoid forwarding bundles to remote external nodes. SGBR performs with better Packet Delivery Ratio (PDR) and reduced delivery delays and flooded bundle copies compared to some common DTN routers.

3 Critics

The nature-inspired optimization approaches are exposed to a huge challenge under VDTNs on adapting SCF forwarding on sparse vehicular mobility scenarios. Considering the different VDTN routing categories figuring flooding-based, prediction-based, geography-based, and knowledge-based routing, swarm-based metaheuristics can be adapted to SCF routing and it have been proven in the realized nature-inspired geography-based routing protocols including the hybrid greedy-v2v DTN routing. According to our study, the swarm-based optimization algorithms have not been involved in VDTN routing yet, except human-inspired behaviors which have been tried for VDTN routing, this opens a wide research area that can be effective for supporting sparse VANET routing.

We notice that every version of DTN routers has its own flaws; we detail in the following points the major drawbacks of discussed VDTN routing literature by each category figuring classic DTN routing, DTN-enhanced routers, geographic DTN/VDTN routers, knowledge-based VDTN routers and bio-inspired DTN routers:

For personalizing each classic protocol's flaws, we spread few cases below:

- In SnW, flooding a static number of copies for each message is not adapted to each destination since it can either not suffice to reach distant destinations or causing unnecessary network congestion for reachable destinations.

 For DTN enhanced versions:

- The hybridization of ProPHET and Epidemic found in Improved ProPHET does not consider certain influencing parameters such as the network density, message flooding speed (number of copies versus time) or message moving speed (hop count versus time). Thus, the switching factors (Hm and Nm) need to be dynamic to exploit ProPHET and Epidemic without altering forwarding delays and delivery ratios.
- Grouping notion suggested in G-ER is a similar approach of cluster-based routing and can be fluent to numerous constraints like: group longevity against either drastic change of network density or sparse areas for longer-time, frequent intra-group updates especially for larger group and unhandled conformity of buffer information freshness, etc.
- Inter-grouped bundles dissemination priority of G-ER is not adapted to the remaining distance to the destination, where far destinations needs fast replication while too close ones need longer carrying time.
- Sorting bundles whose destination is known on distance is not effective to find the optimum transmission priority, other parameters like hop count, buffer residence time or bundle's size should be considered for priority calculation.

VDTN routing including DTN-originated protocols adjusted for VANETs improved numerous aspects in intermittent network routing, but still considerable progression still to be realized:

- The tendency of few geography-based protocols to balance for either single-copy multicopy forwarding as the case of GeOpps and GeoVDM respectively. GeoSpray came with the hybridization of the two modes supported by the notion of geographic NP which is a context for enhancement in this chapter.
- Single-copy geography-based routing (GeOpps) and other protocols lacks exploiting further unknown routing path alternatives, whereas multi-copy forwarding replicates too many bundles copies which wastes network resources especially during late stages of routing cycle.
- Position prediction of vehicle proposed in VDTN-ToD necessitates too frequent verifications which entails a congested computational complexity for VDTN nodes.
- METD metric, suggested in GeOpps and GeoSpray, suffer inaccuracy when exposed to numerous constraints like long urban trajectories, traffic lights, NP recalculation, static vehicles, disconnected streets, among others.
- Majority of geography-based routing protocols are GPS-dependent and intersection-related algorithms restricting forwarding policies on urban mobility models where routing decision is made on intersections and junctions.

The bio-inspired DTN routing is relatively emergent and still to spread particularly for VANETs. Thus, realized protocols as AntProPHET, SGBR and CGrAnt.

- The realized human-based DTN routing algorithms are task-specific approaches concentrating on the flooding enhancement or next-SCF node selection, while the bio-inspired approaches should have a global impact on the basic VDTN routing tasks.
- The use of human-based behavior is exposed to be improved by numerous bio-metaheuristics having better stochastic abilities in tracking better relays nodes and managing buffer cache.

4 Swarm-VDTN Solution Details

We considered the limitation of discussed VDTN literature to define a general strategy for managing VDTN routing under heterogeneous sparse mobility scenarios removing dependency from Road-Side Units (RSUs), avoiding intersection-based routing decision and impacting routing parts related to next-SCF node selection particularly the buffer priority management, vehicle's position prediction and contact exchanges. The proposed solution seeks with integration of nature-inspired optimization approaches, introducing new notion which consists of a swarm-assisted hybrid VDTN router combining both geography-based and knowledge-based routing aspects:

- Combining the qualities gathered from the GPS-implemented VDTN routers, prediction-based forwarding the mobilization of knowledge of network topology information and quality of relay vehicles in flooding customization and geographic forwarding strategies.
- Exploiting the stochastic behavior of swarm computation to improve numerous aspects in VDTN forwarding particularly the next-SCF vehicle selection, the optimal number of bundle's replicated copies and the buffer time of stored bundles.

It is clear that MANET-adjusted, classic and enhanced VDTN-destined protocols suffer from numerous lacks especially network-flooded routing, bandwidth consumption, buffer capacity limitations, etc. VDTN protocols are free from energy limitations and distances. Meanwhile, it is important to define dissemination priority of bundles which has a direct impact on delivery delays, since the more important swarm-evaluated bundles are forwarded or replicated in first priority. That's further to the optimal exploitation of contact opportunities in routing decision which helps in reducing flooded bundle copies. This is done by:

- Reducing the number of disseminated bundle copies by avoiding basic flooding in early stages using a natural particle-inspired technique, when minimum network density allows to avoid the conventional replication-based routing.

- Finding local-optimum relay nodes using an intelligent approach as the swarm-based technique, which does not select forcibly vehicles having the closest position, highest speed or lowest direction angle toward the destination.
- Switching the geography-based routing by providing location-assisted information of nodes' position using GPS when required in particular stages of forwarding phase.
- Extending the probabilistic DTN forwarding using the notion intelligent nature-inspired routing approaches supported by the routing and neighborhood historic of each node.

The proposed Hybrid Swarm-Geographic Vehicular DTN Routing (HSGVDR) solution whose details are given in this section is given to respond all these suggestions:

4.1 Solution Principles

HSGVDR is a swarm-inspired metaphor approach combined with a hybrid geographic knowledge-based VDTN routing. HSGVDR is conceived mainly to optimize next-SCF vehicle selection to cope with intermittent connectivity and constraining vehicular mobility conditions by implementing an intelligent metaphor-inspired metaheuristic in this case the Firefly Algorithm (FA). HSGVDR is mobility-independent so as it's not restricted to urban scenario and conceived for a variety of vehicular mobility patterns.

This protocol exploits the advantages provided by both knowledge-based and geography-based routing information, where the FA is implemented for the knowledge-based routing mode while the geography-based forwarding is based on an evaluation metric called the Reliable Estimated Time of Delivery (RETD). A switching system between the geographic forwarding and the topology-based mode is performed.

HSGVDR's SCF forwarding is based on two strategies:

- Swarm-inspired knowledge-based approach: this mode is used in the early stages of the routing cycle to localize better positioned vehicle toward each buffered bundle's destination. This phase is based on the knowledge of vehicle historic and flooding information.
- Geography-based approach: this mode is enabled in the late stages of routing cycle to complete the swarm-based strategy and adapted dynamically in each vehicle for each stored bundle depending on the bundle's swarm fitness and the vehicle's geographic information figuring GPS-calculated path and direction.

The sequence of two methods in the framework of the next-SCF vehicle selection are illustrated in the Fig. 1 below:

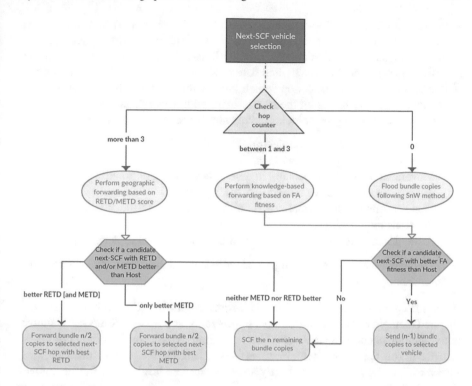

Fig. 1 HSGVDR's solution plan

4.2 Firefly Algorithm (FA)

We describe in this subsection the main properties that we need in our implemented bio-inspired approach that fit most with the characteristics of VDTN vehicles:

- The up-to-date/reactive movement of particles (speed and direction) following natural indicators (river soil, pheromone, flagella brightening). This property can be found in the water-drops (Intelligent Water-Drop) [33], PSO (Particle Swarm Optimization) [34], FA (Firefly Algorithm) [35], etc.
- The progressive grouping behavior of particles in nature during their displacements. This property is noticed in elephants (Elephant Herding Optimization) [36], ant swarm (ACO) [37], lions groups (Lion Optimization Algorithm) [38], etc.
- The intelligent distribution/dispersion of particles toward zones/directions following the changes on natural substances as pheromone, river soil, shining flagella, etc.

We believe in involving the bio-inspired optimization to improve the performances of VDTN routing by modifying several aspects tuning around the orientation of bundles dissemination, VDTN relay nodes selection and by the adaptability of bundles' replication/forwarding balance. These points are detailed in the below underlines:

- Prediction of more or less reliable suggested GPS-traced VDTN paths, by considering the properties of selected vehicles. With aid of swarm-inspired particles movement, such operation allows selecting longer-time or slower-speed paths but with longer-expected longevity and better progress toward bundles' destinations.
- Prediction of potential next-hop vehicle's carry time using its historic of bundle's average store duration using the swarm-inspired fitness evaluation.
- Anticipation of grouping vehicles having similar behaviors in term of:

  - Speed variation, traversed trajectories.
  - Closeness of targeted destinations.
  - Movement directions especially in case of less routing alternatives such as freeways, main city roads or closed network subareas.
  - Vehicles moving in normalized mobility models as traffic lights or serial of junctions/intersections.
  - Particular categories of vehicles as taxi cars and same-company buses.

- Tracking network zones with higher or more stable density to relay buffered bundles stacking in local optima.

In this contribution, we introduce the Firefly Algorithm (FA) which is a swarm-inspired metaphor conceived for global optimization problems. FA models the interactive shining phenomena that occurs between fireflies based on their light brightness which controls their movement. Fireflies uses light attraction to bring other fireflies, preys, etc. It's noticed through firefly swarms that less brightening fireflies moves toward brighter ones.

The FA is implemented to support the knowledge-based VDTN routing phase since we believe its adaptive quality to mimic the increasing forwarding quality of the SCF principle used for DTN routing. Fireflies offers the progressive characteristic of passing from a solution to a better one. This property is identical to SCF hops when forwarding or replicating bundles from source to destination vehicles. For VDTN routing, the FA seeks the anticipation of near-optimum relay SCF vehicles that speed up delivering bundles according to a list of predefined brightness parameters that seek optimizing replication/forwarding balance moreover to buffer residence time. In our solution, a modified FA is implemented to be adapted with the intermittent routing that characterizes VDTN networks.

In the FA, the light intensity [Eq. 1.7] of a given firefly (i) is proportional to its distance to another firefly (j):

$$I = \frac{I_S}{|r^2|} \tag{1.7}$$

Where: I_S the initial light intensity while r represents the distance (i, j).

The brightness value (β) reflects the intensity value (i) and is calculated based on the initial brightness (β_0) as mentioned in Eq. 1.8:

$$\beta = \beta_0 \times \exp^{-\gamma r} \tag{1.8}$$

Where γ is a predefined absorption factor and r the predefined brightness parameter.

The illustration of the FA in Fig. 2 below:

As illustrated in Fig. 2, the FA process includes six steps:

1. Initialization phase: constitutes the initial population of the firefly candidate solutions.
2. Fitness evaluation phase: a periodic fitness calculation of all candidate solution after each iteration.
3. Solution ranking: a classification of the evaluated candidate fireflies.
4. Local best extraction: selection of the best candidate from the current iteration.
5. Global best update: upgrade the global best solution from the personal best values.
6. Stopping criteria evaluation: checks the meeting of required criteria for ending the FA process.

4.3 Routing Operations of HSGVDR Protocol

We detail in this section the forwarding mechanisms adopted for HSGVDR in both strategies demonstrating the functions of the bio-inspired approaches selected for this solution:

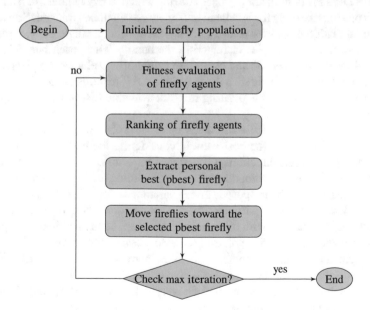

Fig. 2 Firefly Algorithm (FA) flowchart

4.3.1   Knowledge-Based Forwarding of HSGVDR

We introduce a GPS-assisted knowledge-based strategy for HSGVDR that is restricted to the early stages of routing operation. This restriction is due to the limitations of geography-based routing especially for remote destinations which is exposed to the inaccuracy of GPS-suggested paths. Consequently, incorrect METD calculation to estimate erroneous NPs can occur. Indeed, it is difficult to predict the accurate delivery time for faraway destinations due to quick mobility, frequent GPS recalculation, radio obstacles and traffic lights. Thus, it is preferable to create comfortable relay positions which will allow to apply METD more accurately. That's done with the aid of an intelligent stochastic approach the case of bio-swarm computation algorithms.

The knowledge-based routing phase is built around a progressive FA-inspired SCF selection on each node towards each bundle's destination supported by the freshness of neighborhood information collected through contact exchanges. The evolutive attractiveness between fireflies is applied on VDTN vehicles since they can share similar mobility behaviors as in speed, acceleration and deceleration varieties, directions and frequented paths where attractivity between nodes can be predicted through common targeted bundles' destinations.

Each firefly VDTN node (FV) holding a bundle seeks a next hop vehicle which has the best Attractiveness Value (AV) which is extracted relatively from a predefined Brightness Value (BV) calculated for the bundle's destination according to a quality fitness based on the below-mentioned predefined parameters' factors of the HSGVDR's FA optimization process during which, every bundle's carrier makes decision on either keeping the bundle in buffer with a calculated transmission ranking following a predefined buffer priority policy, forwarding a single copy, replicating a calculated number of copies or rejecting the bundle. Thus, each bundle's AV is calculated for the host vehicle and candidate neighboring vehicles. For each bundle, the selection decision is made seen it's the highest AV towards the bundle's destination which reflects the vehicle offering the best available forwarder to the specified destination.

Both BV an AV are calculated on the bases of probabilistic indicators on each contacts update including the host vehicle considering the node's forwarding performances (speed range, directions, GPS-traced trajectories) and its historic routing reputation figuring neighborhood reliability, forwarding efficiency and buffer management consistency. The proposed fitness parameters for BV calculation evaluate each candidate SCF node's participation degree on forwarding or replicating bundles, connectivity stability with other vehicles, current state information as position and direction and the historic evolution tendency toward final targets.

It is worth noting that AV tendency keeps an increasing evolution until delivering the bundle to its destination as the FA's challenge is to speed up the bundle's AV progress which concludes in shortened delivery delays and lower bundle loss rate.

Each VDTN node's brightness is calculated following Eq. 1.9:

$$\beta = \beta_0 \times \exp^{\frac{-\gamma}{fitness}} \tag{1.9}$$

Where:

- β is the up-to-date vehicle's brightness for the stored bundle copy
- β_0 is the initial brightness of the bundle which is set for all nodes carrying a copy of this bundle
- Fitness is the brightness fitness of the current hop's neighbor for the given bundle
- The absorption factor of our solution is calculated basing on the fitness value

We extract the vehicle attractiveness (α) [Eq. 1.10] for a given bundle from the Brightness value (β):

$$\alpha = \alpha_0 \times \exp^{-\gamma r^2} \tag{1.10}$$

Thus:

$$\alpha = \alpha_0 \times \exp^{\frac{-\gamma}{fitness^2}} \tag{1.11}$$

As mentioned above the HSGVDR introduces three knowledge-based routing metrics and one geography-based metric on which the next SCF-hop candidates are evaluated based on their attractiveness toward the bundle's destination, in this case:

1. Vehicle Efficiency (VE):
This parameter evaluates the forwarding abilities of the vehicle including the historic factors of replicated/forwarded bundles and buffered bundles information.
The VE includes: number of Relayed Bundles ($NbRB$), number of Delivered Bundles ($NbDB$), Average Lifespan of Buffered Bundles ($ALBB$) and Average Number of Bundles Copies ($ANBC$).
The $RRFB$ calculation formula is:

$$NbRB = NFB + NRB \tag{1.12}$$

Where: NFB (number of forwarded bundles) and NCB (number of replicated bundles) The VE calculation formula is:

$$VE = (0.1 \times ALBB) - (0.6 \times (NbRB + NbDB)) + (0.3 \times ANBC) \tag{1.13}$$

Considering:

- $ALBB$ indicates how much average time bundles reside in buffer; lower values show less congested buffer cache.
- $NbRB$ indicates the number of forwarded bundles including forwarded bundles (1-copy disseminations) and replicated bundles (n-copies disseminations); higher values show better delivery abilities.
- $NbDB$ indicates the number of delivered bundles; higher values show better delivery abilities.

- $ANBC$ indicates the average copies of bundles in buffer; higher values show better node positioning toward destinations since number of copies reduces in late forwarding stages.

2. Vehicle Popularity (VP):
This parameter evaluates the vehicle's neighborhood stability which is a major indicator of its relaying quality.
The VP includes: Mean Degree (MD) and Average Lifespan of Active Neighbors ($ALAN$).
The MD considers the historic maximum number of active contacts (HMD) and the current degree (CD) whose formula is:

$$MD = HMD + CD \tag{1.14}$$

The VP calculation formula is:

$$VP = (0.6 \times ALAN) + (0.4 \times MD) \tag{1.15}$$

Considering:

- $ALAN$ indicates how good a given node is linked to the network seen its active neighborhood size.
- MD indicates how long a given node is linked to the network seen its average active neighborhood longevity

3. Geographic Forwarding Quality to the destination (GFQ):
This parameter reflects the vehicle's geographic positioning quality for supporting both VE and VP parameters. The GFQ includes: Relative Speed to Destination (RSD). Geographic Advancement to Destination (GAD).
The RSD is calculated based on the Direction Angle to Destination (DAD) as follows:

$$RSD = DAD \times CD \tag{1.16}$$

The GAD is calculated based on the candidate's METD and Euclidean distance to destination (ED) as follows:

$$GAD = \frac{ED}{METD} \tag{1.17}$$

The GFQ calculation formula is:

$$FQD = RSD \times GAD \tag{1.18}$$

Considering:

- RSD indicates ED to destination.

- GAD indicates angle formed between candidate neighbor's direction and destination's direction.

For this geography-based indicator, we estimate the profitability of a potential next-hop seen its position and behavior toward the destination which is measured by GPS position, speed difference to current carrier and candidate direction.

Based on the three predefined metrics, the fitness value [Eq. 1.19] of each neighboring vehicle to potentially hold the bundle from the current hop is deduced:

$$fitness = (\alpha \times VE) + (\beta \times VP) + (\theta \times FQD) \qquad (1.19)$$

Considering: $\alpha + \beta + \theta = 1$.
The estimated weights are affected to the predefined parameters' of the fitness formula reflecting the importance of each factor [Eq. 1.20]:

$$fitness = (0.5 \times VE) + (0.3 \times VP) + (0.2 \times FQD) \qquad (1.20)$$

We assign a predefined value for the absorption factor (γ) considering $\gamma = 0.2$.
We deduce the final formulae for calculating BV (β) of a bundle (b) in a given vehicle buffer [Eq. 1.21]:

$$\beta_b = \beta_{b0} \times \exp^{-\frac{\gamma}{fitness}} \qquad (1.21)$$

$$\beta_b = \beta_{b0} \times \exp^{-\frac{\gamma}{(0.5 \times VE) + (0.3 \times VP) + (0.2 \times FQD)}} \qquad (1.22)$$

Consequently, we extract the attractiveness value (α) of b as in Eq. 1.23:

$$\alpha_b = \alpha_{b0} \times \exp^{-\frac{\gamma}{fitness^2}} \qquad (1.23)$$

$$\alpha_b = \alpha_{b0} \times \exp^{-\frac{\gamma}{((0.5 \times VE) + (0.3 \times VP) + (0.2 \times FQD))^2}} \qquad (1.24)$$

In our solution, we refer by the initial BV (BV_0) to the bundle's current brightness in interactivity with its destination. An equal predefined constant is given to each vehicle as an initial brightness value β_0 [Eq. 1.25] and attractiveness value α_0 for all its buffered bundles ($Init\_Av$). The evolution of this value depends on each bundle's destination and the quality of its next SCF carrier nodes:

$$\beta_0 = \alpha_0 = Init_{Av} \qquad (1.25)$$

The $Init\_Av$ is affected only by the bundle generator node, whereas it will be updated by each node which will handle this bundle as next-SCF vehicle.
The pseudo-code shown in Algorithm 1 summarizes the Firefly-based SCF selection process.

Algorithm 1. HSGVDR pseudocode

1: Define CH: the current hop of bundle.
2: Define $Bdle$: the SCF message (bundle).
3: Define $Cand_{SCF}$: a candidate next-SCF vehicle of Bdle.
4: Define $Best_{NextHop}$: the best $CandSCF$ vehicle.
5: Define $Source_{Msg}$: the source vehicle of $Bdle$.
6: Define $Dest_{Msg}$: the destination vehicle of $Bdle$.
7: Define $Msg_{BRIGHTNESS}$: the stored $Bdle$ brightness fitness.
8: Define $Msg_{ATTRACTIVENESS}$: the stored $Bdle$ attractiveness fitness.
9: Define $Best_{METD}$: the vehicle with best stored METD value for $Bdle$.
10: Define $Best_{RETD}$: the vehicle with best stored RETD value for $Bdle$.
11: Define N_{LIST}: the available next-hop candidates.
12: Define Nb_{COPLES}: number of remaining copies of $Bdle$.
13: Declare constant $Init_{Brightness}$: the initial firefly brightness value
14: Declare constant $Init_{Attractiveness}$: the initial firefly attractiveness value
15: Init $Best_{NextHop}$ to CH;
16: **if** New bundle $Bdle$ forwarding request for $Dest_{Msg}$ **then**   ▷ $CH = Source_{Msg}$
17:     Init $Msg_{BRIGHTNESS}$ to $Init_{Brightness}$;
18:     Init $Msg_{ATTRACTIVENESS}$ to $Init_{Attractiveness}$;
19: **else**   ▷ CH = intermediary SCF vehicle
20:     **if** Hop count ≤ 3 **then**   ▷ Perform FA-based SCF forwarding
21:         Check N_{LIST} for $Bdle$;
22:         **while** N_{LIST} is empty **do**
23:             Enable SCF mode on CH of $Bdle$;
24:             Update active contacts of CH;
25:         **end while**
26:         **for** $Cand_{SCF}$ appear in N_{LIST} **do**
27:             $Msg_{BRIGHTNESS}$ = Calculate_brightness ($Bdle$, $Cand_{SCF}$, $Dest_{Msg}$);
28:             $Msg_{ATTRACTIVENESS}$ = Calculate_attractiveness ($Bdle$, $Cand_{SCF}$, $Dest_{Msg}$);
29:             **if** $Msg_{ATTRACTIVENESS}$ > Calculate_attractiveness ($Bdle$, CH, $DestMsg$) **then**
30:                 Update $Best_{NextHop}$ = $Cand_{SCF}$;
31:                 **if** Nb_{COPIES} of $Bdle$ > 1 **then**
32:                     Forward (Nb_{COPIES}-1) of $Bdle$ to $Best_{NextHop}$;
33:                 **else**
34:                     Enable SCF mode on CH of $Bdle$;
35:                 **end if**
36:             **else**
37:                 Enable SCF mode on CH of $Bdle$;
38:             **end if**
39:         **end for**
40:     **else**   ▷ Perform geographic SCF forwarding
41:         $Best_{RETD}$ = Calculate_RETD ($Bdle$, $Dest_{Msg}$);
42:         **if** $Best_{RETD} \neq CH$ **then**
43:             Forward (Nb_{COPIES}-1) of $Bdle$ to $Best_{RETD}$;
44:         **end if**
45:         $Best_{METD}$ = Calculate_METD ($Bdle$, $Dest_{Msg}$);
46:         **if** $Best_{METD} \neq CH$ **then**
47:             Forward (Nb_{COPIES}-1) of $Bdle$ to $Best_{METD}$;
48:         **else**
49:             Enable SCF mode on CH of $Bdle$;
50:         **end if**
51:     **end if**
52: **end if**

4.3.2 Geography-Based Forwarding of HSGVDR

HSGVDR sets a probabilistic-based geography-assisted routing which is an enhancement of GeoSpray and GeoVDM restricted for close destination respecting the above-mentioned conditions, this is by introducing the RETD as a new metric for finding the best path toward bundles' destinations; the RETD seeks improving the METD proposed in GeoSpray and GeoVDM seeking a reasonable delay with optimized delivery insurance. This routing mode uses GPS to collect vehicular position and recuperate METD values.

RETD's calculation seeks finding the best Reliable Point (RP) which is not forcibly the NP towards the destination (D) from current hop. RP considers further traffic parameters, namely:

- The distance between the candidate vehicle and the nearest point (NP) to the destination ($Dist_{Host,NPi}$).
- The distance between the NP and the destination ($Dist_{Host,NPi}$)
- The candidate vehicle's average path speed (Avg_{SPEED_PATH}).

The Delivery Reliability (DR) value is introduced to estimate the probability of vehicle to traverse its preselected path with the estimated speed. DR is calculated based on the predefined values $Dist_{Host,NPi}$ and $Dist_{NPi,D}$ [Eq. 1.26].

$$DR_i = \frac{Dist_{Host,NPi}}{Dist_{NPi,D}} \tag{1.26}$$

The DR value reflects the reliability degree of every candidate path based on the distance separating it from the NP it can reach. Less this distance is higher the DR is.

As introduced in the METD, the estimated time of arrival between the candidate vehicle (i) and selected NP_i ETA($Host, NP_i$) [Eq. 1.27], and between the latter and the destination ETA(NP_i, D) [Eq. 1.28] is based.

$$ETA_{Host,NPi} = Avg_{SPEED\_PATH} \times Dist_{Host,NPi} \tag{1.27}$$

$$ETA_{NPi,D} = Avg_{SPEED\_PATH} \times Dist_{NPi,D} \tag{1.28}$$

For each candidate NP(NP_i), the ETA is recuperated from GPS to calculate the RETD [Eq. 1.29].

$$RETD_i = DR_i \times (ETA_{Host,NPi} + ETA_{NPi,D}) \tag{1.29}$$

The RP is the candidate vehicle with highest RETD value [Eq. 1.30]:

$$RP = Max(NPs, RETDs), D \tag{1.30}$$

Then the select RP handles the bundle until the NP when another SCF selection is performed to find a new candidate node getting to a closer point to the destination.

This approach came to recover knowledge-based forwarding limits from one side, and find an effective exploitation to GPS-collected information to solve geography-based forwarding drawbacks. In VDTN routing case we mention the quick mobility of destination vehicles, recurrent location information solicitations and updates which entails usually inaccurate long routing sequences. That's further to the nature of urban areas which force considering speed and direction of vehicles.

5 Simulation and Results

5.1 Configuration of the Mobility Scenario

The proposed HSGVDR protocol is implemented using the ONE simulator, a specific network simulator for DTNs and supports VDTNs as well. The simulation tests are executed on an i5-processor Acer laptop.

HSGVDR (SwarmVDTN) is compared to SnW, GeoSpray, ProPHET and ER protocols which are available in ONE 1.4.1 version.

The simulation measured a number of VDTN performance indicators, namely:

- Average delivery delay (latency): represents the end-to-end transmission delay from bundle's source to its destination considering SCF time.
- Delivery probability: represents the PDR.
- Number of flooded bundle copies: represents the average number copies replicated for all sent bundles.
- Number of dropped bundle copies: represents the average number of copies lost for all sent bundles.
- Overheads ratio: represents the ratio of the difference between number of relayed copies and delivered copies on the number of delivered bundles.

Table 1 details the configuration of the performed simulations. The tested Helsinki scenario as shown in Fig. 3 considers three vehicular types of nodes, namely cars, buses and taxis. Each category has the adequate mobility model and the speed range which fit with its engine capacities. It is worth noting that the set mobility areas of all three categories' allow cars, buses and taxis to come cross during the simulation time since the corresponding movement trajectories have shared areas where cars, buses and taxis can exchange bundle copies. This configuration is set to ensure the variety of mobility and different condition of bundles' forwarding.

The mobility scenario, as shown in Fig. 3, used for HSGVDR's simulation experiments fits with the urban mobility patterns which offer all possible mobility trajectories and motions.

Table 2 details the density distribution for each density level: The partition of nodes is split to four categories from the three vehicular types and differentiating the

Table 1 Configuration settings of simulation tests

VDTN simulator	ONE	
City scenario	Helsinki downtown (Fig. 3)	
Simulated area's size	4.5 × 3.4 Km	
Simulation time	18000 s	
Vehicles speed ranges	Slow cars	12 to 24 Km/h
	Fast cars	20 to 50 Km/h
	Buses	0 to 30 Km/h
	Taxis	5 to 40 Km/h
Density levels (number of nodes)	Distribution of vehicles (Table 2)	
Mobility models	Cars	Shortest path map-based movement
	Buses	Bus movement
	Taxis	Map route movement
Average transmission range	25 m	
Pause time	0 to 120 s	
Bundle TTL	300 min	
Buffer sizes	Cars	30 MB
	Buses	50 MB
	Taxis	10 MB

Table 2 Distribution of vehicles

Density level	30 nodes	45 nodes	60 nodes	75 nodes	90 nodes
Fast cars	14 nodes	21 nodes	29 nodes	37 nodes	44 nodes
Slow cars	9 nodes	13 nodes	17 nodes	21 nodes	26 nodes
Bus vehicles	4 nodes	6 nodes	8 nodes	9 nodes	11 nodes
Taxi cars	3 nodes	5 nodes	6 nodes	8 nodes	9 nodes

cars into slow cars subcategory and fast cars subcategory. It is worth noting that the proposed distribution respects the real mobility distribution respecting for instance that the number cars are clearly higher than buses and taxis. Also, this configuration allows to judge the stability of HSGVDR performances comparing to other VDTN routers.

5.2 Discussion of the Results

The obtained statistics of of the tested scenario are analysed in this section below for each performance indicator:

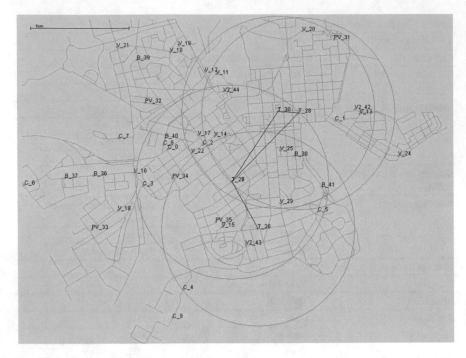

Fig. 3 Simulation of Helsinki city mobility model

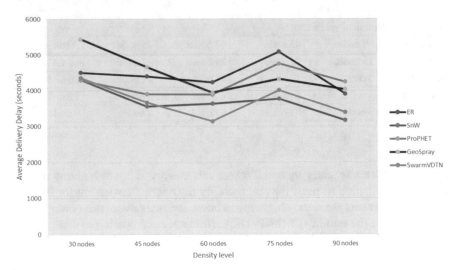

Fig. 4 Average delivery delay

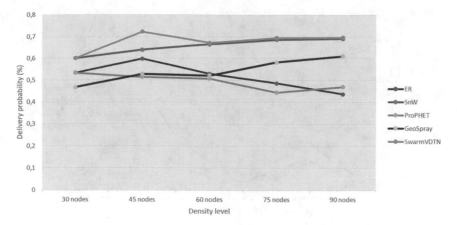

Fig. 5 Delivery probability

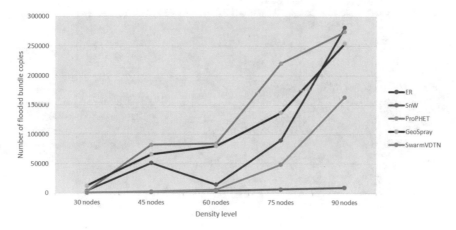

Fig. 6 Number of flooded bundles

- Graph shown in Fig. 4 demonstrates that HSGVDR offers reduced average delivery delays and performs better in middle sparse areas while it increases in either too low or high density levels. This is due mainly to lack of opportunities and congestion between close nodes respectively. Hence, HSGVDR tends to perform better in middle-range sparse networks.
- Graph shown in Fig. 5 shows that the HSGVDR offers better average delivery probability comparing to other routers while it performs closely to SnW router when density level increases. Noticing that the Swarm router shows some inconsistency in average delay and slight deterioration when density increases.
- The obtained results of flooded bundle copies as shown in Fig. 6 indicates the efficiency of HSGVDR in more sparse areas while it offers average performances in more dense areas.

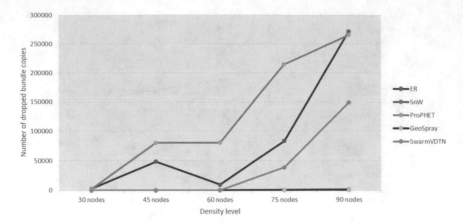

Fig. 7 Number of dropped bundles

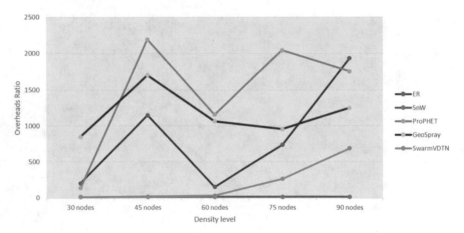

Fig. 8 Ratio of overheads

- The obtained results of dropped bundle copies as shown in Fig. 7 indicates that the number of lost copies get controlled by the number of flooded copies.
- The obtained results of overheads ratio as shown in Fig. 8 indicates the impact of number of flooded copies on increasing overheads levels. Consequently, HSGVDR get an optimized overhead ratio comparing to some multi-copy DTN router such as SnW and ER.

6 Conclusion and Perspectives

In this chapter the HSGVDR protocol for DTN-based VANETs is proposed based on firefly-inspired optimization approach. This bio-inspired VDTN router came as an extension to the human-based VDTN routing scheme to enhance the quality of routing in VDTNs with the aid of bio-inspired approaches and the geography-based forwarding. The latter is involved to support a better localization of next-SCF during advanced routing stages to follow the stochastic selection of relay nodes by the Firefly-inspired algorithm in early stages of forwarding cycle.

The HSGVDR includes a geography-based forwarding based on the METD concept introduced in GeOpps and GeoSpray to support the firefly-based SCF vehicle selection. Th si mechanism brought effective results in term of delivery delay and delivery probability optimization according to the simulated scenario.

Considering the overall simulation process results, the proposed swarm-based VDTN router opens for further enhancements notably the extension of emerging bio-inspired metaheuristics, the hybrid swarm-inspired optimization and the enhanced swarm-inspired optimization approaches in the framework of the next-SCF vehicle selection. Also, the improvement of probabilistic VDTN routing and the customization of buffer management to manage vehicles with limited energy resources in order to reduce bundle copies' loss particularly in case of drastic density changes.

Acknowledgements We present our special tributes to Professor Vasco Nuno da Gama de Jesus Soares for his great help in geographic VDTN routing concepts using ONE simulator.

References

1. Sobin, C.C., Raychoudhury, V., Marfia, G., Singl, A.: A survey of routing and data dissemination in delay tolerant networks. J. Netw. Comput. Appl. **67**, 128–146 (2016)
2. Ali, S., Qadir, J., Baig, A.: Routing protocols in delay tolerant networks - a survey. In: Paper Presented at the 6th International Conference on Emerging Technologies (ICET) (2010)
3. Soares, V.N.G.J., Isento, J.N., Dias, J.A., Silva, B.M., Rodrigues, J.J.P.C.: Solutions for Vehicular Communications: a Review (2010). https://repositorio.ipcb.pt/handle/10400.11/551
4. Ahmed, S.H., Kang, H., Kim, D.: Vehicular delay tolerant network (VDTN): routing perspectives. In: Paper Presented at the 12th IEEE Consumer Communications and Networking Conference (CCNC 2015 Workshops), Las Vegas, 2015 (2015)
5. Pereira, P.R., Casaca, A., Rodrigues, J.J.P.C., Soares, V.N.G.J., Triay, J., Cervello-Pastor, C.: From delay-tolerant networks to vehicular delay-tolerant networks. IEEE Communications Surveys & Tutorials (2014)
6. Tsuru, M., Takai, M., Kaneda, S., Agussalim, R., Aina Tsiory, R.: Towards practical store-carry-forward networking: examples and issues. IEICE Trans. Commun. **100**(1), 2–10 (2017)
7. Mahajan, R.M., Bang, A.O.: Study of various geographic position based routing protocols in VANETs. In: International Journal of Advent Research in Computer and Electronics in CONVERGENCE 2015 Conference (2015)
8. Bitam, S., Mellouk, A., Zeadally, S.: Bio-inspired routing algorithms survey for vehicular ad hoc networks. IEEE Commun. Surv. Tutorials **17**(2), 843–867 (2014)
9. Boussaïd, I., Lepagnot, J., Siarry, P.: A survey on optimization metaheuristics. Inform. Sci. **237**, 82–117 (2013)

10. Wang, T., Cao, Y., Zhou, Y., Li, P.: A survey on geographic routing protocols in delay/disruption tolerant networks. Int. J. Distrib. Sens. Netw. **12**(2), 3174670 (2016)
11. Vahdat, A., Becker, D.: Epidemic Routing for Partially-Connected Ad Hoc Networks. Technical report, Duke Computer Science (2000)
12. Spyropoulos, T., Psounis, K., Raghavendra, C.S.: Spray and wait: an efficient routing scheme for intermittently connected mobile networks. In: Proceedings of the 2005 ACM SIGCOMM Workshop on Delay-Tolerant Networking, August 2005, pp. 252–259 (2005)
13. Lindgren, A., Doria, A., Davies, E., Grasic, S.: Probabilistic Routing Protocol for Intermittently Connected Networks (2012). https://tools.ietf.org/html/rfc6693
14. Burgess, J., Gallagher, B., Jensen, D., Levine, B.N.: MaxProp: routing for vehicle-based disruption-tolerant networks. In: Proceedings of 25th IEEE International Conferences on Computer Communications (INFOCOM 2006), Barcelona (2006)
15. Sandulescu, G., Nadjm-Tehrani, S.: Opportunistic DTN routing with window-aware adaptive replication. In: Proceedings of the 4th Asian Conference on Internet Engineering (AINTEC), Bangkok (2008)
16. Yashaswini, K.N., Prabodh, C.P., et al.: Spray and wait protocol based on prophet with dynamic buffer management in delay tolerant network. Int. J. Recent Innov. Trends Comput. Commun. **5**(6), 1034–1037 (2017)
17. Xie, L.F., Chong, P.H.J., Guan, Y.L., Chong Ng, B.: G-ER: group-epidemic routing for mobile ad hoc networks with buffer sharing mechanism. In: Paper Presented at the 7th IEEE International Wireless Communications and Mobile Computing Conference. Istanbul (2011)
18. Han, S.D., Chung, Y.W.: An improved PRoPHET routing protocol in delay tolerant network. Scientific World Journal (2015)
19. Benamar, N., Singh, K.D., Benamar, M., El Ouadghiri, D., Bonnin, J.-M.: Routing protocols in vehicular delay tolerant networks: a comprehensive survey. Comput. Commun. **48**(15), 141–158 (2014)
20. Kang, H., Ahmed, S.H., Kim, D., Chung, Y.-S.: Routing protocols for vehicular delay tolerant networks: a survey. Int. J. Distrib. Sens. Netw. **11**(3), 325027 (2015)
21. Agarwal, R.K., Mathuria, M., Sharma, M.: Study of routing algorithms in DTN enabled vehicular ad-hoc network. Int. J. Comput. Appl. **159**(7), 8887 (2017)
22. Isento, J.N.G., Rodrigues, J.J.P.C., Dias, J.A.F.F., Paula, M.C.G., Vinel, A.: Vehicular delay-tolerant networks - a novel solution for vehicular communications. IEEE Intell. Transp. Syst. Mag. **5**(4), 10–19 (2013)
23. Ruiz, P.M., Cabrera, V., Martinez, J.A., Ros, F.J.: BRAVE: beacon-less routing algorithm for vehicular environments. In: 7th International Conferences on Mobile Ad-hoc and Sensor Systems (IEEE MASS 2010), San Francisco (2010)
24. Leontiadis, I., Mascolo, C.: GeOpps: geographical opportunistic routing for vehicular networks. In: 2007 International Symposium on a World of Wireless, Mobile and Multimedia Networks, Espoo (2007)
25. Soares, V.N.G.J., Rodrigues, J.J.P.C., Farahmand, F.: GeoSpray: a geographic routing protocol for vehicular delay-tolerant networks. Inform. Fusion **15**(2014), 102–113 (2014)
26. Cherif, A.H., Boussetta, K., Diaz, G., Fedoua, D.: Improving the performances of geographic vdtn routing protocols. In: Paper Presented at the IEEE 16th Annual Mediterranean Ad Hoc Networking Workshop (Med-Hoc-Net), Budva, 2017 (2017)
27. Vieira, A.S.S., Filho, J.G., Celestino Jr., J., Patel, A.: VDTN-ToD: routing protocol VANET/DTN based on trend of delivery. In: Paper Presented at the 9th Advanced International Conferences on Telecommunications (AICT 2013) (2013)
28. Yu, D., Ko, Y.-B.: FFRDV: fastest-ferry routing in DTN-enabled vehicular ad hoc networks. In: (2009) Paper Presented at the 11th IEEE International Conferences on Advanced Communication Technology (ICACT 2009), Phoenix Park (2009)
29. Vendramin, A.C.B.K., Munaretto, A., Delgado, M.R., Viana, A.C.: CGrAnt: a swarm intelligence-based routing protocol for delay tolerant networks. In: Proceedings of 14th Annual Conferences on Genetic and evolutionary computation (GECCO 2012), Philadelphia (2012)

30. Ababou, M., Elkouch, R., Bellafkih, M., Ababou, N.: AntProPHET: a new routing protocol for delay tolerant networks. In: Proceedings of IEEE 2014 Mediterranean Microwave Symposium (MMS 2014), Marrakech (2014)
31. Ababou, M., Elkouch, R., Bellafkih, M., Ababou, N.: BeeAntDTN: a nature inspired routing protocol for delay tolerant networks. In: Proceedings of IEEE 2014 Mediterranean Microwave Symposium (MMS 2014), Marrakech (2014)
32. Abdelkader, T., Naik, K., Nayak, A., Goel, N., Srivastava, V.: SGBR: a routing protocol for delay tolerant networks using social grouping. IEEE Trans. Parallel Distrib. Syst. **24**(12), 2472–2481 (2013)
33. Shah-Hosseini, H.: The intelligent water drops algorithm: a nature-inspired swarm-based optimization algorithm. Int. J. Bio-Inspired Comput. **1**(1–2), 71–79 (2009)
34. Kennedy, J., Eberhart, R.: Particle swarm optimization. In: Proceedings of IEEE International Conferences on Neural Networks (ICNN 1995), Perth (1995)
35. Yang, X.S.: Firefly algorithms for multimodal optimization. In: Paper Presented at the International Symposium on Stochastic Algorithms (SAGA 2009), pp. 169–178, Sapporo (2009)
36. Wang, G.-G.: A new metaheuristic optimisation algorithm motivated by elephant herding behaviour. Int. J. Bio-Inspired Comput. **8**(6), 394–409 (2016)
37. Dorigo, M., Birattari, M., Stutzle, T.: Ant colony optimization. IEEE Comput. Intell. Mag. **1**(4), 28–39 (2006)
38. Yazdani, M., Jolai, F.: Lion Optimization Algorithm (LOA): a nature-inspired metaheuristic algorithm. J. Computat. Des. Eng. **3**, 24–36 (2016)

SnLocate: A Location-Based Routing Protocol for Delay-Tolerant Networks

Elizabete Moreira, Naercio Magaia, Paulo Rogério Pereira,
Constandinos X. Mavromoustakis, George Mastorakis, Evangelos Pallis,
and Evangelos K. Markakis

Abstract Delay-Tolerant Networks (DTNs) are networks where there are no permanent end-to-end connections, that is, they have a variable topology, with frequent partitions in the connections. Given the dynamic characteristics of these networks, routing protocols can take advantage of dynamic information, such as the node's location, to route messages. Geolocation-based routing protocols choose the node that moves closer to the location of the message destination as the message carrier. However, such protocols suffer from obsolete location information due to node mobility and network partitions. In this chapter and conversely to the state-of-the-art, an epidemic-based decentralized localization system (i.e., DTN-Locate) and a hybrid location-based routing (i.e., SnLocate) are proposed. The former is used for disseminating node's localization information meanwhile the latter to create and route

E. Moreira · P. R. Pereira
INESC-ID, Instituto Superior Técnico, Universidade de Lisboa, 1000-029 Lisbon, Portugal
e-mail: prbp@inesc.pt

P. R. Pereira
INOV, Instituto Superior Técnico, Universidade de Lisboa, 1000-029 Lisbon, Portugal

N. Magaia (✉)
LASIGE, Departamento de Informática, Faculdade de Ciências da Universidade de Lisboa,
1749-016 Lisbon, Portugal
e-mail: ndmagaia@ciencias.ulisboa.pt

C. X. Mavromoustakis
Department of Computer Science, University of Nicosia, Nicosia, Cyprus
e-mail: mavromoustakis.c@unic.ac.cy

G. Mastorakis
Department of Management Science and Technology, Hellenic Mediterranean University,
Agios Nikolaos, Crete, Greece
e-mail: gmastorakis@hmu.gr

E. Pallis · E. K. Markakis
Department of Electrical and Computer Engineering, Hellenic Mediterranean University,
Heraklion, Crete, Greece
e-mail: pallis@pasiphae.eu

E. K. Markakis
e-mail: markakis@pasiphae.eu

© Springer Nature Switzerland AG 2021   459
N. Magaia et al. (eds.), *Intelligent Technologies for Internet of Vehicles*, Internet of Things,
https://doi.org/10.1007/978-3-030-76493-7_15

multiple copies of a message, using geographic mechanisms to disseminate them. Besides, a novel distributed contention mechanism is also proposed. The performance evaluation shows that the SnLocate protocol has a higher delivery rate and lower latency than other geographic and non-geographic routing protocols considered.

Keywords Delay-Tolerant Networks · Routing protocols · Localization system · Geographic routing

1 Introduction

Delay-Tolerant Networks (DTNs) [1] are characterized by inconstant network topology and the nonexistence of paths from one end to another for data routing between a source-destination pair. Their main characteristics in data transmission include sporadic connectivity, prolonged and unpredictable delays, and high error rates. Because of regular network partitions, traditional routing protocols are not able to discover paths, which results in unsuccessful transmissions. DTN nodes (i.e., vehicles, pedestrians) need to cooperate among them to route messages [2]. They store messages in their buffers and carry them. If the carrier finds a better forwarder to the destination of the message, it transfers it. This process is repeated until the message destination is ultimately encountered.

Designing suitable routing protocols and metrics, allowing selecting the most suitable node to transfer messages, still remains one of the biggest challenges in DTNs. Given that in such networks, nodes may not get in contact often, selecting the next message carrier, i.e., the node with the message, is an essential choice.

This chapter focuses on leveraging location and navigation information of the nodes to aid routing in DTNs. The use of location-based routing metrics requires that the message carrier knows its location and the location of the message's destination, and the position of the node with which it communicates. For instance, using the Global Positioning System (GPS), one can obtain location information. The latter can be exchanged nodes get in contact. However, a location system is required to locate the destination of the message. Current routing protocols that rely on geographic information assume that the nodes in the network have the accurate location information of other nodes (e.g., GeoSpray [3]), or even a motionless destination, not contemplating the related delays necessary to get such information nor destination mobility. Nevertheless, location information may soon become obsolete because of node mobility and partitions in the DTN.

Additionally, most DTNs routing protocols using geographical information obtain such information from navigation systems like the destination of the messages or the expected time to reach it. Although mobile nodes are nowadays equipped with GPS, navigation systems are not so common neither knowing their destination on the map, given the need for the user's action and to the fact that it could most probably change during the trip. Hence, some protocols highly depend on navigation systems, which can affect their performance directly.

In this chapter and conversely to the state-of-the-art, an epidemic-based decentralized localization system (i.e., DTN-Locate) is proposed and used for disseminating node's localization information. DTN-Locate allows each network node to maintain a structure with previously known location information, direction, and speed of movement of other nodes. Such information is updated upon an encounter between them. Please note that apart from the overhead, flooding is considered as the upper-bound in terms of information dissemination. In addition, a hybrid location-based routing protocol (also known as SnLocate) is proposed. SnLocate creates multiple copies of a message and uses geographic mechanisms to disseminate messages. Specifically, it has two phases of operation, where it first disseminates a restricted number of message copies in the network and then uses location information via the DTN-Locate system to route the created replicas towards their destination. Besides, a novel distributed contention-based mechanism is proposed aiming to address the overhead of greedy-based routing protocols. It performs better than other routing protocols that use or not geographic information considered during the evaluation in terms of delivery rate and latency.

The remainder of the chapter is as follows. Section 2 presents related work. Section 3 introduces our DTN-based localization system. Section 4 describes the SnLocate protocol. Section 5 presents the performance evaluation of both the localization system and the routing protocol. Lastly, Sect. 6 presents concluding remarks and forthcoming work.

2 Related Work

Many DTN routing protocols can be found in the literature [4, 5]. Mainly, the following criteria are considered: the number of replicas created, and the information used to forward messages. For instance, Greedy-DTN [4], MoVe [6], and V-GRADIENT [7] are geographic routing protocols using a single-copy strategy, i.e., a single message copy exists in the network. On the other hand, Epidemic [8], Spray-and-Wait (SnW) [9], and PRoPHET [10] use a multi-copy strategy, where several replicas exist.

Motion Vector (MoVe) uses the information about the nodes' relative velocities to estimate the closest distance to get to the messages' destination, based on their current trajectories. The one with the closest estimated trajectory to the destination is considered the forwarder.

Greedy-DTN forwards a replica to the destination closest node. It is a DTN adaptation of GPSR, which uses geographical information in Ad-Hoc networks [11]. In dense networks, a hop-by-hop approach is used, and the message is sent to nodes that reduce the distance to the destination. If no neighboring node closer to the message destination in comparison to the current carrier exists, the latter keeps the message until a suitable carrier is found.

V-GRADIENT dynamically alters the methods employed in message dissemination within the geographic region of interest by monitoring parameters such as node density, buffer occupancy, among others. It manages congestion and delivers messages results in improved performance metrics such as delivery ratio and latencies, at the cost of a minor overhead increase.

Epidemic creates an unlimited number of replicas on the network. Each node keeps a list containing the identifier of each message stored in its buffer, also known as the summary vector. Upon an encounter between two nodes, they exchange their summary vectors, which allows each node only to receive messages that it does not have.

SnW restricts the number of replicas created to a given value. It has the SnW phases. In the former phase, L message replicas coming from the source node are firstly disseminated to L different relays. In the latter phase, each of the L replica carriers will forward it merely to its destination if it was not found in the former phase. Two versions exist differing on how L replicas are disseminated. In the Source version, which is the simplest version, the source node forwards all L replicas to the first L different nodes it finds. On the other hand, in the Binary version, if any node having $n > 1$ replicas and encounters another node without a copy, it hands over $\lfloor n/2 \rfloor$ copies also keeping $\lceil n/2 \rceil$ replicas. The node switches to the Wait phase if only one replica remains.

PRoPHET replicates messages based on the encounter history of the nodes. It uses the contact history of the node to assess the delivery predictability, that is, a probabilistic metric that tells how probable a given node is to deliver a message to its destination. Encounters increase the delivery predictability metric meanwhile the opposite decreases it. A node replicates a message to another one if the delivery predictability of the latter to the destination of the message is higher.

Conversely to the single-copy protocols considered, the multi-copy ones may perform sub-optimally in some DTN environments, e.g., vehicular ones, due to the inexistence of specific features. On the contrary, even though Greedy-DTN and MoVe use geographical information, they suffer from obsolete information in sparse environments. On the other hand, our proposed protocol improves delivery probability with reduced latency, although using a multi-copy strategy, and considers well-established geographic information.

3   The DTN Localization System

Here, the DTN-Locate system is presented. It aims to address which suffer from obsolete location information due to node mobility and network partitions. It leverages geographic information for routing in DTNs.

Node ID	Location	Direction	Speed	Info Time

Fig. 1 The structure of GeoInfo

3.1 The Principle of Operation

Given the inexistence of end to end paths, only upon the encounter between nodes communication occur. Therefore, centralized localization systems cannot provide timely information.

The DTN-Locate system is a decentralized system that uses epidemic principles of operation for disseminating localization information. Each node keeps a dictionary with the last known location information of encountered nodes, and through information exchange upon an encounter, this information is updated. The structure of the localization information, also known as GeoInfo, is shown in Fig. 1. For node X, it consists of X's identification, the coordinates of its last location position, its direction of movement angle, speed, and time at which the information was created.

Node X uses its GPS receiver to get its location and time information. With this information, the localization system determines other parameters such as speed and angle of movement direction.

3.2 Dictionary Update

Figure 2 presents the dictionaries' exchange algorithm amongst nodes. Please note that every single network node uses the DTN-Locate system. Initially, nodes update their dictionaries to ensure the removal of old information, therefore enabling to save nodes' limited resources such as storage space. The information age parameter can be configured. Upon an encounter, each node X updates its GeoInfo record and exchanges its dictionary with node Y. Upon receiving X's dictionary, X updates its

---

Algorithm for updating dictionaries

- Delete GeoInfos from the dictionary with old information
- Get node X current GeoInfo and put it in the dictionary of X
- Send the dictionary of X to a neighbor node Y
- Receive the dictionary of Y
- For each i in the received dictionary distinct from X
  - if i is in the dictionary of X
    - If neighbor's information is more recent, update i
  - else
    - Add i to the dictionary of X

Fig. 2 Algorithm for exchanging dictionaries between nodes when contact occurs

dictionary in this manner: for each record received, X verifies if it already has it in its dictionary. If that is the case, X only updates the localization information if its acquisition time in Y's dictionary is more recent. If not in the dictionary, it adds it.

4   The SnLocate Protocol

SnLocate is a geographic routing protocol for DTNs using both single- and multi-copy the routing protocols strategy. It is based on our preliminary work proposed in [12] that mainly focused on vehicular environments, conversely to DTN, that is more comprehensive. It uses the Spray phase concept of the Binary Spray-and-Wait protocol, where a fixed number of replicas are spread across the network. The Spray phase of the binary version reduces the replicas distribution time [9]. Nevertheless, in the wait phase, and conversely to Spray-and-Wait, SnLocate forwards the message to nodes moving in the direction of the message destination.

4.1   The Principle of Operation

SnLocate operation consists of the following phases:

4.1.1   Spray: The First Phase of Operation

In the first phase, the source node sets the maximum number of message copies in the message header. Any node having many message copies that finds another one with none gives half of the copies to it, thus staying with half. A node changes to the second phase of the protocol when only one copy is left, and the message is not delivered in this phase.

4.1.2   Locate: The Second Phase of Operation

Here, each node X geographically forwards the remaining single message copy to its destination. Upon an encounter with another node, the carrier evaluates if the other one compared to itself is heading in a nearer direction to the destination. If that is the case, the message is transferred to node Y. When nodes are relatively close, hence in contact, and due to DTNs' sparsity, the message is forwarded to y going nearer direction wise to the destination. If both X and Y have a similar angle of the direction of movement relative to the destination, it is selected to carry the message the node minimizing the distance to the destination. Therefore, SnLocate is a greedy protocol by taking advantage of existing multi-hop communication opportunities, aiming to reduce latency, compared to a node merely carrying it.

4.2 The Direction of Movement

The location of the node and destination of the message, its direction of movement are considered to determine the direction of the movement angle of the nodes relative to the destination. Let $\vec{v_X}$ denote direction of movement vector of node X, and $\overrightarrow{XD}$ denote the vector initiating at X to the known location of the destination D. The direction of the movement angle of X relative to D is given by

$$\theta = \cos^{-1}\left(\frac{\overrightarrow{XD}.\vec{v_X}}{|\overrightarrow{XD}||\vec{v}|_X}\right) \tag{1}$$

When contact is made with a neighboring node, the message carrier calculates the angle of the direction of movement relative to the destination for itself and the neighboring node. If the neighbor has a smaller angle of movement relative to the destination, the message is forwarded and is cleared from its buffer. Otherwise, the node keeps the message stored in its buffer. If the two nodes have the same direction of movement angle in relation to the destination, the vehicle that minimizes the distance to the destination is selected to keep the message. By being greedy, it is possible to take advantage of multi-hop communication opportunities if they exist, which are expected to have less latency than the message being simply carried by nodes.

Let the location of node X be given by coordinates $(x_X; y_X)$ and the location of the message destination given by the coordinates $(x_D; y_D)$, the distance from X to D is

$$d = \sqrt{(x_D - x_X)^2 + (y_D - y_X)^2} \tag{2}$$

Figure 3 presents two decision situations for forwarding the message using SnLo-cate. Node X has a message, of which the destination is node D and node Y is its neighbor. In Fig. 3-(a), node X does not forward the message to its neighbor Y because its angle of direction of movement in relation to D is less than the angle of Y and, therefore, is going further in the direction destination. In Fig. 3-(b), the angle of the direction of movement with respect to D is equal to X and Y, but the distance from Y to D is less than the distance from X to D. Thus, the message is forwarded to Y and X clears it from its buffer.

4.3 The SnLocate Algorithm

Figure 4 presents the SnLocate algorithm. Upon an encounter between nodes X and Y, they exchange their location dictionaries. Subsequently, X verifies in its list of stored messages the existence of any message destined to Y. In the Spray phase, X

Fig. 3 The situation in which node X does not forwards **a** and forwards **b** a message to node Y

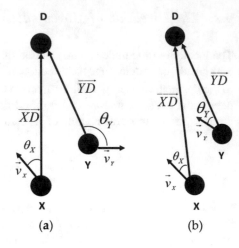

(a) (b)

The SnLocate algorithm
- Node X establishes communication with node Y
  /* DTN-Locate dictionaries are exchanged between X and Y (according to Figure 2) */
- X updates the GeoInfo in its dictionary with current data and sends to Y its updated DTN-Locate dictionary
- X receives DTN-Locate dictionary of Y and updates its dictionary with the latest received information
  /* Delivery of messages whose destination is Y */
- For each message stored in the buffer of X and destined for Y
  - X delivers the message to Y and deletes it from its buffer
  /* Replication/forwarding of the remaining messages */
- For each message i in the buffer of X
  - If i's number of copies L is greater than 1
    - X delivers to Y $\lfloor L/2 \rfloor$ copies of i and keeps $\lceil L/2 \rceil$ copies
  - Else
    - If devAvgHC > μ
      - X accesses its DTN-Locate dictionary to get the GeoInfo for X, Y, and destination node D.
      - If the GeoInfo of the destination is in the dictionary
        - X calculates its distance and the angle of the direction of its movement relative to D
        - X calculates the distance and the angle of the direction of the movement of Y relative to D
        - If the angle of Y is smaller than the angle of X
          - X delivers i to Y and deletes it from its buffer
        - Else
          - If the angle of X is equal to the angle of Y, and the calculated distance of Y is smaller than that of X
            - X delivers i to Y and deletes it from its buffer
    - Else X keeps i in its buffer and moves on to the next message

Fig. 4 The SnLocate routing algorithm when a contact occurs between nodes X and Y

checks which message m among the stored ones has the number of replicas higher than one in order to replicate it to Y. Then, X updates it in m's header to half of the current value and delivers a copy to Y. Lastly, in the second phase (Locate), X check every m only having a single copy in its header to assess if Y is a better forwarder.

If the standard deviation of the average hop count is then the Threshold (which is explained in Sect. 4.4), then X determines its movement direction angle and its distance relative to D. Then, X makes the same computations for Y. Y only receives the message if and only if its movement direction angle relative to D is inferior to X's angle. If X and Y have equal angles, the latter only receives the message if its distance from D is inferior to that of X. X removes every m from its buffer upon successful delivery.

4.4 The Distributed Contention Mechanism

A well-known problem of greedy-based approaches is its high overhead if compared to others, and the goal of the proposed contention mechanism is to restrict the number of retransmissions considering the number of hops of successfully delivered messages. Specifically, the proposed contention mechanism consists of a distributed consensus protocol that aims to determine the average hop count of successfully delivered messages in the network.

Let each node X be composed of the following tuple T < *timestamp, averageHop-Count* >. This tuple is disseminated and/or updated during the GeoInfo dissemination process. The *timestamp* corresponds to the time of reception of successfully delivered message m or of the update of *averageHopCount*. First, each X receiving m, creates and disseminated T to every node Y it encounters. Every Y upon receiving T from X checks if the timestamp of T is higher than the one it has. If that is the case, it updates its *averageHopCount* using an Exponential Weighted Moving Average.

$$avgHopCount = (1 - \alpha)^* avgHopCount + \alpha^* aHopCount \qquad (3)$$

where *aHopCount* and *avgHopCount* are the accumulated hop count information received from X and of Y, respectively.

Upon update of the *averageHopCount*, the standard deviation (i.e., *devAvgHC*) is also computed as it is an indication of the convergence of the consensus protocol.

$$devAvgHC = devAvgHC^*(1 - \phi) + |avgHopCount - aHopCount|^*\phi \quad (4)$$

Please note that α and ϕ are two positive parameters aiming to give more weight to recent information. The goal of our contention mechanism is to restrict the number of transmissions of the greedy phase of SnLocate, after which the current message carrier only delivers the message to its destination. In other words, the greedy mechanism happens as long as the *devAvgHC* is above a given Threshold (μ).

5   Performance Evaluation

This section evaluates both the routing protocol and the localization system. The Opportunistic Network Environment [13] (ONE) simulator is used to perform simulations.

5.1   The Simulation Model

Different simulation scenarios consisting of a synthetic mobility model are considered. Figure 5 shows the Helsinki city map that is used. Table 1 summarizes the simulation parameters explained in the following sections.

5.1.1   The DTN-Locate System

The simulation model consisted of two scenarios, i.e., pedestrians and cars, with the 50, 150, and 300 nodes. Pedestrians' speed of movement ranges from 2 and 6 km/h; meanwhile, cars' speed ranges from 10 to 50 km/h. The Shortest Path Map-Based Movement (SPMBM) model is used.

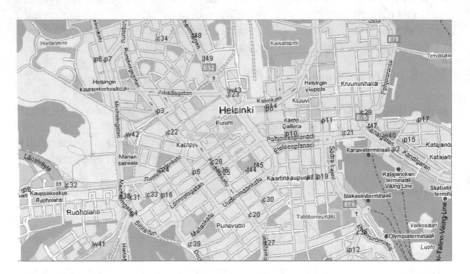

Fig. 5   The Helsinki city map with dimensions of 4.5 km by 3.4 km

Table 1 Simulation parameters and respective values

Map	Helsinki city downtown [4500 m × 3400 m]	
Movement model	[12; 84] h	
Movement model	SPMBM	RMBM
Type of nodes	Vehicles (cars)	Buses
Speed interval	10–50 km/h	25–36 km/h
Buffer size	20 MB	50 MB
Nodes' wait time	0–120 s	10–30 s
Message size interval	500 KB–1 MB	
Message generation interval	25–35 s	
Data rate	4.5 Mb/s	
Transmission range	30 m	
Initial number of copies	6	
TTL	[60; 120; 180; 240; 300] min	
Number of nodes	[50; 150; 300] nodes	

5.1.2 The SnLocate Protocol

The simulation model scenario consisted of 150 nodes, i.e., 105 cars and 45 buses. The simulation duration is set to 12 h. Vehicles and buses have a buffer size of 20 and 50 MB, respectively. Vehicles and buses move on roads using the according to the SPMBM and to the Route Map-Based Movement (RMBM) models, respectively. Vehicles and buses speed ranges from 10 to 50 km/h, with pauses of 0 to 120 s, and 25 to 36 km/h, with pauses of 10 to 30 s, respectively. The wireless communication interface range and transmission rate are of 30 m and 4.5 Mb/s, respectively. The message generation interval ranges from 25 to 35 s. The message size ranges from 500 KB to 1 MB. PRoPHET parameters [9] are set as follows: $Pinit = 0.75$, $\beta = 0.25$ and $\gamma = 0.98$. Both SnW and SnLocate use the binary version with the initial number of copies set to 6 [9]. α and ϕ were set to 0.125 and 0.25, respectively [14]. The values of μ varied from 0.01 to 6.

5.2 Simulation Results

5.2.1 The DTN-Locate System

Aiming at assessing the error associated with the information acquisition speed, the information age, the position and direction errors, and the relative speed errors, a set of simulations are performed using different movement pattern seeds. The data collected every fifteen minutes for four hours from all nodes for each metric during the simulation is averaged. For each metric evaluated, an average of the data collected from all nodes in the simulation was calculated. Please note that the overhead associated with the exchange of GeoInfo was omitted as we considered that the size of the control information exchanged to be negligible as compared with the data.

The dictionary size in relation to time is the quantity of GeoInfos in the node's dictionary and enables understanding of how the propagation of information occurs. The age of information enables knowing when the information was created and, therefore, in what way it is obsolete. The latter is the difference in seconds between the current time and the time when the information is created. The distance in meters between the actual position and the position in the dictionary corresponds to position error. The speed relative error, which is in percentage, enables analyzing the speed information related error in the dictionaries.

Figure 6 presents the average dictionaries size in relation to time for both scenarios. For all node densities of Fig. 6-(a), at 900 s, node dictionaries are practically filled with information about all nodes in the network. The dissemination of information is fast, and after 15 min, the nodes have location information for all other nodes in the

Fig. 6 The average number of nodes in the dictionary for the pedestrians **a** and cars **b** scenarios

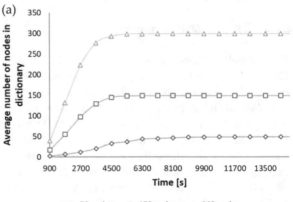

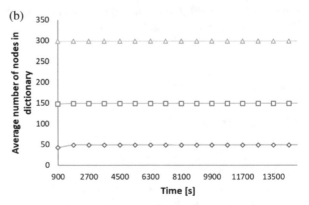

network in their dictionaries. By observing the graphs of the average dimension of the dictionaries of the scenario of the movement of pedestrians and cars, Figs. 6-(a) and 6-(b), respectively, the time of acquisition of information is shorter in the latter scenario. In the pedestrian scenario with 300 nodes, dictionaries are complete only at 5400 s, while in the car scenario with the same node density, they are complete at 900 s. Therefore, one may conclude that the high movement speed of cars increases the frequency of contacts between the nodes in the network, which increases the speed of dissemination of information on the network.

Figure 7 shows the average age of the information present in the dictionaries of the nodes. By comparing the graphs with 50, 150, and 300 nodes, one may conclude that low node density scenarios have more outdated information. The average age of the information, for example, at 7200 s, is 744, 394, and 303 s for the 50, 150, and 300 node density scenarios, respectively. The node dictionary is updated upon

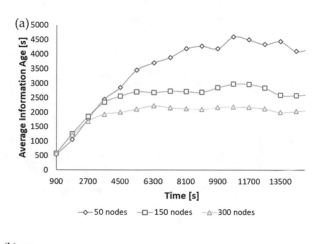

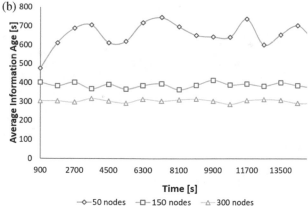

Fig. 7 The average information age for the pedestrians **a** and cars **b** scenarios, in seconds

an encounter between two nodes. In environments with few nodes, they rarely meet as compared with environments with many nodes.

Consequently, the information in the dictionary is not updated as often in low-density networks. By comparing the average age of the information in the scenario of pedestrians and cars, Figs. 7-(a) and 7-(b), respectively, the average age of information is lower in the car scenario. For example, at 5400 s, for 300 nodes, in the pedestrian scenario, the average age of information is 2117 s, while in the car scenario, it is 300 s.

Figure 8 presents the average position error for both scenarios, with different node densities. For the car scenario (Fig. 8-(a)), by comparing all node densities' graphs, it appears that the average position error is smaller and inversely proportional to the node density. The average relative speed error for both scenarios is presented in Fig. 9. Also, for the car scenario (Fig. 9-(a)), at 4500 s, the average relative error considering the following node densities of 50, 150, and 300 nodes is 59, 50

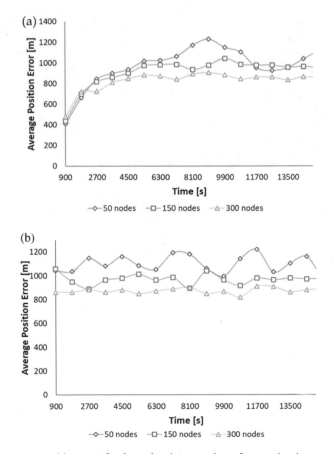

Fig. 8 The average position error for the pedestrians **a** and cars **b** scenarios, in meters

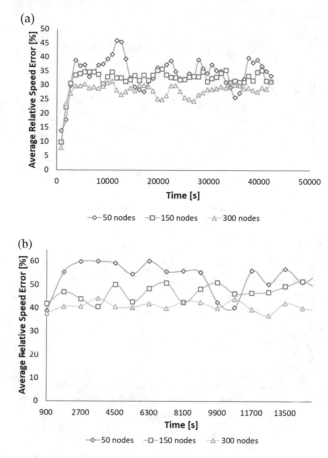

Fig. 9 The average relative speed error for the pedestrians **a** and cars **b** scenarios, in percentage

and 40%, respectively. Similar to position error, one may conclude that the average relative velocity error is lower in scenarios with a higher node density.

The information in the dictionaries has low accuracy. For instance, the average position error for information with a 5 min average age in a 300 nodes' scenario is 850 m (Fig. 7). This will affect routing protocols significantly if they use such outdated information. The relative speed error is considerable, given that nodes often stop and alter the speed. Please note that these results correspond to an average over all nodes. Intuitively, near each node, its errors may be smaller, as the information is disseminated via an epidemic update mechanism.

Figure 10 shows the average direction error in degrees for both scenarios and node densities. By comparing the node density graphs, one may conclude that, on average, the error is higher in low node density configuration. The average direction error fluctuates over time in all scenarios with in the 300, 150 and 50 nodes configurations (Fig. 10-(b)), it does not exceed 96, 100, and 104°, respectively. The average errors

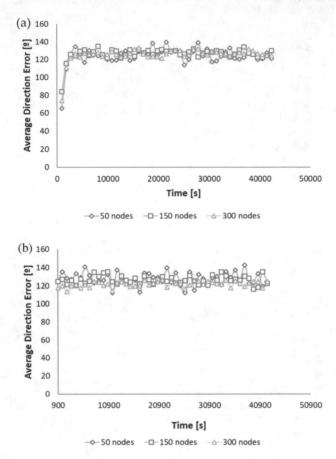

Fig. 10 The average direction of movement error for the pedestrians **a** and cars **b** scenarios, in degrees

associated with the information are smaller in dense node configurations because the information is more recent in high node density configurations than in scenarios of low node density.

By assessing the average direction of movement error in the pedestrian and car scenarios, Figs. 10-(a) and 10-(b), respectively, it appears that this error is not significantly different between the scenarios. However, the information is more recent in the car scenario, and the movement of cars is restricted to roads. For example, at 3600 s, for the 300 nodes configuration, the average direction error is 100 and 96° in the pedestrian and car scenario, respectively. The map topology used also affects the average direction of movement error because if the displacement, for example in the case of cars, is on a highway, there are few changes in the direction of movement, while in a city center, the existence of short streets and many intersections, results in frequent changes in the direction of movement.

In summary, the error associated with the DTN-Locate system depends on the network density and the speed at which the nodes move. In networks with higher density, there is a greater probability of contact between nodes, and the information is disseminated more quickly through it. The increase in the speed of the nodes increases the frequency of contacts, therefore increasing the information propagation speed. These factors influence the information age, which is higher in sparse networks and lower in dense networks. Information age is smaller in scenarios with high node mobility. However, despite speed allowing information to spread rapidly, it also causes information to become outdated. Since the information in the dictionaries has low accuracy, the SnLocate protocol does not make corrections to nodes positions based on speed or direction of movement. This means that, as can be deducted from the algorithm in Fig. 4, nodes do not need to exchange or maintain speed information in their dictionaries and only need to exchange the movement direction information with their neighbours without storing it in their dictionaries. Improvement on location accuracy might be made using maps, which is left for future work.

5.2.2 The SnLocate Protocol

SnLocate's evaluation consisted of comparing it to Epidemic, PRoPHET, Spray-and-Wait (i.e., multi-copy non-geographic routing protocols), and MoVe and Greedy-DTN (i.e., geographic routing protocols), similarly to [14, 15]. Two setups are considered, namely an optimal and realistic one. In the former, nodes have precise knowledge of where all the network nodes; meanwhile, in the latter, nodes use DTN-Locate to assess the most recent known location as they do not know where the destinations of the messages are.

Different Time-To-Live (TTL) values are considered trying to change the network load. TTL gives the validity period of a message. These values range from 60 to 300 min, with increments of 60 min. The following performance evaluation metrics considered are delivery rate, average delay, average hop count, and overhead. The average delivery ratio can be defined as the ratio between the number of messages successfully received and sent. The average delivery latency can be defined as the average time necessary for delivering messages. The overhead metric corresponds to the number of message transmissions for each created message. The average hop count is the average number of hops taken by a message until it reaches its destination. Each simulation is performed eight times, utilizing distinct movement generator seeds, and the results were averaged for statistical confidence. Please note that all performance graphs have 95% confidence intervals plotted, but they are so small that can be hardly seen, which proves the very good confidence in the simulation results.

Average Delivery Rate. Figure 11-(a) presents the average message's delivery rate for both optimal and realistic setups. SnLocate has the highest average delivery rate for different TTL values. The second protocol with the highest average delivery rate is Spray-and-Wait. These two protocols have an equal dissemination phase, in which

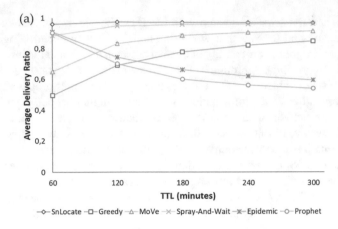

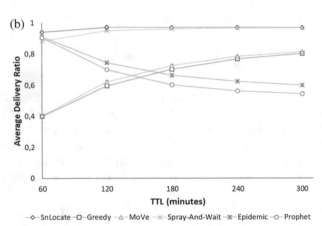

Fig. 11 Average delivery rate for the optimal **a** and realistic **b** setup

they place multiple copies of the messages circulating in the network, increasing the probability that eventually, one of these copies will find the destination of the message. However, in the wait phase, SnW waits until it finds the message destination, meanwhile SnLocate routes replicas to nodes that are moving in the direction of the destination. The delivery rate of Greedy-DTN and MoVe grow with the increase of the TTL as only a single replica exists, and with extra delivery time, the right path is more likely to be found. Epidemic and PRoPHET behave differently from the remaining protocols varying the TTL. They do not restrict the number of replicas in the network; thus, network congestion and message drops increase with the increase of TTL results, also degrading the delivery rate. In Fig. 11-(b), SnLocate's average delivery rate with DTN-Locate is different only for the 60 min of TTL from the optimal setup and remains almost the same for the other TTL values. The single-copy protocols present more significant differences between the average delivery rates in these setups. SnLocate first disseminates a restricted number of replicas and then routes such copies utilizing geographic routing metrics, while MoVe retains just a replica. DTN-Locate is a decentralized location system; therefore, each node's dictionary information and associated errors are different due to the decentralized nature of our localization system. Therefore, different replicas in the network are dependent on unique circumstances imposed by the system.

Average Delivery Latency. Figure 12 presents the message delivery latency for optimal and realistic setups. Mainly, latency increases because of the increase in network congestion with the increase of TTL. Greedy-DTN and MoVe present higher latency if compared to multi-copy ones. Message replication is used to spread message copies in the network, hence improving their odds of reaching their destination. One can conclude that our protocol presents the lowest latency as a result of its hybrid nature. To be precise, its first phase enables disseminating a restricted number of replicas that can delve into various paths, and its second phase routes the latter in the direction of the destination.

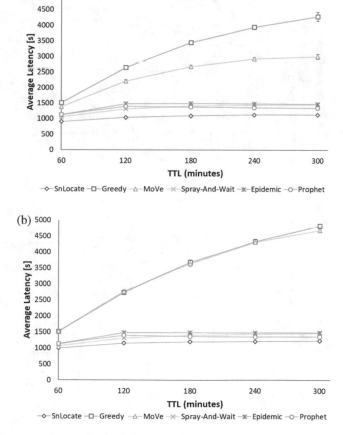

Fig. 12 The average latency for the optimal **a** and realistic **b** setup

Figure 12-(b) shows the latency in the scenario using DTN-Locate. The three geographic routing protocols have higher latency when using DTN-Locate. The inaccurate node location information causes more latency in message delivery. In the optimal setup, the metrics are calculated by always considering the current position, and as a result, the message is being carried towards the destination. In the real setup that uses DTN-Locate, the message suffers more deviations when they are transported to outdated positions.

Average Overhead Ratio. Figure 13 shows the average overhead ratio of all considered protocols for the optimal and realistic setups. One may conclude that replication-based protocols have the highest average overhead, except Spray-and-Wait. This type of protocols registers an increase in overhead with the increase in TTL. On the other hand, single-copy protocols have no significant variation in overhead with increasing TTL of messages. Spray-and-Wait's average overhead ratio does not vary with the TTL. By limiting the number of message copies to 6, the number of transmissions

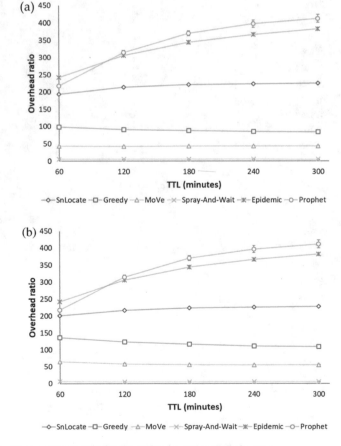

Fig. 13 Average overhead ratio for the optimal **a** and realistic **b** setup

is very close to this value as the protocol enters the Wait phase. SnLocate presents a higher average overhead if compared with Spray-and-Wait because, in the second phase of operation, it makes geographic forwarding of messages, which increases the number of transmissions and, thus, the overhead. This is the price to pay for improving the latency. For scenarios where nodes are not energy constrained, as is the case of vehicle nodes, the overhead increase is not a problem, and the latency gains are more important.

However, the SnLocate protocol makes fewer transmissions than the Epidemic and PRoPHET protocols, as, unlike these two protocols, it limits the number of message copies on the network. Epidemic and PRoPHET have a high average overhead since copies of messages already delivered continue to circulate on the network. These messages are only removed if they are discarded due to TTL expiration or due to congestion.

Greedy-DTN and MoVe have a higher average overhead than Spray-and-Wait. Geographic protocols use location information and movement direction of nodes to calculate their routing metrics. In many cases, the node that has the message and the neighboring node with which it establishes contact are very close. Thus, for example, if the routing metric is the distance to the destination, the value of the distance to the destination of the nodes in contact is very similar. In these situations, cycles may occur; that is, the message may be transferred between the same nodes several times

Average Hop Count. Figure 14 presents the average message hop count, i.e., the number of hops messages take until they are delivered to their destination in the ideal and normal scenarios. One can conclude that the average number of hops is higher in geographic single-copy protocols. Geographic protocols have a greedy behavior, so they are always forwarding to any node that is closest to the destination, which increases messages' hop count. The average hop count of single copy protocols such as Greedy-DTN and MoVe, increases with the increase of the TTL, conversely to message replication protocols. Since replication protocols have several copies of messages circulating in the network, this increases the probability of finding the destination using the shortest path.

Although Spray-and-Wait and SnLocate are replication protocols, the former presents the lowest average number of hops. This is mainly due to the difference in the second phase of operation as both have the maximum number of copies of messages equal to 6 and use the Binary version of Spray-and-Wait in the first phase to distribute the copies, limiting their average number of hops to 3. However, SnLocate forwards messages to nodes moving towards the destination, meanwhile Spray-and-Wait only delivers them directly to the destination. Nonetheless, one may also conclude that the messages delivered by SnLocate had, on average, 3.5 times more hops than those of Spray-and-Wait. This value is informative as it would enable restricting the number of SnLocate retransmissions hence reducing the protocol's overhead. Epidemic and PRoPHET presented low average hop count, although their average delivery rates decrease significantly with the increase in the TTL of messages, as shown in Fig. 11. Due to congestion, messages are only delivered if the source and destination are relatively close.

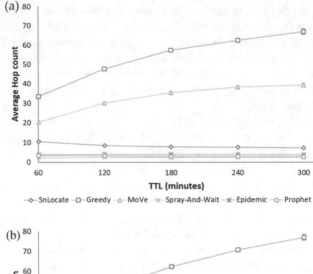

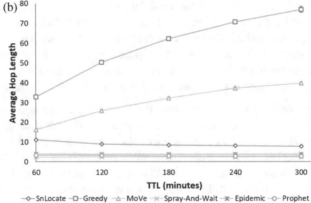

Fig. 14 Average hop count for the optimal **a** and realistic **b** setup

5.2.3 The Distributed Contention Mechanism

Figure 15 highlights the tradeoffs between the average delivery and overhead ratios as function of μ for TTL values of (a) 60 min and (b) 120 min in the scenario considered in this work. Please note that the simulation duration was set to 3.5 days. We only considered the TTL values of 60 and 120 min as for higher values the delivery ratio does not vary much tending to 1.

From the below figure, it is possible to see that the overhead drops to 4.3 and 4.9 times of its initial values for losses of 3.9 and 1.1% on the delivery ratio for TTL values of 60 and 120 min, respectively. Nonetheless, with this value of μ (i.e., 1.0), gains of 2.7 and 5.7% were obtained comparatively to SnW (i.e., the second-best protocol regarding delivery ratio) for TTL values of 60 and 120 min, respectively.

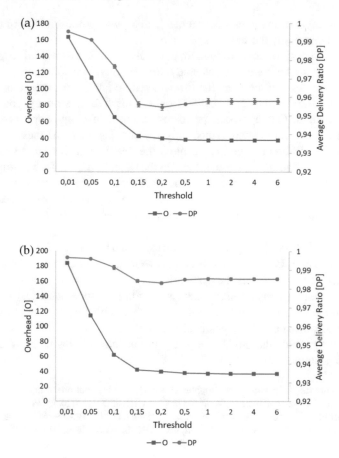

Fig. 15 The average delivery and overhead ratios as function of the μ for TTL values of **a** 60 min and **b** 120 min

In summary, if the goal is to maximize the average delivery ratio, as is the case of most vehicular applications, μ should be set to its minimum. Conversely, if resource-constrained devices are involved, μ should be set accordingly and taking into account that for values greater than 1.0 no overhead reduction will be attained.

6 Conclusions and Future Work

In this chapter, SnLocate, which is a geographic routing protocol, and DTN-Locate, which is a DTN localization system, are proposed for routing in DTNs. The former leverages geographical location information and nodes' moving direction to aid the routing process; meanwhile, the latter gives location information about nodes.

Besides, a novel distributed contention mechanism was proposed aiming to restrict the overhead caused by the greedy approach.

Based on the performance evaluation, one can conclude that our protocol outperforms other ones considered in this study. Specifically, our approach presents the highest average delivery rate and the lowest average latency in both setups considered, namely, the optimal and realistic ones. Our hybrid strategy enables it to have good results. Our approach performs better than single-copy protocols as restricts messages' replication. Thus, only a limited number of message replicas exist, exploring different directions through the destination. It enabled to increase the probability of finding the destination. On the other hand, the forwarding replicas to nodes moving in the direction of the destination enhances its latency contrasted to other multi-copy protocols. The price to pay is an increase of the overhead, which is not a problem for scenarios where nodes are not energy-constrained, like vehicle nodes.

Moreover, single copy protocols are more impacted by the inaccuracy of the DTN-Locate information. Each message is subject to its own context, and the associated errors reduce as it approaches the destination.

Extra evaluations considering scenarios with higher node densities, speeds, and city maps are left as future work, aiming at understanding the protocols scalability as well as the performance of the localization system. SnLocate will also be compared to other routing protocols, and DTN-Locate will be integrated with maps to enhance its location accuracy.

Acknowledgements This work was supported by Portuguese national funds through FCT, Fundação para a Ciência e a Tecnologia, under project UIDB/50021/2020, by Portuguese national funds through FITEC - Programa Interface, with reference CIT "INOV - INESC Inovação - Financiamento Base" and by LASIGE Research Unit, ref. UIDB/00408/2020 and ref. UIDP/00408/2020.

References

1. Pereira, P.R., Casaca, A., Rodrigues, J.J.P.C., Soares, V.N.G.J., Triay, J., Cervelló-Pastor, C.: From delay-tolerant networks to vehicular delay-tolerant networks. IEEE Commun. Surv. Tutorials **14**, 1166–1182 (2012). https://doi.org/10.1109/SURV.2011.081611.00102
2. Magaia, N., Rogerio Pereira, P., Correia, M.P.: Security in delay-tolerant mobile cyber physical applications. In: Rawat, D.B., Rodrigues, J.J.P.C., Stojmenovic, I. (Eds.) Cyber-Physical Systems: From Theory to Practice, pp. 373–394. CRC Press (2015). https://doi.org/10.1201/b19290-22
3. Soares, V.N.G.J., Rodrigues, J.J.P.C., Farahmand, F.: GeoSpray: a geographic routing protocol for vehicular delay-tolerant networks. Inf. Fusion. **15**, 102–113 (2014). https://doi.org/10.1016/j.inffus.2011.11.003
4. Tornell, S.M., Calafate, C.T., Cano, J.C., Manzoni, P.: DTN protocols for vehicular networks: an application oriented overview. IEEE Commun. Surv. Tutorials **17**, 868–887 (2015). https://doi.org/10.1109/COMST.2014.2375340
5. Wang, T., Cao, Y., Zhou, Y., Li, P.: A survey on geographic routing protocols in delay/disruption tolerant networks. Int. J. Distrib. Sens. Netw. **12**(2), 3174670 (2016). https://doi.org/10.1155/2016/3174670

6. LeBrun, J., Chuah, C.N., Ghosal, D., Zhang, M.: Knowledge-based opportunistic forwarding in vehicular wireless ad hoc networks. In: IEEE Vehicular Technology Conference, pp. 2289–2293 (2005). https://doi.org/10.1109/vetecs.2005.1543743

7. Nascimento, H., Pereira, P.R., Magaia, N.: Congestion-aware geocast routing in vehicular delay-tolerant networks. Electronics **9**, 477 (2020). https://doi.org/10.3390/electronics9030477

8. Vahdat, A., Becker, D.: Epidemic routing for partially connected ad hoc networks, Technical report CS-200006, Duke University (2000)

9. Spyropoulos, T., Psounis, K., Raghavendra, C.S.: Spray and wait. In: Proceeding 2005 ACM SIGCOMM Work. Delay-Tolerant Networks – WDTN 2005, pp. 252–259. ACM Press, New York, USA (2005). https://doi.org/10.1145/1080139.1080143

10. Lindgren, A., Doria, A., Schelén, O.: Probabilistic routing in intermittently connected networks. In: Springer Berlin Heidelberg, pp. 239–254 (2004). https://doi.org/10.1007/978-3-540-27767-5_24

11. Karp, B., Kung, H.T.: GPSR. In: Proceedings of the 6th Annual International Conference on Mobile Computing and Networking – MobiCom 2000, pp. 243–254. Association for Computing Machinery (ACM), New York, USA (2000). https://doi.org/10.1145/345910.345953

12. Moreira, E., Magaia, N., Pereira, P.R.: Spray and locate routing for vehicular delay-tolerant networks. In: Proceedings of the 2018 16th International Conference on Intelligent Transportation Systems Telecommunications ITST 2018, Institute of Electrical and Electronics Engineers Inc. (2018). https://doi.org/10.1109/ITST.2018.8566757

13. Keränen, A., Ott, J., Kärkkäinen, T.: The ONE simulator for DTN protocol evaluation. In: Proceedings of the 2nd International Conference on Simulation Tools and Techniques (ICST), p. 55 (2009). https://doi.org/10.4108/ICST.SIMUTOOLS2009.5674

14. Magaia, N., Borrego, C., Pereira, P.R., Correia, M.: ePRIVO: an enhanced PRIvacy-preserving opportunistic routing protocol for vehicular delay-tolerant networks. IEEE Trans. Veh. Technol. **67**, 11154–11168 (2018). https://doi.org/10.1109/TVT.2018.2870113

15. Magaia, N., Borrego, C., Pereira, P., Correia, M.: PRIVO: a privacy-preserving opportunistic routing protocol for delay tolerant networks. In: 2017 IFIP Networking Conference (IFIP Networking) Workshops, pp. 1–9. IEEE (2017). https://doi.org/10.23919/IFIPNetworking.2017.8264835

Assisted Living (Invited)

New Ambient Assisted Living Technology: A Narrative Review

Costas S. Constantinou, Tirsan Gurung, Hosna Motamedian,
Constandinos Mavromoustakis, and George Mastorakis

Abstract This chapter presents the results of a narrative literature review study of acceptability of new assistive and information technology, including Internet of Things (IoT) technologies and artificial intelligence, by older adults (65 or older). The study followed a careful search strategy in specific databases and was based on inclusion/exclusion criteria and keywords. The search strategy resulted in 28 articles, which reflected the research aim, and were reviewed on the basis of an interpretive approach and critically appraised in accordance with the 'critical assessment skills programme' guidelines. This study is an important contribution to scholarship because, unlike other literature reviews, it explored both assistive and information technology and looked for overarching reasons why older adults may accept new technology. The results showed that older adults accept technology when they have a good sense of control over the devices and their lives, the technology is useful, has characteristics which are not threatening, and other compromising factors such as financial cost, restricting health conditions and inappropriate physical environment are not present. Based on these findings, we propose the N-ACT principles whereby technology developers should consider users' Needs, Adjustable technology and personalised service, users' Control over technology and their lives and Trust in technology.

Keywords Assistive and information technology · Acceptability · Narrative review · The N-ACT principles

C. S. Constantinou (✉) · T. Gurung · H. Motamedian
University of Nicosia Medical School, Nicosia, Cyprus
e-mail: constantinou.c@unic.ac.cy

C. Mavromoustakis
University of Nicosia, Nicosia, Cyprus

G. Mastorakis
Hellenic Mediterranean University, Heraklion, Greece

© Springer Nature Switzerland AG 2021
N. Magaia et al. (eds.), *Intelligent Technologies for Internet of Vehicles*, Internet of Things,
https://doi.org/10.1007/978-3-030-76493-7_16

1 Introduction

"I am not going to use a technology that cannot make a difference in my life" (participant during the pilot of vINCI technology in Cyprus).

The above expressed attitude captures the importance of acceptability of technology developed for use by older adults, and it was communicated during the piloting of vINCI[1] technology in Cyprus. vINCI (Clinically-validated INtegrated Support for Assistive Care and Lifestyle Improvement: the Human Link) aims to construct, pilot and validate a set of technologies which will provide end-users information and feedback about a number of aspects in their daily life based on integrating IoT (Internet of Things), monitoring technologies and artificial intelligence (Fig. 1). vINCI's design reflect a microservice architecture[2], whereby small, autonomous but interconnected services are at play. During the pilot phase, 20 older adults in Cyprus were invited to use insoles, watches and a tablet for seven days in order to collect feedback and understand their willingness to use the technology in the long run.

During the vINCI pilot the participant above appeared more critical than others and clearly referred to the usefulness of the technology. However, her expressed attitude reflects the essence of designing technology for people in many more respects. It pinpoints the importance of a technology for people's needs and for contributing to people with specific positive outcomes. It also highlights other aspects of the

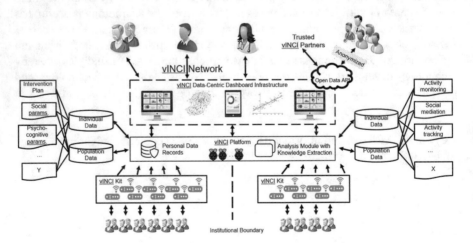

Fig. 1 The vINCI architecture

[1] vINCI is approved under the AAL (Active and Assisted Living) programme of the EU and it is coordinated by the National Institute for Research and development in Informatics (ICI Bucharest). The following countries participate in the project: Romania, Italy, Cyprus, Poland, and Slovenia. For more information about the project visit vINCI - AAL Programme (aal-europe.eu).

[2] Microservices vs Monolithic architecture|by Bhagwati Malav|Hash#Include|Medium.

relationship between people and technology, such as control and acceptance for long-term use. Moreover, it reflects research evidence suggesting that new technology is not always accepted and, as a result, used by end-users [1, 2].

The development of new technology for older adults is growing rapidly [3–5] and has resulted from the fact that contemporary western societies are faced with the challenge of their ageing populations and the need to promote active ageing, supporting older adults live independently in their community [6–8]. As a result, many research projects rely on the combination of technologies on the basis of Internet of Things (IoT) and artificial intelligence so to offer such technology that addresses end-users' needs and reflects their capabilities. One of the fundamental principles of offering technological assistance to older adults is the user-centred approach to designing, constructing, testing and finalising technology [9, 10]. Yet, there is evidence to suggest that end-users do not always eagerly accept and use new technology and technologies which have been designed for them do not always reflect their needs [1, 2].

Based on the identified importance of the acceptability of newly developed technology [1, 2, 9], this study aims to explore acceptability of new assistive and information technology based on review of the relevant literature. The word "new" is very important to clarify here as it does not refer to any new technology but to the technology which is new to end-users. Gaining insights into this question would help technology developers better understand the importance of involving end-users to the development process from the design stage to the end result and formulate a process which would reflect the characteristics of the populations they are targeting. In order to address the research aim, we designed a narrative literature review, which is described in detail in the next section. Although there are other literature reviews on acceptability of technology, we did not identify either a systematic or a narrative literature review which combined both assistive and information technologies, which were new to older adults. Therefore, it is imperative to understand how older adults accept new technology and find overarching patterns which could potentially help technology developers and researchers.

2 Methodology

The methodology used for this narrative review was based on guidelines by Ferrari [11] and the SANRA scale for quality assessment of narrative reviews [12], which clarify that narrative reviews should include a clear research aim, justification and a search strategy. The aim of this narrative review is to explore the use of new assistive and information technology, including Internet of Things (IoT) and technology of artificial intelligence, by older adults (65 or older) and focus more specifically on how older adults accept such new technology. Because we did not have a fixed research hypothesis, as per the guidelines for narrative reviews [11], our inclusion and exclusion criteria were the following: peer-reviewed articles, theses and dissertations, literature reviews, conference papers, reports and books/textbooks that included older people/adults and new technology use, published between 2009–2020. We had three overarching criteria. First, the technology should not necessarily be new in general but new to end-users. Second, sources should present research evidence on older adults' acceptability of new technology. Third, technology designed to address specific medical conditions should not be included. This is because a technology designed to improve a medical condition could potentially bias end-users in favour of this technology.

Based on the criteria above, the following databases were searched: MEDLINE, Scopus, Google scholar, PubMed, EBSCO Academic Search Premier, CINAHL, and Web of Science. In order to achieve a focused search and address the aim of the project, we used specific keywords and these were: assistive technology, information technology, assisted living technology, Internet of Things (IoT), artificial intelligence, robotics, new technology, devices, use, facilitators, barriers, perceptions, effectiveness, and quality of life. All these were in combination and/or in combination with the words "older adults", "older people, "the elderly", "people 65+". As per Fig. 2, the initial search generated 156 sources. The process of excluding duplicates resulted

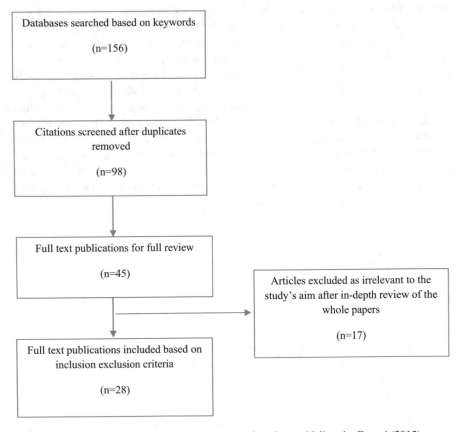

Fig. 2 Flowchart on the literature selection process, based on guidelines by Ferrari (2015)

in 98 sources. Based on reading the abstracts of these sources in accordance with our inclusion/exclusion criteria, 45 sources were selected for in-depth review. The detailed review resulted in the selection of 28 articles as they clearly reflected the research aim of this study.

For the analysis of the articles, we relied on an interpretive approach in order to better understand how older adults accept new assistive and information technology and on Thomas and Harden's [13] thematic synthesis. Thematic synthesis consists of three stages. First, the articles were read multiple times in order to gain familiarity in the methods and the findings. Second, the findings in each article were coded based on the research aim of this study. That is, acceptability including individual factors, facilitators, barriers and any other factor which related to the research aim. Third, the codes were grouped together in order to construct overarching themes, eventually setting up a codebook which helped organising and interpreting the results.

For the critical appraisal of the articles included, the CASP (Critical Assessment Skills Programme) was used as a guide [14]. More specifically, due to the absence of specific appraisal guidelines for narrative literature reviews [11], the CASP checklists for systematic reviews and qualitative studies were consulted. As a result, the critical appraisal of the articles is presented under results in the form of a narrative text instead of a systematic presentation of information in a table.

3 Results

The coding of the papers resulted in two main themes, namely facilitators and barriers, and individual and external factors.

3.1 Facilitators and Barriers

Facilitators and barriers were grouped together because very often they refer to the same aspects but from different angles. Facilitators and barriers have been further organised in three subthemes, namely sense of control, usefulness and characteristics of the technology.

3.1.1 Sense of Control

This aspect largely pertains to who is having the control over the user's life. This perceived sense of control was among the most prevailing themes in the articles included in this study. More specifically, one of the main facilitators for deciding to use the technology is when the technology is not intrusive and causes older adults to maintain their autonomy [15], and they can have control over the technology and their lives [16]. For instance, Vandemeulebroucke, de Casterle and Gastmans'

[16] study of older adults' acceptability of social assistive robots found that older adults wanted a more "boss-employee" relationship with robots whereby they would take all decisions. Such autonomy is associated with other components of perceived control and these are privacy and trust [15, 17]. That is, violation of privacy by technology such as cameras or sensors or robots entering the older adult's personal space are a huge barrier for technology acceptability and long-term use [15, 18, 19]. One of the main factors which affects feelings of trust and concerns about privacy is lack of familiarity. This means that those who were less familiar with technology due to limited prior exposure were more likely to raise these issues [19–22]. Interestingly, such attitudes were also influenced by prior personal negative experience with technology [19]. Furthermore, friends' or other family members' prior experience with technology also had an impact [19], showing that older adults shape their attitudes towards technology through interaction with other people.

Interestingly, the sense of control reached out to encompass the person's sociality. Older adults would be reluctant to use technology that would reduce or substitute social interaction with other people, including peers and family members [15, 17]. In fact, they would accept technology that would help them to socialise more; they would reject a technology which would be their main companion [15, 17, 23]. This relates to their need to maintain their independence not only in healthcare or care but in holistic terms to cover them as biopsychosocial beings [16, 24–28].

In addition to the aspects of control presented above, older adults would also like to have some control over the development of new technology. More specifically, they would like to be consulted for the development of technology and the technology must satisfy their needs in an individualised ecosystem of care and support [18, 39]. This means that older adults would be reluctant to use a technology which was developed without their input and especially if this technology did not reflect and address their needs. Addressing their needs reflects usefulness of developed technology and this is discussed below.

3.1.2 Usefulness

Perceived usefulness of technology was another dominant subtheme and took many forms in the reviewed articles. Older adults clarified in many studies that they would be interested in using technology if its capabilities reflected their needs or if it was adjusted to meet their changing needs [15, 17, 19, 20, 29–31]. A main factor that would cause more older adults to use technology would be technology's capability of making a difference in their lives and improving their condition and quality of life [15, 17, 19, 20, 29–31]. The technology should be able to help older adults, not only enhance their sociality but also to maintain it or increase it in a way that their life would continue as before [15, 17, 32, 41, 42]. In order for the older adults to be convinced, technology needed to be reliable with no functioning or failing issues [15, 33–35].

Perceived usefulness of technology as a criterion for use reflected back to the perceived sense of control. That is, a technology would be acceptable if it adjusted to older adults' changeable needs which would eventually improve their quality of living, and enhanced older adults' sense of control over their life but also over the technology itself. That is, older adults would decide which technology would be able to help them; hence exerting control and power over the use of technology.

3.1.3 Characteristics of the Technology

Characteristics of the technology are also very important for older adults and can potentially have an impact on their final decision for use. The appearance of technology is an interesting characteristic with mixed results. In one study older adults preferred machine-like, instead of human-like appearance because older adults considered human-like appearance as more threatening in the sense the technology could be seen as making decisions for them or as sharing responsibility [32, 36]. For example, Boradbent, Stafford and MacDonald [36] found that older adults preferred robots that did not look like humans and were small. In another study, there was no clear preference between human- and machine-like technology [16]. Apart from technology's appearance, other communication characteristics were also understood as important. That is, older adults tended to prefer the technology projecting female voice [16, 36], and being responsive and expressing emotions so that to feel understood. In addition to communication characteristics, older adults would be more eager to use technology if it was not complicated and was easy to use [17–19, 35].

Interestingly, these preferred characteristics of technology relate back to perceived sense of control. More specifically, preference for machine-like appearance, expression of emotions and female voice may be understood as characteristics of vulnerability which older adults may view as less threatening than characteristics such as human-like appearance with male voice. Also, technological complexity places the two parties (older adults and technology) in a power relationship, sending the message that technology cannot easily be under control or that it is more capable and smarter than humans.

3.2 Individual and Other External Factors

The above facilitators and barriers were largely technology-specific. However, a number of other factors were identified in the reviewed literature which had to do with individual characteristics and other external factors such as the physical environment.

Age was a factor that older adults considered in the sense that the older they were the less likely to be open to the use of new technology [26, 36, 37]. Age was directly associated with skills and technology literacy [20, 37, 38] but also with older adults' limited capabilities due to health condition [38, 40]. In other words, when older adults' wellbeing and capabilities were compromised they were reluctant to accept and use new technology. Lack of skills and health conditions were coupled with the issue of financial cost of new technology and restrictions in or lack of appropriate physical environment [16, 18].

Another factor which was explored by a couple of studies was that of gender. Interestingly, gender had an impact on both the sense of control and usefulness. For example, women were more interested in knowing about the interactive skills of the technology, whereas men focused more on what the technology can do [36]. In addition, Fand and Chang [32] study of older adults' attitudes toward wearable devices showed that men were more embarrassed with technology on the arm and neck, while women did not have an issue with these bodily locations. Men were more concerned than women regarding social reactions while wearing the devices.

The individual and external factors discussed above mirror again the sense of control presented earlier in this section. That is, factors that compromise control, such as skills or capabilities, cost and inappropriate physical environment, as well as characteristics which could potentially be interpreted by male older adults as weakening their sense of control, such as being perceived by others as vulnerable, were likely to raise barriers for accepting and using new technology.

3.3   Critical Appraisal of Articles Reviewed

The 28 articles reviewed were evaluated on the basis of their relevance to the research aim of this study and scientific robustness. All studies discussed new assistive or information technology, such as wearable devices, home monitoring technology, censors, tablets, computers, robots, telecare devices, digital cards etc., which aimed to help older adults maintain their dependence and active life in the community.

Based on CASP guidelines, all articles had either a clear aim or research question(s) and hypotheses depending on the nature of research and their importance ranged between moderate to high. This was gauged on the basis of how valuable each research was in accordance with the following: discussion of contribution to scholarship, identification of new areas in research, and generalisability or transferability of the findings. The sample sizes in the studies ranged from a few participants to a few hundreds and the research methods varied widely. That is, from randomised control trial, to use of questionnaires, to experiments and qualitative interviews and focus groups. The majority of them employed cross-sectional research designs and collected data with the use of closed and standardised questionnaires. In spite of methodological variation, there have been specific findings which were identified across many studies. These were, control, autonomy, independence, privacy, reliability, ease of use and usefulness. This indicates that the experience of new technology can be similar across different groups of older adults and that it was studied based on thought-out research protocols.

The review of these 28 studies has also revealed some important limitations which can open new directions in research. First, the employment of qualitative methods to understand acceptability by older adults is scarce. Nonetheless, qualitative research is very important for exploring in depth how older adults understand new technology and why some facilitators and barriers are more important than others. Interpretivist phenomenological analysis could provide important insights from in-depth inquiry and understanding of older adults' experience with new technology. Second, most studies have not explored the relationship between acceptability and older adults' social characteristics. More specifically, although there are a couple of studies which looked into gender, more research is needed in this direction in order to develop technology which reflects men's and women's experience and needs or if it is adaptable to each user's needs. The same applies to education, which very few studies have touched up on. Importantly enough, other social characteristics have not been researched adequately or at all. Social class and ethnic and cultural backgrounds do not seem they have been part of the equation. However, there is a lot of evidence to suggest how social class has an impact of illness behaviour and how people understand health and medicine [43]. This means socioeconomic status does not merely affect affordability of new technology but also people's perception and eventually long-term use of it for health and wellbeing purposes. Moreover, culture plays a critical role in people's understanding of health and illness [44], let alone technological devices which could potentially work in partnership with people to improve their health and wellbeing.

4 Discussion

This study reviewed 28 articles which explored the acceptability of the use of new assistive and information technology. "New" refers to the technology which older adults had not used before and the technology was not designed to address a specific health condition. This was important for our search strategy because a technology designed to improve a specific health condition would possibly bias older adults to accept it and use in the long-term because they would expect and hope that their health would improve. This study differs from other systematic reviews on the use of technology by older adults [2, 21, 24, 26, 28, 32] in the sense that it focuses on technology which is new to older adults, it does not address a health condition and it combines both assistive and information technology. The objective was to find overarching reasons of acceptability of new technology by older adults which would help technology designers and developers. The findings revealed that a technology which allows older adults to maintain or even enhance their control over their life, including autonomy and independence, that is useful enough to make a difference in their life and that has such characteristics which are not threatening is more appealing to older adults. In addition, when older adults have at least some technology skills and their health is not restricting then they are more open to accept and use technology.

Based on these findings, we propose the N-ACT principles for the developers of new technology, which go beyond the user-centred model for designing technology or other models such as the CAT (Comprehensive Assistive Technology) model [10]. N-ACT stands for Needs, Adjustability, Control, and Trust. Needs relate to perceived usefulness of new technology and older adults' concern that the new technology should make a difference for them. However, the technology should not only meet older adults' needs but it should also be able to accommodate their changing needs. This means that it should be adjustable in order to help older adults constantly and provide personalised service; hence, **A**djustability. Regardless of technology's perceived usefulness, it cannot be accepted if it is the one that has full control. In other words, older adults should have the **C**ontrol over both technology and their life; technology should be a tool for humans and not the other way around. This indicates that older adults should be well-trained in order to enhance their technological literacy and familiarity and make the technology easier to understand and use. Other factors, which could potentially compromise their sense of control, such as financial cost, health conditions and physical environment, should be considered. All these three components, namely needs, adjustability, and control, have to be demonstrated in practice and in a reliable way in order for older adults to gain **T**rust in the technology which will eventually cause them to accept and use in a long-term manner.

Thereafter, technology developers as well as other scientists involved should consider the following N-ACT questions during the process of designing, constructing and piloting new technology for older adults:

What are the needs of the older adults we want to develop technology for?

Is the technology going to be adjustable and provide personalised care and service?

Are the older adults empowered enough? Have they been trained adequately and is the technology causing them to maintain or enhance control over their lives? Have other external factors which could potentially compromise their sense of control been addressed?

Do older adults trust that the technology is indeed useful for them, it is reliable and that it is not taking control away from them?

Based on the findings of this study, more research is needed in order to explore the N-ACT principles in more depth, to validate them and assess their contribution to the design, development and refinement of new technology not only for older adults but for all people. In addition, as per the critical appraisal of the 28 articles included in this review, the N-ACT principles should be studied in relation to people's social characteristics such as gender, culture, occupation and socioeconomic status.

5 Conclusion

In this study, 28 articles were reviewed and appraised in order to better understand how older adults accept new assistive and information technology. The study showed that older adults accept technology when they have a good sense of control over the devices and their lives, the technology is useful, has characteristics which are not threatening, and other compromising factors such as financial cost, restricting health conditions and inappropriate physical environment do not interfere their relationship with technology. These findings constitute an important contribution to scholarship as they can help technology developers to design and construct such technology which will eventually be used by the designated end-users. Technology developers could consider the N-ACT principles (Needs, Adjustability, Control, Trust) during the development process before the technology is piloted, refined and marketed.

References

1. Ianculescu, M., Alexandru, A.: Silver digital patient, a new emerging stakeholder in current healthcare. Study case: proactiveageing. Stud. Inform. Control **25**(4), 461–469 (2016). ISSN 1220-1766
2. Voicu, R.A., Dobre, C., Băjenaru, L., Ciobanu, R.C.: Human physical activity recognition using smartphone sensors. Sensors **19**(3), 458 (2019). ISSN 1424-8220, https://doi.org/10.3390/s19 030458
3. Khosravi, P., Ghapanchi, A.H.: Investigating the effectiveness of technologies applied to assist seniors: a systematic literature review. Int. J. Med. Inform. **85**(1), 17–26 (2016). https://doi.org/10.1016/j.ijmedinf.2015.05.014. Accessed 10 Oct 2020
4. Abdi, J., Al-Hindawi, A., Ng, T., Vizcaychipi, M.: Scoping review on the use of socially assistive robot technology in elderly care. BMJ Open **8**, e018815 (2018). https://doi.org/10.1136/bmjopen-2017-018815. Accessed 12 Oct 2020
5. Bong, W., Bergland, A., Chen, W.: Technology acceptance and quality of life among older people using a TUI application. Int. J. Environ. Res. Publ. Health **16**, 4706 (2019). https://doi.org/10.3390/ijerph16234706. Accessed 12 Oct 2020
6. Khosravi, P., Ghapanchi, A.H.: Investigating the effectiveness of technologies applied to assist seniors: a systematic literature review. Int. J. Med. Inform. **85**(1), 17–26 (2016). https://doi.org/10.1016/j.ijmedinf.2015. Accessed 10 Oct 2020
7. Broekens, J., Hecrink, M., Rosendal, H.: Assistive social robots in elderly care: a review. Gerontechnology **8**(2), 94–103 (2009). https://doi.org/10.4017/gt.2009.08.02.002.00. Accessed 14 Oct 2020
8. Bobillier Chaumon, M., Michel, C., Tarpin Bernard, F., Croisile, B.: Can ICT improve the quality of life of elderly adults living in residential home care units? From actual impacts to hidden artefacts. Behav. Inform. Technol. **33**, 574–590 (2013). https://doi.org/10.1080/0144929x.2013.832382. Accessed 09 Oct 2020
9. Bajenaru, L., Marinescu, I.A., Dobre, C., Prada, G.I., Constantinou, C.S.: Towards the development of a personalized healthcare solution for elderly: from user needs to system specifications. In: 2020 12th International Conference on Electronics, Computers and Artificial Intelligence (ECAI) 2020 Jun 25, pp. 1–6. IEEE (2020)
10. Hersh, M.A.: A user-centred approach for developing advanced learning technologies based on the comprehensive assistive technology model. In: Seventh IEEE International Conference on Advanced Learning Technologies (ICALT 2007) 2007 Jul 18, pp. 919–920. IEEE (2007)
11. Ferrari, R.: Writing narrative style literature reviews. Med. Writ. **24**(4), 230–235 (2015). https://doi.org/10.1179/2047480615Z.000000000329. Accessed 10 Sep 2020
12. Baethge, C., Goldbeck-Wood, S., Mertens, S.: SANRA—a scale for the quality assessment of narrative review articles. Res. Integrity Peer Rev. **4**(1), 1–7 (2019)
13. Thomas, J., Harden, A.: Methods for the thematic synthesis of qualitative research in systematic reviews. BMC Med. Res. Methodol. **8**(1), 45 (2008). https://doi.org/10.1080/13607863.2017.1286455
14. Critical Appraisal Skills Programme. CASP (systematic review, qualitative studies) Checklist. Available at: CASP CHECKLISTS - CASP - Critical Appraisal Skills Programme (casp-uk.net) (2019). Accessed 11 Nov 2020. Accessed 10 Sep 2020
15. Vassli, L., Farshchian, B.: Acceptance of health-related ICT among elderly people living in the community: a systematic review of qualitative evidence. Int. J. Hum. Comput. Interact. **34**, 99–116 (2017). https://doi.org/10.1080/10447318.2017.1328024. Accessed 07 Oct 2020

16. Vandemeulebroucke, T., de Casterlé, B.D., Gastmans, C.: How do older adults experience and perceive socially assistive robots in aged care: a systematic review of qualitative evidence. Aging Mental Health **22**(2), 149–167 (2017). https://doi.org/10.1080/13607863.2017.128 645512. Accessed 17 Oct 2020

17. Shareef, M., Kumar, V., Dwivedi, Y., Kumar, U., Akram, M., Raman, R.: A new health care system enabled by machine intelligence: elderly people's trust or losing self control. Technol. Forecast. Soc. Chang. **162**, 120334 (2020). https://doi.org/10.1016/j.techfore.2020.120334. Accessed 04 Oct 2020

18. Claes, V., Devriendt, E., Tournoy, J., Milisen, K.: Attitudes and perceptions of adults of 60 years and older towards in-home monitoring of the activities of daily living with contactless sensors: an explorative study. Int. J. Nurs. Stud. **52**, 134–148 (2015). https://doi.org/10.1016/j.ijnurstu.2014.05.010

19. Louie, W.-Y.G., McColl, D., Nejat, G.: Acceptance and attitudes toward a human-like socially assistive robot by older adults. Assistive Technol. **26**(3), 140–150 (2014). https://doi.org/10.1080/10400435.2013.869703. Accessed 14 Oct 2020

20. Fischer, S.H., David, D., Crotty, B.H., Dierks, M., Safran, C.: Acceptance and use of health information technology by community-dwelling elders. Int. J. Med. Inform. **83**(9), 624–635 (2014). https://doi.org/10.1016/j.ijmedinf.2014.06.005. Accessed 14 Oct 2020

21. Orellano-Colón, E.M., Mann, W.C., Rivero, M., Torre, M., Jutai, J., Santiago, A., et al.: Hispanic older adult's perceptions of personal, contextual and technology-related barriers for using assistive technology devices. J. Racial Ethnic Health Disparities **3**(4), 676–686 (2015). https://doi.org/10.1007/s40615-015-0186-8. Accessed 20 Oct 2020

22. Bradwell, H., Edwards, K., Winnington, R., Thill, S., Jones, R.: Companion robots for older people: importance of user-centred design demonstrated through observations and focus groups comparing preferences of older people and roboticists in South West England. BMJ Open **9**, e032468 (2019). https://doi.org/10.1136/bmjopen-2019-032468

23. Macedo, I.: Predicting the acceptance and use of information and communication technology by older adults: an empirical examination of the revised UTAUT2. Comput. Hum. Behav. **75**, 935–948 (2017). https://doi.org/10.1016/j.chb.2017.06.013

24. González, A., Ramírez, M., Viadel, V.: Attitudes of the elderly toward information and communications technologies. Educ. Gerontol. **38**, 585–594 (2012). https://doi.org/10.1080/03601277.2011.595314. Accessed 12 Oct 2020

25. Steele, R., Lo, A., Secombe, C., Wong, Y.K.: Elderly persons' perception and acceptance of using wireless sensor networks to assist healthcare. Int. J. Med. Inform. **78**(12), 788–801 (2009). https://doi.org/10.1016/j.ijmedinf.2009.08.001. Accessed 20 Oct 2020

26. Tural, E., Lu, D., Cole, D.: Factors predicting older adults' attitudes toward and intentions to use stair mobility assistive designs at home. Prev. Med. Rep. **18**, 101082 (2020). https://doi.org/10.1016/j.pmedr.2020.101082. Accessed 20 Sep 2020

27. Dahler, A., Rasmussen, D., Andersen, P.: Meanings and experiences of assistive technologies in everyday lives of older citizens: a meta-interpretive review. Disabil. Rehabil. Assistive Technol. **11**, 619–629 (2016). https://doi.org/10.3109/17483107.2016.1151950

28. Singh, D., Psychoula, I., Kropf, J., Hanke, S., Holzinger, A.: Users' perceptions and attitudes towards smart home technologies. In: Mokhtari, M., Abdulrazak, B., Aloulou, H. (eds) Proceedings of Smart Homes and Health Telematics, Designing a Better Future: Urban Assisted Living. ICOST (2018). https://doi.org/10.1007/978-3-319-94523-1_18. Accessed 07 Oct 2020

29. Vichitvanichphong, S., Talaei-Khoei, A., Kerr, D., Ghapanchi, A.H.: Adoption of assistive technologies for aged care: a realist review of recent studies. In: Proceedings of 2014 47th Hawaii International Conference on System Sciences (2014). https://ieeexplore.ieee.org/stamp/stamp.jsp?tp=&arnumber=6758941. Accessed 12 Oct 2020

30. Vaportzis, E., Giatsi Clausen, M., Gow, A.J.: Older adults perceptions of technology and barriers to interacting with tablet computers: a focus group study. Front. Psychol. **8**, 1687 (2017). https://doi.org/10.3389/fpsyg.2017.01687. Accessed 10 Oct 2020

31. Gramstad, S., Storli, L., Hamran, T.: Do i need it? do i really need it?" elderly peoples experiences of unmet assistive technology device needs. Disabil. Rehabil. Assistive Technol. **8**(4), 287–293 (2012). https://doi.org/10.3109/17483107.2012.699993. Accessed 13 Oct 2020

32. Fang, Y., Chang, C.: Users' psychological perception and perceived readability of wearable devices for elderly people. Behav. Inform. Technol. **35**, 225–232 (2016). https://doi.org/10.1080/0144929X.2015.1114145. Accessed 21 Sep 2020
33. Hawley-Hague, H., Boulton, E., Hall, A., Pfeiffer, K., Todd, C.: Older adults' perceptions of technologies aimed at falls prevention, detection or monitoring: a systematic review. Int. J. Med. Inform. **83**(6), 416–426 (2014). https://doi.org/10.1016/j.ijmedinf.2014.03.002. Accessed 10 Oct 2020
34. Yusif, S., Soar, J., Hafeez-Baig, A.: Older people, assistive technologies, and the barriers to adoption: a systematic review. Int. J. Med. Inform. **94**, 112–116 (2016). https://doi.org/10.1016/j.ijmedinf.2016.07.004. Accessed 09 Oct 2020
35. Holzinger, A., Searle, G., Pruckner, S., Steinbach-Nordmann, S., Kleinberger, T., Hirt, E., et al.: Perceived usefulness among elderly people: experiences and lessons learned during the evaluation of a wrist device. In: Proceedings of the 4th International ICST Conference on Pervasive Computing Technologies for Healthcare (2010). https://ieeexplore.ieee.org/document/5482228. Accessed 12 Oct 2020
36. Broadbent, E., Stafford, R., MacDonald, B.: Acceptance of healthcare robots for the older population: review and future directions. Int. J. Soc. Robot. **1**, 319–330 (2009). https://doi.org/10.1007/s12369-009-0030-6
37. Amaro, F., Gil, H.: ICT for elderly people: « Yes, "They" Can!. In: Proceedings of 2011 e-CASE & e-Tech International Conference (2011). https://core.ac.uk/download/pdf/62717788.pdf. Accessed 12 Oct 2020
38. Kim, E-H., Stolyar, A., Lober, W.B., Herbaugh, A.L., Shinstrom, S.E., Zierler, B.K., et al.: Challenges to using an electronic personal health record by a low-income elderly population. J. Med. Internet Res. **11**(4), 44 (2009). https://doi.org/10.2196/jmir.1256. Accessed 14 Oct 2020
39. Charness, N., Boot, W.: Aging and information technology use. Curr. Dir. Psychol. Sci. **18**, 253–258 (2009). https://doi.org/10.1111/j.1467-8721.2009.01647
40. Astell, A., McGrath, C., Dove, E.: That's for old so and so's!: does identity influence older adults' technology adoption decisions? Ageing Soc. **40**, 1550–1576 (2019). https://doi.org/10.1017/S0144686X19000230
41. Lai, C.K., Chung, J.C., Leung, N.K., Wong, J.C., Mak, D.P.: A survey of older Hong Kong people's perceptions of telecommunication technologies and telecare devices. J. Telemedicine Telecare **16**(8), 441–6 (2010). https://doi.org/10.1258/jtt.2010.090905. Accessed 07 Oct 2020
42. Dogruel, L., Joeckel, S., Bowman, N.: The use and acceptance of new media entertainment technology by elderly users: development of an expanded technology acceptance model. Behav. Inform. Technol. **34**, 1052–1063 (2015). https://doi.org/10.1080/0144929X.2015.1077890. Accessed 05 Oct 2020
43. Marmot, M.: Social determinants of health inequalities. Lancet **365**(9464), 1099–1104 (2005)
44. Helman, C.: Culture, Health and Illness. CRC Press, Boca Raton (2007)

Piloting Intelligent Methodologies for Assisted Living Technology Through a Mixed Research Approach: The VINCI Project in Cyprus

Costas S. Constantinou, Constandinos Mavromoustakis, Anna Philippou, and George Mastorakis

Abstract This chapter presents the procedure and results of piloting vINCI, a new ambient assisted technology. vINCI aims to enhance older adults' active life and quality of living by measuring end-users' physical, psychological and social state and providing them with information and feedback about any necessary corrective measures they need to take. To achieve this, vINCI has been based on a microservices architecture and integrated IoT (Internet of Things) monitoring technologies and artificial intelligence. The interconnected devices which are capturing end-users' biopsychosocial state are tablets (the vINCI app), smart insoles and smart watches. In order to ensure that end-users would accept and use such set of technologies, this study employed a mixed research methodology to understand any acceptability factors. The results indicated that clarity of instructions, comfort of technology, ease to use, usefulness, and a sense of safety, control, familiarity and normalisation were very important features of vINCI, which could cause participants to accept and use such technology. The study highlights the importance of a mixed research method for gauging acceptability in order to ensure that the end-users' experience with new technology during pilots is fully captured and understood.

Keywords Artificial intelligence · AAL approaches · Assistive technology · Mixed methodology · The vINCI project

C. S. Constantinou (✉)
University of Nicosia Medical School, Nicosia, Cyprus
e-mail: constantinou.c@unic.ac.cy

C. Mavromoustakis
University of Nicosia, Nicosia, Cyprus

A. Philippou
Daily Centre for Elderly People, Strovolos Municipality, Nicosia, Cyprus

G. Mastorakis
Hellenic Mediterranean University, Heraklion, Greece

1 Introduction

The development of new technology for older adults is growing rapidly and this is largely because contemporary societies are faced with the challenge of supporting their ageing population to stay active and healthy in the community [1–3]. Yet there is evidence to suggest that new technology is not always accepted and, as a result, used by end-users [4, 5]. End-users have pointed out that technologies which have been designed for them do not always reflect their needs and have not been adequately tested for acceptability before finalised [4, 5]. A number of studies have highlighted specific facilitators and barriers for using new technology. More specifically, a forthcoming narrative review [6], which explored the factors affecting acceptability of new assistive and information technology by older adults, showed that older adults accepted new technology when they had control over the devices and their lives, the technology could make a difference in their life, and it did not have threatening characteristics. The study also showed that acceptance was subject to the absence of other compromising factors, such as ill health, financial cost and restricting physical environment. This narrative review focused on technology which was new to the end-users. Such principle captures the essence of this chapter, which focuses on the vINCI project which is about the construction of new assistive technology for older adults and describes how this technology was piloted in a way to ensure that acceptability by users would be well understood before the technology was finalised and made available in the market. Before explaining the process followed, let us first describe the vINCI technology.

vINCI* stands for Clinically-validated INtegrated Support for Assistive Care and Lifestyle Improvement: the Human Link and it aims to enhance older adults' active life and quality of living by constructing, piloting and validating a set of technologies, which will provide end-users information and feedback about any necessary corrective measures they need to take. To achieve this, vINCI has been based on a microservices architecture and integrated IoT (Internet of Things) monitoring technologies and artificial intelligence. The interconnected devices which are measuring end-users' biopsychosocial state are tablets (the vINCI app), smarts insoles and smart watches. vINCI is an example of microservices architecture, whereby small, autonomous but interconnected services work together in synergies [7]. vINCI architecture is shown in more detail in Fig. 1.

On the basis of the above architecture, end-users will have the opportunity to score their quality of life electronically (tablet or phone) including psychological and physical wellbeing, sociality, use of services etc. Two validated questionnaires (WHOQoL and IPAQ) will be integrated in the vINCI app which end-users should complete. Based on the selected answers by end-users vINCI would then proceed with calculations, which would result in providing feedback to the end-users about the quality of life and physical and mental wellbeing, and about the actions they should take in case their quality of life and physical activity fell below a threshold. In order to make the calculation and provide the feedback, vINCI would also draw data about end-users' physical mobility and other physical indicators (i.e. steps, heart

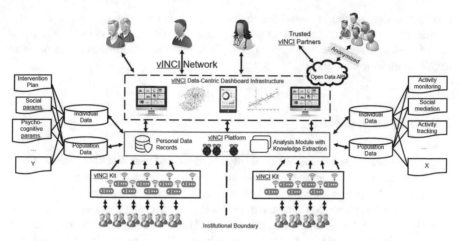

Fig. 1 The vINCI architecture

functioning, location) measured through smart shoe insoles and arm watches. The purpose is to provide end-users a tool of gaining control over their health and living and actively engage in improving their quality of life. End-users will have the option to connect other people involved in their healthcare such as their family member or informal carers but also their personal doctor. One of the main objectives of the vINCI technology is to actively involve older people in the monitoring and enhancing of their physical, psychological and social wellbeing. Such technology is innovative in the sense that it approaches the end-user's health holistically, reflecting WHO's definition of health as biopsychosocial wellbeing [8], but also the modern medical curricula which no longer focuses on treating the disease but more importantly on treating the patient as a social being [9]. In order to achieve this the vINCI technology was prepared in a draft version based on the existing literature about the needs of older adults and was piloted in order to understand whether the end-users would accept such new technology. Below we describe the methodology we have used for piloting vINCI in Cyprus for acceptability purposes.

2 Methodology

2.1 Mixed Method

The methodology used for piloting the vINCI project in Cyprus for acceptability purposes is that of mixed research methods whereby participants used the technology for a period of seven days and then completed questionnaires with closed and open questions, which led to short interviews. According to Bryman [10], a mixed research method is used for many reasons, namely triangulation, comple-mentarity, development, initiation and expansion. In this study, a mixed method was used for triangulation (corroboration), complementarity (elaboration) and expansion (exploration in more depth).

2.2   The Instrument

The instrument for measuring acceptability was designed to reflect vINCI technology and consisted of 15 likert-scale statements (five statements per device). The state-ments were: 1) The instructions of how to use this devise were clear, 2) Using the device was comfortable, 3) Using the device was easy, 4) Using the device was useful, and 5) Overall, I am satisfied with this device. These statements were answered on the basis of a likert-scale, such as "strongly disagree", "disagree", "neither agree nor disagree", "agree", "strongly disagree". Internal consistency of the closed question-naire was calculated by using the Cronbach's alpha coefficient (α). Cronbach's alpha reliability coefficient normally ranges between 0 and 1. The closer Cronbach's alpha coefficient is to 1.0 the greater the internal consistency of the questionnaire. An α of 0.6–0.7 indicates an acceptable level of reliability, and 0.8 or greater a very good level [11]. Analysis of our data showed that the value of Cronbach's alpha coefficient for internal consistency for the 15-items acceptability questionnaire was 0.735, which indicated an acceptable level of reliability. This shows that the questionnaire reliably measured acceptability of the vINCI technology.

For each device there were two open questions asking participants to reflect on what was good about the device and on any suggestions for improvement. However, these were not fixed questions but were open enough to allow space for more questions by the researcher turning them eventually into short interviews.

2.3   Sampling, Recruitment and Procedure

The convenient and purposive sample methods were used to recruit participants for this pilot study [12]. The sample was convenient because older adults were found in a convenient place [12], that of a Day Centre in Strovolos, Nicosia. Sampling was also purposive in the sense that the participants had to have specific characteristics, such as being older than 65 and being active in the community, and they were recruited for a specific purpose [12], that of using technology for some period and providing feedback. In order to recruit the participants and begin the pilot study, an ethics approval was granted by the Cyprus National Bioethics Committee.

The procedure of the pilot is shown in Fig. 2. That is, each device was used separately for a period of seven days and then participants were asked to complete the acceptability questionnaires. Participants' comments were sent to the developers

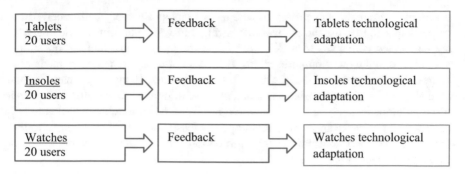

Fig. 2 Process of acceptability pilot

of vINCI technology for any technological adaptation and the technology would then be piloted again.

2.4 Analysis

The analysis of data was done through the SPSS (Statistical Package for the Social Sciences), whereby means were calculated and statistically compared. Qualitative comments were coded inductively and analyzed on the basis of the General Inductive Approach [13]. Thomas [13p.242] described the process of inductive coding as follows: "label the segments of texts to create categories → reduce overlap and redundancy among categories → create a model incorporating most important categories". Because the qualitative data generated were largely for triangulation, complementarity and expansion we did not aim to develop a model but an interpretive explanation which would reflect the purpose of employing a mixed research methodology.

3 Results

3.1 Tablets

Tablets would host the vINCI app and participants were given the opportunity to complete WHOQOL-Bref (Quality of Life) [14] and IPAQ (Physical exercise) [15] questionnaires, and were explained how the information inserted would be processed along with other information from the insoles and watches in order to receive feedback about their biopsychosocial condition. Out of a maximum score of 5, participants scored clarity of instruction with 4.85, comfort of device with 4.80, ease of use with 4.55, usefulness with 4.65, and they overall scored tablets with 4.55. These scores show that participants had a positive experience.

Qualitative comments showed that the use of tablets was very well-received by the participants. Because the use was very straight forward, participants did not have any elaborated comments to make. They liked the instructions, that the questions were clear and the questionnaires short. They did not raise any issues or made any suggestions for the device itself. This is possibly because many of the participants were already familiar with tablets and had positive prior experience. However, some of the participants clarified that they would not like a tablet to tell them what to do. They would prefer the app giving them options, based on the information form the questionnaires, the insoles and watches, to choose from so that intervention would reflect their needs at the time.

3.2   Insoles

Participants were provided with the insoles, which they should use for seven days and during their daily routine. Therefore, they started using the insoles at the Day Centre because they had visited the Centre for an activity they wanted to do (e.g. dancing lessons, signing, computer lessons etc.). When their commitment at the Centre ended they continued wearing it at their home and anywhere else they went for other activities, such as going to a supermarket, visiting friends and relatives, driving a car and so forth. When the 7-days period lapsed, the participants returned the insoles and were asked to provide feedback. They scored the insole as follows: clarity of instructions received 4.75, comfort of device 4.55, ease of use 4.75, usefulness 4.55, and overall evaluation 4.30.

Unlike the tablets, participants provided more elaborated feedback on the insoles. The feedback from the participants was generally positive. Participants found the insoles to be very thin and that they did not feel them in their shoe. The general feeling was comfort. They found the instructions clear, the insoles easy to use and useful for what they will be doing in the end. Participants clarified that they understood that the use of the insoles will be for their own good and they would be happy to use them permanently. Interestingly, all participants explained that using such a device which would make the specific measures and provide feedback to the users about the next steps and actions to be taken would give a context of safety and would enhance their confidence in daily living independently. Participants also clarified that the insoles would not cause them to feel different and that it would not be a context of stigmatisination. On the other hand, the participants identified an areas which needed to be considered for improving the device. The main issue was that the dock did not stick well to the shoe. This was because the dock with the tab inserted became heavy and it got off easily.

3.3 Watches

The use of arm watches was similar as the use of insoles in the sense that participants were provided with the devices, which they used for seven days in their daily life. Participants' experience with watches was also positive as the scored were: clarity of instructions: 4.85, comfort of device: 4.40, ease of use: 4.45, usefulness: 4.45, overall evaluation: 4.40.

Participants wrote very positive comments on the watches. They explained that the watch was easy to wear, it could safely stay on their wrist without them worrying if it could go off, it was light enough and they often forgot that they were wearing it. They did not have any negative comments on the device itself and they clarified that they would not feel stigmatised while wearing the device possibly because of their familiarity with watches in general or with smart watches.

3.4 Comparing Scores Across the Three Devices

Table 1 shows the scores for each device per question. Comparing these scores in graph 1 indicates that the devices have been scored similarly. This suggests that the vINCI devices gave a relatively consistent experience and were accepted in a similar way.

Although all three devices were scored positively, it was imperative to check if there were any statistically significant differences in terms of how the devices were scored per question. The results from all five questions showed no important difference in scores on the devices, as the statistical testing revealed no statistical significance (see Table 2). This means that the information about the devices as well as the devices themselves were understood as being of similar clarity, comfort, ease to use, usefulness, and importance.

Table 1 Scores on devices per question

	Tablets (T)	Insoles (I)	Watches (W)
Q1	4.85	4.75	4.85
Q2	4.80	4.55	4.40
Q3	4.55	4.75	4.45
Q4	4.65	4.55	4.45
Q5	4.55	4.30	4.40

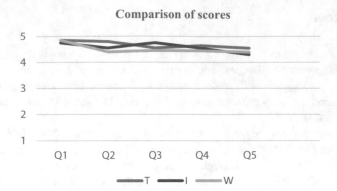

Graph 1 Comparison of scores among the devices

Table 2 Statistical comparison of how the devices were scored per question

	Asymptomatic significance (2-sided)	Extact sig. (2-sided)	Extact sig. (1-sided)
T1, I1	0.71	0.140	0.140
I1, W1	0.71	0.140	0.140
T1, W1	0.335	0.404	0.404
T2, I2	0.436	0.617	0.396
I2, W2	0.714	1.000	0.535
T2, W2	0.292	0.347	0.292
T3, I3	0.795	1.000	0.604
I3, W3	0.795	1.000	0.604
T3, W3	0.640	0.92	0.80
T4, I4	0.279	0.374	0.272
I4, W4	0.621	0.90	0.78
T4, W4	0.279	0.374	0.272
T5, I5	0.095	0.157	0.119
I5, W5	0.111	0.161	0.137
T5, W5	0.582	0.67	0.465

Statistical significance < 0.05

3.5 Acceptability of VINCI Technology

Integrating the scores and the qualitative comments as per the purpose of a mixed research method, acceptability of the vINCI technology has been affected by a number of parameters which are shown in Fig. 3. These parameters are clarity of instruction use and purpose of the device, how comfortable the devices are on the human body and to the senses, how easy they are to use, how useful they are for older adults and their health or life, the devices give a sense of safety for the future,

Fig. 3 Parameters affecting acceptability of vINCI technology

and older adults have control over the decision making with regard to corrective actions required following processing of information from the WHOQOL and the IPAQ questionnaires, as well as, the insoles and the watches. Moreover, older adults appreciated that the vINCI technology was familiar as it was employing devices used in daily life (i.e. tablets, watches, very thin unnoticeable insoles) and the technology would cause them to continue their life as normal as before with the potential of improvement.

4 Discussion

This pilot study has relied on a mixed research methodology (closed questionnaire, and open questions/short interviews) in order to gauge older adults' acceptability of new ambient assisted technology, namely the vINCI which has been based on a microservices architecture. The quantitative and qualitative data showed that older adults demonstrated high rates of acceptance of the vINCI technology.

The findings of this pilot as presented above share some interesting similarities and differences with other acceptability studies of new assistive technology. More specifically, usefulness and ease to use have been found by many other studies [16–18]. Also, other studies revealed that older adults would like to have control over the devices and their life [16, 17]. There are some differences with other studies as well. That is, in this study the participants did not raise any questions about physical characteristics (size, appearance, voice etc.) of the technology [19, 20]. This does not mean that they did not find such characteristics important but possibly they were happy with the fact that tablets and watches fell inside their cognitive template of understanding and familiarity, while insoles were so thin and comfortable to use. Furthermore, participants in this study were not concerned about being noticed or stigmatised for using assistive technology [19]. Again this has possibly to do with the fact that tablets and watches cannot cause a user to stand out in the sense that they are devices which are widely used for other purposes by the general population and the insole was very thin and almost unnoticeable where it was placed.

This study has confirmed Constantinou et al.'s review of the literature [6] which showed that when technology addressed older adults' needs, it was easy to use, it was useful and it did not take control away from end-users then it was likely to be accepted and used. Moreover, this study has showed that a mixed research methodology is an appropriate means to gauge end-users' understanding and acceptability of new technology as quantitative data alone may cause researchers to miss out important aspects of the end-users' perspective.

5 Conclusion

This study employed a mixed research method to pilot the vINCI technology which is based on a microservices architecture which combines virous technologies in order to provide the designate service to end-users. The results showed that participants well received vINCI technology because the information received was clear, the technology was comfortable, easy to use, useful, and it should provide a sense of safety, control, familiarity and normalisation. This pilot study makes an important contribution to scholarship, provides useful insights to technology developers, it has confirmed other studies about acceptability of technology and highlighted the importance of employing both quantitative and qualitative methods during the pilot in order to capture all relevant parameters to the appropriate depth; thus, adequately address the issue of involving the end-users appropriately or of finalising a technology which does actually address end-users' needs.

Note
*vINCI is approved under the AAL (Active and Assisted Living) programme of the EU and it is coordinated by the National Institute for Research and development in Informatics (ICI Bucharest). The following countries participate in the project: Romania, Italy, Cyprus, Poland, and Slovenia. For more information about the project visit vINCI - AAL Programme (aal-europe.eu)

References

1. Khosravi, P., Ghapanchi, A.H.: Investigating the effectiveness of technologies applied to assist seniors: A systematic literature review. Int. J. Med. Inf. **85**(1), 17–26 (2016). https://doi.org/10.1016/j.ijmedinf.2015.05.014
2. Abdi, J., Al-Hindawi, A., Ng, T., Vizcaychipi, M.: Scoping review on the use of socially assistive robot technology in elderly care. BMJ Open **8**(e018815), 2020 (2018). https://doi.org/10.1136/bmjopen-2017-018815
3. Bong, W., Bergland, A., Chen, W.: Technology acceptance and quality of life among older people using a TUI application. Int. J. Environ. Res. Public Health, **16,** 4706 (2019). doi:10.3390/ijerph16234706

4. Ianculescu, M., Alexandru, A.: Silver digital patient, a new emerging stakeholder in current healthcare. study case: ProActiveAgeing. Stud. Inf. Control **25**(4), 461–469 (2016)
5. Voicu, R.A., Dobre, C., Băjenaru, L., Ciobanu, R.C.: Human physical activity recognition using smartphone sensors. Sensors **19**(3), 458, 1–19 (2019). https://doi.org/10.3390/s19030458
6. Constantinou, S.C., Gurung, T., Motamedian, H., Mavromoustakis, C., Mastorakis, G.: New ambient assisted technology: a narrative review. In: Magaia, N., Mastorakis, G., Mavromoustakis, X.C., Pallis, E., Markakis, K.E. (eds.) Intelligent Technologies for Internet of Vehicles (Series on Internet of Things - Technologies, Communications and Computing). Springer (forthcoming)
7. Bhagwati, M.: Microservices vs Monolithic Architecture (2017). Available at: Microservices vs Monolithic architecture|by Bhagwati Malav|Hash#Include|Medium
8. Rokho, K.: WHO and wellbeing at work. Available at: World Health Organization and wellbeing at work (hsl.gov.uk) (2012)
9. Schiffman, F.: Treating the patient, not the disease: fred schiffman on humanism in medicine. Oncology **31**(4), 246 (2017)
10. Bryman, A.: Integrating quantitative and qualitative research: how is it done? Qual. Res. **6**(1), 97–113 (2006)
11. Gliem, J.A., Gliem, R.R.: Calculating, interpreting, and reporting Cronbach's alpha reliability coefficient for Likert-type scales (2003). Midwest Research-to-Practice Conference in Adult, Continuing, and Community Education
12. Bowling, A.: Research methods in health: investigating health and health services. McGraw-hill education, UK, 1 July 2014
13. Thomas, D.R.: A general inductive approach for analyzing qualitative evaluation data. Am. J. Eval. **27**(2), 237–246 (2006)
14. WHOQoL-Bref questionnaire. Available at: 76.pdf (who.int)
15. IPAQ questions. Available at: Downloadable questionnaires - International Physical Activity Questionnaire (google.com)
16. Vassli, L., Farshchian, B.: Acceptance of health-related ICT among elderly people living in the community: a systematic review of qualitative evidence. Int. J. Hum.-Comput. Interact. **34**, 99–116 (2017). https://doi.org/10.1080/10447318.2017.1328024
17. Shareef, M., Kumar, V., Dwivedi, Y., Kumar, U., Akram, M., Raman, R.: A new health care system enabled by machine intelligence: elderly people's trust or losing self control. Technol. Forecast. Soc. Chang. **162**, (2020). https://doi.org/10.1016/j.techfore.2020.120334
18. Louie, W.-Y.G., McColl, D., Nejat, G.: Acceptance and attitudes toward a human-like socially assistive robot by older adults. Assistive Technol. **26**(3), 140–150 (2014). https://doi.org/10.1080/10400435.2013.869703
19. Fang, Y., Chang, C.: Users' psychological perception and perceived readability of wearable devices for elderly people. Bchav. Inf. Technol. **35**, 225–232 (2016). https://doi.org/10.1080/0144929X.2015.1114145
20. Broadbent, E., Stafford, R., MacDonald, B.: Acceptance of healthcare robots for the older population: review and future directions. Int. J. Soc. Robot. **1**, 319–330 (2009). https://doi.org/10.1007/s12369-009-0030-6

Printed in the United States
by Baker & Taylor Publisher Services